Engineering Electromagnetics

Engineering Electromagnetics

C.L. Wadhwa
Formerly Professor and Head
Department of Electrical Engineering
Delhi College of Engineering, Delhi
and
Formerly Dean
Faculty of Technology
University of Delhi, Delhi

PUBLISHING FOR ONE WORLD

NEW AGE INTERNATIONAL (P) LIMITED, PUBLISHERS
New Delhi • Bangalore • Chennai • Cochin • Guwahati
Hyderabad • Kolkata • Lucknow • Mumbai
Visit us at **www.newagepublishers.com**

Published by New Age International (P) Ltd., Publishers
First Edition : 2012

Branches:

- No. 37/10, 8th Cross (Near Hanuman Temple), Azad Nagar, Chamarajpet, **Bangalore**-560 018. Tel.: (080) 26756823, Telefax: 26756820, E-mail: bangalore@newagepublishers.com
- 26, Damodaran Street, T. Nagar, **Chennai**-600 017. Tel.: (044) 24353401, Telefax: 24351463 E-mail: chennai@newagepublishers.com
- CC-39/1016, Carrier Station Road, Ernakulam South, **Cochin**-682 016. Tel.: (0484) 2377004, Telefax: 4051303, E-mail: cochin@newagepublishers.com
- Hemsen Complex, Mohd. Shah Road, Paltan Bazar, Near Starline Hotel, **Guwahati**-781 008. Tel.: (0361) 2513881, Telefax: 2543669, E-mail: guwahati@newagepublishers.com
- No. 105, 1st Floor, Madhiray Kaveri Tower, 3-2-19, Azam Jahi Road, Nimboliadda, **Hyderabad**-500 027. Tel.: (040) 24652456, Telefax: 24652457, E-mail: hyderabad@newagepublishers.com
- RDB Chambers (Formerly Lotus Cinema) 106A, Ist Floor, S.N. Banerjee Road, **Kolkata**-700 014. Tel.: (033) 22273773, Telefax: 22275247, E-mail: kolkata@newagepublishers.com
- 16-A, Jopling Road, **Lucknow**-226 001. Tel.: (0522) 2209578, 4045297, Telefax: 2204098, E-mail: lucknow@newagepublishers.com
- 142C, Victor House, Ground Floor, N.M. Joshi Marg, Lower Parel, **Mumbai**-400 013. Tel.: (022) 24927869, Telefax: 24915415, E-mail: mumbai@newagepublishers.com
- 22, Golden House, Daryaganj, **New Delhi**-110 002. Tel.: (011) 23262370, 23262368, Telefax: 43551305, E-mail: sales@newagepublishers.com

ISBN: 978-81-224-3356-2

₹ 275.00

C-12-01-6117

Printed in India at Ajit Printers, Delhi
Typeset at Pagiteck Graphics, Delhi.

PUBLISHING FOR ONE WORLD

NEW AGE INTERNATIONAL (P) LIMITED, PUBLISHERS
4835/24, Ansari Road, Daryaganj, New Delhi - 110002
Visit us at **www.newagepublishers.com**

Preface

Engineering Electromagnetics has been written as a basic course in Electrical Engineering, Electronic and Communication Engineering, Computer Engineering, Information Technology etc. In view of the rapid growth in wireless technology and several specialized branches of electrical engineering a student may not have the opportunity to take advanced courses in field theory. On the other hand field theory is a very important subject having a wide variety of applications. Hence a basic course in electromagnetics must be offered at the undergraduate level in electrical engineering.

Vector analysis being the language of electromagnetics, first chapter is devoted entirely for this aspect which includes vector algebra and vector calculus. Various coordinate systems have been explained. Various properties of vector in vector algebra have been discussed. Gradient of a scalar, divergence and curl of a vector have been explained. Laplace's and Poisson's equations have been used to solve boundary value problems. Uniqueness theorem has been developed.

Chapter 2 deals in Coulomb's law and electric field intensity due to different charge distributions has been explained.

Chapter 3 discusses concept of Electric flux density and explains Gauss' law and its application to electric fields.

Chapter 4 deals into electrostatic field energy and potentials. The concept of potential, absolute potential and potential difference has been explained. Potential due to several charge distributions has been discussed in detail.

Chapter 5 discusses properties of conductors, insulators and semiconductors. Also this chapter discusses boundary conditions when the interfaces are conductor free space, dielectric free space and dielectric-dielectric.

Chapter 6 deals into dielectric materials and capacitances. Expressions for capacitances of various devices have been developed. Method of images has been discussed.

In Chapter 7 Poisson's and Laplace's equations have been derived and their applications to boundary value problems have been explained. Uniqueness theorem and its significance has been explained.

Chapter 8 describes steady magnetic fields. Biot-Savart law, Ampere's circuital law and Stoke's Theorem have been explained and their applications have been discussed. The concept of magnetic flux and magnetic flux density has been made clear. Also, the physical concept of scalar and vector magnetic potential has been explained.

Chapter 9 deals in magnetic forces, magnetic materials and inductance. Magnetic force on charged particles and combined effect of magnetic and electric force on charged particles have been explained. Magnetic force on a planar coil has been discussed. The concepts of permeability, leakage flux, fringing of flux, self inductance, mutual inductance hysteresis and eddy current loss of magnetic circuits have been explained.

Chapter 10 describes Faraday's law and Maxwell's equations for time varying fields and in phasor form. The concept of retarded potential and displacement current has been discussed.

Chapter 11 describes propagation of signals on transmission lines. Transmission line equations have been derived and their solutions sought in phasor form. Wave reflection and refraction at discontinuities of transmission lines have been discussed and voltage standing wave ratio explained.

Chapter 12 deals in the uniform plane waves. Here equation for plane wave has been derived and propagation of these waves in good conductors and dielectric materials has been explained. The concept of Poynting theorem and wave polarization has been discussed.

A suitable number of problems have been solved in each chapter to help understand the relevant theory. At the end of each chapter unsolved problems with their answers have been suggested for further practice. At the end, a large number of multiple choice questions have been added to help the reader to test himself. An extensive bibliography will help the reader to locate detailed information on various topics of his interest.

Any constructive suggestion for the improvement of the book will be gratefully acknowledged.

Last but not the least, I wish to express my gratitude to my wife Usha and daughter Meenu for their patience and encouragement during the preparation of the book.

C.L. Wadhwa

Contents

Chapter 1

Vector Analysis

1.1 INTRODUCTION

The rapid progress made in electrical engineering during the past few decades has been possible due to the fact that an electrical engineer is able to assess, analyse and predict with greater accuracy the performance of various systems with the help of a simple powerful tool known as circuit theory. In fact circuit theory has revolutionalised the process of analysis in all branches of engineering be it civil, mechanical, electronics and communication, computer engineering etc. It has been possible to represent an electrical analog of a real physical system in terms of electric circuit elements and analysis made accurately. Circuit theory is a simplified approximation of a more complex and exact field theory.

Electromagnetic field theory deals directly with electric and magnetic field vectors E and H which are due to the electric charge and current respectively whereas circuit theory deals with voltages and currents that are integrated effects ($V = \int E \cdot dl$ and $I = \int H \cdot dl$) of electric and magnetic fields. Following approximations are made while writing circuit equations:

(*i*) The distributed capacitance effects or displacement currents from one conductor to another have been ignored.
(*ii*) The inductance is assumed to be independent of frequency.
(*iii*) The radiation effects are ignored.

These approximations are quite excellent at power frequency.

However, for an electronics and communication engineer, who has to deal with transmission of signals at ultra high frequencies, the approach of circuit theory is just not acceptable to him. In the overall design of a communication system, the engineer can use circuit theory to design the terminal equipments but between the output terminals of the transmitter and the input terminals of the receiver *i.e.* the communication channel, the circuit theory does not help him and he has to take recourse to field theory. At UHF these lines act as circuit elements. It is seen that for line lengths less than a quarter of a wavelength the shorted section is equivalent to an inductance and the open section to a capacitor and these inductances and capacitances are a function of frequency of the operating system.

When we connect voltages to the transmission line conductors, current is supplied through these conductors to the load. This is a highly simplified version of what actually is happening. The voltage gives rise to an electric field emanating radially from the surface of the conductor and, magnetic field due to current flow through the conductor is circular and surrounds the pair conductor. Thus the two fields E and H intersect each other at right angle and hence the intersection of the two fields gives rise to flow of energy, the wires acting as guides for the flow of energy. A part of the energy is radiated into the space. In fact at VHF the entire power may be radiated into the space and thus the circuit will act as an antenna. The radio interference of communication lines due to power lines is a well known example of effects of electromagnetic field. Similarly many electromagnetic devices are inadvertently coupled to other systems. Video display units of television and computer may emit sufficient radiations which could be picked up, decoded and the screen displays reproduced at sufficient distance. Any device that limits electromagnetic radiations is coupled to the rest of the world. Therefore, the motivation and study of electromagnetic field is basically very strong.

The study of electromagnetic field is more difficult as compared to circuit theory as field require more variables to describe them completely. In a circuit the current and voltages may be a function of time at the most. However, fields depend upon both time and space: Fields have to be dealt with third dimensional perspective besides time. A good knowledge of 3-dimensional geometry or solid geometry and vector's will make things much simpler for the study of electromagnetics and also it will give better insight into the various phenomenon taking place in the system. A brief study of vector analysis in three dimensional space is given here, before we embark upon the study of electromagnetic fields.

1.2 VECTOR ANALYSIS

Vector analysis is a concise language or mathematical short hand which greatly facilitate the analysis and better understanding of the electric and magnetic fields.

Scalar: A scalar quantity is characterised only by its magnitude and algebraic sign. Physical quantities which are termed as scalars are mass, temperature, time, volume, energy etc. The letter x, y, z or a, b, c we use in algebra are scalar quantities.

Vector: A vector quantity, however, has both the magnitude and direction. Force, velocity (not speed), displacement, acceleration, temperature gradient etc. are examples of vector quantities. Vector quantities are represented in bold-face or heavy type letters e.g. A, B, X, Y etc. A vector is represented geometrically by a line with an arrow. The line has an origin (starting point) and an end point (at the arrow). The orientation of the line and the arrow indicates the direction of the vector. The length of the line is proportional to the magnitude of the vector.

It is to be noted that voltage is also a scalar quantity, although the complex representation of a sinusoidal voltage produces a complex scalar or phasor which requires two real numbers for it's representation such as amplitude and phase angle or real part and imaginary part.

We shall mostly be concerned in electromagnetics with scalar and vector fields. A scalar or vector field is defined as some function of that scalar/vector which connects an arbitrary origin to a general point in space: we usually associate some function with a field e.g. a force on a compass needle in the earth's magnetic field. The field concept is invariably related to a region. If at each point of a region there is a corresponding value of some physical function, the region is called a field.

If the value of the physical function at each point in the region is a scalar quantity, the field is said to be scalar field. However, if the value of the physical function at each point in the region is a vector quantity, the field region is said to be vector field.

Having defined scalar/vector quantities and their corresponding fields we next move on to carry out basic operation of addition, subtraction, multiplication and division of vectors.

1.2.1 Addition of Vectors

Suppose there are two vectors A and B as shown in Fig. 1.1(a).

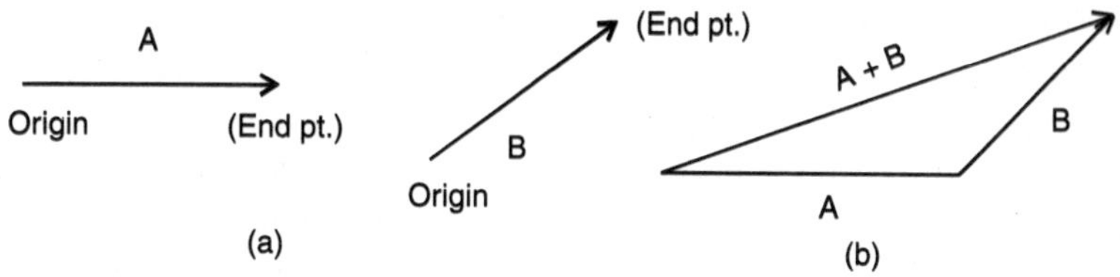

Fig. 1.1 (*a*) Two vectors *A* and *B* (*b*) Addition of *A* and *B*

To obtain $A+B$, graphically we place *A* with magnitude and orientation as shown in Fig. 1.1(*a*) and we place origin of *B* at the end point of *A* and draw *B* with its orientation and magnitude as in Fig. 1.1(*b*) and join the origin of *A* and end point of *B* by the straight line. This vector $A + B$ has its origin at origin of *A* and end point at end point of *B*.

It can be seen that addition of two vectors $A + B$ follows law of parallelogram *i.e.* *A* and *B* represent the two sides of the parallel and the diagonal of the parallelogram represents the addition of two vectors.

Let us now find the addition of vectors *B* and *A* *i.e.* $B + A$. Again we draw vector *B* with its magnitude and orientation as shown in Fig. 1.1(*a*) and place the origin of *A* at the end point (arrow) of *B* as shown in Fig. 1.2. The resultant $B + A$ as can be seen has the same magnitude and direction as $A + B$ in Fig. 1.1(*b*), which means

$$A + B = B + A \qquad \text{...(1.1)}$$

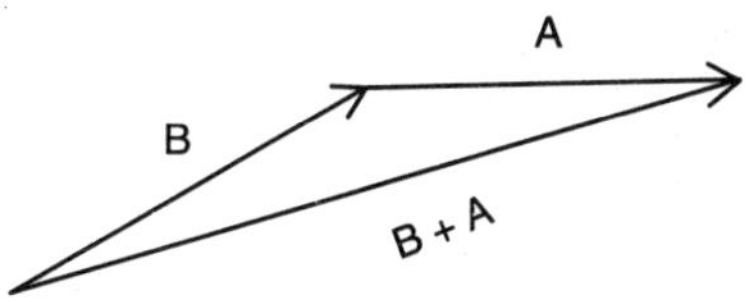

Fig. 1.2 Addition of *B* and *A*

When the order of the operation may be reversed with no effect on the result, the operation is said to follow law of commutation.

1.2.2 Addition of Three Vectors—Associative Law

Consider three vectors *A, B, C* as shown in Fig. 1.3(*a*) with their orientation and magnitude. We first obtain $(A + B) + C$ as shown in Fig. 1.3(*b*) and then $A + (B + C)$ as shown in Fig. 1.3(*c*). We find that $(A + B) + C = A + (B + C)$ which shows that the sum of three vectors is independent of the way in which vector *B* is associated with vectors *A* or *C* *i.e.* the vector addition obeys the associative law.

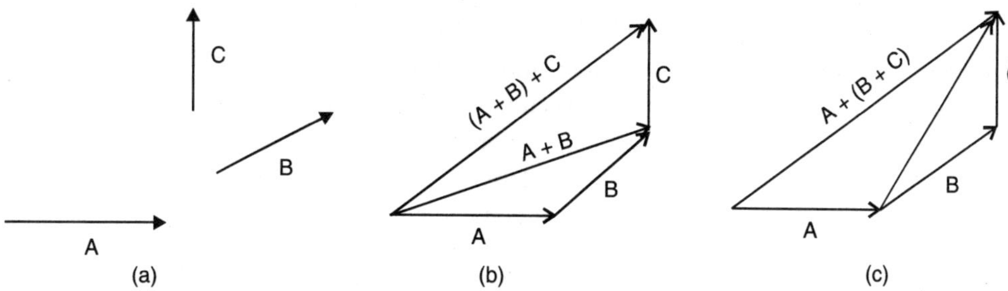

Fig. 1.3 Addition of three vectors—Associative law

1.2.3 Subtraction of Two Vectors

Consider the vectors A and B as shown in Fig. 1.4(*a*). The subtraction could be considered as $A - B = A + (-B)$ *i.e.* addition of the vectors after reversing the direction of the vector B *i.e.* the line vectors now will be as shown in Fig. 1.4(*b*) using the same procedure of addition *i.e.* we place tail end of $(-B)$ on the arrow head of A and keep the orientation and magnitude of $-B$ as in Fig. 1.4(*b*) and finally, the result is obtained by joining the tail end of B to the arrow head of A. It can be shown easily (Fig. 1.4*d*) that $A - B = - B + A$ which means that the subtraction operation on two vectors is commutative.

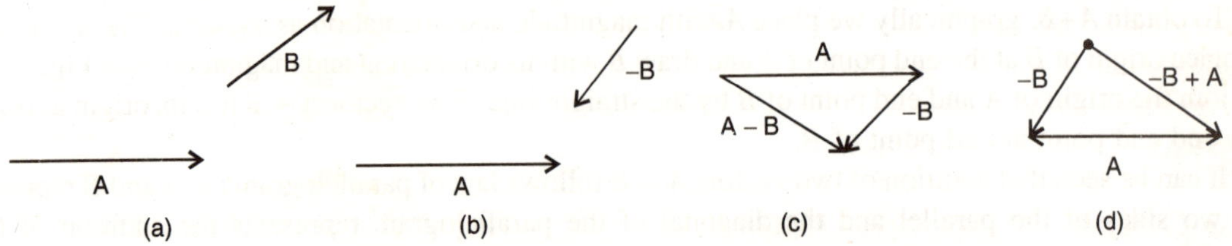

Fig. 1.4 Subtraction of the vectors A and B *i.e.* $A - B$

It is to be noted here with caution that vectors A, B and C all refer to the same point in space. The displacements shown in Fig. 1.1 through 1.4 are only for the convenience of illustrating graphically the addition and subtraction operations.

1.2.4 Multiplication or Division of a Vector by a Scalar

When a vector is multiplied or divided by a scalar, the magnitude of the original vector is changed by an amount equivalent to the multiplying or dividing scalar. However the direction of resultant vector remains same as the original vector.

1.2.5 Multiplication of Two Vectors

In vector analysis there are two types of vector multiplications. These are known as scalar or dot product of two vectors and vector or cross product of two vectors. It is known as scalar as the final result of multiplication of two vectors is a scalar quantity and it is known as dot product as it is indicated with a dot between the two vectors *i.e.* say if A and B are to be multiplied as dot product it is denoted as $A \cdot B$. The dot product is defined as the multiplication of the magnitude of A and the magnitude of B and cosine of the smaller angle between A and B when we move from A to B *i.e.* $A \cdot B = |A|\,|B| \cos\theta$ where θ is the minimum angle between A and B.

Vector product or cross product means the product results in a vector quantity and cross means it is denoted by a cross between the two vectors e.g. cross product of A and B is denoted as $A \times B$ and it is given as

$$A \times B = |A|\,|B| \sin\theta \cdot \bar{a}_n \qquad \text{...(1.2)}$$

i.e. the product of magnitude of the two vectors and further multiplied by the sin of the minimum angle θ between the two vectors when we move from A to B and $\bar{a}_n$ represent the direction of the resultant vector which is perpendicular to the plane containing the two vectors A and B. Here a_n is a unit vector and subscript n stands for normal *i.e.* the resultant is perpendicular to the plane containing A and B and is along that one of the two possible perpendiculars which is in the direction of advance of a right handed screw as A is turned into B. If A and B are supposed to be in the plane of the paper, and the order of A and B is as shown in Fig. 1.5 then a_n is normal to the plane of the paper and

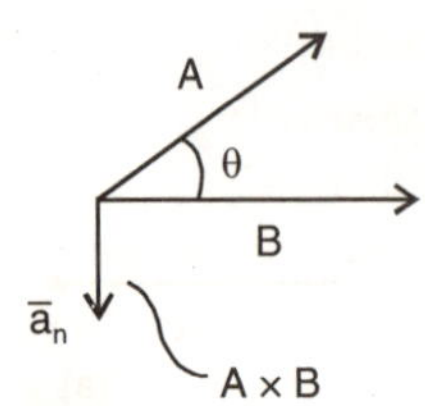

Fig. 1.5 Cross product of A and B

is into the paper. However if we have $B \times A$ we find that the magnitude of the product remains the same but this time the direction is normal to the plane of the paper but coming out of the paper. Hence

$$A \times B = -B \times A \qquad \text{...(1.3)}$$

This means cross product of two vectors is not commutative. However, we know that

$$A \cdot B = B \cdot A \qquad \text{...(1.4)}$$

Hence dot product of two vectors is commutative.

There are few more operations like divergence of a vector or curl of a vector (calculus of vectors). We will study them in this chapter a little later. We have to study the electric and magnetic field in a three dimensional space and, therefore, we should learn how to specify a vector or a point in a three dimensional space. A vector can be specified by giving some specific length, or its projection or components along some axis, direction and/or angles. We will study here three different coordinate systems.

1.3 COORDINATE SYSTEMS

The coordinate systems are:

(*i*) Rectangular Coordinate System
(*ii*) Circular Cylindrical Coordinate System
(*iii*) Spherical Coordinate System

We are going to use right handed orthogonal coordinate system and the above mentioned three systems come under this category. We will try to learn the terminology right handed and orthogonal with reference to the rectangular coordinate system and we will study this system first in detail as this happens to be the simplest of the three systems.

In order to locate a point in three dimensional system we need three coordinates of the point. In cartesian coordinates, the three coordinates are x, y and z. Right handed system means, if we stretch, the thumb, the forefinger and the central finger at right angle to each other and the minimum angle between two fingers is 90°. If we choose forefinger as x-axis and rotate through 90° then central finger shows the y-axis and if we rotate x through y, then thumb represents the z-axis.

Orthogonal system means, to locate a point say (x, y, z) in three dimensional space, three surfaces/planes $x = c_1$, $y = c_2$, $z = c_3$ will intersect each other orthogonally and this orthogonal intersection defines the point (x, y, z). We know that when two planes intersect we get a straight line and when three planes intersect we get a point in a three dimensional space. It is to be noted that these planes must intersect each other orthogonally or they should be mutually perpendicular to each other. Origin in the three dimensional space is obtained by intersection of $x = 0$, $y = 0$ and $z = 0$ planes *i.e.* $y - z$, $z - x$ and $x - y$ planes. In fact it is a one step forward as the three vectors are of simple nature, each one is always directed along one of the coordinate axes *i.e.* the component, vectors have magnitude which depends upon the original vector, however, the direction is fixed. Because of this we make use of what is known as unit vector concept. A unit vector is a vector with unit magnitude and is directed along the coordinate axes in increasing direction of the coordinate value. $\bar{a}_x$ is a unit vector along x-axis and in the increasing direction of x. Therefore, if A_x, A_y and A_z are the magnitudes of the three components of A, the vector A can be completely defined in terms of the components and unit vectors as

$$A = a_x A_x + a_y A_y + a_z A_z \qquad \text{...(1.5)}$$

Now unit vector is given as $\dfrac{\bar{A}_x}{|\bar{A}_x|}$ in x-direction *i.e.*

$$\bar{a}_x = \frac{\bar{A}_x}{|\bar{A}_x|}. \text{ Similarly } \bar{a}_y = \frac{\bar{A}_y}{|\bar{A}_y|} \text{ and } \bar{a}_z = \frac{\bar{A}_z}{|\bar{A}_z|}$$

Suppose two points P and Q are given with coordinate (x_1, y_1, z_1) and (x_2, y_2, z_2) as shown in Fig. 1.6(*b*). The line joining the points P and Q from the origin represents vector $\bar{A}$ and $\bar{B}$ respectively.

Our objective is to determine the vector directed from point P to Q. The coordinates of points P and Q are used to write the two position vectors A and B

$$\bar{A} = \bar{a}_x x_1 + \bar{a}_y y_1 + \bar{a}_z z_1$$

$$\bar{B} = \bar{a}_x x_2 + \bar{a}_y y_2 + \bar{a}_z z_2$$

Hence $\bar{B} - \bar{A} = \bar{a}_x (x_2 - x_1) + \bar{a}_y (y_2 - y_1) + \bar{a}_z (z_2 - z_1)$

As can be seen from the Fig. 1.6(*b*), vector $\bar{A}$ does not extend outward from the origin. We also know that two vectors having the same magnitude and direction are equal and therefore, to help us in our visualisation process we are free to slide any vector over to the origin before determining its component vectors. Parallelism, however, must be maintained while sliding the vectors.

We next consider the properties of the unit vectors associated with dot and cross products especially in cartesian coordinate system.

$$\begin{array}{lll} \bar{a}_x \cdot \bar{a}_x = 1 & \bar{a}_x \cdot \bar{a}_y = 0 & \bar{a}_x \cdot \bar{a}_z = 0 \\ \bar{a}_y \cdot \bar{a}_y = 1 & \bar{a}_y \cdot \bar{a}_x = 0 & \bar{a}_y \cdot \bar{a}_z = 0 \\ \bar{a}_z \cdot \bar{a}_z = 1 & \bar{a}_z \cdot \bar{a}_x = 0 & \bar{a}_z \cdot \bar{a}_y = 0 \end{array} \quad ...(1.6)$$

If P is a point with coordinates x, y, z and we generate a differential voltage by increasing at the coordinates differentially *i.e.* the new coordinates are $x + dx$, $y + dy$ and $z + dz$. The volume of the differential elements which is a parallelopipe is $dx\,dy\,dz$ and the surfaces are $dx\,dy$, $dy\,dz$ and $dz\,dx$ and length of the element joining the two points is $\sqrt{dx^2 + dy^2 + dz^2}$.

So far we have described scalar quantities e.g. volume, surface or length in cartesian coordinate system. Next we consider vectors in terms of coordinate system.

1.3.1 Rectangular Coordinates

To describe a vector in the rectangular coordinate system consider a vector A starting from the origin and has its arrow head in space. The most obvious way to describe A is by giving the three component vectors lying along the three coordinate axes such that the vector sum of the three components equals the original vector. Let

$$\bar{A} = \bar{A}_x + \bar{A}_y + \bar{A}_z \quad ...(1.7)$$

where A_x is vector along x-axis and in the increasing direction of x, similarly for A_y and A_z. At first sight it appears that we have now three vectors instead of one and we have complicated the analysis. But this is not. However in cross product, since angle between the two unit vectors is 90° and sin 90° = 1. Hence

$\bar{a}_x \times \bar{a}_y = 1 \cdot \bar{a}_z, \quad \bar{a}_z \times \bar{a}_x = \bar{a}_y, \quad \bar{a}_x \times \bar{a}_x = 0$

However, $\bar{a}_y \times \bar{a}_x = -\bar{a}_z, \quad \bar{a}_x \times \bar{a}_z = -\bar{a}_y$

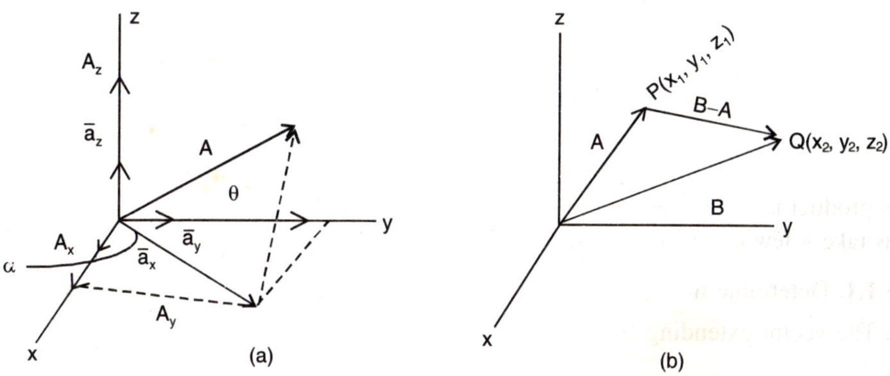

Fig. 1.6 Decomposition of vector A along three axes

For positive value the cyclic order x, y, z should be maintained.

To obtain the components of $\bar{A}$ as shown in Fig. 1.6 following procedure can be followed. A subtends an angle θ and projection of A on xy plane subtends an angle ψ with x-axis, then

$$A_x = |A|\sin\theta\cos\psi,\ A_y = |A|\sin\theta\sin\psi \text{ and } A_z = |A|\cos\theta$$

However, if vector A is given and unit vectors $\bar{a}_x, \bar{a}_y$ and $\bar{a}_z$ are known, then

$$A_x = \bar{A}\cdot\bar{a}_x = |A|\,|\bar{a}_x|\cos\alpha$$

$$A_y = \bar{A}\cdot\bar{a}_y = |A|\,|\bar{a}_y|\cos\beta \qquad ...(1.8)$$

and $$A_z = \bar{A}\cdot\bar{a}_z = |A|\,|\bar{a}_z|\cos\gamma$$

where α, β and γ are the angles subtended by the vector with x, y and z axes respectively and cos α, cos β and cos γ are known as direction cosine and their values are such that

$$\cos^2\alpha + \cos^2\beta + \cos^2\gamma = 1 \qquad ...(1.9)$$

We next find out the scalar product of the vectors in cartesian coordinates

Let $$\bar{A} = \bar{a}_x A_x + \bar{a}_y A_y + \bar{a}_z A_z$$

and $$\bar{B} = \bar{a}_x B_x + \bar{a}_y B_y + \bar{a}_z B_z$$

Hence $\bar{A}\cdot\bar{B} = (\bar{a}_x A_x + \bar{a}_y A_y + \bar{a}_z A_z)\cdot(\bar{a}_x B_x + \bar{a}_y B_y + \bar{a}_z B_z)$. We know that $\bar{a}_x\cdot a_x = 1$ and $\bar{a}_x\cdot\bar{a}_y = 0$. Using similar relation in the above equation we have

$$\bar{A}\cdot\bar{B} = A_x B_x + A_y B_y + A_z B_z \qquad ...(1.10)$$

We next find out cross product of two vectors taking components of the vectors into consideration.

$$\bar{A}\times\bar{B} = (a_x A_x + a_y A_y + a_z A_z)\times(a_x B_x + a_y B_y + a_z B_z)$$

Again making use of the relations after cross product

$$\bar{a}_x\times\bar{a}_x = 0,\quad \bar{a}_x\times\bar{a}_y = \bar{a}_z,\quad \bar{a}_y\times\bar{a}_z = \bar{a}_x \text{ and } \bar{a}_y\times a_x = -\bar{a}_z \text{ we have}$$

$$\bar{A} \times \bar{B} = \begin{vmatrix} \bar{a}_x & \bar{a}_y & \bar{a}_z \\ A_x & A_y & A_z \\ B_x & B_y & B_z \end{vmatrix} \qquad \text{...(1.11)}$$

Cross product is thus expressed in terms of a determinant.

Let us take a few examples to understand what all we have studied so far.

Example 1.1. Determine the unit vector extending from the origin towards the point $P(1, 2, 3)$.

Solution. The vector extending from the origin to point P.

$$P = a_x + 2a_y + 3a_z$$

The magnitude of the vector P is $\sqrt{1^2 + 2^2 + 3^2} = \sqrt{14}$ and hence the desired unit vector

$$\bar{a}_p = \frac{\bar{P}}{|\bar{P}|} = \frac{\bar{a}_x}{\sqrt{14}} + \frac{2\bar{a}_y}{\sqrt{14}} + \frac{3\bar{a}_z}{\sqrt{14}} \quad \textbf{Ans.}$$

Example 1.2. Two points $P(2, -4, 1)$ and $Q(0, -2, 0)$ are given in cartesian coordinates. Say corresponding to position P the position vector A is A and to position Q it is B. Determine the vector joining the line PQ with arrow head at Q.

Solution. The position vector $A = 2\bar{a}_x - 4\bar{a}_y + a_z$

$$B = -2a_y$$

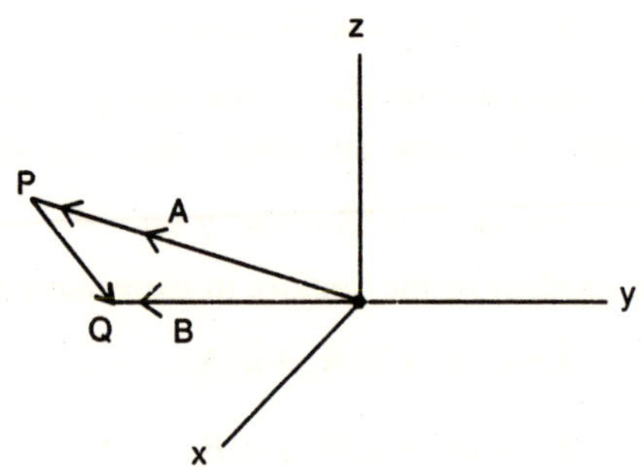

Fig. E 1.2

Hence $$B - A = -2a_y - 2a_x + 4a_y - a_z$$

$$= -2\bar{a}_x + 2\bar{a}_y - \bar{a}_z$$

The magnitude of vector $|B - A| = \sqrt{2^2 + 2^2 + 1^2} = 3$

Hence $$\bar{a}_{B\cdot A} = -\frac{2}{3}\bar{a}_x + \frac{2}{3}\bar{a}_y - \frac{\bar{a}_z}{3} \quad \textbf{Ans.}$$

Example 1.3. Vectors $\bar{A} = 6\bar{a}_y + \bar{a}_z$ and $\bar{B} = 2\bar{a}_x + 4\bar{a}_y$ are position vectors from the origin. Determine $\bar{A} \cdot \bar{B}, \bar{A} \times \bar{B}$ and the angle between the vectors.

Solution.

$$\bar{A}\cdot\bar{B} = \left(6\bar{a}_y + \bar{a}_z\right)\cdot\left(2\bar{a}_x + 4\bar{a}_y\right) = 24$$

and
$$\bar{A}\times\bar{B} = \left(6\bar{a}_y + \bar{a}_z\right)\times\left(2\bar{a}_x + 4\bar{a}_y\right) = -12\bar{a}_z + 2\bar{a}_y - 4\bar{a}_x$$

Now
$$A\cdot B = |A|\,|B|\cos\theta$$

$$24 = \sqrt{37}\cdot\sqrt{20}\cos\theta \text{ or } \cos\theta = \frac{24}{6.08\times 4.47} = 0.883$$

$\theta = 28°$ **Ans.**

Using cross product

$$|A\times B| = |A|\,|B|\sin\theta$$

or
$$\sin\theta = \frac{|A\times B|}{|A|\,|B|} = \frac{\sqrt{12^2+2^2+4^2}}{\sqrt{37}\cdot\sqrt{20}} = \frac{\sqrt{164}}{\sqrt{37\times 20}}$$

$$= 0.47$$

$\theta = 28.08°$ **Ans.**

Example 1.4. Given $\bar{A} = 3\bar{a}_y + 5\bar{a}_z$ and $\bar{B} = 2\bar{a}_x + 4\bar{a}_y$. Determine the projection of $\bar{A}$ on $\bar{B}$.

Solution. Projection of A on B

$$\frac{A\cdot B}{|B|} = \frac{\left(3\bar{a}_y + 5\bar{a}_z\right)\cdot\left(2\bar{a}_x + 4\bar{a}_y\right)}{|B|}$$

$$= \frac{12}{|B|} = \frac{12}{\sqrt{20}} = \frac{6}{\sqrt{5}}$$ **Ans.**

Example 1.5. Given $\bar{A} = \bar{a}_x + \bar{a}_y$, $\bar{B} = 2\bar{a}_x + 3\bar{a}_z$ and $c = 2\bar{a}_y + \bar{a}_z$. Determine $(A\times B)\times C$ and compare it with $A \times (B \times C)$.

Solution.

$$A\times B = \begin{vmatrix} \bar{a}_x & \bar{a}_y & \bar{a}_z \\ 1 & 1 & 0 \\ 2 & 0 & 3 \end{vmatrix} \begin{aligned} &= \bar{a}_x(3-0) - \bar{a}_y(3-0) + \bar{a}_z(-2) \\ &= 3\bar{a}_x - 3\bar{a}_y - 2\bar{a}_z \end{aligned}$$

$$(A\times B)\times C = \begin{vmatrix} \bar{a}_x & \bar{a}_y & \bar{a}_z \\ 3 & -3 & -2 \\ 0 & 2 & 1 \end{vmatrix} \begin{aligned} &= \bar{a}_x(-3+4) - \bar{a}_y(3-0) + \bar{a}_z 6 \\ &= \bar{a}_x - 3\bar{a}_y + 6\bar{a}_z \end{aligned}$$

Now
$$B\times C = \begin{vmatrix} \bar{a}_x & \bar{a}_y & \bar{a}_z \\ 2 & 0 & 3 \\ 0 & 2 & 1 \end{vmatrix} \begin{aligned} &= -6\bar{a}_x - \bar{a}_y(2) + \bar{a}_z 4 \\ &= -6\bar{a}_x - 2\bar{a}_y + 4\bar{a}_z \end{aligned}$$

$$A \times (B \times C) = \begin{vmatrix} \bar{a}_x & \bar{a}_y & \bar{a}_z \\ 1 & 1 & 0 \\ -6 & -2 & 4 \end{vmatrix} = 4\bar{a}_x - 4\bar{a}_y + 4\bar{a}_z \quad \textbf{Ans.}$$

Hence it is important to keep in mind the parentheses while performing triple cross product.

1.4 CYLINDRICAL COORDINATE SYSTEM

As mentioned earlier, depending upon the kind of charge distribution e.g. point charges, line charges, surface charges etc. or current distribution e.g. linear, circular etc., a particular type of coordinate system is found more handy as compared to others. We have so far studied rectangular or cartesian coordinate system. The other two systems are circular cylindrical coordinate and spherical coordinate systems. We next study circular cylindrical coordinate system or simply cylindrical coordinate system.

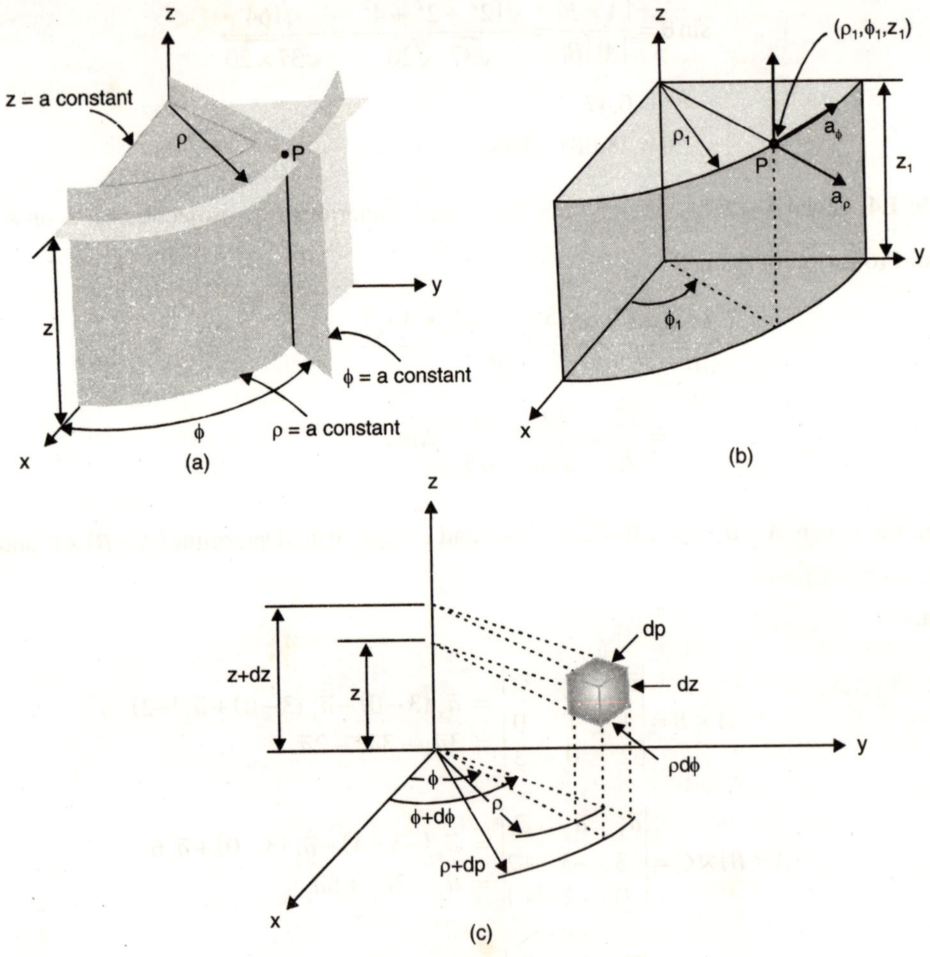

Fig. 1.7 (*a*) The three mutually perpendicular surfaces of the circular cylindrical coordinate system. (*b*) The three unit vectors of the circular cylindrical coordinate system. (*c*) The differential volume unit in the circular cylindrical coordinate system; *dp*, ρ*d*ϕ, and *dz* are all elements of length.

We are omitting the word circular for simplicity. In fact we have elliptic cylindrical, parabolic cylindrical, hyperbolic cylindrical coordinate system which we will not deal with here. Again to locate a point in three dimensional space using cylindrical coordinate we use three coordinates ρ, ϕ, Z. Of course we don't set up three axes as with rectangular coordinate, rather we consider any point as the intersection of three mutually perpendicular surfaces.

The cylindrical coordinate system is the three dimensional version of the polar coordinate in two dimensional plane surface. In the two dimensional polar coordinates a point is located in a plane by giving it's distance ρ from the origin and the angle ϕ between the line from the point to the origin (*i.e.* ρ) and an arbitrary radial line taken as $\phi = 0$ or the arbitrary line could be x-axis. In three dimensional space point P is obtained by specifying third coordinate along z-axis.

The three surfaces in cylindrical coordinates are

ρ = constant, a circular cylinder
ϕ = constant, a plane and
z = constant a plane

The three surfaces are shown in Fig. 1.7. These three surfaces intersect each other orthogonally and result in a point with general coordinate $|\rho, \phi, z|$.

Next we define unit vectors in cylindrical coordinates. We know that in rectangular coordinates, unit vector corresponding to x-axis is a vector that is normal to the plane x = constant and is in the increasing direction of x and of course its magnitude is unity.

In cylindrical coordinates, however, such axes don't exist and hence we will talk general view *i.e.* we will define unit vector is one which is normal to the surface for which that coordinate is constant and is in the increasing direction of that coordinate e.g. a_p is a unit vector at a point $\rho(\rho_y, \theta, Z)$ which is directed radially outward (*i.e.* in the increasing direction of ρ) and is normal to the cylindrical surface $\rho = \rho_1$ and this lies in the planes defined by the other two coordinates *i.e.* $\phi = \phi$, and $Z = Z_1$.

Similarly $\bar{a}_\phi$ is normal to ϕ and is directed in the increasing value of ϕ and lies in the plane $Z = Z_1$ and is tangent to the cylindrical surface $\rho = \rho_1$. $\bar{a}_z$ is normal to the surface $Z = Z_1$ and is in the increasing direction of Z and lies in the plane $\phi = \phi_1$ and tangent to $\rho = \rho_1$.

In rectangular co-ordinates the directions of unit vectors remain unchanged whereas in cylindrical coordinates $\bar{a}_z$ remains unchanged, however, $\bar{a}_\rho$ and $\bar{a}_\phi$ change their direction as ϕ changes. Therefore, while performing differentiation and integration operation $\bar{a}_\rho$ and $\bar{a}_\phi$ must not be treated constant. The unit vectors are in cylindrical co-ordinates and mutually perpendicular to each other as each one of them lies in two surfaces/planes whereas it is perpendicular to the third surface/plane and we may define a right handed cylindrical coordinate system as one in which $\bar{a}_\rho \times \bar{a}_\phi = \bar{a}_z$.

Similarly taking in the sequence ρ, ϕ, z, $a_\phi \times a_z = a_\rho$ and so on.

Next we find out relationship between the rectangular and cylindrical coordinate system for which refer to Fig. 1.8.

From Fig. 1.8 it is clear that

$$x = \rho\cos\phi,\ y = \rho\sin\phi \text{ and } z = z \qquad \ldots(1.12)$$

From the above relations we can express cylindrical coordinates in terms of rectangular co-ordinates

$$\rho = \sqrt{x^2 + y^2},\ \phi = \tan^{-1}\frac{y}{x} \text{ and } z = z \qquad \ldots(1.13)$$

$\rho \geq 0$

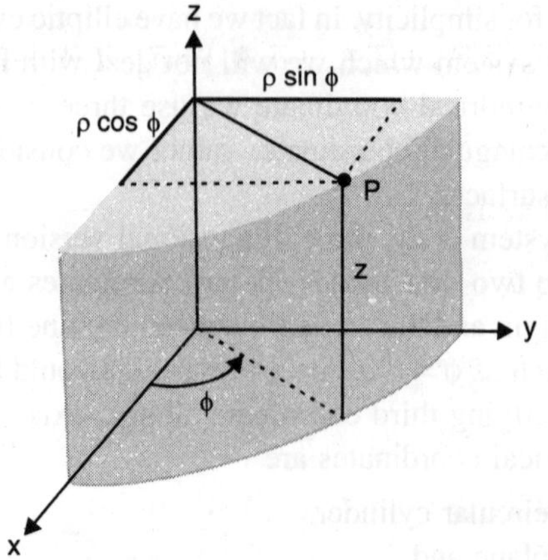

Fig. 1.8 Relationship between the rectangular and cylindrical coordinate system

Next we find out expressions for differential lengths, surfaces and volume when each coordinate is incremented infinitesimally. Let ρ, φ and z be incremented by differential elements $d\rho$, $d\phi$ and dz. With this, three change in lengths are $d\rho$, $\rho d\phi$ and dz and various differential surfaces will be $\rho d\rho d\phi$, $\rho d\phi dz$ and $d\rho dz$. In terms of differential volume, two cylinders of radius ρ and $\rho + d\rho$, the two radial planes at angles φ and $\phi + d\phi$ and the two horizontal planes at heights z and $z + dz$. All these six components form a differential volume as shown in Fig. 1.8. The volume is $\rho d\rho d\phi dz$ units. The length of the line vector between points P and Q $(\rho + d\rho, \phi + d\phi, z + dz)$ is

$$d_L = \sqrt{d\rho^2 + (\rho d\phi)^2 + z^2}$$

Using equations (1.12) and (1.13) the scalar functions given in one co-ordinate system can be transformed into the other systems.

However, a vector function in one coordinate system requires two steps to transform it into another system.

The first step is same as with scalar functions and the second step involves the use of dot products of the unit vectors as obtained below to change components. The two steps could be taken in either order.

Suppose we are given a vector in rectangular coordinates

$$A = \bar{a}_x A_x + \bar{a}_y A_y + \bar{a}_z A_z$$

where each component A_x, A_y, A_z are given in x, y, z coordinates. We now find a vector in cylinder coordinate

$$A = \bar{a}_\rho A_\rho + \bar{a}_\phi A_\phi + \bar{a}_z A_z \qquad \text{...(1.14)}$$

Here each component is given in (ρ, ϕ, z).

We know that

$$A_\rho = \bar{a}_\rho \cdot \bar{A},\ A_\phi = \bar{a}_\phi \bar{A}$$

Substituting rectangular components for $\bar{A}$ we have

$$A_\rho = \bar{a}_\rho\left(\bar{a}_x A_x + \bar{a}_y A_y + \bar{a}_z A_z\right) = \bar{a}_\rho \cdot \bar{a}_x A_x + \bar{a}_\rho \cdot \bar{a}_y A_y + \bar{a}_\rho \cdot \bar{a}_z A_z$$

$$A_\phi = \bar{a}_\phi\left(\bar{a}_x A_x + \bar{a}_y A_y + \bar{a}_z A_z\right) = \bar{a}_\phi \cdot \bar{a}_x A_x + \bar{a}_\phi \cdot \bar{a}_y A_y + \bar{a}_\phi \cdot \bar{a}_z A_z$$

and $\qquad A_z = \bar{a}_z \cdot A = A_z$

Now angle between $\bar{a}_\rho$ and $\bar{a}_x$ is ϕ

a_ρ and $\bar{a}_y$ is $90° - \phi$

and $\qquad a_\rho$ and a_z is $90°$

Hence $\qquad \bar{a}_\rho \cdot \bar{a}_x = \cos\phi,\ \ a_\rho \cdot a_y = \sin\phi$ and $\bar{a}_\rho \cdot \bar{a}_z = 0$

Therefore $\qquad A_\rho = A_x \cos\phi + A_y \sin\phi \qquad$ ...(1.15)

and $\qquad A_\phi = -A_x \sin\phi + A_y \cos\phi \qquad$ ...(1.16)

The dot products of the unit vectors in the two coordinate systems are summarised in Table 1.1.

Table 1.1 Dot products of unit vectors in cartesian and cylindrical coordinate system

	$\bar{a}_\rho$	$\bar{a}_\phi$	$\bar{a}_z$
$\bar{a}_x$	$\cos\phi$	$-\sin\phi$	0
$\bar{a}_y$	$\sin\phi$	$\cos\phi$	0
$\bar{a}_z$	0	0	1

Example 1.6. Transform the vector $A = y\bar{a}_x + x\bar{a}_y + z\bar{a}_z$ into cylindrical coordinates.

Solution. The new components A_ρ, A_ϕ and $\bar{a}_z$ are obtained as follows:

$$A_\rho = A \cdot \bar{a}_\rho = y\bar{a}_x \cdot \bar{a}_\rho + x\bar{a}_y \cdot \bar{a}_\rho + z\bar{a}_z \cdot \bar{a}_\rho$$

$$= y\cos\phi + x\sin\phi$$

$$= \rho\sin\phi\cos\phi + \rho\cos\phi\sin\phi = 2\rho\sin\phi\cos\phi$$

$$= \rho\sin 2\phi$$

$$A_\phi = A \cdot \bar{a}_\phi = y\bar{a}_\phi \cdot \bar{a}_x + x\bar{a}_y \cdot \bar{a}_\phi + z\bar{a}_z \cdot \bar{a}_\phi$$

$$= -y\sin\phi + x\cos\phi$$

Next replacing y and x in terms of cylindrical coordinates

$$A_\phi = -\rho\sin^2\phi + \rho\cos^2\phi = \rho\left(\cos^2\phi - \sin^2\phi\right) = \rho\cos 2\phi$$

and $\qquad A_z = A_z$

Hence $\qquad A = \rho\sin 2\phi \cdot \bar{a}_\rho + \rho\cos 2\phi \cdot \bar{a}_\phi$ **Ans.**

Example 1.7. Determine the distance between (4, 3π/2, 0) and (4, π/2, 0) in cylindrical coordinates.

Solution. Since the position vectors are given in cylindrical coordinates, there is no direct way to obtain the length so we convert cylindrical to cartesian coordinates.

From equation (1.13)

$$x_1^2 + y_1^2 = 16 \quad \phi = \frac{3\pi}{2} = \tan^{-1}\frac{y_1}{x_1} \quad x_1 = 0 \text{ and}$$

since ϕ is in third quadrant y is negative and hence

$$y_1^2 = 16 \text{ or } y_1 = -4 \text{ and } z_1 = 0$$

Hence $(0, -4, 0)$ or $A = -4\,\bar{a}_y$

For the second position vector B

$$x^2 + y^2 = 16 \text{ and since } \phi \text{ is in first quadrant } y \text{ is +ve}$$

Hence $B = 4\bar{a}_y + 10\bar{a}_z$

Hence length $B - A = 4\bar{a}_y + 10\bar{a}_z + 4\bar{a}_y = 8\bar{a}_y + 10\bar{a}_z$

Hence distance $\sqrt{8^2 + 10^2} = \sqrt{164} = \sqrt{4 \times 41} = 2\sqrt{41}$ units **Ans.**

Example 1.8. Given cartesian coordinate vector

$$A = y\bar{a}_x + x\bar{a}_y + \frac{x^2}{\sqrt{x^2 + y^2}}a_z$$

Transform into cylindrical coordinate system.

Solution. In the first step we transform the scalar variables from cartesian to cylindrical coordinates using equation (1.12).

$$A = \rho\sin\phi\bar{a}_x + \rho\cos\phi\bar{a}_y + x\cos\phi\bar{a}_z$$

$$= \rho\sin\phi\bar{a}_x + \rho\cos\phi\bar{a}_y + \rho\cos^2\phi\bar{a}_z$$

$$A_\rho = A\cdot\bar{a}_\rho = \rho\sin\phi\bar{a}_x\cdot\bar{a}_\rho + \rho\cos\phi\bar{a}_y\cdot\bar{a}_\rho + \rho\cos^2\phi\bar{a}_z\cdot\bar{a}_\rho$$

$$= \rho\sin\phi\cos\phi + \rho\cos\phi\sin\phi$$

$$= 2\rho\sin\phi\cos\phi = \rho\sin 2\phi \text{ (Using Table 1.1)}$$

$$A_\phi = A\cdot\bar{a}_\phi = \rho\sin\phi\bar{a}_x\cdot\bar{a}_\phi + \rho\cos\phi\bar{a}_y\cdot\bar{a}_\phi + \rho\cos^2\phi\bar{a}_z\cdot a_\phi$$

$$= -\rho\sin^2\phi + \rho\cos^2\phi = \rho\left(\cos^2\phi - \sin^2\phi\right) = \rho\cos 2\phi$$

$$A_z = A\cdot\bar{a}_z = A_z = \rho\cos^2\phi$$

Hence $A = \rho\sin 2\phi\bar{a}_\rho + \rho\cos 2\phi\bar{a}_\phi + \rho\cos^2\phi\bar{a}_z$ **Ans.**

1.5 THE SPHERICAL COORDINATES

Next we study the spherical coordinate system. Here a point P is specified in three dimensional space by the coordinates (r, θ, ϕ) where r is the distance of the point P measured from the origin, θ is the angle between r and the z-axis and ϕ is the angle between x-axis and the projection of r on the $z = 0$ plane or xy-plane as shown in Fig. 1.9. r = constant represents a sphere with centre at the origin and r the radius. θ = constant represents a cone with z-axis as the axis of the cone and origin as the vertex of the cone. The intersection of the cone θ = constant and sphere r = constant is mutually perpendicular everywhere and is a circle of radius $r \sin \theta$. The surface ϕ = constant is a plane passing through $\theta = 0$ or the z-axis. The coordinates of any point P in space is a result of intersection of three mutually perpendicular surfaces, a sphere, a cone and a plane.

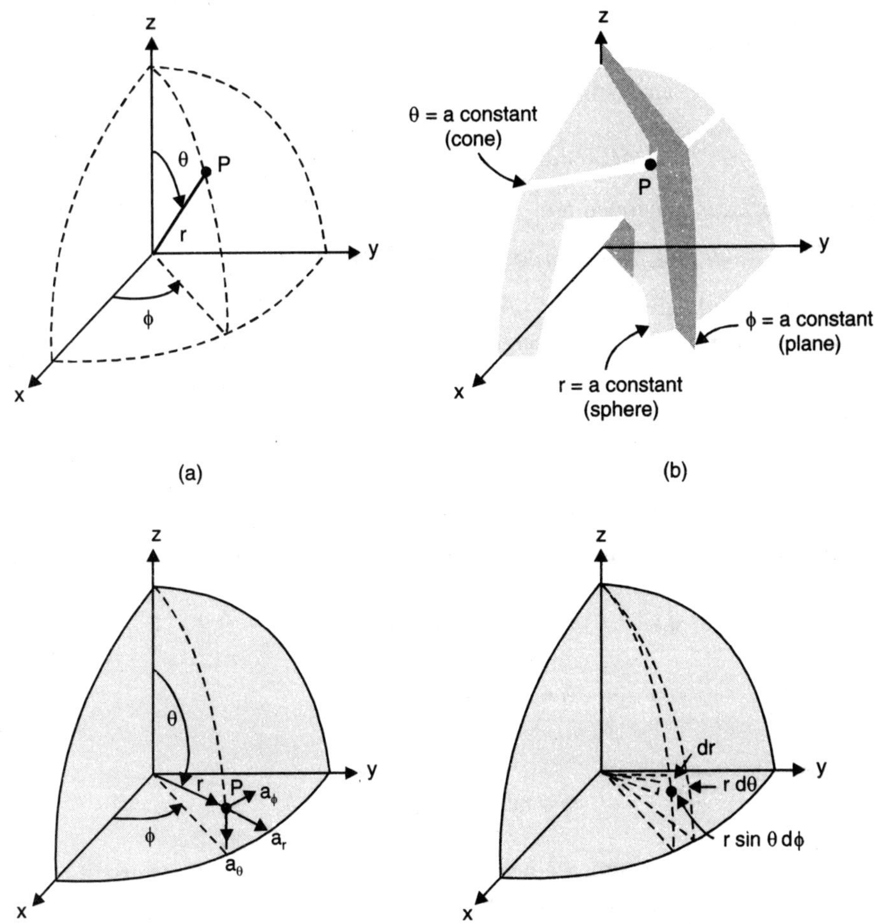

Fig. 1.9 *(a)* The three spherical coordinates. *(b)* The three mutually perpendicular surfaces of the spherical coordinate system. *(c)* The three unit vectors of spherical coordinates: $a_r \times a_\theta = a_\phi$. *(d)* The differential volume element in the spherical coordinate system.

We next define unit vectors in spherical coordinates. $\bar{a}_r$ is a unit vector which is directed in the increasing direction of r, is perpendicular to the sphere r = constant and lies in surface of θ = constant (a cone) and ϕ = constant (a plane). Similarly $\bar{a}_\theta$ is a unit vector directed in the increasing value of θ, is normal to the conical surface θ = constant and lies in the plane ϕ = constant and is tangent to the sphere r = constant.

Similarly $\bar{a}_\phi$ is a unit vector normal to the plane ϕ = constant directed in the increasing value of ϕ and is tangent to both r = constant the sphere and θ = constant a cone. The three unit vectors are shown in Fig. 1.9(*b*). The unit vectors are again mutually perpendicular and follow right handed coordinate system *i.e.* $\bar{a}_r \times a_\theta = a_\phi$, the sequence to be followed is r, θ and ϕ.

Next we consider differential lengths, surfaces and volumes generated as a result of giving differential increments in the three coordinates. When we give differential increments to r, the differential length generated is dr. Similarly when θ is changed to $\theta + d\theta$, the change in length is $rd\theta$ and when ϕ is changed to $\phi + d\phi$, the change in length is $r \sin\theta\, d\phi$ where $r \sin\theta$ is the projection of r on the xy plane or $z = 0$ plane. Therefore when all the three coordinates are incremented, the change in length or the distance between the points P and Q is

$$dL = \sqrt{dr^2 + r^2 d\theta^2 + r^2 \sin^2 \theta d\phi^2} \qquad \text{...(1.17)}$$

The differential surfaces generated are

$$dr\, r\, d\theta,\ dr\, r \sin\theta\, d\phi \text{ and } rd\theta,\ r \sin\theta\, d\phi$$

and the differential volume is given as $dr \cdot rd\theta\ r\sin\theta d\phi$ or $r^2 \sin\theta\ dr\ d\theta\ d\phi$ units.

For transformation from spherical to rectangular coordinates and vice versa, following scalar functions are obtained, see Fig. 1.9(*a*)

$$x = r \sin\theta \cos\phi$$

$$y = r \sin\theta \sin\phi \qquad \text{...(1.18)}$$

and

$$z = r \cos\theta$$

The spherical coordinates in terms of rectangular coordinates are related as follows:

$$r = \sqrt{x^2 + y^2 + z^2} \qquad r \geq 0$$

$$\theta = \cos^{-1} \frac{z}{\sqrt{x^2 + y^2 + z^2}} \qquad \text{...(1.19)}$$

and

$$\phi = \tan^{-1} \frac{y}{x}$$

For transformation of unit vectors, we need to find out dot product of vectors in rectangular and spherical coordinates, we will find these with the help of Fig. 1.9, since the dot product of any spherical unit vector with any rectangular unit vector is the component of unit spherical vector along the rectangular vector, the dot product with a_x are found to be

$$\bar{a}_x \cdot a_r = \sin\theta \cos\phi$$

$$\bar{a}_x \cdot a_\theta = \cos\theta \cos\phi \qquad \text{...(1.20)}$$

$$\bar{a}_x \cdot a_\phi = -\sin\phi$$

Similarly other dot products can be obtained. Table 1.2 gives dot products of unit vectors in spherical and rectangular co-ordinate systems.

Table 1.2

	a_r	a_θ	a_ϕ
$\bar{a}_x$	$\sin\theta\cos\phi$	$\cos\theta\cos\phi$	$-\sin\phi$
$\bar{a}_y$	$\sin\theta\sin\phi$	$\cos\theta\sin\phi$	$\cos\phi$
$\bar{a}_z$	$\cos\theta$	$-\sin\theta$	0

Example 1.9. Determine the vector directed from $\left(10, \frac{3\pi}{4}, \frac{\pi}{6}\right)$ to $\left(5, \frac{\pi}{4}, \pi\right)$ where the end points are given in spherical coordinates.

Solution. Using the relations $x = r \sin\theta \cos\phi$, $y = r \sin\theta \sin\phi$ and $z = r \cos\theta$ we have

$$x_1 = 10\sin\frac{3\pi}{4}\cos\frac{\pi}{6} = +10\cos\frac{\pi}{4}\cos\frac{\pi}{6} = +10\times\frac{1}{\sqrt{2}}\cdot\frac{\sqrt{3}}{2}$$

$$= \frac{10\sqrt{3}}{2\sqrt{2}} = 5\sqrt{\frac{3}{2}}$$

$$y_1 = 10\sin\frac{3\pi}{4}\sin\frac{\pi}{6} = 10\cos\frac{\pi}{4}\sin\frac{\pi}{6} = 10\frac{1}{2\sqrt{2}} = \frac{5}{\sqrt{2}}$$

$$z_1 = 10\cos\frac{3\pi}{4} = -10\sin\frac{\pi}{4} = -\frac{10}{\sqrt{2}}$$

$$(x_1, y_1, z_1) = \left(+5\sqrt{\frac{3}{2}}, \frac{5}{\sqrt{2}}, \frac{10}{\sqrt{2}}\right)$$

$$x_2 = 5\sin\frac{\pi}{4}\cos\pi = -\frac{5}{\sqrt{2}}$$

$$y_2 = 5\sin\frac{\pi}{4}\sin\pi = 0$$

$$z_2 = 5\cos\frac{\pi}{4} = \frac{5}{\sqrt{2}}$$

$$(x_2, y_2, z_2) = \left(-\frac{5}{\sqrt{2}}, 0, \frac{5}{\sqrt{2}}\right)$$

Hence the vector $\left(-\frac{5}{\sqrt{2}} - 5\sqrt{\frac{3}{2}}\right)\bar{a}_x - \frac{5}{\sqrt{2}}\bar{a}_y + \left(\frac{5}{\sqrt{2}} + \frac{10}{\sqrt{2}}\right)\bar{a}_z$

$$= -9.65\bar{a}_x + 3.53\bar{a}_y + 10.608\bar{a}_z \quad \textbf{Ans.}$$

Example 1.10. Transform the cartesian coordinate vector $\left(\frac{xy}{z}\right)\bar{a}_x$ into spherical coordinate vector.

Solution. Let $\bar{A} = \frac{xy}{z}\bar{a}_x$

$$A_r = \bar{A}\cdot\bar{a}_r = \frac{xy}{z}\bar{a}_x\cdot\bar{a}_r = \frac{r\sin\theta\cos\phi\cdot r\sin\theta\sin\phi}{r\cos\theta}\cdot\sin\theta\cos\phi$$

$$= \frac{r\sin^3\theta\sin\phi\cos^2\phi}{\cos\theta}$$

$$A_\theta = \bar{A}\cdot\bar{a}_\theta = \frac{xy}{z}\bar{a}_x\cdot\bar{a}_\theta = \frac{r\sin\theta\cos\phi\cdot r\sin\theta\sin\phi}{r\cos\theta}\cdot\cos\theta\cos\phi$$

$$= r\sin^2\theta\sin\phi\cos^2\phi$$

$$A_\phi = \bar{A}\cdot\bar{a}_\phi = \frac{xy}{z}\bar{a}_x\cdot\bar{a}_\phi = -\frac{r\sin\theta\cos\phi\cdot r\sin\theta\sin\phi}{r\cos\theta}\cdot\sin\phi$$

$$= \frac{-r\sin^2\theta\sin^2\phi\cos\phi}{\cos\theta}$$

Hence $$A = \frac{r\sin^3\theta\sin\phi\cos^2\phi}{\cos\theta}\bar{a}_r + \left(r\sin^2\theta\sin\phi\cos^2\phi\right)\bar{a}_\theta - \frac{r\sin^2\theta\sin^2\phi\cos\phi}{\cos\theta}\cdot\bar{a}_\phi$$ **Ans.**

Example 1.11. Determine the components of $B = 2\bar{a}_x - \bar{a}_y + 2\bar{a}_z$ in spherical coordinates at $(1, \pi/2, \pi)$.

Solution. $B_r = B\cdot\bar{a}_r = \left(2\bar{a}_x - \bar{a}_y + 2\bar{a}_z\right)\cdot\bar{a}_r$

$$= 2\bar{a}_x\cdot\bar{a}_r - \bar{a}_y\cdot\bar{a}_r + 2\bar{a}_z\cdot\bar{a}_r$$

Using Table 1.2.

$$B_r = 2\sin\theta\cos\phi - \sin\theta\sin\phi + 2\cos\theta$$

$$= 2\sin\frac{\pi}{2}\cos\pi - \sin\frac{\pi}{2}\sin\pi + 2\cos\frac{\pi}{2}$$

$$= -2$$

$$B_\theta = B\cdot\bar{a}_\theta = \left(2\bar{a}_x - \bar{a}_y + 2\bar{a}_z\right)\cdot\bar{a}_\theta = 2\bar{a}_x\cdot\bar{a}_\theta - \bar{a}_y\cdot\bar{a}_\theta + 2\bar{a}_z\bar{a}_\theta$$

$$= 2\cos\theta\cos\phi - \cos\theta\sin\phi - 2\sin\theta$$

$$= -2$$

$$B_\phi = \bar{B}\cdot\bar{a}_\phi = \left(2\bar{a}_x - \bar{a}_y + 2\bar{a}_z\right)\cdot\bar{a}_\phi = 2\bar{a}_x\cdot\bar{a}_\phi - \bar{a}_y\bar{a}_\phi + 2\bar{a}_z\cdot\bar{a}_\phi$$

$$= -2\sin\phi - \cos\phi + 0$$

$$= 1$$

Hence, the three components are $B_r = -2,\ B_\theta = -2,\ B_\phi = 1$

$$B = -2\bar{a}_r - 2\bar{a}_\theta + \bar{a}_\phi \quad \textbf{Ans.}$$

1.6 DIFFERENTIATION OF VECTOR

Differentiation of a scalar function $f(x)$ is defined as

$$\frac{df(x)}{dx}_{Lt\,\Delta x\to 0} = \frac{f(x+\Delta x) - f(x)}{\Delta x} \qquad \ldots(1.21)$$

However, if function $f(x)$ is a function of more than one variable, we define what is known as partial derivative of $f(x, y, z)$ with respect to one variable at a time *i.e.*

$$\frac{\partial f}{\partial x}_{Lt\,\Delta x\to 0} = \frac{f\left[\{(x+\Delta x), y, z\} - f(x, y, z)\right]}{\Delta x} \qquad \ldots(1.22)$$

and if all the three variables are incremented by Δx, Δy, Δz, respectively, then change in function

$$\Delta f = \frac{\partial f}{\partial x}\Delta x + \frac{\partial f}{\partial y}\Delta y + \frac{\partial f}{\partial z}\Delta z \qquad \ldots(1.23)$$

where the derivatives are partial derivatives as given above. The differentiation of a vector function $\bar{A}(x,y,z)$ is also defined on the same line as the scalar function and is given as

$$\frac{\partial \bar{A}}{\partial x}_{Lt\,\Delta x\to 0} = \frac{\bar{A}(x+\Delta x, y, z) - \bar{A}(x, y, z)}{\Delta x} \qquad \ldots(1.24)$$

Similarly $\frac{\partial \bar{A}}{\partial y}$ and $\frac{\partial \bar{A}}{\partial z}$ are defined and the differential increment in $\bar{A}$ from point $P(x, y, z)$ to the point $Q(x + \Delta x,\ y + \Delta y,\ z + \Delta z)$ is given as

$$d\bar{A} = \frac{\partial \bar{A}}{\partial x}\Delta x + \frac{\partial \bar{A}}{\partial y}\Delta y + \frac{\partial \bar{A}}{\partial z}\Delta z \qquad \ldots(1.25)$$

Since $\bar{A}(x+\Delta x, y, z) - \bar{A}(x, y, z)$ is a vector, the derivatives $\frac{\partial \bar{A}}{\partial x}$ is also a vector. However it's direction is different from that of $\bar{A}$. Similarly $\frac{\partial \bar{A}}{\partial y}$ and $\frac{\partial \bar{A}}{\partial z}$ are also oriented in different direction as compared to $\bar{A}$. The component vectors of the vector function $\bar{A}(x, y, z)$ are given as follows:

$$\bar{A}(x, y, z) = A_x(x, y, z)\,\bar{a}_x + A_y(x, y, z)\bar{a}_y + A_z(x, y, z)\bar{a}_z$$

and

$$d\bar{A} = dA_x\bar{a}_x + dA_y\bar{a}_y + dA_z\bar{a}_z \qquad \ldots(1.26)$$

Here $\bar{a}_x, \bar{a}_y, \bar{a}_z$ are unit vectors along x, y, z axes respectively and are independent of x, y and z.

In cylindrical and spherical coordinate system the expression for the differential of vector A as in equation (1.26) in cartesian coordinate system does not hold good *i.e.* $d\bar{A} \neq dA_r a_r + d\bar{A}_\phi a_\phi + dA_z\bar{a}_z$.

In order to derive correct expressions for these two coordinate system, the differentiation rule.

$$d(fA) = fdA + dfA \qquad \text{...(1.27)}$$

where f is a scale function

In cylindrical coordinate system

$$\bar{A} = A_r\bar{a}_r + A_\phi\bar{a}_\phi + A_z\bar{a}_z$$

Hence
$$d\bar{A} = d\left(A_r\bar{a}_r + A_\phi\bar{a}_\phi + A_z\bar{a}_z\right) \qquad \text{...(1.28)}$$

$$= A_r d\bar{a}_r + a_r dA_r + a_\phi dA_\phi + A_\phi d\bar{a}_\phi + a_z dA_z \qquad \text{...(1.29)}$$

Similarly in spherical coordinates

$$\bar{A} = A_r\bar{a}_r + A_\theta\bar{a}_\theta + A_\phi\bar{a}_\phi \qquad \text{...(1.30)}$$

we have
$$d\bar{A} = d\left[A_r\bar{a}_r + A_\theta\bar{a}_\theta + A_\phi\bar{a}_\phi\right]$$

$$= A_r d\bar{a}_r + dA_r\bar{a}_r + A_\theta d\bar{a}_\theta + dA_\theta\bar{a}_\theta + dA_\phi\bar{a}_\phi + A_\phi da_\phi \qquad \text{...(1.31)}$$

However, if $\bar{A}$ is also a function of time t in addition to x, y, z we have

$$d\bar{A} = dA_x\bar{a}_x + dA_y\bar{a}_y + dA_z\bar{a}_z \qquad \text{...(1.32)}$$

where
$$dA_i = \frac{\partial A_i}{\partial x}\Delta x + \frac{\partial A_i}{\partial y}\Delta y + \frac{\partial A_i}{\partial z}\Delta z + \frac{\partial A_i}{\partial t}\Delta t, \qquad i = x, y, z \qquad \text{...(1.33)}$$

It is to be noted that the rules for the differentiation of dot and cross product of vectors are given as

$$d(\bar{A}\cdot\bar{B}) = d\bar{A}\cdot\bar{B} + \bar{A}\cdot d\bar{B}$$

$$d(A\times B) = d\bar{A}\times\bar{B} + \bar{A}\times d\bar{B} \qquad \text{...(1.34)}$$

1.7 GRADIENT

The gradient of a scalar function results in a vector function. The magnitude of this vector function at any point in the scalar field region is the maximum rate of increase of the scalar function at that point. Suppose potential at any point in the scalar field is V then gradient of V is in the direction of the maximum rate of change of the potential V and in a direction opposite to that of the electric field. Suppose V is the scalar potential at any point $P(x, y, z)$ and $V + dV$ is the potential at point $Q(x + \Delta x, y + \Delta y, z + \Delta z)$ an infinitesimal distance away from P. Hence the change in potential

$$dV = \frac{\partial V}{\partial x}\Delta x + \frac{\partial V}{\partial y}\Delta y + \frac{\partial V}{\partial z}\Delta z \qquad \text{...(1.35)}$$

$$= \left(\frac{\partial V}{\partial x}\bar{a}_x + \frac{\partial V}{\partial y}\bar{a}_y + \frac{\partial V}{\partial z}\bar{a}_z\right)\cdot\left(\Delta x\bar{a}_x + \Delta y\bar{a}_y + \Delta z\bar{a}_z\right)$$

$$= \left(\frac{\partial}{\partial x}\bar{a}_x + \frac{\partial}{\partial y}\bar{a}_y + \frac{\partial}{\partial z}\bar{a}_z\right)V\cdot\left(\Delta x\cdot\bar{a}_x + \Delta y\bar{a}_y + \Delta z\bar{a}_z\right)$$

$$= \nabla V\cdot d\bar{l} \qquad \text{...(1.36)}$$

where ∇ is known as 'del' or 'nabla' and is a vector operator defined as

$$\nabla = \frac{\partial}{\partial x}\bar{a}_x + \frac{\partial}{\partial y}\bar{a}_y + \frac{\partial}{\partial z}\bar{a}_z \qquad ...(1..37)$$

Suppose P and Q points lie on the equipotential surface at infinitesimal distance $d\bar{l}$ then

$$\nabla V \cdot d\bar{l} = 0$$

Since the dot product is zero, the two vectors are normal to each other. As $d\bar{l}$ lies in the equipotential surface ∇V lies perpendicular to the equipotential surface *i.e.* the gradient of a potential function is a vector field which is everywhere normal to the equipotential surfaces. It is to be noted that the operator ∇ as shown in equation (1.37) has a particularly concise form in rectangular coordinates. It must be mentioned that it consists of derivatives over displacements along the coordinate axes *i.e.* changes with respect to distances and not changes in coordinates.

In cylindrical coordinates when r is changed it is dr and along $\bar{a}_r$, when ϕ is changed by $d\phi$, the change in displacement is $rd\phi$ and along $\bar{a}_\phi$, similarly to z-axis it is dz along $\bar{a}_z$.

Hence
$$dV = \frac{\partial V}{\partial r}dr + \frac{\partial V}{rd\phi}\cdot rd\phi + \frac{\partial V}{\partial z}dz \qquad ...(1.38)$$

and the increment displacement $d\bar{l}$ is $d\bar{l} = \bar{a}_r dr + rd\phi\bar{a}_\phi + \bar{a}_z dz$

Hence

$$dV = \left(\frac{\partial V}{\partial r}\bar{a}_r + \frac{1}{r}\frac{\partial V}{\partial \phi}\bar{a}_\phi + \frac{\partial V}{\partial z}\bar{a}_z\right)\cdot\left(dr\,\bar{a}_r + rd\phi\bar{a}_\phi + \bar{a}_z dz\right)$$

$$= \nabla V \cdot d\bar{l}$$

Thus in cylindrical coordinates

$$\nabla V = \frac{\partial V}{\partial r}\bar{a}_r + \frac{1}{r}\frac{\partial V}{\partial \phi}\bar{a}_\phi + \frac{\partial V}{\partial z}\bar{a}_z \qquad ...(1.39)$$

Similarly in spherical coordinate, the coordinates are r, θ and ϕ. When these are increments, the change in displacements $a_r d_r$, $rd\theta$, $r\sin\theta d\phi$ and hence the del operator in spherical coordinates is

$$\nabla = \left(\frac{\partial}{\partial r}\bar{a}_r + \frac{\partial}{rd\theta}\bar{a}_\theta + \frac{\partial}{r\sin\theta}\bar{a}_\phi\right) \qquad ...(1.40)$$

Example 1.12. Determine the rate at which the scalar function $V = r^2 \sin 3\phi$ in cylindrical coordinates increases in the direction of the vector $\bar{A} = \bar{a}_r + \bar{a}_\phi$ at the point $\left(1, \frac{\pi}{6}, 0\right)$.

Solution. We are required to find out $\nabla V \cdot \frac{\bar{A}}{|A|}$ at $\left(1, \frac{\pi}{6}, 0\right)$.

Now
$$\nabla V = \frac{\partial}{\partial r}\left(r^2 \sin 3\phi\right)\bar{a}_r + \frac{1}{r}\frac{\partial}{\partial \phi}\left(r^2 \sin 3\phi\right)\bar{a}_\phi + \frac{\partial}{\partial z}\left(r^2 \sin 3\phi\right)\bar{a}_z$$

$$= 2r\sin 3\phi\bar{a}_r + \frac{3}{r}\cdot r^2\cos 3\phi\bar{a}_\phi + 0$$

$$= 2r\sin 3\phi\bar{a}_r + 3r\cos 3\phi\bar{a}_\phi$$

$$\nabla V\cdot\bar{A} = (2r\sin 3\phi\bar{a}_r + 3r\cos 3\phi\bar{a}_\phi)\cdot(\bar{a}_r + \bar{a}_\phi)$$

$$= 2r\sin 3\phi + 3r\cos 3\phi$$

Now $\quad |\bar{A}| = \sqrt{1^2+1^2} = \sqrt{2}$

Hence $\quad \nabla V\cdot\frac{\bar{A}}{|\bar{A}|} = \sqrt{2}r\sin 3\phi + \frac{3}{\sqrt{2}}r\cos 3\phi$

At $\quad r = 1$ and $\phi = \frac{\pi}{6}$ $\nabla V\cdot\frac{\bar{A}}{|\bar{A}|} = \sqrt{2} + 0 = \sqrt{2}$ **Ans.**

1.8 LINE INTEGRAL

An important property of the dot product or scalar product involves the line integral. The line integral evaluates along a specified path the integral of the dot product of a vector and the differential displacement vector tangential to the path of integration. Suppose we consider a path from point a to point b as shown in Fig. 1.10.

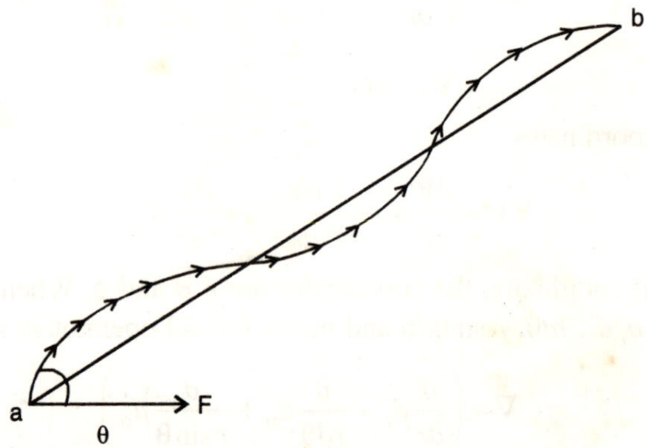

Fig. 1.10 Curve indicating the path of integration divided into several infinitesimal segments and a straight line joining the points a and b

A force F as shown in Fig. 1.10 acts in this region on some mass located at a and to be shifted to point b. In order to find total work done by the force when the particle is moved along the curved path from a to b, we divide the path from a to b into a number of infinitesimal segment of lengths dl_1, dl_2, ..., dl_n. Since the segments are very small, we could treat them as straight line and define their orientations by the corresponding differential vectors $d\bar{l}_1$, $d\bar{l}_2$ ... and within each differential length, and we can assume the force F to be constant. Consider first infinitesimal length dl, and force F_1 and let the angle between the force F_1 and dl_1 is θ_1. The work done by this force is the product of component of force F_1 along dl_1 and the length dl_1 *i.e.* $dW_1 = F_1 \cos\theta_1 dl_1$ or $dW_1 = \bar{F} \cdot d\bar{l}_1$. Similarly the work done by the forces for all the other infinitesimal segments is $F_2 d\bar{l}_2$, $F_3 d\bar{l}_3$...... and so on. Hence total work done from a to b is

$$W = F_1 \cdot dl_1 + F_2 \cdot dl_2 + ... = \sum_i \bar{F}_i \, d\bar{l}_i \qquad ...(1.41)$$

Equation (1.41) assumes constant force vector for any infinitesimal segment which holds good only in the limit when the length of the segment tends to zero *i.e.* it shrinks to a point. In this limit equation (1.41) becomes an integral

$$W = \int_a^b \bar{F} \cdot d\bar{l} \qquad ...(1.42)$$

Here the integration is carried out along the path from a to b irrespective of the shape of the path, it could be a straight line between a and b or any other sharp curvilinear shape. The integral on the right hand side of equation (1.42) is known as a line integral. If the line integral is evaluated around a closed path, it is known as the "circulation" around that path and we write it with a small circle associated with the integral sign as shown here

$$W = \oint_C \bar{F} \cdot d\bar{l} \qquad ...(1.43)$$

However, if the integration is carried out from b to a as in Fig. 1.10 the work done will be

$$W = \int_b^a F \cdot dl = -\int_a^b F \cdot dl \qquad ...(1.44)$$

Here W is the work done in moving against F and is the negative of equation (1.42).

It is to be noted that if the path of integration is a closed loop, we observe the convention of keeping the enclosed area to the left as we travel about the contour *i.e.* we evaluate the integral along the path as we move in a counterclockwise direction.

Example 1.13. Suppose $F = y\bar{a}_x$, determine $\oint \bar{F} \cdot d\bar{l}$ around the rectangular path shown in Fig. E 1.13.

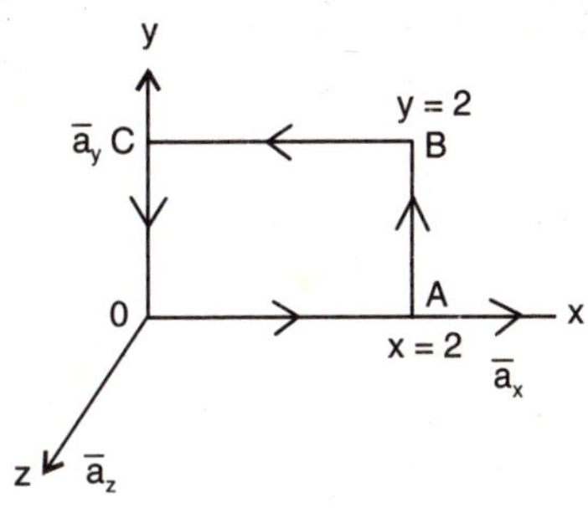

Fig. E 1.13

Solution. Moving from the origin and going in the anticlockwise direction with respect to the rectangular area we have

$$\int_{0}^{2} y\bar{a}_x \cdot \bar{a}_x dx + \int_{0}^{2} ya_x \cdot a_y dy + \int_{0}^{2} y\bar{a}_x \cdot (-a_x dx) + \int_{0}^{2} y\bar{a}_x(-\bar{a}_y dy)$$

$$\quad y=0 \qquad\qquad x=2 \qquad\qquad y=2 \qquad\qquad x=0$$

$$[yx]_0^2 + 0 + (-yx)_0^2 + 0$$

Here in the first bracket $y = 0$ hence $(0.x)_0^2 = 0$

In the third $y = 2$, hence $(-2x)_0^2 = -4$ **Ans.**

While traversing along the arrow, it should be seen that if we are in the positive direction then limits of the integration should be from minimum to maximum. Here from 0 to A since $\bar{a}_x$ is positive therefore the lower limit for $x = 0$ and upper 2. Next from A to B again we are going along +ve $\bar{a}_y$ hence the limits are from minimum to maximum *i.e.* $y = 0$ and $y = 2$.

From B to C we either take $\bar{a}_x$ as negative and hence the limits will be from minimum to maximum *i.e.* $x = 0$ to $x = 2$ or if we take $\bar{a}_x$ as positive, then the limits will be from maximum to minimum. Therefore, the direction should be included either in the unit vector or in the limits, but care should be taken not to do both *i.e.* not to introduce a double negative.

As mentioned earlier the line integral of a vector field depends only on the end points and is independent of the path chosen. The path could be a straight line joining the two points or a curvilinear path. Any field for which the line integral around a closed path is zero, is called a conservative or lamellar field.

Not all fields are conservative *i.e.*, for conservative field

$$\oint F \cdot dl = 0 \qquad \text{...(1.45)}$$

Example 1.14. Given an electric field $E = y\bar{a}_x + x\bar{a}_y + z\bar{a}_z$, determine the line integral of the field between the points (1, 0, 1) to (0.8, 0.6, 1).

Solution. The differential length in cartesian coordinates $d\bar{l} = \bar{a}_x dx + \bar{a}_y dy + \bar{a}_z dz$

Hence
$$E \cdot dl = (y\bar{a}_x + x\bar{a}_y + z\bar{a}_z) \cdot (\bar{a}_x dx + \bar{a}_y dy + \bar{a}_z dz)$$
$$= y\,dx + x\,dy + z\,dz$$

Hence
$$\int E \cdot dl = \int_{1}^{0.8} y\,dx + \int_{0}^{0.6} x\,dy + \int_{1}^{1} z\,dz$$

The equations, of the straight lines passing through the general two points are given as

$$x - x_B = \frac{x_A - x_B}{z_A - z_B}(z - z_B)$$

$$y - y_B = \frac{y_A - y_B}{x_A - x_B}(x - x_B) \qquad \text{...}(\alpha)$$

$$z - z_B = \frac{z_A - z_B}{y_A - y_B}(y - y_B)$$

Any two of these equations for planes passing through the line are sufficient to define the line. From the second equation

$$y - 0 = \frac{(0.6 - 0)}{0.8 - 1}(x - 1) = -\frac{0.6}{0.2}(x - 1) = -3(x - 1)$$

$$y = -3(x - 1) \text{ and } x - 1 = -\frac{y}{3} \text{ or } x = 1 - y/3$$

and from the third equation

$$z - 1 = \frac{0}{0.6}(y - 0) \text{ or } z = 1$$

Hence rewriting the line integral

$$\int E \cdot dl = \int_1^{0.8} -3(x - 1)\,dx + \int_0^{0.6}\left(1 - \frac{y}{3}\right)dy +$$

$$= \left[-\frac{3x^2}{2} + 3x\right]_1^{0.8} + \left[y - \frac{y^2}{6}\right]_0^{0.6}$$

$$= \frac{-3 \times 0.64}{2} + 2.4 + \frac{3}{2} - 3 + 0.6 - \frac{0.36}{6}$$

$= 0.48$ **Ans.**

Again if we want to find out the line integral between the same two points but along the shorter arc of the circle $x^2 + y^2 = 1$, $z = 1$.

We have from this $x = \sqrt{1 - y^2}$ and $y = \sqrt{1 - x^2}$

Substituting in the line integral we have

$$\int E \cdot dl = \int_1^{0.8} \sqrt{1 - x^2}\,dx + \int_0^{0.6} \sqrt{1 - y^2}\,dy + 0$$

$$= \frac{1}{2}\left[x\sqrt{1 - x^2} + \sin^{-1} x\right]_1^{0.8} + \frac{1}{2}\left[y\sqrt{1 - y^2} + \sin^{-1} y\right]_0^{0.6}$$

$$= \frac{1}{2}[0.8 \times 0.36 + 0.927 - 0 - 1.57] + \frac{1}{2}[0.6 \times 0.8 + 0.644 - 0 - 0] = 0.48$$

Hence, it can be seen that the line integral between any two points of the field *E*, irrespective of the path chosen is same.

Example 1.15. Determine the work done in moving a 10 μC charge from (0, 0, 0) to (2, –1, 4) through the field $E = 2xyz\bar{a}_x + x^2z\bar{a}_y + x^2y\bar{a}_z$ V/m via the path of the straight line $x = -2y$, $z = 2x$

Solution. The line integral

$$-Q\int E \cdot dl = -Q\int\left(2xyz\bar{a}_x + x^2z\bar{a}_y + x^2y\bar{a}_z\right)\cdot\left(\bar{a}_x dx + \bar{a}_y dy + \bar{a}_z dz\right)$$

$$= -Q\int_0^2 2xyz\,dx + \int_0^{-1} x^2 z\,dy + \int_0^4 x^2 y\,dz$$

$$= \int_0^2 2x\left(-\frac{x}{2}\right)2x\,dx + \int_0^{-1} 4y^2(-4y)\,dy + \int_0^4 \frac{z^2}{4}\left(-\frac{z}{4}\right)dz$$

$$= \int_0^2 -\frac{x^3}{2}dx + \int_0^{-1} -16y^3 dy + \int_0^4 -\frac{z^3}{16}dz$$

$$= -10\left[\left(-\frac{x^4}{8}\right)_0^2 + \left(-\frac{16y^4}{4}\right)_0^{-1} + \left(-\frac{z^4}{64}\right)^4\right]$$

$$= -10[-2-4-4] = 100\ \mu\text{J} \quad \textbf{Ans.}$$

Example 1.16. A force field F is in the x-direction and increases linearly with distance. Determine the work done by the force F in moving an object from point $x = 2$ to $x = 10$.

Solution. Since the force F increases linearly with distance along x-direction $\bar{F} = a_x x$

Hence work done in moving an object from $x = 2$ to $x = 10$ is

$$\int_2^{10} \bar{a}_x x \cdot \bar{a}_x dx = \int_2^{10} x dx = \left[\frac{x^2}{2}\right]_2^{10} = 48 \text{ joule} \quad \textbf{Ans.}$$

where F is in newton and displacement in metre.

Example 1.17. Determine the line integral of E from $B(0, 0, 0)$ to $A(2, 1, 3)$ if $E = -4xy\bar{a}_x - (4y + 2x^2)\bar{a}_y + 8z\bar{a}_x$. Use the path (*i*) straight line segments (0, 0, 0) to (2, 0, 0) to (2, 1, 0) to (2, 1, 3) (*ii*) direct-straight line.

Solution. Using equation (1.12) for the straight line segments

$$x - x_B = \frac{x_A - x_B}{z_A - z_B}(z - z_B)$$

$$y - y_B = \frac{y_A - y_B}{x_A - x_B}(x - x_B)$$

$$z - z_B = \frac{z_A - z_B}{y_A - y_B}(y - y_B)$$

$$x - 0 = \frac{2-0}{3-0}(z-0) \text{ or } x = \frac{2}{3}z$$

$$y - 0 = \frac{1-0}{2-0}(x-0) = \frac{x}{2}$$

Now $$E \cdot dl = \left\{-4xy\bar{a}_x - \left(4y+2x^2\right)\bar{a}_y + 8z\bar{a}_z\right\} \cdot \left(dx\bar{a}_x + dy\bar{a}_y + dz\bar{a}_z\right)$$

$$= -4xy\,dx - 4y\,dy - 2x^2dy + 8z dz$$

Hence $$\int E \cdot dl = \int_0^2 -4xy\,dx - \int_0^1 4y\,dx - \int_0^1 2x^2dy + \int_0^3 8z\,dz$$

$$= \int_0^2 -4 \cdot x \cdot \frac{x}{2}dx - \left[2y^2\right]_0^1 - \int_0^1 8y^2dy + 36 = -\int_0^2 2x^2dx - 2 - \frac{8}{3} + 36$$

$$= -\left[\frac{2}{3}x^3\right]_0^2 + \frac{94}{3} \quad \text{or} \quad \frac{94}{3} - \frac{16}{3} = \frac{78}{3} = 26\text{ V}$$

(*i*) Using the straight line segments (0, 0, 0) to (2, 0, 0)

$y = 0$ potential is zero

(2, 0, 0) to (2, 1, 0)

$$x - 2 = 2 - 2\ (\ )\ \text{or}\ x = 2$$

$$y - 0 = \frac{1-0}{0}\ (\ )\ \text{or}\ dx = 0$$

$$z - 0 = 0(\ ) = 0$$

$$(2, 1, 0)\ \text{to}\ (2, 1, 3)$$

$$x - 2 = \frac{2-2}{3}(z-0)\ \text{or}\ x = 2$$

$$y - 1 = \frac{0}{0} \qquad dx = 0$$

$$z - 0 = 3.$$

$$F \cdot dl = -4xy\,dx - 4y\,dy - 2x^2dy + 8z\,dz$$

$$\int F \cdot dl = -\int_0^1 4y\,dy - \int_0^1 8\,dy = -2 - 8 = -10$$

$$= -\int_1^1 4ydy - \int_1^1 8\,dy + \int_0^3 8z\,dz = \left(4z^2\right)_0^3$$

$$= 36\text{ V}.$$

Hence net potential = 36 – 10 = 26 V **Ans.**

1.9 SURFACE INTEGRAL

If the flux density of any physical quantity is given over a specified surface we can find out the amount of flux of that quantity crossing that surface using the surface integral of that quantity. Let us consider that the current density at all points on a particular surface s is known. Since current is due to the flow of charge, the current density at a point has both the magnitude and direction. Figure 1.11 shows the division of a surface s into an infinitesimal areas to evaluate the flux of the vector J crossing the surface. The infinitesimal areas are $s_1, s_2, \ldots, s_n$. Since each of these areas is very very small, we can treat them as plane surfaces and define their orientations by their corresponding normal vectors $a_{n_1}, a_{n_2}, \ldots, a_{n_n}$. Also, the current density in each small areas could be considered constant. Also since the differential areas are infinitesimal these could be considered as plane surface.

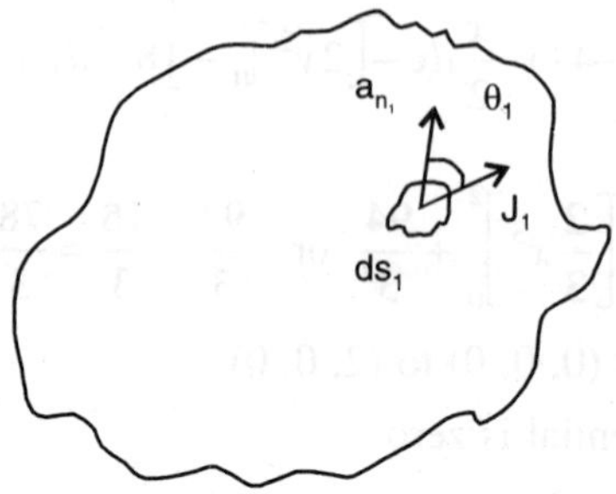

Fig. 1.11 Surface charge distribution

Let us consider infinitesimal area $ds_1\ a_{n_1}$ and its associated current density vector $\bar{J}_1$ as shown in Fig. 1.11 and let θ, be the angle between J_1 and $ds_1\ a_{n_1}$. Hence the current crossing the differential area is $\bar{J}_1 ds\,\bar{a}_{n_1} = J_1 ds\cos\theta_1$.

Similarly, it is possible to find out the current contributed by all the differential surfaces and the total current is the sum of the currents contributed by the various differential surface *i.e.*,

$$I = \Sigma J_n \cdot ds_i \cdot \bar{a}_{n_i} \quad \text{or} \quad \sum_n J_n \cdot \bar{d}s_n \qquad \ldots(1.46)$$

Here $\qquad \bar{d}s_n = ds\,\bar{a}_n$

Equation (1.46) is approximate as the assumptions the current density is constant over the infinitesimal area is time only if $ds \to 0$. Under this situation to make equation (1.46) exact, the summation sign is replaced by the integral sign a_n thus,

$$I = \int_s \bar{J} \cdot d\bar{s} = \int_s J \cdot ds\bar{a}_n \qquad \ldots(1.47)$$

where the integration is performed over the entire surface s. The integral on the right side of equation (1.47) is known as surface integral. This integral is a double integral as $d\bar{s}$ involves two differential lengths. If the surface is closed we call it a closed surface integral and indicate it with a small circle on the integral sign as shown here

$$I = \oint_S \bar{J} \cdot d\bar{s} \qquad \ldots(1.48)$$

Also, the normal vectors to the differential surfaces are chosen to be pointing away from the closed surface so that the integral represents the flux emanating from the volume.

Example 1.18. Water flowing in x-direction of a 3-dimensional space has a rate of flow given as $Q_x = 3yz$ litres/min/m^2. Determine the total flow of water through the rectangular area with corners (0, 0, 0), (0, 4, 0), (0, 0, 3) and (0, 4, 3) m.

Solution. Since the water is flowing along x-direction the discharge $\bar{Q}_x = \bar{a}_x 3yz$ and the differential surface $ds = dy\,dz\,\bar{a}_x$

Hence
$$\int Q_x \cdot ds = \int_0^3 \int_{y=0}^4 3yz\,dy\,dz = \left(\frac{3y^2}{2}\right)_0^4 \int_0^3 z\,dz$$

$$= 24 \cdot \left(\frac{z^2}{2}\right)_0^3 = 24 \times \frac{9}{2} = 108 \text{ litres/min.} \quad \textbf{Ans.}$$

Example 1.19. A 20 μC point charge is located at the origin. Calculate the electric flux passing through (*i*) that portion of the sphere where $r = 20$ cm, bounded by $0 \le \theta \le \pi$ and $0 \le \phi \le \pi/2$ (*ii*) the closed surface $\rho = 0.8$ m, $z = \pm\, 0.5$ m.

Solution. (*i*) In spherical coordinate the differential surface when θ and φ are incremented is given as $rd\theta \cdot r \sin\theta d\phi$. Now the electric flux density at any radius r is

$$D = \frac{q}{4\pi r^2} = \frac{20}{4\pi r^2}\bar{a}_r$$

$$D \cdot ds = D \cdot r^2 \sin\theta d\theta d\phi$$

or
$$\psi = \frac{q}{4\pi r^2}\int_0^{\pi/2}\int_0^{\pi} r^2 \sin\theta d\theta d\phi$$

$$= \frac{q}{4\pi}\int_0^{\pi/2}\int_0^{\pi} \sin\theta d\theta d\phi$$

$$= \frac{q}{4\pi}[\cos\theta]_0^{\pi}\int_0^{\pi/2} d\phi = \frac{20}{4\pi}(12)\cdot \pi/2$$

$$= 5\mu\text{C} \quad \textbf{Ans.}$$

(*ii*) In this section the surface is cylindrical and the differential surface $d\rho dz$.

Hence
$$\oint_S D \cdot ds = \int_0^{0.8}\int_{-0.5}^{5} \frac{q}{4\pi\rho^2}\cdot d\rho\,dz$$

$$= -\left[\frac{20}{4\pi\rho}\right]_0^{0.8}\int_{-0.5}^{0.5} dz = \frac{25}{4\pi} \quad \textbf{Ans.}$$

1.10 VOLUME INTEGRAL

Consider certain volume having some substance with varying density. To find out the amount of quantity present in the volume we make use of volume integral. Suppose the density of the substance within the volume is a function of co-ordinates (x, y, z) or (r, ϕ, z) or (r, θ, ϕ). To obtain the total mass contained in the volume we divide the volume into a large number of infinitesimal volumes $dv, dv_2, \ldots, dv_n$. Within each infinitesimal volume we assume the density to be constant and say is given by $d\rho_1, d\rho_2, \ldots, d\rho_n$. Hence the mass contained within each volume is given by the product of the infinitesimal volume and the density in that volume. The total mass m is then the sum of the infinitesimal masses *i.e.*

$$m = \rho_1 dv_1 + \rho_2 dv_2 + \ldots\ldots = \sum_i \rho_i\, dv_i \ldots\ldots \qquad \ldots(1.49)$$

Equation 1.49 gives only the approximate value of mass m since the density within differential volumes is not exactly constant unless of course when the differential volume $dv \to 0$ *i.e.* the volume shrinks to a point in which case the summation becomes integral *i.e.*

$$m = \int_v \rho dv \ldots \qquad \ldots(1.50)$$

where the integration is carried out throughout the volume V, as indicated by the letter V associated with the integral sign. The integral in the right side of equation (1.50) is known as volume integral. The volume integral is a triple integral since dv is the product of three differential lengths.

Example 1.20. The density of mass of a spherical body of radius R centered at the origin is given by

$$\rho(r, \theta, \phi) = \frac{\rho v}{r^2}$$

where ρv is a constant. Determine the mass of the body.

Solution. The differential length in spherical coordinates are dr, $rd\theta$, $r \sin\theta d\phi$ and hence the differential volume in spherical co-ordinates is

$$dv = r^2 \sin\theta\, dr\, d\theta\, d\phi$$

Hence
$$m = \int_0^R \int_0^\pi \int_0^{2\pi} \frac{\rho v}{r^2} r^2 \sin\theta\, dr\, d\theta\, d\phi$$

$$= \int_0^R \int_0^\pi \int_0^{2\pi} \rho v \sin\theta\, dr\, d\theta\, d\phi$$

$$= \int_0^\pi \int_0^{2\pi} \rho v R \sin\theta\, d\theta\, d\phi$$

$$= \rho v R \int_0^\pi \int_0^{2\pi} \sin\theta\, d\theta\, d\phi = \rho v[-\cos\theta]_0^\pi \int_0^{2\pi} d\phi$$

$$= 2\rho v \cdot 2\pi = 4\pi\rho v \quad \textbf{Ans.}$$

Example 1.21. The density of a mass of substance is given as $\rho = 3\,xyz$ kg/m^3. Determine the total mass in a volume bounded by the planes $x = 0$, $x = 3$, $y = 0$, $y = 4$, $z = 0$, $z = 5$.

Solution. The differential volume in rectangular coordinates is $dx\, dy\, dz$, hence the mass of the differential volume is $\rho\, dx\, dy\, dz$ and the mass bounded by the various planes is the volume integral of the volume bounded by the planes.

Hence
$$M = \int_0^3 \int_0^4 \int_0^5 3xyz\, dx\, dy\, dz$$

$$= \left[\frac{3x^2}{2}\right]_0^3 \int_0^4\int_0^5 yz\,dy\,dz$$

$$= \frac{27}{2}\left(\frac{y^2}{2}\right)_0^4\left(\frac{z^2}{2}\right)_0^5 = \frac{27}{2}\times\frac{16}{2}\times\frac{25}{2} = 27\times 2\times 25$$

$= 1350$ kg. **Ans.**

1.11 DIVERGENCE AND DIVERGENCE THEOREM

A vector field changes from point to point in space in mainly two different ways. The indicators showing the changes are known as Divergence and curl of a vector. The divergence of a vector is a scalar quantity whereas the curl of a vector at a point is a vector quantity. In this section we will deal with divergence of a vector when the divergence of a vector is non zero that region is said to have sources or sinks. If the divergence is positive the region contains sources and if it is negative, it contains sinks. In static electric fields sources are associated with +ve charges Q and sinks are associated with –ve charges $-Q$ and if the divergence is zero at a point in space, there is neither source nor sink present in that region. So positive value of divergence, source and positive charge correspond to one condition while negative value of divergence, sink and negative charge correspond to the other condition. Electric flux ψ by convention originates from the positive charges and terminates at the negative charges. Thus a region which contains positive charge contains the source of ψ.

Divergence of the vector field D at any point P is defined

as
$$\text{div}\, D = \underset{\Delta V \to 0}{\text{Lt}} \frac{\oint D \cdot ds}{\Delta v} \quad \ldots(1.51)$$

Here the integration is over the surface of an infinitesimal volume Δv when $\Delta v \to 0$.

1.11.1 Divergence in Cartesian Coordinate System

The expression for divergence of a vector can be obtained in any coordinate system. However, since Cartesian coordinate system is the simplest of all system, we consider here a differential cuboid with edges Δx, Δy and Δz parallel to the x, y, z axes respectively as shown in Fig. 1.12(a) at any point P.

The vector field D is defined at P, the center of the cuboid with the lowest values of the coordinates x, y and z

$$\bar{D} = \bar{a}_x D_x + \bar{a}_y D_y + \bar{a}_z D_z$$

In order to obtain $\oint D \cdot ds$ for the cuboid, all six faces must be covered. On each face, the direction of ds is outward. Now, in order to evaluate the integral over the closed surface, the integral must be broken up into six integrals, one over each face

$$\oint_S D \cdot ds = \int_{\text{front}} + \int_{\text{back}} + \int_{\text{left}} + \int_{\text{right}} + \int_{\text{top}} + \int_{\text{bottom}}$$

Let us consider the first integral. Since the surface element is very small, $\bar{D}$ is almost constant over the portion of the entire closed surface. Let D_{ox} be the flux density at point P along $\bar{a}_x$. Hence the

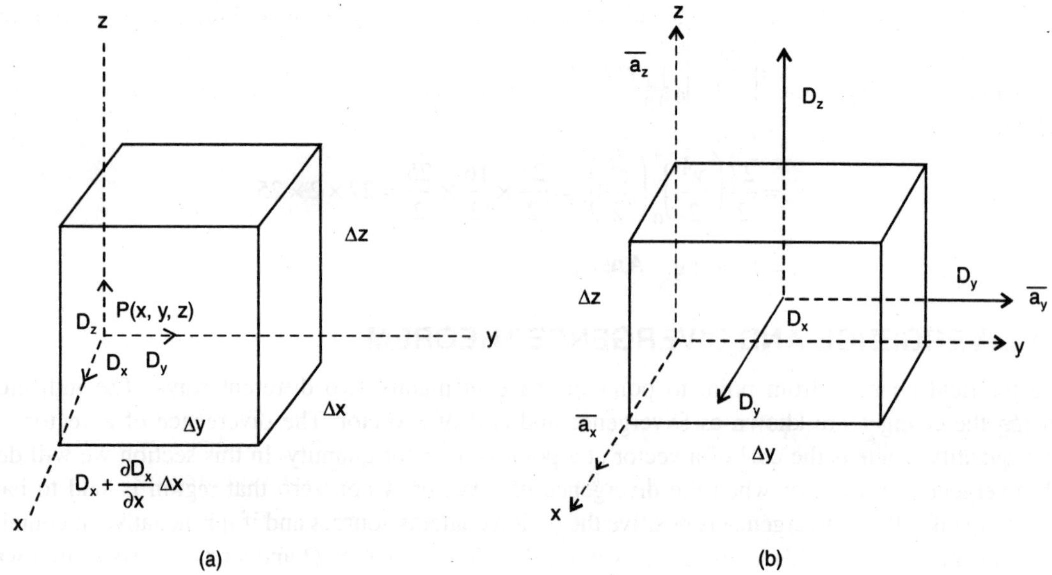

Fig. 1.12 Divergence in Cartesian coordinates

approximate value of D_x normal to front surface which is at a distance of $\frac{\Delta x}{2}$ from the point P is $D_{x\text{front}}\left(x+\frac{\Delta x}{2}\right)$

$$D_{x\text{front}} \approx D_{ox} + \frac{\Delta x}{2}\frac{\partial D_x}{\partial x} + \ldots$$

neglecting the higher terms of Taylor's series.

Therefore, for the front face the integral is

$$\int_{\text{front}} = \int D \cdot ds \text{ where } D = D_{ox} + \frac{\Delta x}{2}\frac{\partial D_x}{\partial x} \text{ and area } ds = dy\,dz\,\bar{a}_x$$

Hence
$$\left(D_{ox} + \frac{\Delta x}{2}\frac{\partial D_x}{\partial x}\right)\cdot \Delta y\,\Delta z\,\bar{a}_x$$

Similarly $D_{x\text{back}} = D_{ox} - \frac{\Delta x}{2}\frac{\partial D_x}{\partial x}$ and area $ds = -\Delta y\,\Delta z\,\bar{a}_x$. Hence the back integral is

$$\left(D_{ox} - \frac{\Delta x}{2}\frac{\partial D_x}{\partial x}\right)\cdot(-\Delta y\,\Delta z\,\bar{a}_x) = \left(-D_{ox} + \frac{\Delta x}{2}\frac{\partial D_x}{\partial x}\right)\Delta y\,\Delta z$$

Combining the two integral *i.e.* front and back we have

$$\int_{\text{front}} + \int_{\text{back}} \approx \frac{\partial D_x}{\partial x}\Delta x\,\Delta y\,\Delta z = \frac{\partial D_x}{\partial x}\Delta v$$

Here with its negative sign is attached as the back area is in the negative direction of $\bar{a}_x$ and Δv is the differential volume of the cuboid.

Similarly, following the same procedure we can write down expressions for the remaining four integrals

$$\int_{\text{right}} + \int_{\text{left}} \approx \frac{\partial D_y}{\partial y}\Delta v \text{ and } \int_{\text{top}} + \int_{\text{bottom}} \approx \frac{\partial D_z}{\partial z}\Delta v$$

Hence if Q is the charge located at point P

$$\int D \cdot ds = Q = \left(\frac{\partial D_x}{\partial x} + \frac{\partial D_y}{\partial y} + \frac{\partial D_z}{\partial z}\right)\Delta v$$

Now divergence of D, by definition is

$$\nabla \cdot D_{\Delta V \to 0} = \frac{\oint_s D \cdot ds}{\Delta v} \qquad \text{...(1.52)}$$

and from the expression above we have

$$\nabla \cdot D_{\Delta V \to 0} = \frac{\oint D \cdot ds}{\Delta v} = \frac{\partial D_x}{\partial x} + \frac{\partial D_y}{\partial y} + \frac{\partial D_z}{\partial z} = \frac{Q}{\Delta v}$$

and the charge enclosed by the differential volume Δv is

$$\left(\frac{\partial D_x}{\partial x} + \frac{\partial D_y}{\partial y} + \frac{\partial D_z}{\partial z}\right)\Delta v$$

and hence the volume charge density at point P is obtained by dividing this charge by the differential volume Δv.

i.e.
$$\rho_v = \frac{\partial D_x}{\partial x} + \frac{\partial D_y}{\partial y} + \frac{\partial D_z}{\partial z} = \nabla \cdot D$$

Hence the divergence of the vector flux density D is the outflow of flux from a differential volume per unit volume as the volume reduces to zero and this is also known as volume charge density.

As mentioned earlier a positive value of divergence at any point in a region represents the presence of source whereas a negative value *a* sink.

The expression

$$\text{div}\,\bar{D} = \frac{\partial D_x}{\partial x} + \frac{\partial D_y}{\partial y} + \frac{\partial D_z}{\partial z} \qquad \text{...(1.53)}$$

is of a form which does not involve charge density, rather it is as a consequence of applying definition of divergence to a differential volume element in rectangular coordinates.

If, however, the differential volumes are expressed in cylindrical coordinates ($d\rho \cdot \rho d\phi \cdot z$) or spherical coordinates ($r^2 \sin\theta\, dr.\, d\theta \cdot d\phi$), the expressions for divergence involving the components of the vector in the particular co-ordinate system and the partial derivatives with respect to the variables of that system would have been obtained. The divergence in cylindrical coordinate system is given as

$$\operatorname{div} \bar{D} = \frac{1}{\rho}\frac{\partial}{\partial p}\left(\rho D_\rho\right) + \frac{1}{\rho}\frac{\partial D_\phi}{\partial \phi} + \frac{\partial D_z}{\partial z} \quad ...(1.54)$$

and in spherical coordinates

$$\operatorname{div} \bar{D} = \frac{1}{r^2}\frac{\partial}{\partial r}\left(r^2 D_r\right) + \frac{1}{r\sin\theta}\frac{\partial}{\partial\theta}\left(\sin\theta D_\theta\right) + \frac{1}{r\sin\theta}\frac{\partial D_\phi}{\partial\phi} \quad ...(1.55)$$

It is to be noted that div D is a dot productor of two vectors, $\nabla \cdot D$ where ∇ is a vector $\left(\frac{\partial}{\partial x}\bar{a}_x + \frac{\partial}{\partial y}\bar{a}_y + \frac{\partial}{\partial z}\bar{a}_z\right)$ and $D = \bar{a}_x D_x + \bar{a}_y D_y + \bar{a}_z D_z$, the divergence of a vector is a scalar quantity.

Example 1.22. Suppose vector field $D = 6x^2\left(\sin\frac{\pi x}{3}\right)\bar{a}_x$. Determine div D at $x = 1$.

Solution. Since D is a function of x and

$$\operatorname{div} D = \frac{\partial D}{\partial x} = \frac{\partial}{\partial x}\left\{6x^2 \sin\frac{\pi x}{3}\right\}$$

$$= 6x^2 \cdot \left(\cos\frac{\pi x}{3}\right)\frac{\pi}{3} + \left(\sin\frac{\pi x}{3}\right)12x$$

$$= 6 \cdot \left(\cos\frac{\pi}{3}\right)\frac{\pi}{3} + \left(\sin\frac{\pi}{3}\right)12$$

$$= \frac{6\pi}{6} + 12 \times \frac{\sqrt{3}}{2} = 3.14 + 6\sqrt{3}$$

$$= 3.14 + 10.39 = 13.53 \quad \textbf{Ans.}$$

Example 1.23. Determine the divergence of a field directed radially and varying inversely with the radius within a cylinder.

Solution. Since the field is directed radially *i.e.* it is along $\bar{a}_r$ and since it varies inversely with radius it is proportional to $\frac{1}{r}$. Therefore the field F is described as

$$\bar{F} = \frac{k}{r}\bar{a}_r$$

It is a case of cylindrical co-ordinates, therefore F_ϕ and F_z are zeros and $F_r = \frac{k}{r}\bar{a}_r$

Therefore, $\operatorname{div} F = \frac{1}{r}\frac{\partial}{\partial r}\left(r \cdot \frac{k}{r}\right) = \frac{1}{r}\frac{\partial}{\partial r}(k) = 0$ **Ans.**

Example 1.24. Let $\nabla \cdot D = \frac{dD_x}{dx} = 50\,\text{pC/m}^3$. Determine D at $x = 5$ cm and Q in the volume $0 \le x \le 10$, $5 \le y \le 15$, $0 \le z \le 12$ cms.

Solution. (*i*) Since $\frac{dD_x}{dx} = 50$. Hence $D_x = 50x = 50 \times \frac{5}{100}$

$$= 2.5 \text{ pc/m}^2 \quad \textbf{Ans.}$$

(*ii*) Since
$$Q = \left(\frac{\partial D_x}{\partial x} + \frac{\partial D_y}{\partial y} + \frac{\partial D_z}{\partial z}\right)\Delta x \, \Delta y \, \Delta z$$

and as
$$\frac{\partial D_y}{\partial y} = 0 \text{ and } \frac{\partial D_z}{\partial z} = 0$$

Hence
$$Q = \frac{\partial D_x}{\partial x}\Delta x \, \Delta y \, \Delta z = 50 \times \frac{10}{100} \times \frac{10}{100} \times \frac{12}{100}$$

$$= 60 \times 10^{-3} \times 10^{-12} = 60 \times 10^{-15} \text{C} \quad \textbf{Ans.}$$

Example 1.25. Determine $\nabla \cdot A$ if $A = xy\bar{a}_x + yz\bar{a}_y + zx\bar{a}_z$.

Solution. $\nabla \cdot A = \frac{\partial}{\partial x}(xy) + \frac{\partial}{\partial y}(yz) + \frac{\partial}{\partial z}(zx)$

$$= y + z + x \quad \textbf{Ans.}$$

Example 1.26. Determine divergence of $A = \sin\theta \, a_\theta + \sin\phi \, \bar{a}_\phi$ at $\left(1, \frac{\pi}{2}, \pi\right)$.

Solution. This is a case of spherical coordinates

where
$$\nabla \cdot D = \frac{1}{r^2}\frac{\partial}{\partial r}\left(r^2 D_r\right) + \frac{1}{r\sin\theta}\frac{\partial}{\partial\theta}\left(\sin\theta D_\theta\right) + \frac{1}{r\sin\theta}\frac{\partial D_\phi}{\partial\phi}$$

Here in the given problem $D_r = 0$ and hence the expression becomes

$$\nabla \cdot A = \frac{1}{r\sin\theta}\frac{\partial}{\partial\theta}\left(r\sin^2\theta\right) + \frac{1}{r\sin\theta}\frac{\partial}{\partial\phi}(\sin\phi)$$

$$= \frac{1}{r\sin\theta} \cdot r \cdot 2\sin\theta\cos\theta + \frac{1}{r\sin\theta} \cdot \cos\phi$$

$$= 2\cos\theta + \frac{\cos\phi}{r\sin\theta}$$

Therefore at $r = 1, \theta = \frac{\pi}{2}$ and $\phi = \pi$

$$\nabla \cdot A = 2\cos\frac{\pi}{2} + \frac{\cos\pi}{1 \cdot \sin\pi/2}$$

$$= 0 + (-1) = -1 \quad \textbf{Ans.}$$

As mentioned earlier a vector field changes from point to point in space. One of the indicators of this change is divergence of a vector at a point which has been discussed in the last section. The second indicator of this change is curl of a vector which will be discussed in the next section whereas divergence of a vector is a dot product of two vectors and hence is a scalar quantity, curl of a vector is the cross product of two vectors and hence is a vector quantity.

1.12 CURL OF A VECTOR

To obtain curl of a vector we apply Ampere's circuital law to the perimeter of a differential surface element. Our objective here is to obtain the point form of Ampere's circuital law.

Because of simplicity we again select cartesian coordinate system and an incremental surface bounded by the sides Δx and Δy as shown here in Fig. 1.13.

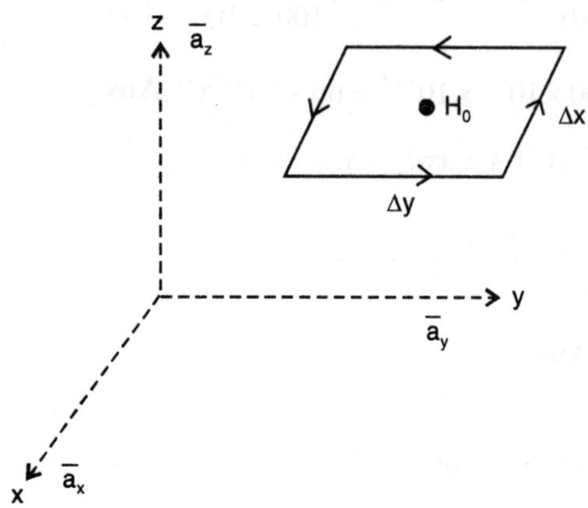

Fig. 1.13 Curl of a vector

Suppose there is some current which gives rise to magnetic field intensity at the center of the differential rectangle as H_0 such that

$$H_0 = H_{ox}\bar{a}_x + H_{oy}\bar{a}_y + H_{oz}\bar{a}_z$$

Our path of travel for integration will be from front to right to back to left differential length along the arrow. The magnetic field intensity along the front to element is

$$H_{y_f} = H_{oy} + \frac{\partial H_y}{\partial x}\frac{\Delta x}{2}$$

Hence, it is $\frac{\partial H_y}{\partial x}$ as we are finding out H at a distance of $\frac{\Delta x}{2}$ from the center point, hence it is partial derivatives of H_y with respect to x and since, it is at a distance of $\frac{\Delta x}{2}$. In fact it is Taylor's series expansion and we have truncated the series with first derivative only and ignored the higher derivative

i.e.
$$H_{yf} = H_y\left(x + \frac{\Delta x}{2}\right) = H_{0y} + \frac{\partial H_y}{\partial x}\frac{\Delta x}{2} + ...$$

Thus
$$(\bar{H}\cdot d\bar{L})_f = \left(H_{oy} + \frac{1}{2}\frac{\partial H_y}{\partial x}\Delta x\right)\Delta y$$

Along the right side of the differential element since the unit vector is $(-\bar{a}_x)$

$$(\bar{H}\cdot\Delta\bar{L})_{\text{right}} \approx H_{xr}(-\Delta x) = -\left(H_{ox} + \frac{1}{2}\frac{\partial H_x}{\partial y}\Delta y\right)\Delta x$$

Taking the remaining two differential lengths and combining the results we have

$$\oint \bar{H}\cdot d\bar{l} = \left(\frac{\partial H_y}{\partial x} - \frac{\partial H_x}{\partial y}\right)\Delta x\,\Delta y$$

However, according to Ampere's circuital law this integral must equal the current enclosed by the differential surface. If J is the current density and in this case using right hand thumb rule the current density will be along positive z-axis and hence the current density could be written as $J_z\bar{a}_z$ and the differential current $\Delta I = J_z\,\Delta x\,\Delta y$ and hence

$$\oint \bar{H}\cdot d\bar{L} \simeq \left(\frac{\partial H_y}{\partial x} - \frac{\partial H_x}{\partial y}\right)\Delta x\,\Delta y = J_z\,\Delta x\,\Delta y$$

or

$$\frac{\int \bar{H}\cdot d\bar{L}}{\Delta x\,\Delta y} \simeq \frac{\partial H_y}{\partial x} - \frac{\partial H_x}{\partial y} \simeq J_z$$

and as the differential area tends to zero *i.e.* it shrinks to a point, the above expression becomes an equality *i.e.*

$$\underset{\text{Lt}\,\Delta x\Delta y\to 0}{\frac{\oint H\cdot dL}{\Delta x\,\Delta y}} = \frac{\partial H_y}{\partial x} - \frac{\partial H_x}{\partial y} = J_z \qquad \text{...(1.56)}$$

In order to obtain J_x and J_y, the x and y components of the current densities we choose closed paths which are oriented perpendicular to each of the remaining two coordinate axes and we finally obtain

$$\underset{\text{Lt}\,\Delta y\,\Delta z\to 0}{\frac{\oint H\cdot dL}{\Delta y\,\Delta z}} = \frac{\partial H_z}{\partial y} - \frac{\partial H_y}{\partial z} = J_x$$

and

$$\underset{\text{Lt}\,\Delta z\,\Delta x\to 0}{\frac{\oint \bar{H}\cdot d\bar{L}}{\Delta z\,\Delta x}} = \frac{\partial H_x}{\partial z} - \frac{\partial H_z}{\partial x} = J_y$$

From the above it is clear that the curl of a vector is a vector and any component of the curl (say J_x) is given by limit of the quotient of the closed line integral of the vector about a differential path in a plane normal to that component desired ($\Delta y\ \Delta z$) and the area enclosed when the differential area tends to a point. This definition of curl is applicable to all coordinate system and is not limited only to cartesian coordinate system.

Combining all the three components of the curl of H *i.e.* J_x, J_y and J_z, the curl of H in cartesian coordinate system is given as

$$\text{Curl } H = \left(\frac{\partial H_z}{\partial y} - \frac{\partial H_y}{\partial z}\right)\bar{a}_x + \left(\frac{\partial H_x}{\partial z} - \frac{\partial H_z}{\partial x}\right)\bar{a}_y + \left(\frac{\partial H_y}{\partial x} - \frac{\partial H_x}{\partial y}\right)\bar{a}_z \qquad \text{...(1.57)}$$

This result may be written in the form of a determinant as

$$\text{Curl } H = \begin{vmatrix} \bar{a}_x & \bar{a}_y & \bar{a}_z \\ \dfrac{\partial}{\partial x} & \dfrac{\partial}{\partial y} & \dfrac{\partial}{\partial z} \\ H_x & H_y & H_z \end{vmatrix}$$

and may also be written in terms of vector product operator

$$\text{Curl } H = \nabla \times H \qquad \ldots(1.58)$$

$$= \left(\frac{\partial}{\partial x}\bar{a}_x + \frac{\partial}{\partial y}\bar{a}_y + \frac{\partial}{\partial z}\bar{a}_z\right) \times \left(H_x\bar{a}_x + H_y\bar{a}_y + H_z\bar{a}_z\right)$$

$$= \frac{\partial H_x}{\partial x}\bar{a}_x \times \bar{a}_x + \frac{\partial H_y}{\partial x}\bar{a}_x \times \bar{a}_y + \frac{\partial H_z}{\partial x}\bar{a}_x \times \bar{a}_z + \frac{\partial H_x}{\partial y}\bar{a}_y \times \bar{a}_x + \frac{\partial H_y}{\partial y}\bar{a}_y \times \bar{a}_y + \frac{\partial H_z}{\partial y}\bar{a}_y \times \bar{a}_z$$

$$+ \frac{\partial H_x}{\partial z}\bar{a}_z \times \bar{a}_x + \frac{\partial H_y}{\partial z}\bar{a}_z \times \bar{a}_y + \frac{\partial H_z}{\partial z}\bar{a}_z \times \bar{a}_z$$

Using the relations

$$\bar{a}_x \times \bar{a}_x = \bar{a}_y \times \bar{a}_y = \bar{a}_z \times \bar{a}_z = 0$$

and

$$\bar{a}_x \times \bar{a}_y = \bar{a}_z, \quad a_y \times a_x = -\bar{a}_z$$

The above nine terms reduce to six terms, three with +ve sign and three with –ve sign and the expression reduces to that given by the determinant.

i.e.

$$\text{Curl } H = \nabla \times H = \left(\frac{\partial H_z}{\partial y} - \frac{\partial H_y}{\partial z}\right)\bar{a}_x + \left(\frac{\partial H_x}{\partial z} - \frac{\partial H_z}{\partial x}\right)\bar{a}_y + \left(\frac{\partial H_y}{\partial x} - \frac{\partial H_x}{\partial y}\right)\bar{a}_z \qquad \ldots(1.59)$$

The expressions for curl H in cylindrical and spherical coordinates are given here under

$$\nabla \times H = \frac{1}{\rho}\left(\frac{\partial H_z}{\partial \phi} - \frac{\partial H_\phi}{\partial z}\right)\bar{a}_\rho + \left(\frac{\partial H_\rho}{\partial z} - \frac{\partial H_z}{\partial \rho}\right)\bar{a}_\phi$$

$$+ \left(\frac{1}{\rho}\frac{\partial(\rho H_\phi)}{\partial \rho} - \frac{1}{\rho}\frac{\partial H_\rho}{\partial \phi}\right)\bar{a}_z \qquad \text{(Cylindrical)} \qquad \ldots(1.60)$$

$$\nabla \times H = \frac{1}{r\sin\theta}\left(\frac{\partial(H_\phi \sin\theta)}{\partial \theta} - \frac{\partial H_\theta}{\partial \phi}\right)\bar{a}_r + \frac{1}{r}\left(\frac{1}{\sin\theta}\frac{\partial H_r}{\partial \phi} - \frac{\partial(rH_\phi)}{\partial r}\right) \cdot a_\theta$$

$$+ \frac{1}{r}\left(\frac{\partial(rH_\theta)}{\partial r} - \frac{\partial H_r}{\partial \theta}\right)\bar{a}_\phi \qquad \text{(Spherical)} \qquad \ldots(1.61)$$

The curl of a vector may be described as circulation per unit area and it is defined at a point. The curl of $\bar{E}$ the electric field intensity must be zero $\nabla \times E = 0$ for the circulation is zero. However, the curl of H is not zero as the circulation of H per unit area is the current density according to Ampere's circuital law.

Example 1.27. Determine curl of the vector $A = \dfrac{-\bar{a}_x y + \bar{a}_y x}{\left(x^2 + y^2\right)^{1/2}}$.

Solution. Now
$$\nabla \times A = \begin{vmatrix} a_x & a_y & a_z \\ \dfrac{\partial}{\partial x} & \dfrac{\partial}{\partial y} & \dfrac{\partial}{\partial z} \\ A_x & A_y & \dfrac{A_z}{0} \end{vmatrix}$$

Here $A_z = 0$

$$\nabla \times A = \bar{a}_z \left\{ \frac{\partial}{\partial x}\left[\frac{x}{\left(x^2 + y^2\right)^{1/2}}\right] + \frac{\partial}{\partial y}\left[\frac{y}{\left(x^2 + y^2\right)^{1/2}}\right]\right\}$$

$$\frac{\partial}{\partial x}\left\{\frac{x}{\left(x^2 + y^2\right)^{1/2}}\right\} = \frac{\left(x^2 + y^2\right)^{1/2} \cdot 1 - x \cdot \dfrac{1}{2} \cdot \dfrac{1}{\left(x^2 + y^2\right)^{1/2}} 2x}{x^2 + y^2}$$

$$= \frac{x^2 + y^2 - x^2}{\left(x^2 + y^2\right)^{3/2}} = \frac{y^2}{\left(x^2 + y^2\right)^{3/2}}$$

Similarly
$$\frac{\partial}{\partial y}\frac{y}{\left(x^2 + y^2\right)^{1/2}} = \frac{x^2}{\left(x^2 + y^2\right)^{3/2}}$$

Hence
$$\nabla \times A = \frac{y^2 + x^2}{\left(x^2 + y^2\right)^{3/2}} \bar{a}_z = \frac{\bar{a}_z}{\left(x^2 + y^2\right)^{1/2}} \quad \textbf{Ans.}$$

Example 1.28. Suppose $H = 0.5z^2\bar{a}_x$ for $z > 0$ and $H = 0$ elsewhere as shown in Fig. E 1.28. Determine $\oint H \cdot dl$ about a square path with side a centred at $(0, 0, z_1)$ in the $y = 0$ plane where $z_1 > a/2$.

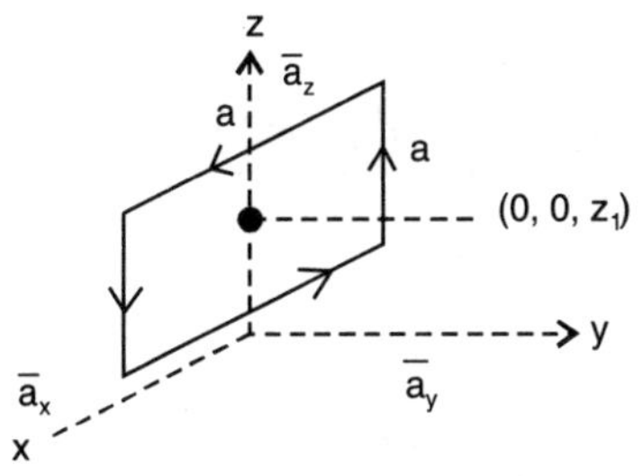

Fig. E 1.28

Solution. Since the differential surface is lying in $y = 0$ plane *i.e.* it is lying in the $x - z$ plane and coinciding with the x and z axes. The direction of the curl of H will be along $\bar{a}_y$ and the magnitude of line integral will be as following:

The value of H along the upper segment will be

$$H_{top} = 0.5\left(z_1 + \frac{a}{2}\right)^2 \text{ and since it is a square } H{\cdot}dl \text{ will be } 0.5\left(z_1 + \frac{a}{2}\right)^2 \cdot a$$

Since H is zero along $\bar{a}_z$ hence $H_{left} \cdot dl = 0$

$$H_{bottom} = 0.5\left(z_1 - \frac{a}{2}\right)^2 \text{ and since its direction is along } \bar{a}_x$$

$$H_{bottom} \cdot d = -0.5\left(z_1 - \frac{a}{2}\right)^2 a$$

Similarly for the right segments $H = 0$ and hence $H \cdot dl = 0$

Therefore
$$\oint H \cdot dl = 0.5\left(z_1 + \frac{a}{2}\right)^2 a + 0 - 0.5\left(z_1 - \frac{a}{2}\right)a + 0$$

$$= z_1 a^2$$

In the limit when area $a^2 \to 0$ we have

$$Lt\, a^2 \to 0 \frac{\oint H \cdot dl}{a^2} = J\bar{a}_y = z_1\bar{a}_y = \nabla \times H$$

The same result can be obtained by using the relation

$$\nabla \times H = \begin{vmatrix} a_x & a_y & a_z \\ \frac{\partial}{\partial x} & \frac{\partial}{\partial y} & \frac{\partial}{\partial z} \\ 0.5z^2 & 0 & 0 \end{vmatrix} = \frac{\partial}{\partial z}(0.5z^2)\bar{a}_y = z_1\bar{a}_y \quad \textbf{Ans.}$$

Example 1.29. Suppose we are given $A = (y\cos ax)\bar{a}_x + (y + e^{2x})\bar{a}_z$. Determine $\nabla \times A$ at (0, 0, 0).

Solution.
$$\nabla \times A = \begin{vmatrix} \bar{a}_x & \bar{a}_y & \bar{a}_z \\ \frac{\partial}{\partial x} & \frac{\partial}{\partial y} & \frac{\partial}{\partial z} \\ y\cos a_x & 0 & y + e^{2x} \end{vmatrix} = \bar{a}_x \frac{\partial(y + e^{2x})}{\partial y} - \bar{a}_y \frac{\partial(y + e^{2x})}{\partial x} - \bar{a}_z \frac{\partial(y\cos a_x)}{\partial y}$$

$$= \bar{a}_x - 2e^{2x}\bar{a}_y - a_z \cos ax$$

At (0, 0, 0)
$$= \bar{a}_x - 2\bar{a}_y - a_z \quad \textbf{Ans.}$$

Example 1.30. Determine J if $H = 3\bar{a}_x + 7y\bar{a}_y + 2x\bar{a}_z$.

Solution.
$$J = \nabla \times H = \begin{vmatrix} \bar{a}_x & \bar{a}_y & \bar{a}_z \\ \frac{\partial}{\partial x} & \frac{\partial}{\partial y} & \frac{\partial}{\partial z} \\ H_x & H_y & H_z \end{vmatrix}$$

$$= \bar{a}_x\left(\frac{\partial H_z}{\partial y} - \frac{\partial H_y}{\partial z}\right) - \bar{a}_y\left(\frac{\partial H_z}{\partial x} - \frac{\partial H_x}{\partial z}\right) + \bar{a}_z\left(\frac{\partial H_y}{\partial x} - \frac{\partial H_x}{\partial y}\right)$$

$$= \bar{a}_x\left(\frac{\partial(2x)}{\partial y} - \frac{\partial(7y)}{\partial z}\right) - \bar{a}_y \cdot \left(\frac{\partial(2x)}{\partial x} - \frac{\partial(3)}{\partial z}\right) + \bar{a}_z\left(\frac{\partial(7y)}{\partial x} - \frac{\partial(3)}{\partial y}\right)$$

$$J = 0 - \bar{a}_y(2) + \bar{a}_z(0) = -2\bar{a}_y \quad \textbf{Ans.}$$

Example 1.31. Determine $\nabla \times F$ if $F = x^2\bar{a}_x + 2yz\bar{a}_y - x^2\bar{a}_z$.

Solution. Using the above expression

$$\nabla \times F = \bar{a}_x\left(\frac{-\partial x^2}{\partial y} - \frac{\partial(2yz)}{\partial z}\right) - \bar{a}_y\left(\frac{\partial(-x^2)}{\partial x} - \frac{\partial(x^2)}{\partial z}\right) + \bar{a}_z\left(\frac{\partial(2yz)}{\partial x} - \frac{\partial x^2}{\partial y}\right)$$

$$= -2y\bar{a}_x + 2x\bar{a}_y + 0 = 2\left(x\bar{a}_y - y\bar{a}_x\right) \quad \textbf{Ans.}$$

Example 1.32. Determine $\nabla \times A$ if $A = 5e^{-r}\cos\phi\,\bar{a}_r - 5\cos\phi\, a_z$ over the xz plane.

Solution.
$$\nabla \times A = \begin{vmatrix} \frac{1}{r}\bar{a}_r & \bar{a}_\phi & \frac{1}{r}a_z \\ \frac{\partial}{\partial r} & \frac{\partial}{\partial \phi} & \frac{\partial}{\partial z} \\ A_r & rA_\phi & A_z \end{vmatrix}$$

$$= \frac{1}{r}\left(\frac{\partial A_z}{\partial \phi} - \frac{\partial A_\phi r}{\partial z}\right)\bar{a}_r - \bar{a}_\phi\left(\frac{\partial A_z}{\partial r} - \frac{\partial A_r}{\partial z}\right) + \frac{1}{r}\bar{a}_z\left(\frac{\partial r A_\phi}{\partial r} - \frac{\partial A_r}{\partial \phi}\right)$$

$$= \frac{1}{r}\left(\frac{\partial}{\partial \phi}(-5\cos\phi)\right)\bar{a}_r - \bar{a}_\phi\left(\frac{\partial}{\partial r}(-5\cos\phi) - 0\right) - \frac{1}{r}\bar{a}_z\frac{\partial}{\partial \phi}5e^{-r}\cos\phi$$

$$= \frac{5\sin\phi}{r}\bar{a}_r + \frac{5}{r}e^{-r}\sin\phi\,\bar{a}_z \quad \textbf{Ans.}$$

This is to be evaluated over the xz plane *i.e.* when $\phi = 0$ to $\phi = \pi$
when $\phi = 0$ $\nabla x = 0$ and also it is zero for $\phi = \pi$.

1.12.1 Stoke's Theorem

From Faraday's law we know that the line integral of the electric field intensity E around a loop of incremental area ΔS enclosing a time varying magnetic flux density B equals the voltage V induced in the loop or mathematically

$$V = \oint E \cdot dl = -\iint_{\Delta S} \frac{\partial B}{\partial t} \cdot d\overline{s} = -\frac{\partial B}{\partial t}\Delta S$$

Dividing both sides by ΔS and taking the limit as $\Delta S \to 0$ around a point yields the curl of E at the point under consideration or

$$\nabla \times E = \underset{\Delta S \to 0}{\text{Lt}} \frac{\oint E \cdot dl}{\Delta S} = -\frac{\partial B}{\partial t}$$

From this it is clear that

$$\oint E \cdot dl = \iint_S (\nabla \times E) \cdot ds \qquad \ldots(1.62)$$

The above equation (1.62) is mathematical statement of Stoke's theorem which states that the line integral of a vector function over a closed surface contour C equals the integral of the curl of that vector function over any surface having C as its boundary.

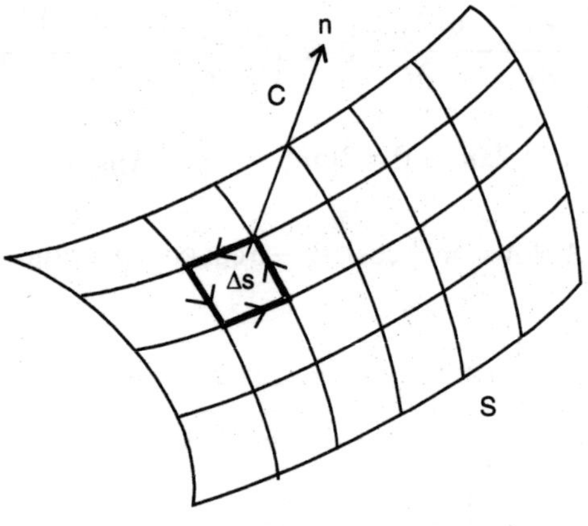

Fig. 1.14 Surface S with differential surfaces ΔS and perimeter C, n is normal to surface ΔS

Consider the surface S with perimeter C and is divided into a large number of differential area ΔS. n is normal to the differential surface. Curl E is the circulation for every ΔS comprising S and sum of all differential areas gives the net result. When we evaluate the closed line integral for each ΔS, some cancellation (line integral) will occur because every interior line is covered once in each direction. The only boundaries where cancellation shall not occur from the outside boundary C, the path enclosing S. Therefore Stoke's theorem relates the line integral of vector with surface integral of the curl of the same vector.

Example 1.33. Verify Stoke's theorem for the vector field $A = \overline{a}_x + y^2\overline{a}_y$ and the flat surface in the yz plane as shown in the Fig. E 1.33 here. Describe the contour C in the anti-clockwise direction.

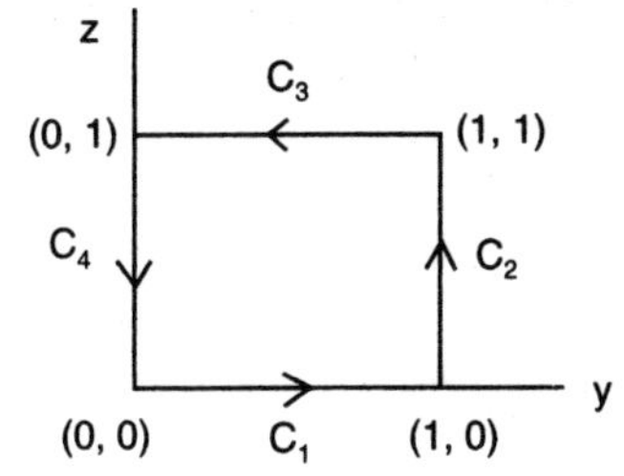

Fig. E 1.33

Solution. We first take the line integral along the contour *C*.

$$\oint A \cdot dl = \int_0^1 0.y^2\, dy + \int_0^1 0\, dz + \int_1^0 1 \cdot y^2 dy + \int_1^0 0\, dz = -\frac{1}{3}.$$

Next we take up the surface integral

$$\int (\nabla \times A) \cdot ds$$

$$\text{New curl of } A \text{ is } \bar{a}_x \frac{\partial F_y}{\partial z} = -\bar{a}_x \frac{\partial}{\partial z}\left(zy^2\right) = -\bar{a}_x y^2$$

and $\quad ds = dydz$

Hence $\quad \int (\nabla \times A) \cdot ds = \int_{y=0}^{1} \int_{z=0}^{1} -y^2 dy\, dz = -\frac{1}{3}$

Hence Stoke's theorem is verified.

1.13 POISSON'S EQUATION AND LAPLACE'S EQUATION

We know that

$$\nabla \cdot D = \nabla \cdot \varepsilon E = \rho$$

or $\quad \nabla \cdot E = \frac{\rho}{\varepsilon}$

Now $\quad E = -\nabla V$

Hence $\quad \nabla \cdot (-\nabla V) = \frac{\rho}{\varepsilon} \qquad \ldots(1.63)$

or $\quad \nabla^2 V = -\frac{\rho}{\varepsilon}$

This is known as Poisson's Equation. For a charge free region $\rho = 0$, *i.e.* volume charge density is zero but allowing point charges and surface charge density to exist at singular location as sources of the field $\nabla^2 V = 0$ then, this is known as Laplace's Equation.

From equation (1.63) it is clear that Laplace's equation is the divergence of the gradient of *V* and since using this fact expression for Laplace's equation in different coordinate system can be obtained.

Cartesian Coordinates

The expression for gradient is

$$\nabla V = \frac{\partial V}{\partial x}\bar{a}_x + \frac{\partial V}{\partial y}\bar{a}_y + \frac{\partial V}{\partial z}\bar{a}_z$$

and the operator

$$\nabla = \frac{\partial}{\partial x}\bar{a}_x + \frac{\partial}{\partial y}\bar{a}_y + \frac{\partial}{\partial z}\bar{a}_z$$

Hence

$$\nabla \cdot \nabla V = \left(\frac{\partial}{\partial x}\bar{a}_x + \frac{\partial}{\partial y}\bar{a}_y + \frac{\partial}{\partial z}\bar{a}_z\right)\cdot \frac{\partial V}{\partial x}\bar{a}_x + \frac{\partial V}{\partial y}\bar{a}_y + \frac{\partial V}{\partial z}\bar{a}_z$$

$$= \frac{\partial^2 V}{\partial x^2} + \frac{\partial^2 V}{\partial y^2} + \frac{\partial^2 V}{\partial z^2} \qquad \text{...(1.64)}$$

Cylindrical Coordinates

$$\nabla V = \frac{\partial V}{\partial r}\bar{a}_r + \frac{1}{r}\frac{\partial V}{\partial \phi}\bar{a}_\phi + \frac{\partial V}{\partial z}\bar{a}_z \qquad \text{...(1.65)}$$

For a general vector field A

$$\nabla \cdot A = \frac{1}{r}\frac{\partial}{\partial r}\cdot(rA_r) + \frac{1}{r}\frac{\partial A_\phi}{\partial \phi} + \frac{\partial A_z}{\partial z}$$

$$\frac{\partial V}{\partial r} \leftrightarrow A_r \quad \frac{\partial V}{\partial \phi} \rightarrow A_\phi \text{ and } \frac{\partial V}{\partial z} \rightarrow A_z$$

$$\nabla^2 V = \nabla \cdot \nabla V = \frac{1}{r}\frac{\partial}{\partial r}\left(r\frac{\partial V}{\partial r}\right) + \frac{1}{z^2}\frac{\partial^2 V}{\partial \phi^2} + \frac{\partial^2 V}{\partial z^2} = 0 \qquad \text{...(1.66)}$$

Spherical Coordinates

$$\nabla V = \frac{\partial V}{\partial r}\bar{a}_r + \frac{1}{r}\frac{\partial V}{\partial \theta}\bar{a}_\theta + \frac{1}{r\sin\theta}\frac{\partial V}{\partial \phi}\bar{a}_\phi$$

and

$$\nabla \cdot A = \frac{1}{r^2}\frac{\partial}{\partial r}\left(r^2 A_r\right) + \frac{1}{r\sin\theta}\frac{\partial}{\partial \theta}\left(A_\theta \sin\theta\right) + \frac{1}{r\sin\theta}\frac{\partial A_\phi}{\partial \phi}$$

so that the Laplace's equation is

$$\nabla^2 V = \frac{1}{r^2}\frac{\partial}{\partial r}\left(r^2\frac{\partial V}{\partial r}\right) + \frac{1}{r^2\sin\theta}\frac{\partial}{\partial \theta}\left(\sin\theta\frac{\partial V}{\partial \theta}\right) + \frac{1}{r^2\sin^2\theta}\frac{\partial^2 V}{\partial \phi^2} = 0 \qquad \text{...(1.67)}$$

Laplace's equations can be used wherever volume charge density is zero. It states that every possible configuration of electrodes or conductors produces a field for which $\nabla^2 V = 0$. All these fields would be different with different potential values and space rates of charge yet for each of them $\nabla^2 V = 0$. If $\rho_v = 0$ every field satisfies Laplace's equation to find one specific field in which we would be interested. Obviously more information is required and we shall find that we must solve Laplace's equations subject to certain boundary conditions.

The solution of Laplace's equation which is a second order equation within different boundary conditions will be different. The solution of Laplace's equations with certain boundary conditions must give the same solution irrespective of which technique or method is used for solution of these equation. If we solve the equation and if we get two solutions as V_1 and V_2 then V_1 must equal to V_2. This is known as uniqueness theorem and is discussed hereunder.

1.14 UNIQUENESS THEOREM

Suppose we have two solutions of Laplace's equations V_1 and V_2 which are of course, functions of the coordinates used. Therefore,

$$\nabla^2 V_1 = 0 \text{ and } \nabla^2 V_2 = 0$$

and hence $$\nabla^2 (V_1 - V_2) = 0 \qquad ...(1.68)$$

Solution. V_1 and V_2 have been obtained with the same boundary condition and say the boundary potential for both is V_b and hence the value of V_1 on the boundary is V_{1b} and that for V_2 it is V_{2b} and these must be identical to V_b.

Hence $$V_{1b} = V_{2b} = V_b$$

or $$V_{1b} - V_{2b} = 0$$

Next we make use of the vector identity

$$\nabla \cdot (V\bar{D}) = V(\nabla \cdot \bar{D}) + \bar{D} \cdot (\nabla V)$$

(This is identical to finding differential of product of two function.)

Here V and $\bar{D}$ are any scalar and vector respectively. We select here $(V_1 - V_2)$ as scalar and $\nabla (V_1 - V_2)$ as vector. Thus

$$\nabla \cdot [(V_1 - V_2)\nabla(V_1 - V_2)] = (V_1 - V_2)[\nabla \cdot \nabla(V_1 - V_2)] + \nabla(V_1 - V_2) \cdot \nabla(V_1 - V_2)$$

We integrate both the sides throughout the volume enclosed by the boundary surfaces specified

$$\int_{V_d} \nabla \cdot [(V_1 - V_2)\nabla(V_1 - V_2)]dv = \int_{V_d} (V_1 - V_2)[\nabla \cdot \nabla(V_1 - V_2)]dv + \int_{V_d} [\nabla(V_1 - V_2)]^2 dv \qquad ...(1.69)$$

According to Divergence theorem we can replace the volume integral on the left hand side of the equation with the closed surface integral over the surface surrounding the volume. This surface is nothing but the one specified earlier where $V_{1b} = V_{2b}$ and hence

$$\int_{V_d} \nabla \cdot [(V_1 - V_2)\nabla(V_1 - V_2)]dv = \oint_S [(V_{1b} - V_{2b})\nabla(V_{1b} - V_{2b})] \cdot d\bar{s} = 0$$

One of the factors in the first-integral on the right side of equation (1.69) *i.e.* $\nabla \cdot \nabla(V_1 - V_2) = \nabla^2 (V_1 - V_2)$ is zero as per equation (1.68) and therefore, the second term on the right side of equation (1.69) should be zero *i.e.*

$$\int_{V_d} [\nabla(V_1 - V_2)]^2 dv = 0$$

There are two possibilities for this integral to be zero. The first is that the integrand is zero and the other possibility is that the integrand is positive for some region of the surface and negative for the other region such that the net integrand is zero. In this case since the integrand is a square term and hence the net integrand can not be negative. Therefore, the first possibility holds good in this case *i.e.*

$$\left[\nabla\left(V_1 - V_2\right)\right]^2 = 0$$

or

$$\nabla\left(V_1 - V_2\right) = 0$$

Now if gradient of a function is zero *i.e.* the function must be constant *i.e.*

$$V_1 - V_2 = \text{constant}$$

Since at a point on the boundary the potential V_b cannot be different for different solutions hence $V_{1b} = V_{2b}$ or $V_{1b} - V_{2b} = 0$. Also at the boundary $V_1 - V_2 = V_{1b} - V_{2b} = 0$ and, therefore, we see that the constant is indeed zero and therefore,

$$V_1 = V_2$$

which means the two solutions are identical.

It is to be noted that uniqueness theorem is applicable not only to Laplace's equation, it is applicable to Poisson's equation as well as, if $\nabla^2 V_1 = -\rho/\varepsilon$ and $\nabla^2 V_2 = -\rho/\varepsilon$ then $\nabla^2(V_1 - V_2) = 0$ as in Laplace's equation. Boundary Condition still require that $V_{1b} = V_{2b}$ or $V_{1b} - V_{2b} = 0$. Thus uniqueness theorem ensures us that whatever be the technique or method used for the solution of Laplace's equation or Poisson's equation subject to a given set of boundary condition, the result will be identical.

Laplace's equations are second order partial differential equations and various methods have been used for solution of these equations. The first and of course the simplest of the method is the direct integration. We will use this technique for solution of one dimensional variation of potential problem in various coordinate system.

1.15 PHYSICAL INTERPRETATION OF GRADIENT, DIVERGENCE AND CURL

The three operations which are carried out with the operator ∇(del) have important physical significance when operated on scalar and vector fields. These are explained as hereunder.

Gradient: The gradient of a scalar function is the maximum space rate of change of that function and is in the increasing direction of the function. Suppose if T represents the temperature at any point in a field then $\nabla T = \text{grad } T$ is a temperature gradient and is the rate of change of temperature with distance. However, if we are moving in a direction where the temperature is decreasing then $\nabla T = -\text{grad } T$. It is evident that although the temperature T is a scalar quantity having magnitude only and no direction, the temperature gradient ∇T is a vector quantity the direction being that in which the temperature changes most rapidly with distance. Like any other vector quantity, this can also be expressed in terms of its components in x, y and z direction. These components are $\frac{\partial T}{\partial x}, \frac{\partial T}{\partial y}$ and $\frac{\partial T}{\partial z}$ respectively and the resultant temperature gradient is the vector sum of these three components *i.e.*

$$\nabla T = \frac{\partial T}{\partial x}\bar{a}_x + \frac{\partial T}{\partial y}\bar{a}_y + \frac{\partial T}{\partial z}\bar{a}_z$$

If we have a positive point charge in space, this gives rise to electric field lines emanating from the charge. As a result a potential field is set up and the voltage decreases as we move away from the charge. Whereas the direction of field lines is away from the charge, the voltage decreases so we go away from the charge, hence the gradient is negative.

As we move away from the charge and hence $E = -\nabla V$ as the maximum rate of change of voltage along the field line is decreasing.

Curl: The concept of curl or rotation of a vector quantity is illustrated in the stream flow problem. Figure 1.15 (*a*) shows a stream on the surface of which floats a leaf in the *x*-*y* plane.

If the leaf is moving along say *y*-axis only, then it is a translational motion and hence curl of velocity at any point is zero. However if there are eddies in the water, the leaf will have both rotational and translational motion and the rotational motion in this case will be along *z*-axis and the direction of rotation will be from *x*-axis to *y*-axis *i.e.* in the anti clockwise direction as shown in Fig. 1.15 (*b*). The rate of rotation or angular velocity at any point is a measure of the curl of the velocity of water at that point and curl is positive as the rotation is from *x* to *y* *i.e.* anticlockwise. In Fig. 1.15 (*b*) it can be seen that a positive value for $\dfrac{\partial V_y}{\partial x}$ will tend to rotate the leaf in anticlockwise direction whereas a positive

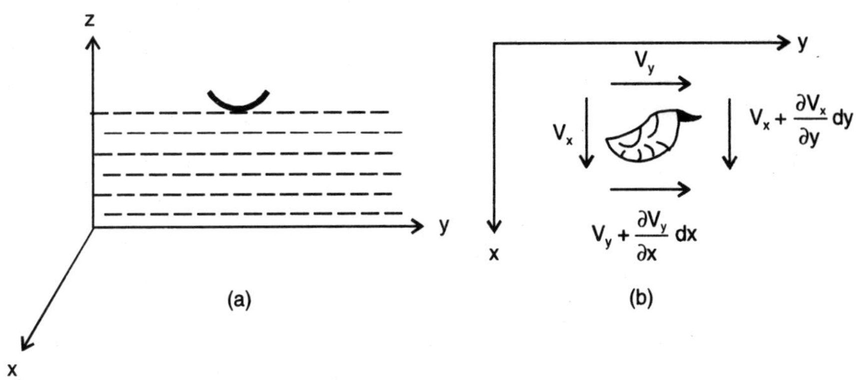

Fig. 1.15 Rotation of a floating leaf

value for $\dfrac{\partial V_x}{\partial y}$ will produce a clockwise rotation. The rate of rotation about the *z*-axis is therefore proportional to the difference between these two quantities. By definition

$$\text{Curl}_z \bar{v} = \nabla_z \times V = \left(\frac{\partial V_y}{\partial x} - \frac{\partial V_x}{\partial y} \right) \bar{a}_z$$

However, in general, the rotation may be about the *x* and *y* axes as well and hence the corresponding components of the curl are given by

$$\text{Curl}_x \bar{v} = \left(\frac{\partial V_z}{\partial y} - \frac{\partial V_y}{\partial z} \right)$$

$$\text{Curl}_y \bar{v} = \left(\frac{\partial V_x}{\partial z} - \frac{\partial V_z}{\partial x} \right)$$

The rotation about any axis can always be expressed as the sum of the component rotations about the *x*, *y* and *z*-axes. Since the rotations (curl) has both the direction and magnitude, this will be a vector sum and the resultant rate of rotation or angular velocity will be proportional to

$$\text{Curl } \bar{v} = \left(\frac{\partial V_z}{\partial y} - \frac{\partial V_y}{\partial z}\right)\bar{a}_x + \left(\frac{\partial V_x}{\partial z} - \frac{\partial V_z}{\partial x}\right)\bar{a}_y + \left(\frac{\partial V_y}{\partial x} - \frac{\partial V_x}{\partial y}\right)\bar{a}_z$$

The direction of the resultant curl is the axis of rotation.

Divergence: Physical significance of divergence can be understood with the help of fluid motion. Consider first the flow of incompressible fluid. Water is more or less an incompressible fluid. Consider a differential rectangular parallelopiped of dimension Δx, Δy, Δz in a flowing incompressible fluid as shown in Fig. 1.16.

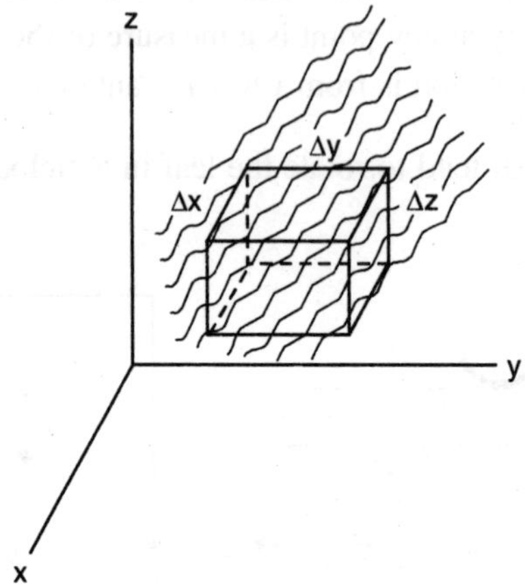

Fig. 1.16 A differential parallelopiped volume $\Delta x\ \Delta y\ \Delta z$

Consider the left surface with area $\Delta x\ \Delta z$ and the right surface opposite to the left surface with area again $\Delta x\Delta z$. It can be seen that whatever volume of water enters the left surface, the same volume leaves the right surface. This holds good for the upper and lower surfaces and so also in case of front and back surfaces *i.e.* the net divergence of water from the differential parallelopipe is zero. However, if we consider compressible fluids like steam, air or any gas we find that there is divergence of the fluid from a differential volume and hence divergence is positive. Sometimes the fluid may converge into the differential volume hence the divergence will be negative. Consider a rubber tube filled with air. If the tube bursts, there is divergence of air from the tube into the atmosphere and hence divergence is positive. However if we consider an incandescent lamp which has partial vacuum within and as it breaks down, the air from the atmosphere rushes into it and hence air converges into the lamp and hence divergence is negative.

From the above it is clear that the divergence of an incompressible fluid at a point is zero. An incompressible fluid can not diverge from nor converge towards a point.

Example 1.34. Determine the potential function for the region between the parallel circular discs. The lower disc is lying in the xy-plane.

Solution. Since V is not a function of r or ϕ
Laplace's equation reduces to

$$\frac{d^2V}{dz^2} = 0 \text{ or } \frac{dV}{dz} = A$$

or

$$V = Az + B$$

If, however, the parallel plates are placed in the *yz*-plane then $V = Ax + B$ and if these are placed in *zx*-plane then $V = Ay + B$.

Example 1.35. Determine the potential function and the electric field intensity for the region between two concentric right circular cylinders where $V = 0$ at $r = 2$ mm and $V = 100$ V at $r = 25$ mm. Neglect fringing. Also obtain expression for capacitance of this system.

Solution. (*i*) We will use here right circular cylindrical coordinates for the system and since cylinder is placed vertical, the potential is constant with ϕ and z and therefore the Laplace's equation reduces to

$$\frac{1}{r}\frac{d}{dr}\left(r\frac{dV}{dr}\right) = 0$$

or
$$\frac{d}{dr}\left(r\frac{dV}{dr}\right) = 0$$

Integrating once we have

$$r\frac{dV}{dr} = A$$

or
$$\frac{dV}{dr} = \frac{A}{r}$$

or
$$V = A\log r + B$$

Using boundary conditions as there are two unknowns

$$0 = A \ln 0.002 + B$$

$$100 = A \ln 0.025 + B$$

$$100 = A(\ln 0.025 - \ln 0.002)$$

$$= A(-3.6888 + 6.214)$$

$$= 2.525A$$

or
$$A = 39.6$$

Now
$$100 = 39.6 \ln 0.025 + B$$

$$= 39.6\ (-3.688) + B$$

or
$$B = 246$$

Hence
$$V = 39.6 \ln r + 246 \text{ volts}$$

$$E = -\frac{dV}{dr} = \frac{39.6}{r}(-\bar{a}_r)\,\text{V/m}. \quad \textbf{Ans.}$$

(*ii*) Calculation of capacitance

Let at $r = a,\ V = V_0$ and at $r = b\ V = 0\ b > a$

$$V = A \ln r + B$$

$$0 = A \ln b + B \quad \text{and} \quad V_0 = A \ln a + B$$

$$V_0 = A\ln\frac{a}{b} \quad \text{or} \quad A = V_0/\ln a/b$$

and $$B = -A\ln b = -\frac{V_0}{\ln a/b}\ln b$$

$$V = \frac{V_0}{\ln a/b}\ln r - \frac{V_0}{\ln a/b}\ln b = \frac{V_0}{\ln a/b}\ln\frac{r}{b} = \frac{V_0 \ln b/r}{\ln b/a}$$

$$E = -\frac{dV}{dr} = \frac{V_0}{r\ln b/a}\bar{a}_r = \frac{V_0}{a\ln b/a}\bar{a}_r$$

$$D = \varepsilon E = \frac{\varepsilon V_0}{a\ln b/a}$$

$$Q = D\cdot \text{Area} = \frac{\varepsilon V_0}{a\ln b/a}\cdot 2\pi a \text{ per unit length}$$

or $$\frac{Q}{V_0} = \frac{2\pi\varepsilon}{\ln b/a}\text{ F/metre} \quad \textbf{Ans.}$$

Example 1.36. Determine the capacitance of a spherical shell with inner radius a and outer radius b. The outer shell is held at $V = 0$.

Solution. Since potential is not a function of θ and ϕ, then Laplace's equation becomes

$$\frac{1}{r^2}\frac{\partial}{\partial r}\left(r^2\frac{\partial V}{\partial r}\right) = 0 \quad \text{or} \quad \frac{\partial}{\partial r}\left(r^2\frac{\partial V}{\partial r}\right) = 0$$

Integrating once $$r^2\frac{dV}{dr} = A \quad \text{or} \quad \frac{dV}{dr} = \frac{A}{r^2}$$

Integrating again $$V = -\frac{A}{r} + B$$

At $$r = b \qquad V = 0 \qquad 0 = -\frac{A}{b} + B$$

$$r = a \qquad V = V_0 \quad \text{and} \quad V_0 = -\frac{A}{a} + B$$

or $$V_0 = \frac{A}{b} - \frac{A}{a} = A\left[\frac{1}{b} - \frac{1}{a}\right]$$

or $$A = \frac{V_0}{\frac{1}{b} - \frac{1}{a}} \text{ and } B = \frac{A}{b} = \frac{V_0/b}{\frac{1}{b} - \frac{1}{a}}$$

$$E = -\frac{dV}{dr} = +\frac{A}{r^2} \text{ or } E = \frac{A}{r^2}$$

$$D = \varepsilon E = \frac{\varepsilon A}{r^2}$$

$$Q = D \cdot 4\pi \cdot r^2 = 4\pi\varepsilon A \quad \text{or} \quad Q = \frac{4\pi\varepsilon V_0}{\frac{1}{b} - \frac{1}{a}}$$

or $$C = \frac{Q}{V_0} = \frac{4\pi\varepsilon}{\frac{1}{b} - \frac{1}{a}} \quad \textbf{Ans.}$$

Example 1.37. Two ϕ = constant planes are insulated along the z-axis as shown. Determine the expression for E between the planes, assuming a potential of 200 V for $\phi = \alpha$ and V is zero at $\phi = 0$. Neglect fringing.

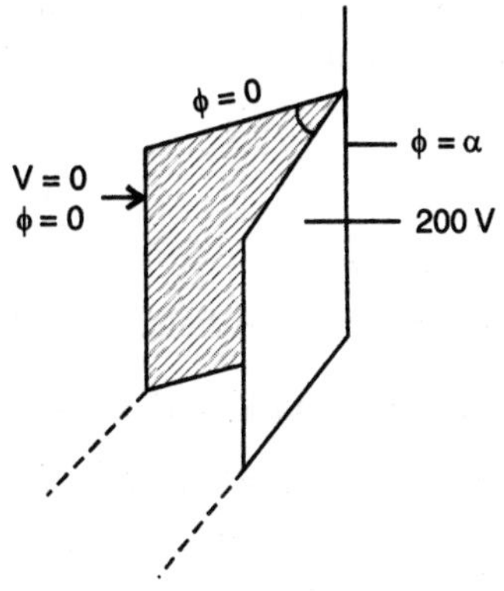

Fig. E 1.37

Solution. Using Laplace equation in cylindrical coordinates and since here potential is constant with r and z, Laplace's equation is

$$\frac{1}{r}\frac{d^2V}{d\phi^2} = 0 \qquad \text{or} \qquad \frac{d^2V}{d\phi^2} = 0$$

Integrating once $\dfrac{dV}{d\phi} = A$ and $V = A\phi + B$

Using boundary condition to obtain A and B we have at $\phi = 0$, $V = 0$ and $\phi = \alpha$, $V = 200$ V

$$0 = A \cdot 0 + B \quad \text{or} \quad B = 0$$

$$200 = A\alpha \quad \text{or} \quad A = \frac{200}{\alpha}$$

Therefore, $$V = \frac{200\phi}{\alpha}$$

Now $$E = -\nabla V = -\frac{1}{r}\frac{d}{d\phi}\left(200\frac{\phi}{\alpha}\right)\bar{a}_\phi$$

$$= -\frac{200}{r\alpha}\bar{a}_\phi \quad \textbf{Ans.}$$

Example 1.38. Two parallel conducting plates occupying the planes $x = 0$ and $x = d$ are kept at potentials $V = 0$ and $V = V_0$ respectively. The region between the two plates consists of dielectric media having permittivity ε_1 for $0 < x < t$ and ε_2 for $t < x < d$ as shown. Determine the potential distributions in the two regions.

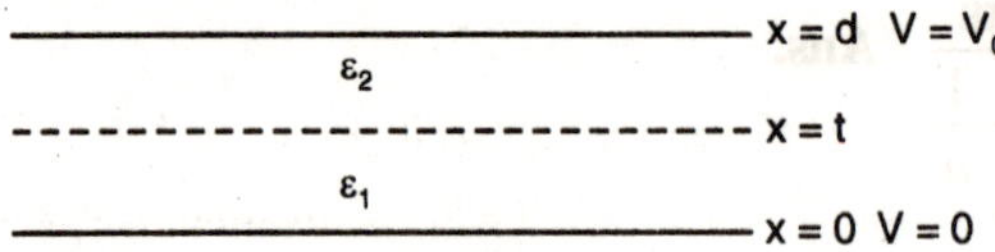

Fig. E 1.38

Solution. In a parallel plate capacitor with different permittivity dielectric materials the normal component of electric flux density will be same. However, since the dielectric permittivities are different hence electric field intensity will be different and so also the potential distribution. The Laplace's equation

$$\frac{\partial^2 V}{\partial x^2} = 0$$

The general solution for this is $V = Ax + B$.

But because the potential distribution across the two dielectric is different, hence let these solutions be

$$V_1 = A_1 x + B_1 \text{ and } V_2 = A_2 x + B_2$$

There are four constants A_1, B_1, A_2, B_2 and hence four boundary conditions are required. Two already are known which are at $x = 0$, $V_1 = 0$, $x = d$, $V = V_2$

At $x = t$ $\quad V_1 = V_2$

and since $D_{n_1} = D_{n_2}$ at $x = t$

$$\varepsilon_1 \frac{\partial V_1}{\partial x} = \varepsilon_2 \frac{\partial V_2}{\partial x}$$

or $\qquad \varepsilon_1 A_1 = \varepsilon_2 A_2$

Making use of the other three conditions

$$0 = B_1 \qquad V_0 = A_1 d + B_2$$

$$A_1 t + B_1 = A_2 t + B_2$$

Solving these four equations for the four constants and substituting the values in the first two equations, we have

$$V_1 = \frac{\varepsilon_2 x}{\varepsilon_2 t + \varepsilon_1 (d - t)} V_0 \qquad 0 < x < t$$

and
$$V_2 = \frac{\varepsilon_2 t + \varepsilon_1 (x - t)}{\varepsilon_2 t + \varepsilon_1 (d - t)} V_0 \qquad t < x < d$$

and the potential at $x = t \dfrac{\varepsilon_2 t}{\varepsilon_2 t + \varepsilon_1 (d - t)} V_0$

So far we have considered one dimensional case for which the potential is a function of $x(V = Ax + B)$. The one dimensional problems for which the potentials are function of y or z are identical to the one when the variable is x. Thus there is only one one-dimensional problem in Cartesian coordinate system. However in cylindrical and spherical coordinate system there are two each one dimensional problems. Hence these are only five different one-dimensional problems in all even though there are nine coordinates.

Example 1.39. Consider the region between the co-axial cones as shown in Fig. E 1.39. A potential V_1 is assumed at θ_1 and $V = 0$ at θ_2. The cone vertices are insulated at $r = 0$. Determine the voltage distribution between the cones.

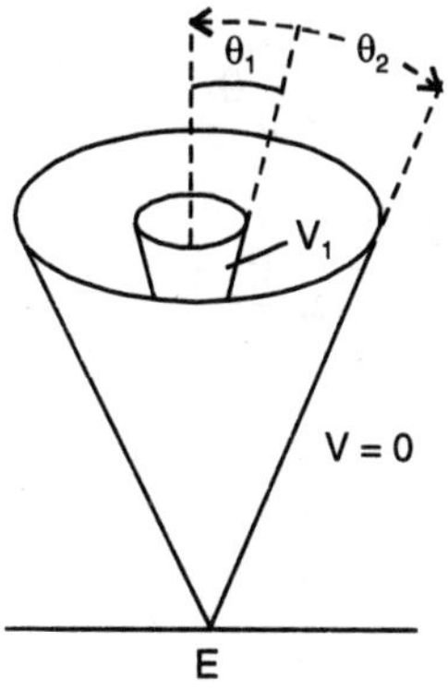

Fig. E 1.39

Solution. Since the potential is constant with r and θ Laplace's equation reduces to

$$\frac{1}{r^2 \sin\theta} \cdot \frac{d}{d\theta}\left(\sin\theta \frac{dV}{d\theta}\right) = 0 \quad \text{or} \quad \frac{d}{d\theta}\left(\sin\theta \frac{dV}{d\theta}\right) = 0$$

Integrating once we have

$$\sin\theta \frac{dV}{d\theta} = A$$

or

$$\frac{dV}{d\theta} = \frac{A}{\sin\theta}$$

Integrating again

$$V = \int \frac{A}{\sin\theta} d\theta + B$$

$$= \int \frac{A \sec\theta/2}{2\sin\frac{\theta}{2}\cos\frac{\theta}{2}\sec\frac{\theta}{2}} d\theta + B$$

$$= \int \frac{A \sec^2\theta/2}{2\tan^2\theta/2} d\theta$$

Now the differential of the denominator with θ is $\frac{1}{2}\cdot 2\sec^2\theta/2 = \sec^2\theta/2$ which is same as the numerator and hence the integration of $\frac{A}{\sin\theta}$ is

$$V = A\ln\tan\frac{\theta}{2} + B$$

Using boundary conditions, we have

$$0 = A\ln\tan\frac{\theta_2}{2} + B \rightarrow B = -A\ln\tan\frac{\theta_2}{2}$$

$$\therefore \qquad V_1 = A\ln\tan\frac{\theta_1}{2} - A\ln\tan\frac{\theta_2}{2}$$

$$= A\left[\ln\tan\frac{\theta_1}{2} - \ln\tan\frac{\theta_2}{2}\right]$$

or

$$A = \frac{V_1}{\ln\tan\frac{\theta_1}{2} - \ln\tan\frac{\theta_2}{2}}$$

$$\therefore \qquad V = \frac{V_1\ln\tan\frac{\theta_1}{2}}{\ln\tan\frac{\theta_1}{2} - \ln\tan\frac{\theta_2}{2}} - \frac{V_1\ln\tan\frac{\theta_2}{2}}{\ln\tan\frac{\theta_1}{2} - \ln\tan\frac{\theta_2}{2}}$$

Hence

$$V = V_1\frac{\ln(\tan\theta_1/2) - \ln(\tan\theta_2/2)}{\ln(\tan\theta_1/2) - \ln(\tan\theta_2/2)} \quad \textbf{Ans.}$$

Next we solve a few problems based on application of Poisson's equation.

Example 1.40. Determine the charge density at (1, π/2, 0) corresponding to

$$D = r\sin\phi\bar{a}_r + r^2\cos\phi\bar{a}_\phi + 2re^{-4z}\bar{a}_z.$$

Solution. We know that

$$\nabla\cdot D = \rho = \frac{1}{r}\frac{\partial}{\partial r}(r^2\sin\phi) + \frac{1}{r}\frac{\partial}{\partial\phi}(r^2\cos\phi) + \frac{\partial}{\partial z}(2re^{-4z})$$

$$= \frac{1}{r}\cdot 2r\sin\phi - \frac{1}{r}\cdot r^2\sin\phi - 8re^{-4z}$$

$$= 2\sin\phi - r\sin\phi - 8re^{-4z}$$

$$= 2 - 1 - 8 = -7 \text{ C/m}^3 \quad \textbf{Ans.}$$

Example 1.41. The volume charge density in a particular system is given by

$$\rho_v = 2\rho_{\max} \sec h\frac{x}{a} \tan h\frac{x}{a}$$

Determine the potential distribution in the system when the boundary conditions are

At $\qquad x \to \pm\infty \qquad E_x \to 0$

$\qquad x = 0 \qquad V = 0$

Solution. Using Poisson's equation

$$\nabla^2 V = -\frac{\rho_V}{\varepsilon}$$

$$= -\frac{2\rho_{\max}}{\varepsilon} \sec h\frac{x}{a} \tan h\frac{x}{a}$$

Since the equation is a function of one variable x as y and z are not present, integrating the equation once, we have

$$\frac{dV}{dx} = \frac{2\rho_{\max} a}{\varepsilon} \sec h\frac{x}{a} + A$$

or

$$\frac{dV}{dx} = -E = \frac{2\rho_{\max} a}{\varepsilon} \sec h\frac{x}{a} + A$$

or

$$E = -\frac{2\rho_{\max} a}{\varepsilon} \sec h\frac{x}{a} + A$$

To obtain A at $x = \pm\infty$ $E \to 0$ and at

$$x = \pm\infty,\ \sec h\frac{x}{a} = 0. \text{ Hence } A = 0 \text{ and}$$

We have

$$\frac{dV}{dx} = \frac{2\rho_{\max} a}{\varepsilon} \sec h\frac{x}{a}$$

Now

$$\sec h\frac{x}{a} = \frac{1}{\cos h\frac{x}{a}} = \frac{2}{e^{x/a} + e^{-x/a}}$$

Let

$$e^{x/a} = \tan\theta$$

$$\frac{1}{a} e^{x/a} dx = \sec^2\theta\, d\theta \text{ or } \tan\theta\, dx = a\sec^2\theta\, d\theta$$

$$dx = a\frac{\sec^2\theta}{\tan\theta} d\theta$$

Hence
$$\sec h\frac{x}{a}=\frac{1}{\tan\theta+\dfrac{1}{\tan\theta}}=\frac{2\tan\theta}{\sec^2\theta}$$

Rewriting
$$dV=\frac{4\rho_{\max}a}{\varepsilon}\tan\theta\cdot\frac{a\sec^2\theta}{\tan\theta}d\theta$$

$$dV=\frac{4\rho_{\max}a^2}{\varepsilon}d\theta$$

$$V=\frac{4\rho_{\max}a^2}{\varepsilon}\theta+B=\frac{4\rho_{\max}a^2}{\varepsilon}\tan^{-1}e^{x/a}+B$$

To obtain B at $x=0$ $V=0$

$$V=0=\frac{4\rho_{\max}a^2}{\varepsilon}\tan^{-1}1+B$$

or
$$B=-\frac{4\rho_{\max}a^2}{\varepsilon}\cdot\frac{\pi}{4}+0$$

Hence
$$V=\frac{4\rho_{\max}a^2}{\varepsilon}\left(\tan^{-1}e^{x/a}-\frac{\pi}{4}\right)\quad \textbf{Ans.}$$

Example 1.42. The spherically symmetric potential distribution in a system is given by

$$V(r)=\frac{Ae^{-\alpha r}}{r}\left(1+\frac{\alpha r}{2}\right)$$

where α and A are positive constant. Determine the spherical charge distribution required to produce this potential, origin to be excluded from calculation.

Solution. In spherical coordinates, Poisson's equation is given as

$$\frac{1}{r^2}\frac{\partial}{\partial r}\left(r^2\frac{\partial V}{\partial r}\right)=-\frac{\rho}{\varepsilon}\qquad r>0$$

as the voltage is independent of ϕ and θ.

For the given V

$$\frac{1}{r^2}\frac{\partial}{\partial r}\left(r^2\frac{\partial V}{\partial r}\right)=\frac{A}{r^2}\frac{\partial}{\partial r}\left[r^2\left(-\frac{e^{-\alpha r}}{r^2}+\frac{\alpha e^{-\alpha r}}{r}-\frac{\alpha^2e^{-\alpha r}}{2}\right)\right]$$

$$=\frac{A}{r^2}\left(\alpha e^{-\alpha r}+\alpha^2re^{-\alpha r}-\alpha e^{-\alpha r}+\frac{\alpha^3}{2}r^2e^{-\alpha r}-\alpha^2re^{-\alpha r}\right)$$

$$=\frac{A}{2}\alpha^3e^{-\alpha r}$$

Hence $$-\frac{\rho}{\varepsilon_0} = \frac{A}{2}\alpha^3 e^{-\alpha r}$$

or $$\rho = -\frac{\varepsilon_0 A}{2}\alpha^3 e^{-\alpha r} \textbf{ Ans.}$$

Finally a few important identities are listed below: Here *A, B, C* and *D* are vector and *U, V* are the scalar quantities.

1. $A \cdot B \times C = B \cdot A \times C = C \cdot A \times B$
2. $(A \times B) \times C = B(A \cdot C) - A(B \cdot C)$
3. $A \times (B \times C) = B(A \cdot C) - C(A \cdot B)$
4. $A \times (B \times C) + B \times (C \times A) + C \times (A \times B) = 0$
5. $(A \times B) \cdot (C \times D) = (A \cdot C)(B \cdot D) - (B \cdot C)(A \cdot D)$
6. $(A \times B) \times (C \times D) = (A \times B \cdot D)C - (A \times B \cdot C)D$
7. $\nabla(A \cdot B) = A \times (\nabla \times B) + B \times (\nabla \times A) + (A \cdot \nabla)B + (B \cdot \nabla)A$
8. $\nabla \cdot (A \times B) = B \cdot \nabla \times A - A \cdot \nabla \times B$
9. $\nabla \times (A \times B) = A\nabla \cdot B - B\nabla \cdot A + (B \cdot \nabla)A - (A \cdot \nabla)B$
10. $\nabla \cdot \nabla \times A = 0$
11. $\nabla(U + V) = \nabla U + \nabla V$
12. $\nabla(UV) = U\nabla V + V\nabla U$
13. $\nabla \times (UA) = \nabla U \times A + U\nabla \times A$
14. $\nabla \times \nabla U = 0$
15. $\nabla \times \nabla \times A = \nabla(\nabla \cdot A) - \nabla^2 A$
16. $\nabla \cdot (UA) = A \cdot \nabla U + U\nabla \cdot A$

PROBLEMS

P 1.1 Comment on the statement "The study of electromagnetic field theory is much more difficult as compared to circuit theory".

P 1.2 Explain what you mean by scalar fields and vector fields. Give examples of each.

P 1.3 Starting from first principle derive expression for gradient of a scalar in (*i*) Cartesian (*ii*) Cylindrical and (*iii*) Spherical coordinate systems. Give the physical significance of gradient of a scalar.

P 1.4 Starting from first principles derive an expression for divergence of a vector in cartesian coordinates and hence the expressions in cylindrical and spherical coordinates.

P 1.5 State and explain Divergence theorem.

P 1.6 What is meant by curl of a vector? Derive an expression for curl of a vector in cartesian coordinates. Hence derive expressions for curl of a vector in cylindrical and spherical coordinates.

P 1.7 State and explain Stoke's theorem.

P 1.8 The equation of a sphere with unit radius and whose center is located at (0, 0, 1) is given in cartesian coordinates by

$$x^2 + y^2 + (z-1)^2 = 1 \cdot$$

Determine its equation in cylindrical and spherical coordinates.

Ans. (*i*) $\rho^2 + (z-1)^2 = 1$ (*ii*) $r = 2\cos\theta$

P 1.9 Determine the rate of change of $V = x^2y + yz^2 + zy^2$ in the direction normal to the surface $x^2y - yz + xz^2 = 5$ at the point (1, 2, 3). **Ans.** 6.98

P 1.10 Determine $\nabla^2\bar{A}$ where $\bar{A} = \bar{a}_x(x^2 - y^2) + \bar{a}_y(y^2 - z^2) + \bar{a}_z(z^2 - x^2)$ **Ans.** $\nabla^2 A = 0$

P 1.11 Show that the vector fields $A = \rho\cos\phi a_\rho + \rho\sin\phi a_\phi + \rho\bar{a}_z$ and $B = \rho\cos\phi a_\rho + \rho\sin\phi a_\phi - \rho\bar{a}_z$ are everywhere perpendicular to each other.

P 1.12 The surfaces ρ = 3, ρ = 5, ϕ = 100°, ϕ = 130°, z = 3 and z = 4.5 define a closed surface. Determine (*a*) the enclosed volume (*b*) the total area of the enclosing surface (*c*) the total length of the twelve edges of the surfaces. **Ans.** (*a*) 6.28 (*b*) 20.7 (*c*) 22.4

P 1.13 Verify Stoke's theorem for the vector field $A = \bar{a}_r + \bar{a}_\theta + \bar{a}_\phi$ the surface being the octant of a sphere $r = 2, 0 \le \theta \le \frac{\pi}{2}, \ 0 \le \phi \le \frac{\pi}{2}$. **Ans.** $\nabla \times A = \pi$

P 1.14 Show that the vector field $A = \dfrac{-y\bar{a}_x + x\bar{a}_y}{x^2 + y^2}$ is curl free at all points except the origin.

P 1.15 The vector field A is given as $A = r\sin\phi\bar{a}_r + 2r\cos\phi\bar{a}_\phi + 2z^2a_z$. Determine $\nabla \cdot A$. **Ans.** $4z$

P 1.16 The vector field A is given as $A = 10\sin^2\phi\ \bar{a}_r + r\bar{a}_\phi + \left[(z^2/r)\cos^2\phi\right]a_z$. Determine $\nabla \cdot A$ at (2, 0°, 5) **Ans.** 5

P 1.17 Given $V = \phi\ z^{1/2}/r$, determine the electric field intensity.

Ans. $\dfrac{\phi z^{1/2}}{r^2}\bar{a}_r - \dfrac{z^{1/2}}{r^2}\bar{a}_\phi - \dfrac{1}{2}\phi\dfrac{1}{r\sqrt{z}}\bar{a}_z$

P 1.18 If the field of a region of space is given by $E = 2\cos\phi\bar{a}_z$. Is this a charge free region?

Ans. Yes

P 1.19 If the charge density $D = \dfrac{\bar{a}_r}{r^2} + \sin\phi \cdot \bar{a}_\phi + z^{-1/3}\bar{a}_z$, determine the charge density at the point (1, 15°, 3). **Ans.** –0.11 C/m^3

P 1.20 If $A = 2x\bar{a}_x + 4x^2yz\bar{a}_y$, determine (*a*) $\nabla \times A$ and (*b*) $\nabla \times \nabla \times A$.

Ans. (*a*) $-4xy^2\bar{a}_x + 4y^2z\bar{a}_z$ (*b*) $8yz\bar{a}_x + 8xy\bar{a}_z$

P 1.21 If $A = -2r\bar{a}_r + 5r\bar{a}_\theta + \sin\theta\bar{a}_\phi$, determine $\nabla \times A$. **Ans.** $\frac{\cos\theta}{r}\bar{a}_r - \frac{\sin\theta}{r}a_\theta + 10\bar{a}_\phi$

P 1.22 The region between two concentric right circular cylinders contains a uniform charge density ρ. Determine expression for V using Poisson's equation. The permittivity is ε.

Ans. $V = -\frac{\rho r^2}{4\varepsilon} + A\ln r + B$

P 1.23 In spherical coordinates $V = 0$ for $r = 10$ cm and $V = 100$ V for $r = 2.0$ m. Assume free space between the concentric spherical shells, determine E and D.

Ans. $E = -\frac{10.5}{r^2}\bar{a}_r$ V/m, $D = \frac{-9.3\times10^{-11}}{r^2}\bar{a}_r$ C/m^2

P 1.24 An electric field is given by $E = xy^2a_x - xz^2a_y + xyza_z$. Determine the x-component of its curl at (2, 1, 3).

Ans. 18

P 1.25 For the spherically conducting shells filled with two dielectrics shown in Fig. P 1.25, determine the potential distribution in the region $c < r < a$ using Laplace equation. c is the radius of inner shell and a the radius of outer shell. b is the radius of the dielectric with permittivity ε_2.

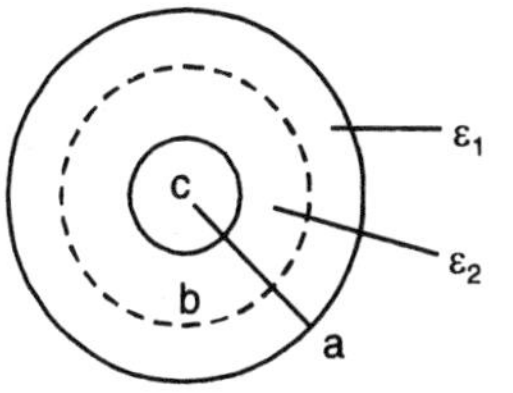

Fig. P 1.25

Ans. $$V_1 = \left[\frac{1/a - 1/r}{(\varepsilon_2/\varepsilon_1)\left(\frac{1}{c}-\frac{1}{b}\right)+\left(\frac{1}{b}-\frac{1}{a}\right)} + 1\right]V_0 \qquad b < r < a$$

$$V_2 = \frac{\frac{1}{c}-\frac{1}{r}}{\left(\frac{1}{c}-\frac{1}{b}\right)+\left(\frac{\varepsilon_1}{\varepsilon_2}\right)\left(1+b-\frac{1}{a}\right)}V_0 \qquad c < r < b$$

P 1.26 Determine the curl of the following vectors:

(*a*) $A = xy\bar{a}_x + yz\bar{a}_y + xz\bar{a}_z$

(*b*) $A = 2r\cos\phi\bar{a}_r + r\bar{a}_\phi$.

Ans. (*a*) $-y\bar{a}_x - z\bar{a}_y - x\bar{a}_z$ (*b*) $(2 + 2\sin\phi)\bar{a}_z$

P 1.27 Find the Laplacian of the following functions:

$(a)\ T = x^2yz^2 \quad (b)\ A = x^2yz\bar{a}_x + xy^2z\bar{a}_y + xyz^2\bar{a}_z$

Ans. $(a)\ 6xyz^2 + 2x^3y\ (b)\ 2\left(yz\bar{a}_x + xz\bar{a}_y + xy\bar{a}_z\right)$

P 1.28 Determine $\nabla \times A$ and $\nabla \cdot A$ at the point $\left(1, \frac{\pi}{2}, \frac{\pi}{6}\right)$ for the following:

$(a)\ A = \left(x^2 + y^2\right)\bar{a}_x + xyz\bar{a}_y + \left(z^2 - x^2\right)\bar{a}_z$

$(b)\ A = r(1 - \cos\theta)\bar{a}_r + r\cos\theta\bar{a}_\theta + \sin\theta\cos\phi\bar{a}_\phi$

Ans. $(a)\ \nabla \times A = -0.433\bar{a}_x + 1.732\bar{a}_y - \bar{a}_z$

$\nabla \cdot A = 1.732$

$(b)\ \nabla \times A = -0.866\bar{a}_\theta + 0.732a_\phi$

$\nabla \cdot A = 1.5$

P 1.29 Evaluate the curl of $A = 5e^{-r}\cos\phi\bar{a}_r - 5\cos\phi\bar{a}_z$ over the xz plane ($\phi = 0, \phi = \pi$).

Ans. $\text{curl}\,A = \frac{5}{r}\sin\phi\bar{a}_r + \frac{5}{r}e^{-r}\sin\phi\bar{a}_z$ At $\phi = 0, \pi, \nabla \times A = 0$

P 1.30 Given $A = r^2\bar{a}_r + 2\sin\theta\bar{a}_\theta$, determine $\nabla \cdot A$ at $(1, \pi, \phi)$.

Ans. $\nabla \cdot A = 4r + \frac{4\cos\theta}{r}$, $\nabla \cdot A = 0$ at $(1, \pi, \phi)$

Chapter 2

Coulomb's Law and Electric Field Intensity

2.1 INTRODUCTION

Historically it has been observed that the Greeks had the knowledge of static electricity as early as 600 B.C. and they are responsible for the term *electricity* derived from their word for amber and they spent many leisure hours rubbing a small piece of amber on their sleeves and observing how the amber piece attracts pieces of fluff and stuff. However, their interest remained only in terms of philosophy and logic and it was considered as a magic or a life force. No experimental set up was developed to asses or measure their force of attraction.

In 1600 A.D. Dr. Gilbert physician to Her Majesty, the Queen of England did some useful experimental work and he showed that not only amber but substances like glass, sulphur etc. These substances not only attract straw and chaff but all metals, woods, leaves, stone etc. also.

Shortly thereafter Charles Coulomb a Colonel in French Army Engineer made elaborate arrangements for experimental set up using a delicate torsion balance designed and developed by himself to asses and measure quantitatively the force exerted between two objects each having a static charge of electricity.

2.2 COULOMB'S LAW

After analysis of his results Coulomb concluded that the force between two very small charged objects separated in vacuum or free space by a distance which is large compared to the size of the objects, is proportional to the charge on each object and inversely proportional to the square of the distance between them or mathematically, it is expressed as

$$F = K\frac{Q_1 Q_2}{r^2} \qquad \text{...(2.1)}$$

where Q_1 and Q_2 are the charges on the two objects separated by a distance r and K is constant of proportionality. Using SI system of units where Q_1 and Q_2 are measured in Coulombs, r is in metres and force F is in Newton, then K is given as

$$K = \frac{1}{4\pi\varepsilon_0}$$

where ε_0 is called the permittivity of free space and has the magnitude measured in Farad per metre

$$\varepsilon_0 = \frac{1}{36\pi} \times 10^{-9} = 8.854 \times 10^{-12} \text{ F/m}$$

Hence Coulomb's law is given as

$$F = \frac{Q_1 Q_2}{4\pi\varepsilon_0 r^2} \qquad \ldots(2.2)$$

or

$$\varepsilon_0 = \frac{Q_1 Q_2}{4\pi r^2 F}$$

It is to be noted that ε_0 is not dimensionless. Rather dimensionally it is given as $C^2/N{\cdot}m^2$.

Consider Fig. 2.1 for further clarification of Coulombs law. Here

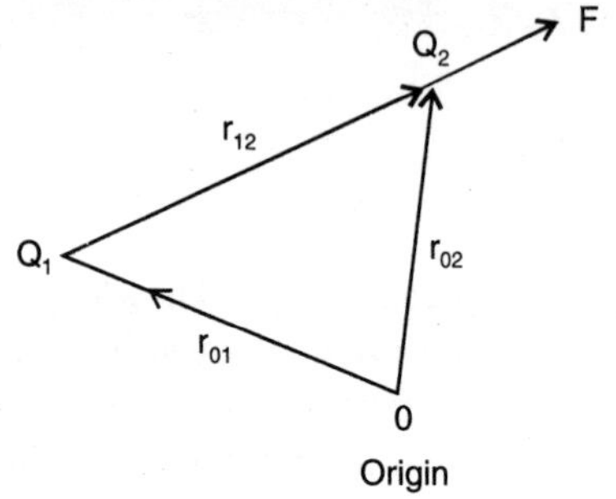

Fig. 2.1

Q_1 and Q_2 are placed at a distance of r_{01} and r_{02} respectively from the origin. Therefore, the distance of Q_2 from Q_1 is $r_{12} = r_{02} - r_{01}$ (from vector addition).

Hence F the force exerted by Q_1 and Q_2 is given by

$$F = \frac{Q_1 Q_2}{4\pi\varepsilon_0 r_{12}^2} \qquad \ldots(2.3)$$

Since force is a vector quantity as it has both magnitude and direction, Coulomb suggested that the force acts along the line joining the two charges and is repulsive if the charges are alike in sign and attractive if they are of opposite in sign. So to define equation (2.3) completely we use a unit vector in the direction r_{12} and we let it write as $\bar{a}_{r_{12}}$. Hence equation (2.3) is completely defined as

$$\bar{F} = \frac{Q_1 Q_2}{4\pi\varepsilon_0 r_{12}^2} \cdot \bar{a}_{r_{12}} \qquad \ldots(2.4)$$

If we know r_{01} and r_{02} in vector form, then the unit vector $\bar{a}_{r_{12}} = \dfrac{r_{02} - r_{01}}{|r_{02} - r_{01}|}$.

Here $\bar{a}_{r_{12}}$ is a unit vector along r_{12}.

We next solve a few problems on Coulomb's law using vector analysis.

Example 2.1. Consider charge $Q_1 = 3\ \mu C$ at $A = (2, 4, 6)$ and $Q_2 = -1\mu C$ at $B = (4, 0, 10)$ in a vacuum. Determine the force experienced by Q_2 due to Q_1.

Solution. $r_{01} = 2\bar{a}_x + 4a_y + 6\bar{a}_z$

$$r_{02} = 4\bar{a}_x + 0\bar{a}_y + 10\bar{a}_z$$

Hence $$r_{12} = r_{02} - r_{01} = 2\bar{a}_x - 4\bar{a}_y + 4\bar{a}_z$$

and $$|r_{12}| = \sqrt{2^2 + 4^2 + 4^2} = 6$$

Hence $$a_{r_{12}} = \frac{2\bar{a}_x - 4\bar{a}_y + 4\bar{a}_z}{6} = \frac{\bar{a}_x - 2\bar{a}_y + 2a_z}{3}$$

Hence force on Q_2 due to Q_1 is given as

$$\bar{F} = -\frac{3\times10^{-6}\times10^{-6}}{4\pi\times\frac{1}{36\pi}\times10^{-9}\times36}\times\left(\frac{\bar{a}_x - 2\bar{a}_y + 2\bar{a}_z}{3}\right)$$

$$= -\frac{3\times10^{-12}\times10^{9}}{4}\left(\frac{\bar{a}_x - 2\bar{a}_y + 2\bar{a}_z}{3}\right)$$

$$= -0.75\times10^{-3}\left(\frac{\bar{a}_x - 2\bar{a}_y + 2\bar{a}_z}{3}\right) \text{ newton} \quad \textbf{Ans.}$$

It is to be noted that the terms within the bracket represents unit vector and these should not be mixed up with the rest of the expression. The magnitude of force between Q_1 and Q_2 is 0.75×10^{-3} newton and the direction is specified by the unit vector. The force on Q_2 may also be considered as three components.

$$-0.25\times10^{-3}\bar{a}_x + 0.5\times10^{-3}\bar{a}_y - 0.5\times10^{-3}\bar{a}_z \text{ newton}$$

The force expressed by Coulomb's law is a mutual force as each of the two charges experiences a force of the same magnitude but in opposite direction. We might as well write

$$F_1 = -F_2 \Rightarrow \frac{Q_1Q_2}{4\pi\varepsilon_0 r_{12}^2}\bar{a}_{r_{21}} = -\frac{Q_1Q_2}{4\pi\varepsilon_0 r_{12}^2}\bar{a}_{r_{12}} \quad \text{...(2.5)}$$

Coulomb's law is linear in terms of charges *i.e.* if Q_1 or Q_2 or both are multiplied by n, the force is also multiplied by n if one of the two charges is multiplied by n. Also, it is true that the force on a charge in the presence of several other charges is the sum of the forces (vector sum) on that charge due to each of the other charges acting alone. However, in Coulomb's law the force is inversely proportional to square of the distances between charges.

Example 2.2. Two point charges of –300 μC and 10μC are located at (2, 4, 5) and (1, 1, 3). Determine the force on the negative charge say Q_1.

Solution. The unit vector along the line from –ve charge to +ve charge in this case is

$$\frac{(1-2)\bar{a}_x + (1-4)\bar{a}_y + (3-5)\bar{a}_z}{1^2 + 3^2 + 2^2} = \frac{-\bar{a}_x - 3\bar{a}_y - 2\bar{a}_z}{14}$$

Hence force from Q_1 to Q_2 is

$$= \frac{-300\times10^{-6}\times10\times10^{-6}}{4\pi\times\frac{1}{36\pi}\times10^{-9}\times14}\left(\frac{-\bar{a}_x - 3\bar{a}_y - 2\bar{a}_z}{14}\right)$$

$$= \frac{-300\times10^{-2}\times9}{14}\left(\frac{-\bar{a}_x - 3\bar{a}_y - 2\bar{a}_z}{14}\right)$$

$$= -1.928\left(\frac{-\bar{a}_x - 3\bar{a}_y - 2\bar{a}_z}{14}\right) \quad \textbf{Ans.}$$

The magnitude of force is 1.928 N and it acts in the direction Q_1 to Q_2.

Example 2.3. Four like point charges of 60 μC each are located at the corners of a square the diagonal of which measures 8 m. Determine the force on a 75 μC point charge located at a height of 3 m from the centre of the square.

Solution. Consider Fig. E 2.3, the force between the charge at location *A* and 75 μC charge will be as shown by the vector *F.* This will have two components one along *z*-axis and another normal to the

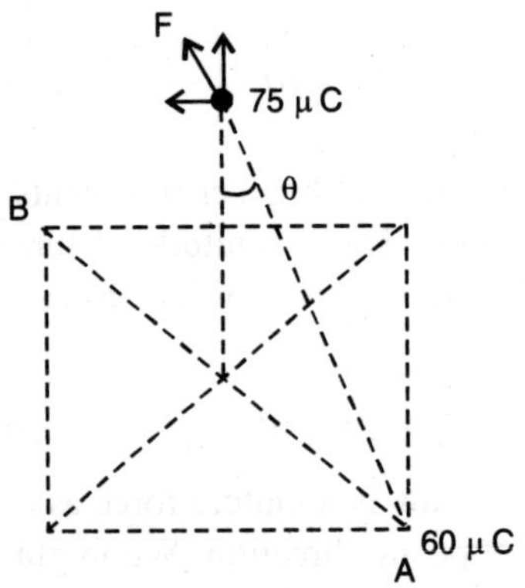

Fig. E 2.3

z-axis. The force due to the charge at *B* will also have two components, one along the *z*-axis as in the previous case and the other normal to *z*-axis but in a direction opposite to previous case and hence due to charges at *A* and *B*, the net force will be $4F\cos\theta$ where $\cos\theta = \frac{3}{5}$. Hence u Coulomb's law, the net force

$$F_{\text{net}} = \frac{4\times60\times75\times10^{-12}\times\frac{3}{5}}{4\pi\times\frac{1}{36\pi}\times10^{-9}\times5^2}\bar{a}_z = \frac{4\times60\times75\times9\times10^{-3}}{25}\bar{a}_z\times\frac{3}{5}$$

$$= 3.89\bar{a}_z \text{ N} \quad \textbf{Ans.}$$

Example 2.4. Charge $Q_1 = 100$ μC located at (2, 4) metres in the *xy* plane experiences a repulsive force of 10 N due to charge Q_2 located at (10, 7) m. Determine Q_2.

Solution. Force $= \dfrac{Q_1 Q_2}{4\pi\varepsilon_0 r^2}\bar{a}_r$

Since the magnitude of force is given, hence

$$F = \frac{Q_1 Q_2}{4\pi\varepsilon_0 r^2} \Rightarrow 10 = \frac{100\times10^{-6}\times Q_2}{4\pi\cdot\dfrac{1}{36\pi}\times10^{-9}\left\{(10-2)^2+(7-4)^2\right\}}$$

$$10 = \frac{9\times100\times10^{-6}Q_2\times10^9}{73}$$

or
$$Q_2 = \frac{73\times10}{9\times10^5} = 81\times10^{-5} = 810\ \mu\text{C}\ \textbf{Ans.}$$

Example 2.5. A circular disc of radius 5 m carries a uniformly distributed charge of 1250 π μC. Determine the force on a 100 μC charge located at a height of 2 m along the axis of the disc.

Solution. Consider a differential surface *ds* at a distance *r* from the centre of the disc with $ds = r\,dr\,d\phi$.

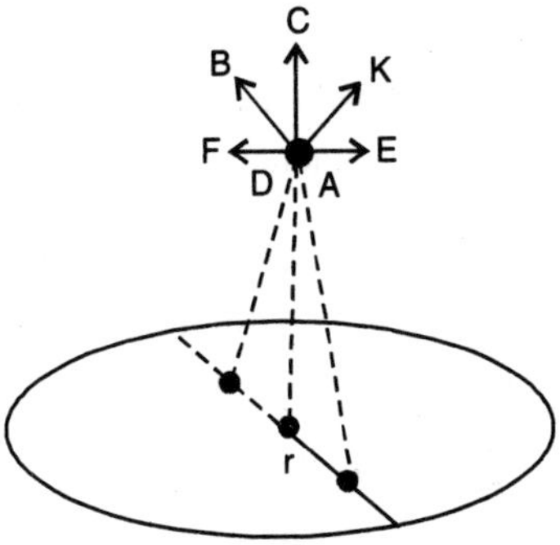

Fig. E 2.5

Since the surface charge density is $\dfrac{1250\pi}{\pi R^2} = \dfrac{1250\pi}{\pi\times25} = 50\mu\text{C}/\text{m}^2$, the charge on the differential surface is $50\cdot ds = 50\,r dr\,d\phi$. Considering this charge as a point charge, the force on charge 100 μC along *AB* is

$$dF = \frac{50\ rdrd\phi\cdot100}{4\pi\cdot\dfrac{1}{36\pi}\times10^{-9}\left(12^2+r^2\right)}$$

A similar differential surface on the opposite side of *r* will have force along AK. Hence the net force is along $\bar{a}_z$ and is given as

$$F = \int_0^{2\pi}\int_0^5 \frac{50\times9\times100\times10^{-9}}{r^2+12^2}\,r\,dr\,d\phi\cdot\frac{5}{\sqrt{r^2+144}}$$

$$F = \int_0^5 45 \times 10^{-6} \cdot 2\pi \cdot \frac{r\,dr}{\left(r^2 + 12^2\right)^{3/2}}$$

Let $r^2 + 12^2 = x^2$

or $2r\,dr = 2x\,dx$

or $rdr = xdx$ and the limits of $x = 12$ to $x = 13$

Hence $F = \int_{12}^{13} 90\pi \times 10^{-6} \cdot \frac{xdx}{x^3} = \int_{12}^{13} 90\pi \times 10^{-6} \cdot \frac{dx}{x^2}$

$$= -9\pi \cdot \left[\frac{1}{x}\right]_{12}^{13} \times 10^{-6} = 9\pi\left[\frac{1}{12} - \frac{1}{13}\right] \times 10^{-6}$$

$$= \frac{9\pi}{156} \times 10^{-6} = 0.18\,\mu\text{C} \quad \textbf{Ans.}$$

2.3 ELECTRIC FIELD INTENSITY

Consider a charge Q_1 fixed in space. Electric field lines emanate radially from or terminate on to Q_1 if it is positive charge or negative charge respectively. For positive charge these lines emanate from the charge and terminate at infinity if it is an isolated charge. If both positive and negative charges of same magnitude are present then, these field lines emanate from the positive charge and terminate on the negative charge.

Now consider a charge Q_1 (positive) fixed in space and if another charge Q_2 is brought in the vicinity of this charge (Q_1) then Q_2 experiences a force and the direction of force is along the line joining two charges. The force due to Q_1 on Q_2 is given as

$$F = \frac{Q_1 Q_2}{4\pi\varepsilon_0 r^2}\bar{a}_r \qquad \ldots(2.6)$$

Writing this force F as force per unit charge $\frac{F}{Q_2}$ we have

$$\frac{F}{Q_2} = \frac{Q_1}{4\pi\varepsilon_0 r^2}\bar{a}_r \qquad \ldots(2.7)$$

It can be seen that the quantity $\frac{F}{Q_2}$ is a function only of Q_1 and its direction is along the direction of force. This quantity describes a vector field and is known as Electric field intensity. We therefore, define electric field intensity as the vector force on a unit positive test charge. A word of caution now! We know that while measuring any quantity, whatever device we bring into the circuit it should not affect the basic quantity being measured as it loses the basic requirement of measurement. Here we are bringing a unit positive test charge for measuring electric field intensity due to charge Q, since unit positive charge is one Coulomb which represents about 6×10^{18} electrons which is a huge amount of charge which will certainly affect the electric field due to Q_1 and hence we would not measure electric field intensity experimentally by finding the force on 1C test charge. It is to be noted that if two unit charge are placed at a distance of 1m, according to Coulombs law, the force experienced by the two charges is 9×10^9 newton or about one million tons. As seen from equation (2.7), the units of electric field intensity is force per unit charge or Newton per Coulomb. The practical units of E are volts/metre. Now

$$E = \frac{\text{Newton}}{\text{C}} = \frac{\text{J/m}}{\text{C}} = \frac{\text{VIt}}{\text{mIt}} = \frac{\text{V}}{\text{m}} \quad \ldots(2.8)$$

Newton is energy/metre and charge is It where I is the current and t is the time. Hence electric field intensity is normally expressed in volt/metre.

Example 2.6. Four 5 μC charges are at the corners of a 4 metre square. The two charges on the left side of the square are positive and on the right side these are negative. Determine the electric field intensity at the centre of the square.

Solution. The configuration of the charges is given in Fig. E 2.6.

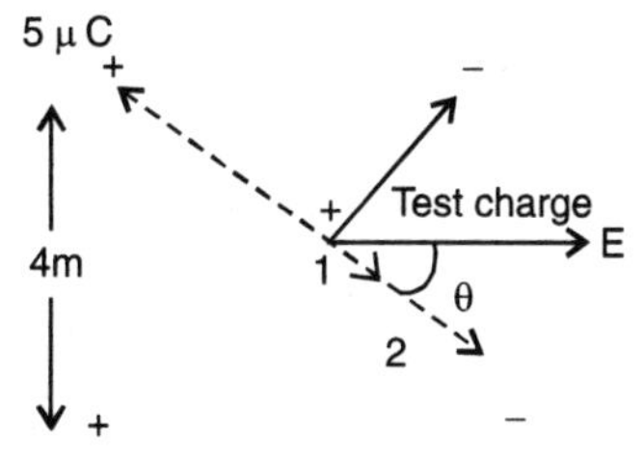

Fig. E 2.6

From the Fig. E 2.6 it is clear that the net field intensity will be $E = 4\, E_0 \cos\theta$ where E_0 is the field intensity between diagonals opposite charges. Here $\cos\theta = 45°$ as the diagonal of a square intersect each other orthogonally.

Now
$$E_0 = \frac{5 \times 10^{-6}}{4\pi \times \frac{1}{36\pi} \times 10^{-9} \left(4\sqrt{2}\right)^2} \times \cos 45°$$

$$= \frac{5 \times 9 \times 10^{-6} \times 10^{9}}{32} \times 0.707 = 0.994 \times 10^3$$

$= 994$ V/m **Ans.**

However, if all the charges were of the same sign, it can be seen that the net electric field intensity at the centre of the square would be zero.

Example 2.7. A 50 μC charge is located at a point (5, 12, 0) m. Determine the electrical field at the origin in (*a*) spherical and (*b*) cartesian coordinates.

Solution. (*a*) **Spherical Coordinates:** The radial distance of the origin from the point charge is $\sqrt{5^2 + 12^2} = 13$ m. The magnitude of electric field intensity at the origin is

$$\frac{50 \times 10^{-6}}{4\pi \cdot \frac{1}{36\pi} \times 10^{-9} \times 13^2} = \frac{50 \times 9 \times 10^3}{169} \text{V/m} \quad \text{or } 2.66 \text{ kV/m}$$

Since, here the field is from charge to the origin, the unit vector is $-\bar{a}_r$ and hence electric field intensity is –2.66 kV/m.

(*b*) **Cartesian Coordinates:** The radial distance in cartesian coordinates is same as in spherical coordinates. However the three components of Electric field intensity would be directed along the three unit vectors. In this case we have unit vectors along x and y direction and the unit vector is given by

$$\frac{-5a_x - 12\bar{a}_y}{13} \text{ or } -\frac{5}{13}\bar{a}_x - \frac{12}{13}\bar{a}_y$$

and hence electric field intensity in cartesian coordinates is given as $2.66\left(-\frac{5}{13}\bar{a}_x - \frac{12}{13}\bar{a}_y\right)\text{kV/m}$.

Example 2.8. Two charges $Q_1 = 5\ \mu\text{F}$ and $Q_2 = 6\ \mu\text{F}$ are located at (0, 4, 0) and (3, 0, 0) respectively. Determine the electric field intensity at (0, 0, 5).

Solution. Charges Q_1 and Q_2 are located along y and x-axis respectively and electric field intensity is required at a point along z-axis.

Vector joining the point (0, 4, 0) and (0, 0, 5) is $-4\ a_y + 5\ a_z$ and the vector joining the points (3, 0, 0) and (0, 0, 5) is $-3a_x + 5a_z$.

The length of the vector $-4\bar{a}_y + 5\bar{a}_z$ is $\sqrt{4^2 + 5^2} = \sqrt{41}$.

Hence electric field intensity due to charge Q_1 is

$$E_1 = \frac{5\times10^{-6}}{4\pi\times\frac{1}{36\pi}\times10^{-9}\cdot41}\left(\frac{-4a_y + 5\bar{a}_z}{\sqrt{41}}\right)$$

$$= \frac{5\times9\times10^3}{41}\left(-0.625\bar{a}_y + 0.78\bar{a}_z\right)$$

$$= 1.097\left(-0.625\bar{a}_y + 0.78\bar{a}_z\right)\times10^3$$

$$= -685\bar{a}_y + 855\bar{a}_z$$

Similarly

$$E_2 = \frac{6\times10^{-6}}{4\pi\times\frac{1}{36\pi}\times10^{-9}}\cdot\frac{1}{34}\cdot\left(\frac{-3a_x + 5a_z}{\sqrt{34}}\right)$$

$$= 1.588\left(-0.51\bar{a}_x + 0.857\bar{a}_z\right)\times10^3$$

$$= -809\bar{a}_x + 1360\bar{a}_z$$

Hence

$$E = E_1 + E_2 = -809\bar{a}_x - 685\bar{a}_y + 2215\bar{a}_z \text{ Volt/m } \textbf{Ans.}$$

2.4 ELECTRIC FIELD INTENSITY DUE TO LINE CHARGE

Consider an infinite length linear charge with ρ_l coulomb/m charge density as shown in Fig. 2.2. The infinite length charge is placed along z-axis. The charge is of cylindrical shape with diameter

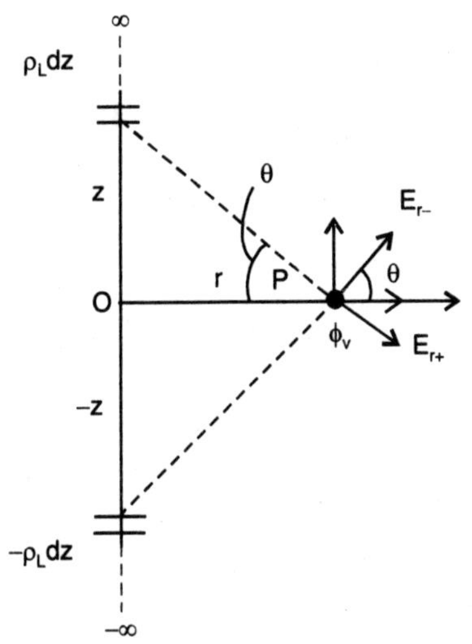

Fig. 2.2

negligibly small and of infinite length. The objective is to obtain electric field intensity at any point P at a distance r from the linear charge. To obtain that we make use of cylindrical coordinates so that symmetry of the linear charge can be taken advantage of for evaluation of E. Consider point P at a distance r from the linear charge where E is required. Since charge is of infinite length any point O on the line charge could be taken as the mid point on the charge such that $OP = r$. Consider a differential charge $\rho_L\, dz$ at a height z from 'O'. Assuming this differential charge as a point charge, electric field intensity at point P is

$$dE_{r+} = \frac{\rho_L dz}{4\pi\varepsilon_0 \left(r^2 + z^2\right)} \bar{a}_{r+} \qquad \ldots(2.9)$$

This intensity can be decomposed into two parts one along $\bar{a}_r$ and another normal downward with respect to $\bar{a}_r$. Similarly it can be seen from the Fig. 2.2 that a symmetrical differential element dz can be taken at a depth z below point O and its charge is $\rho_L dz$. The electric field intensity due to this differential charge at point P will be along E_{r-} as shown in Fig. 2.2 and again this will have two components one along $\bar{a}_r$ and the other normal to $\bar{a}_r$ but upward. Therefore, the net electric field intensity will be along $\bar{a}_r$ and the field intensity along $\bar{a}_r$ due to $\rho_L dz$ is given as

$$dE_r = \frac{\rho_L dz \cos\theta}{4\pi\varepsilon_0 \left(r^2 + z^2\right)} = \frac{\rho_L dz}{4\pi\varepsilon_0 \left(r^2 + z^2\right)} \frac{r}{\left(r^2 + z^2\right)^{1/2}}$$

Hence
$$E_r = \int_{-\infty}^{\infty} \frac{\rho_L r dz}{4\pi\varepsilon_0 \left(r^2 + z^2\right)^{3/2}} \qquad \ldots(2.10)$$

From Fig. 2.2 $\dfrac{z}{r} = \tan\theta$ or $z = r\tan\theta$

$$dz = r\sec^2\theta d\theta$$

Hence $$E_r = \int \frac{\rho_L \cdot r \cdot r\sec^2\theta\, d\theta}{4\pi\varepsilon_0 \left(r^2 + r^2\tan^2\theta\right)^{3/2}} = \frac{\rho_L}{4\pi\varepsilon_0}\int \frac{\sec^2\theta d\theta}{\sec^3\theta}$$

For the limits, since the length of the linear charge is infinite, to cover this length θ should vary from –π/2 to π/2.

Hence $$E_r = \frac{\rho_L}{4\pi\varepsilon_0}\int_{-\pi/2}^{\pi/2} \cos\theta\, d\theta = E_r = \frac{\rho_L}{2\pi\varepsilon_0 r} \qquad \ldots(2.11)$$

In case the length of the linear charge is 2*a*, the limits when the point is at a distance *r* and normal to the charge at its mid-point, would be *a* to –*a*.

i.e. $$E_r = \frac{r\rho_L}{4\pi\varepsilon_0}\int_{-a}^{a} \frac{dz}{\left(r^2 + z^2\right)^{3/2}} = -\frac{\rho_L}{2\pi\varepsilon_0 r\sqrt{\left(\frac{r}{a}\right)^2 + 1}}$$

If $a >> r$ $E_r = \frac{\rho_L}{2\pi\varepsilon_0 r}$ is the magnitude of electric field intensity.

Example 2.9. Two straight line charges with charge density 0.5 μC/m pass through the points *A* and *O* and are parallel to the *z*-axis. Determine the electric field *E* at *P*(5, 5, 0).

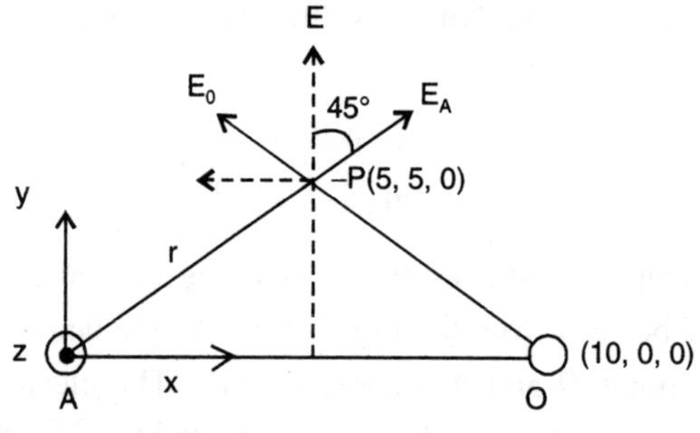

Fig. E 2.9

Solution. The magnitude of the electric field intensity due to either of the two charges is $|E| = \frac{\rho_L}{2\pi\varepsilon_0 r}$.

Hence the resultant electric field is 2|*E*| cos 45°.

Hence $E_r = 2|E| \cos 45°$

$$= \frac{2\times 0.5\times 10^{-6}}{2\pi\cdot\frac{1}{36\pi}\times 10^{-9}\cdot 5\sqrt{2}}\cdot\frac{1}{\sqrt{2}} = \frac{2\times 0.5\times 18\times 10^3}{10}$$

$$= 1800\bar{a}_y \text{ V/m}.$$

Example 2.10. The upper semicircle in Fig. E 2.10 carries a line charge of a density ρ_L C/m while the lower semicircle carries $-\rho_L$ C/m. Determine *E* at the centre.

Solution. Considering an arc of differential length $Rd\theta$ in the upper semicircle, electric field intensity at the centre of the circle is $dE_1 = \frac{\rho_L R d\theta}{4\pi\varepsilon_0 R^2}$ and is directed as shown in the Fig. E 2.10. This will have two components one along $-\bar{a}_x$ and the other along $-\bar{a}_y$.

Similarly the electric field intensity due to $Rd\theta$ in the lower semicircle opposite to the $Rd\theta$ in the upper semicircle is also given by

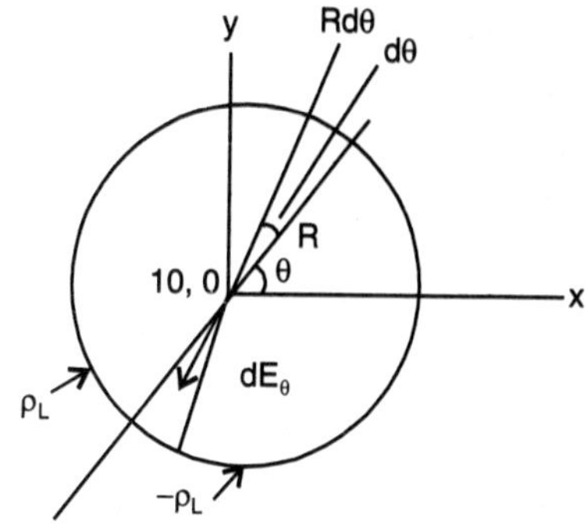

Fig. E 2.10

$$dE_2 = \frac{\rho_L R d\theta}{4\pi\varepsilon_0 R^2}$$

and the direction is same as due to the upper semicircle arc $Rd\theta$. Hence the electric field intensity at the centre due to both the arcs is

$$dE = \frac{\rho_L d\theta}{2\pi\varepsilon_0 R}$$

and is directed along $-\bar{a}_y$ as the x-component cancel each other due to the two arcs $Rd\theta$.

Hence net E at the centre

$$(-\bar{a}_y)\int \frac{\rho_L d\theta}{2\pi\varepsilon_0 R}\sin\theta = \frac{-\bar{a}_y \rho_L}{2\pi\varepsilon_0 R}\int_0^{\pi}\sin\theta\, d\theta$$

$$= \frac{\rho_L}{2\pi\varepsilon_0 R}[-\cos\theta]_0^{\pi}(-\bar{a}_y)$$

$$= -\frac{\rho_L}{\pi\varepsilon_0 R}\bar{a}_y \quad \textbf{Ans.}$$

Example 2.11. A linear charge with charge density $\rho_L = 20\ \mu C/m$ is distributed along z-axis between $z = \pm 10$ m. Determine E at (3, 0, 0) m in Cartesian coordinate system and cylindrical coordinate system.

Solution. Refer to Fig. E 2.11. Electric field intensity due to point charge $\rho_L dz$ at a height z from the origin is given as

$$dE = \frac{20 \times 10^{-6}\, dz}{4\pi \times \frac{1}{36\pi} \times 10^{-9}(9 + z^2)}$$

Since the linear charge is lying symmetrical to the plane $z = 0$, electric field intensity along $\bar{a}_z$ is zero and since the point under consideration is along x-axis we will have resultant electric field intensity along x-axis and it will be $dE \cos\theta$.

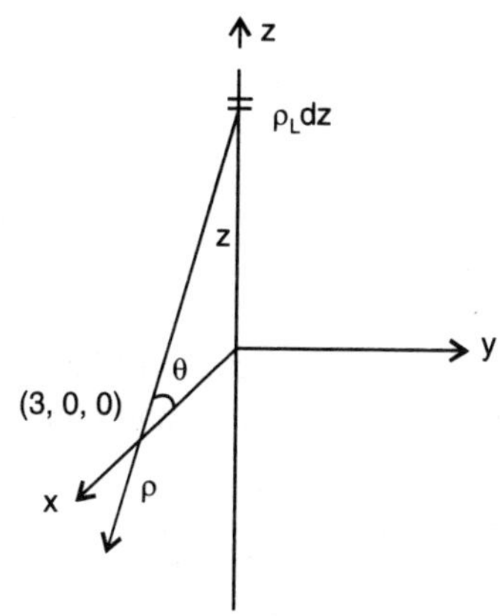

Fig. E 2.11

$$\text{Hence } dE\cos\theta = \frac{20 \times 10^{-6}\, dz}{4\pi \cdot \frac{1}{36\pi} \times 10^{-9}(9 + z^2)} \cdot \frac{3\bar{a}_x}{(9 + z^2)^{1/2}}$$

$$= \frac{180 \times 10^3 dz \times 3\bar{a}_x}{(9+z^2)^{3/2}}$$

$$E = 180 \times 10^3 \int_{-10}^{10} \frac{3dz}{(9+z^2)^{3/2}} \bar{a}_x$$

Let $$z = 3\tan\theta \text{ or } dz = 3\sec^2\theta\, d\theta$$

$$\int \frac{3 \cdot 3\sec^2\theta}{27\sec^3\theta} d\theta = \int \frac{1}{3\sec\theta} d\theta = \frac{1}{3}\int \cos\theta\, d\theta$$

$$= \frac{1}{3}\sin\theta = -\frac{1}{3} \cdot \frac{z}{\sqrt{z^2+9}}$$

$$= -\frac{1}{3}\left[\frac{z}{z^2+9}\right]_{-10}^{10} = -\frac{1}{3}\left[\frac{10}{\sqrt{109}} + \frac{10}{\sqrt{109}}\right] = -\frac{20}{\sqrt{109}} \cdot \frac{1}{3}$$

$$E = 180 \times 10^3 \times \frac{20}{\sqrt{109}} \cdot \frac{1}{3} = 115 \text{V/m}\, \bar{a}_x$$

Hence the resultant

$$E = 2 \times 115 = 230\bar{a}_x \text{ V/m} \quad \textbf{Ans.}$$

So far we have considered the line charge to lie along z-axis. However, we should understand that not all line charges are located along the z-axis. We will illustrate this with the following example.

Example 2.12. A uniform line charge (infinite length) with linear charge density as 3.3 nC/m is located at x = 3 m, y = 4 m. Determine E at the origin.

Solution. Refer to Fig. E 2.12. The magnitude of $E = \frac{\rho_L}{2\pi\varepsilon_0 \bar{r}}$

$$= \frac{3.3 \times 10^{-9}}{2\pi \times \frac{1}{36\pi} \times 10^{-9} \times \sqrt{3^2+4^2}}$$

= 11.88 V/m

Hence $$E_r = 11.88\, \bar{a}_r$$

Since unit vector will have both the components along negative x and y axes and hence

$$\bar{a}_r = \frac{-3\bar{a}_x - 4\bar{a}_y}{\sqrt{3^2+4^2}} = \frac{-3\bar{a}_x - 4\bar{a}_y}{5} = -0.6\bar{a}_x - 0.8\bar{a}_y$$

Hence $\bar{E}_r = 11.88(-0.6\bar{a}_x - 0.8\bar{a}_y) = -7.128\bar{a}_x - 9.504\bar{a}_y$ **Ans.**

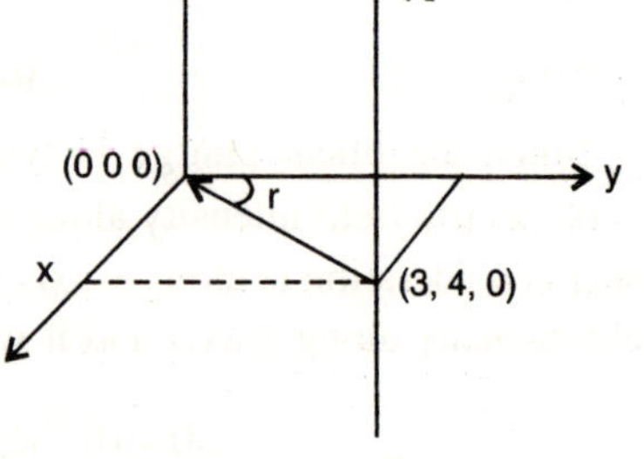

Fig. E 2.12

Example 2.13. In the previous example determine the electric field E at a point (8, 16, 0).

Solution. Refer to Fig. E 2.13. Since electric field is not a function of z in this system, the magnitude of the electric field intensity is given as

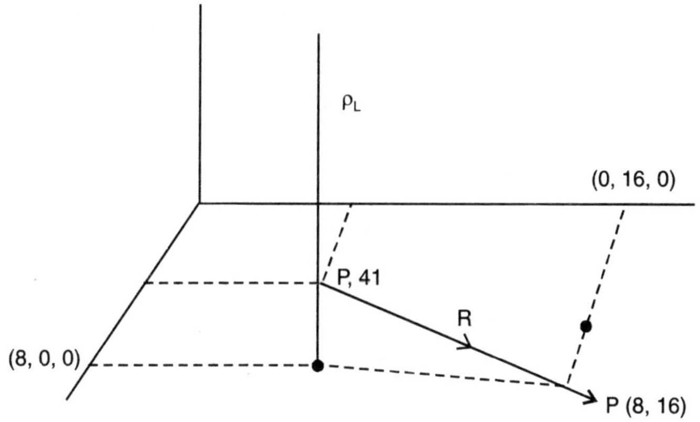

Fig. E 2.13

$$E = \frac{3.3 \times 10^{-9}}{2\pi \times \frac{1}{36\pi} \times 10^{-9} \sqrt{(8-3)^2 + (16-4)^2}}$$

$$= \frac{3.3 \times 18}{13} \approx 4.57 \text{ V/m}$$

The unit vector $$\frac{R}{R} = \frac{(8-3)\bar{a}_x + (16-4)\bar{a}_y}{13}$$

$$= \frac{5\bar{a}_x + 12\bar{a}_y}{13} = 0.3846\bar{a}_x + 0.923\bar{a}_y$$

Here both $\bar{a}_x$ and $\bar{a}_y$ are positive. Hence the electric field intensity at (8, 16, 0) is

$$E = 4.57 \times 0.3846\bar{a}_x + 4.57 \times 0.923\bar{a}_y \text{ V/m}$$

$$= 1.757\bar{a}_x + 4.22\bar{a}_y \text{ V/m } \textbf{Ans.}$$

2.5 ELECTRIC FIELD DUE TO VOLUME CHARGE DISTRIBUTION

Consider a volume enclosed by a regular surface e.g. a cylinder or a sphere etc. Let ρ_v be the uniform volume charge density and to represent a point charge, assume a differential volume Δv the differential charge is given by $\Delta Q = \rho_v \Delta v$ which could be considered a point source charge located at say a distance r' from the assumed origin and let us find out the incremental electric field intensity at a distance r from the origin then the incremental electric field intensity is given by

$$\Delta E(r) = \frac{\Delta Q}{4\pi\varepsilon_0 |r - r'|^2} \cdot \frac{r - r'}{|r - r'|} = \frac{\rho_v \Delta V}{4\pi\varepsilon_0 |r - r'|^2} \frac{r - r'}{|r - r'|} \quad \ldots(2.12)$$

If we sum up the contribution of all the volume charges in a given region, and let the volume elements $\Delta v \to 0$ as the number of these elements tend to infinity, the summation sign is replaced by integral sign. Hence

$$E(r) = \oint_V \frac{\rho_v dv}{4\pi\varepsilon_0 |r - r'|^2} \frac{r - r'}{|r - r'|} \qquad \ldots(2.13)$$

It is to be noted that $\dfrac{r - r'}{|r - r'|}$ is a unit vector and since dv represents differential volume, it is a triple integral as dv, irrespective of system of coordinates, will have three variables and hence while handling any field problem relating to volume charge density we need to perform triple integrals.

Example 2.14. A charge of unit volume charge density is uniformly distributed throughout a sphere of radius a. A redistribution of the charge takes place with density function $\rho_v(r) = k\,(3 - r^2/a^2)$. Determine k.

Solution. The total charge with unit volume charge density is $\dfrac{4}{3}\pi a^3 \cdot 1$.

However with non uniform distribution with volume charge density $\rho_v(r) = k\left(3 - \dfrac{r^2}{a^2}\right)$, the charge density is a function of r. Let us take a sphere with radius r and $(r + dr)$ where we assume the charge density to remain constant as $dr \to 0$, the total charge is:

The volume of the solid shell with radius r and thickness dr is

$4\pi r^2$, dr and hence the charge

$$= \int_0^a 4\pi r^2 \cdot k\left(3 - \frac{r^2}{a^2}\right) dr$$

$$= \int_0^a 4\pi k\left(3r^2 - \frac{r^4}{a^2}\right) dr = 4\pi k\left[\left(r^3 - \frac{r^5}{5a^2}\right)\right]_0^a$$

$$= 4\pi k\left(a^3 - \frac{a^3}{5}\right) = \frac{16}{5}\pi a^3 k = \frac{4}{3}\pi a^3$$

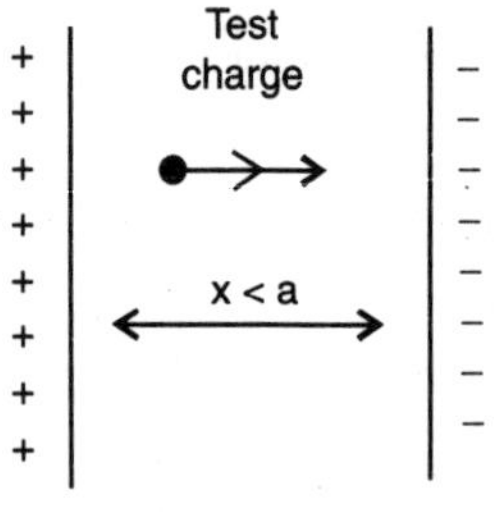

Fig. E 2.14

or $$k = \frac{4}{3} \times \frac{5}{16} = \frac{5}{12} \quad \textbf{Ans.}$$

2.6 ELECTRIC FIELD DUE TO A SHEET OF CHARGE

Next we consider electric field due to an infinite sheet of charge having a uniform charge density of ρ_s C/m^2. Such a charge distribution is normally available on a strip transmission line or on a plate of parallel capacitor. Since we know that charges reside on the surface of the conductors rather than on the interior side of the conductor, this charge density is usually referred to surface charge density. Let us consider a sheet of charge in the yz plane or $x = 0$ plane as shown in Fig. 2.3.

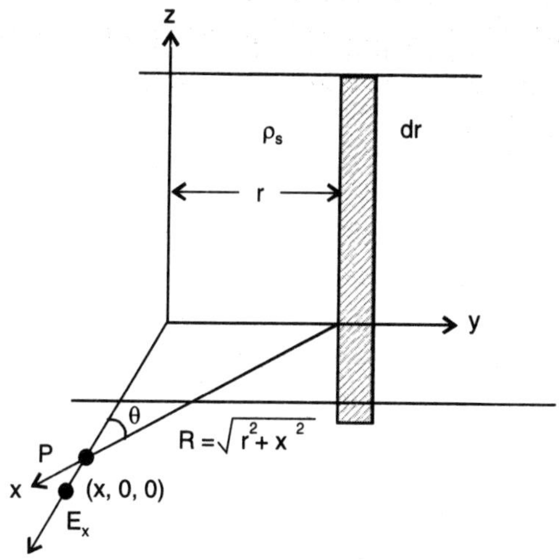

Fig. 2.3 An infinite line of charge in the *yz*-plane

Considering a strip of width dr where $dr \to 0$ and of infinite length at a distance of r from z-axis and along y-axis. The linear charge density is $\rho_2 = \rho_s dr$ and the distance of the general point P along x-axis where we want to find electric field intensity is $R = \sqrt{r^2 + x^2}$. Now looking at the symmetry, since the sheet of charge is lying in the yz plane the electric field cannot vary with y or z and we also see that the y and z components arising from differential elements of charge symmetrically placed with respect to the point (P) at which we evaluate the field will cancel. Therefore only the component along x-axis *i.e.* E_x is present. Considering the infinite strip as an infinite line charge, the electric field intensity due to this line charge at a distance R is given as

$$dE_x = \frac{\rho_s dr}{2\pi\varepsilon_0 \sqrt{(r^2 + x^2)}} \cos\theta = \frac{\rho_s dr}{\sqrt{(r^2 + x^2)} \cdot 2\pi\varepsilon_0} \cdot \frac{x}{\sqrt{r^2 + x^2}}$$

Since $\cos\theta = \dfrac{x}{\sqrt{r^2 + x^2}}$ as seen in Fig. 2.3.

Hence $E_x = \dfrac{\rho_s x}{2\pi\varepsilon_0} \cdot \displaystyle\int_{-\infty}^{\infty} \dfrac{dr}{r^2 + x^2}$

Now from Fig. 2.3 $\dfrac{r}{x} = \tan\theta$ or $r = x\tan\theta$

or $$dr = x \sec^2\theta d\theta$$

$$E_x = \frac{\rho_s x}{2\pi\varepsilon_0} \int_{-\infty}^{\infty} \frac{x \sec^2\theta\, d\theta}{x^2 \sec^2\theta} = \frac{\rho_s}{2\pi\varepsilon_0} \left[\tan^{-1} \frac{r}{x} \right]_{-\infty}^{\infty}$$

$$= \frac{\rho_s}{2\pi\varepsilon_0} \left(\frac{\pi}{2} + \frac{\pi}{2} \right) = \frac{\rho_s}{2\varepsilon_0}.$$

If the point P is taken on negative x-axis, the electric field intensity is then given as $E_x = -\frac{\rho_s}{2\varepsilon_0}$, as the field is always directed away from the positive charge. In order to overcome this problem we associate with the expression a unit vector $\bar{a}_n$ which is normal to the surface and directed outward or away from it. Thus

$$E = \frac{\rho_s}{2\varepsilon_0}\bar{a}_n$$

This result is thrilling as it indicates that the electric field intensity is constant in magnitude and direction irrespective of whether the point under consideration is millions of kms away from the sheet or it is right off the surface.

If, however, a second sheet of charge with negative charge density $-\rho_s$ is placed in the plane at $x = a$ we find out the total field by adding the contribution of each sheet. So we have two infinite sheets separated by a distance a we find out E for three different point (*i*) when $x > a$ (*ii*) $x < a$ and (*iii*) $0 < x < a$. Fig. 2.3 (*a*)

(*i*) $x > a$ Here $E_+ = \frac{\rho_s}{2\varepsilon_0}\bar{a}_x$ and $E_- = -\frac{\rho_s}{2\varepsilon_0}\bar{a}_x$

Hence $E = E_+ + E_- = 0$

(*ii*) $x < a$ *i.e.* the point lies along negative x-axis.

Here $E_+ = \frac{\rho_s}{2\varepsilon_0}\bar{a}_x$, $E_- = -\frac{\rho_s}{2\varepsilon_0}\bar{a}_x$, $E = E_+ + E_- = 0$

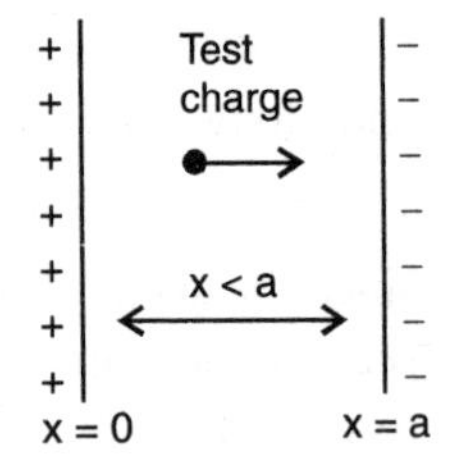

Fig. 2.3 (*a*)

(*iii*) $0 < x < a$ *i.e.* the point lies in between the parallel infinite sheets.

Then $$E_+ = \frac{\rho_s}{2\varepsilon_0}\bar{a}_x$$

$$E_- = \frac{\rho_s}{2\varepsilon_0}\bar{a}_x$$

Hence $$E = \frac{\rho_s}{\varepsilon_0}\bar{a}_x$$

This is a very important result as it gives the electric field between two parallel plates of an air capacitor where the distance between the plates is much smaller than the area of each plate and also that the point under consideration is well away from the edges of the plates of the capacitor. It is to be noted that because of the finite size of the plates of the capacitor, the field outside the parallel plates is not zero but is negligibly small.

Example 2.15. A circular disc of radius R carries a surface charge of uniform density ρ_c C/m^2. Determine the electric field intensity E on the axis of the disc at a distance z from the centre of the disc.

Solution. Consider Fig. E 2.15, taking a differential surface $rdrd\phi$ at a distance of r and angle ϕ from the origin.

In cylindrical coordinates electric field intensity (dE) at a height due to differential element is

$$dE = \frac{\rho_s r\,dr\,d\phi}{4\pi\varepsilon_0\left(r^2+z^2\right)}\,\frac{-r\bar{a}_r + z\bar{a}_z}{\sqrt{r^2+z^2}}$$

Since the disc is symmetrically placed with respect to z-axis, the radial component of dE, and hence we have

$$dE = \frac{\rho_s r\,dr\,d\phi}{4\pi\varepsilon_0\left(r^2+z^2\right)}\cdot\frac{z\bar{a}_z}{\sqrt{r^2+z^2}}$$

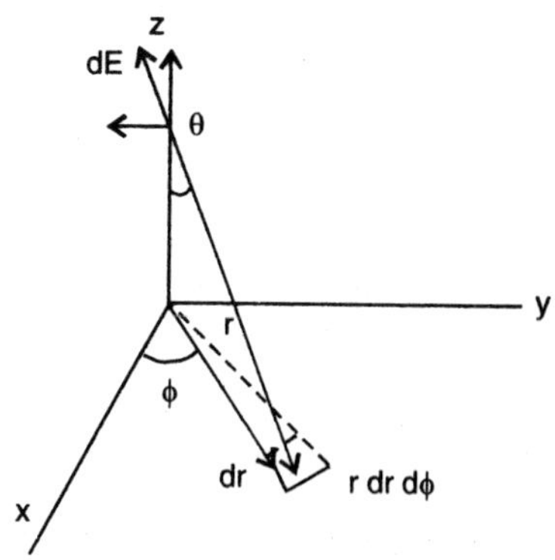

Fig. E 2.15

or

$$E = \frac{\rho_s}{4\pi\varepsilon_0}\int_0^{R}\int_0^{2\pi}\frac{r\,dr\,d\phi\cdot z}{\left(r^2+z^2\right)^{3/2}}\bar{a}_z$$

$$= \frac{\rho_s}{4\pi\varepsilon_0}2\pi\int_0^{R}\frac{r\,dr\,z}{\left(r^2+z^2\right)^{3/2}} = \frac{\rho_s z}{2\varepsilon_0}\int_0^{R}\frac{r\,dr}{\left(r^2+z^2\right)^{3/2}}$$

Let $\quad r = z\tan\theta \quad$ or $\quad dr = z\sec^2\theta\,d\theta$

$$E = \frac{\rho_s z}{2\varepsilon_0}\int_0^{R}\frac{z\tan\theta\cdot z\sec^2\theta\,d\theta}{z^3\sec^3\theta} = \frac{\rho_s}{2\varepsilon_0}\int_0^{R}\sin\theta\,d\theta$$

$$= \frac{\rho_s}{2\varepsilon_0}[-\cos\theta]_0^R = \frac{\rho_s}{2\varepsilon_0}(1-\cos\alpha)\ \text{V/m}\quad \textbf{Ans.}$$

where α is the angle subtended by the z-axis at z to the outer periphery of the disc.

Example 2.16. A sphere of radius 'a' has volume charge density of $\rho_v = 1-\left(\frac{r}{a}\right)^3$ where $r \le a$.

Determine the electric field intensity E at $r = 4a$.

Solution. Since the point where E is required is outside the circle, we can calculate the total charge enclosed by the sphere and assuming this charge to be located at the centre of the sphere, the E can be calculated for $r = 4a$.

Now

$$Q = \oint\rho_v dv = \int_0^a\left\{1-\left(\frac{r}{a}\right)^3\right\}\cdot 4\pi r^2 dr$$

$$= 4\pi\int_0^a\left(r^2-\frac{r^5}{a^3}\right)dr = 4\pi\left(\frac{a^3}{3}-\frac{a^3}{6}\right)$$

$$= 4\pi\cdot\frac{a^3}{6} = \frac{2}{3}\pi a^3$$

Now

$$E = \frac{Q}{4\pi\varepsilon_0(4a)^2} = \frac{\frac{2}{3}\pi a^3}{4\pi\varepsilon_0\cdot 16a^2} = \frac{a}{96\varepsilon_0}\ \text{V/m.}\quad \textbf{Ans.}$$

Example 2.17. Three infinitely long charged lines run parallel to the z-axis as shown in Fig. E 2.17. The lines have uniform linear charge density as shown. Determine E at P.

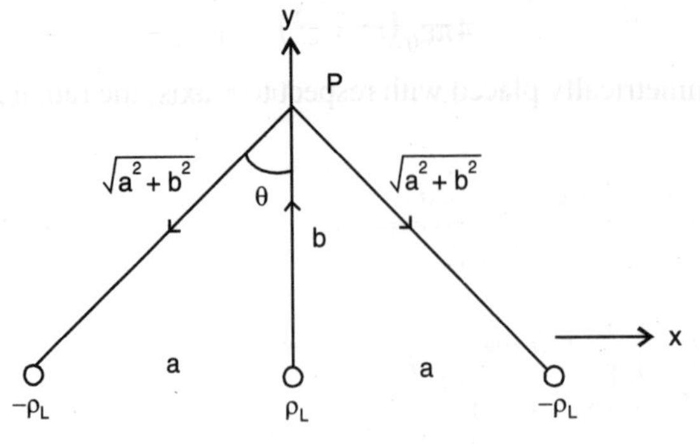

Fig. E 2.17

Solution. The electric field intensity due to an infinite charge at a distance r is given as

$$E = \frac{\rho_L}{2\pi\varepsilon_0 r}$$

Due to symmetry net electric field intensity lies along y-axis as shown in Fig. E 2.17. The net electric field intensity is

$$E = \left\{ \frac{\rho_L}{2\pi\varepsilon_0 b} - \frac{\rho_L \cos\theta}{2\pi\varepsilon_0 \sqrt{a^2+b^2}} - \frac{\rho_L \cos\theta}{2\pi\varepsilon_0 \sqrt{a^2+b^2}} \right\} \bar{a}_y$$

Now from the Fig. E 2.17 $\cos\theta = \dfrac{b}{\sqrt{a^2+b^2}}$

Hence
$$E = \left(\frac{\rho_L}{2\pi\varepsilon_0 b} - \frac{2\rho_L}{2\pi\varepsilon_0} \cdot \frac{b}{\left(\sqrt{a^2+b^2}\right)^2} \right) \bar{a}_y = \frac{\rho_L}{2\pi\varepsilon_0} \left[\frac{1}{b} - \frac{2b}{a^2+b^2} \right] \bar{a}_y$$

$$= \frac{\rho_L}{2\pi\varepsilon_0} \frac{a^2+b^2-2b^2}{b\left(a^2+b^2\right)} \bar{a}_y = \frac{\rho_L}{2\pi\varepsilon_0 b} \frac{a^2-b^2}{a^2+b^2} \bar{a}_y \quad \textbf{Ans.}$$

Example 2.18. A uniform volume charge density of 0.3 μC/m³ is present in a spherical shell of radii 3 cm and outer radius 5 cm. If $\rho_v = 0$ elsewhere, determine (*i*) the charge enclosed by the shell (*ii*) r_1 if half the total charge is located in the region 3 cm $< r < r_1$.

Solution. The charge enclosed by the shell

$$\int_{3\times10^{-2}}^{5\times10^{-2}} 0.3 \times 4\pi r^2 dr = \frac{1.2\pi}{3}\left[5^3 - 3^3\right] \times 10^{-6}$$

$$= 123\,\text{pC}.$$

For the second part

$$61.5\times10^{-12} = 0.4\pi\left(r_1^3 - 3^3\right)\times10^{-12}$$

$$61.5 = 0.4\pi\left(r_1^3 - 3^3\right)$$

or

$$48.9 + 27 = r_1^3$$

$$75.96 = r_1^3 \text{ or } r_1 = 4.23 \text{ cm.}$$

Example 2.19. A uniform line charge of 16 nC/m is located along the line defined by $y = -2$, $z = 5$.
(*a*) Determine E at P (1, 2, 3).

(*b*) Determine E at that point in the $z = 0$ plane where the direction of E is given by $\frac{1}{3}\bar{a}_y - \frac{2}{3}\bar{a}_z$.

Solution. (*a*) Consider point (0, –2, 5) and P(1, 2, 3). The radial vector from (0, –2, 5) and P(1, 2, 3) $\bar{a}_x + 4\bar{a}_y - 2\bar{a}_z$. Here $\bar{a}_x$ component of E will be zero hence we have $4\bar{a}_y - 2\bar{a}_z$. The length of the radial vector is $\sqrt{4^2+2^2} = \sqrt{20}$, the unit vector is $\dfrac{4\bar{a}_y - 2\bar{a}_z}{\sqrt{20}}$

Hence

$$E = \frac{16\times10^{-9}}{2\pi\cdot\dfrac{1}{36\pi}\times10^{-9}\cdot\sqrt{20}} = \frac{16\times18}{\sqrt{20}}\frac{\left(4\bar{a}_y - 2\bar{a}_z\right)}{\sqrt{20}}$$

$$= \frac{4}{5}\times18\left(4\bar{a}_y - 2\bar{a}_z\right) = 57.6\bar{a}_y - 28.8\bar{a}_z \textbf{ Ans.}$$

(*b*) The length of the vector $= \sqrt{\left(\frac{1}{3}\right)^2 + \left(\frac{2}{3}\right)^2} = \frac{\sqrt{5}}{3}$

Hence

$$E = \frac{16\times10^{-9}\times3}{2\pi\cdot\dfrac{1}{36\pi}\times10^{-9}\sqrt{5}}\cdot\left(\frac{1}{\sqrt{5}}\bar{a}_y - \frac{2}{\sqrt{5}}\bar{a}_z\right)$$

$$= \frac{16\times18\times3}{5}\left(\bar{a}_y - 2\bar{a}_z\right) \textbf{ Ans.}$$

Example 2.20. Two infinite uniform sheets of charge each with density ρ_s are located at $x = \pm2$ m. Determine E in all regions.

Solution. Fig. E 2.20 represents the location of the sheet of charge.

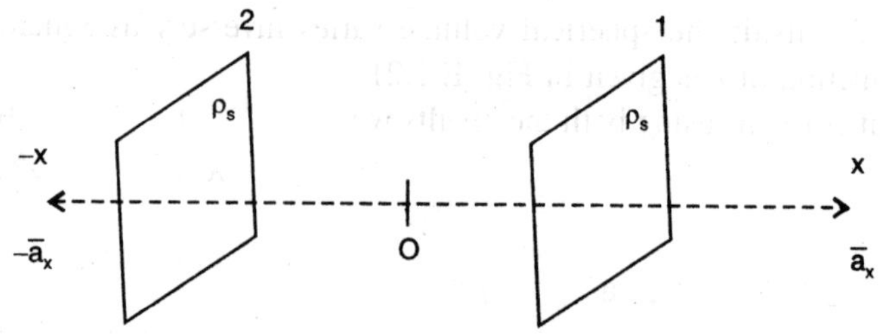

Fig. E 2.20

Here only parts of the two sheets have been shown. Since the sheets are of infinite size, the electric field E is independent of the distance between them *i.e.* $E = \dfrac{\rho_s}{\varepsilon_0}$.

For $x < -2$ *i.e.* the point is on the left side of sheet 2 and the direction of E due to both the sheets will be along $-\bar{a}_x$ and hence

$$E = \frac{-\rho_0}{\varepsilon_0}\bar{a}_x \quad \textbf{Ans.}$$

If the point lies in between the plates, the electric field intensity E due to the two sheets will be in opposition and hence $E = 0, -2 \le x < 2$. **Ans.**

For $x > 2$

$$E = \frac{\rho_0}{\varepsilon_0}\bar{a}_x \quad \textbf{Ans.}$$

Example 2.21. A spherical volume of radius R has uniformly distributes volume charge density of ρ_v. Determine E everywhere.

Solution. (*i*) Field intensity for $r < R$.

The charge enclosed by the radius r is $\dfrac{4}{3}\pi r^3 \rho_v$ and assuming this charge to be concentrated as point charge at the centre of the sphere, then $E = \dfrac{Q}{4\pi\varepsilon_0 r^2} = \dfrac{4}{3}\pi r^3 \rho_v \cdot \dfrac{1}{4\pi\varepsilon_0 r^2}$

$$= \frac{\rho_v r}{3\varepsilon_0}$$

i.e. E varies linear as r for $r \le R$

(*ii*) Field intensity at $r \ge R$

Total charge $\dfrac{4}{3}\pi R^3 \rho_v = Q$

Hence,
$$E = \frac{Q}{4\pi\varepsilon_0 r^2} = \frac{4}{3}\pi R^3 \rho_v \cdot \frac{1}{4\pi\varepsilon_0 r^2} = \frac{\rho_v R^3}{3\varepsilon_0} \cdot \frac{1}{r^2}$$

i.e. the electric field E outside the spherical volume varies inversely as square of the distance. The variation of E as a function of r is given in Fig. E 2.21.

It can be seen that at $r = R$, using both the results we get

$$r < R \quad E = \frac{\rho_v r}{3\varepsilon_0} = \frac{\rho_v R}{3\varepsilon_0} \text{ at } r = R$$

$$r > R \quad E = \frac{\rho_v R^3}{3\varepsilon_0} \cdot \frac{1}{R^2} = \frac{\rho_v R}{3\varepsilon_0}$$

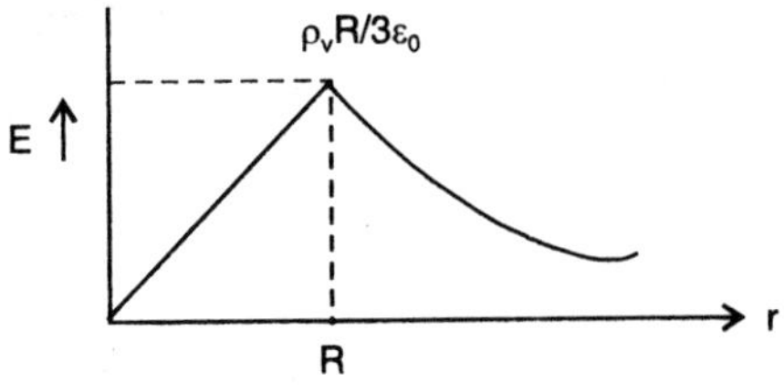

Fig. E 2.21

Example 2.22. A circular flat ring of inner radius 2 m and outer 5 m has surface charge density $\rho_s = 100/r$ μC/m². Determine the electric field intensity on the axis of the ring 15 m away from the centre.

Solution. Consider Fig. E 2.22 for the solution.

Consider differential surface $rdrd\phi$ and the differential charge $dQ = rdrd\phi \cdot \rho_s$ which could be considered as point charge and hence electric field intensity due to this point charge at a vector distance R is given by

$$dE = \frac{r\,dr\,d\phi\rho_s}{4\pi\varepsilon_0 R^2}$$

The component of dE along z-axis is $dE \cos\theta$ where

$\cos\theta = \frac{z}{R} = \frac{15}{R}$ and due to symmetry the other component of dE vanishes and hence

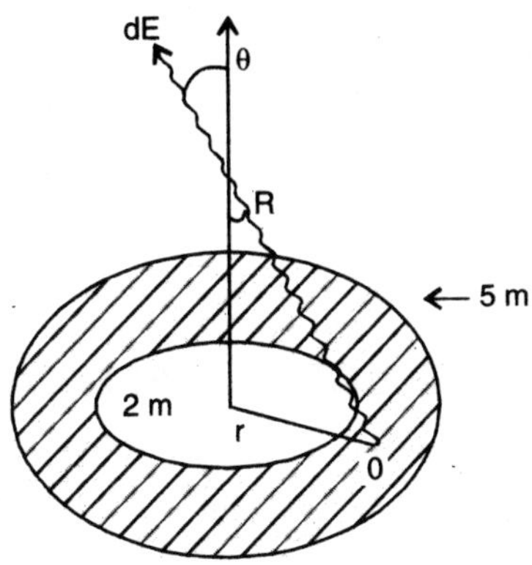

Fig. E 2.22

$$dE = \frac{\rho_s r\,dr\,d\phi}{4\pi\varepsilon_0 R^2} \cdot \frac{15}{R} \text{ where } R = \sqrt{r^2 + z^2}$$

Hence,

$$dE = \frac{\rho_s r \partial r\, d\phi}{4\pi\varepsilon_0} \cdot \frac{z}{\left(r^2 + z^2\right)^{3/2}}$$

$$= \frac{100}{r} \cdot 9 \times 10^9 \times 10^{-6} \cdot \frac{15}{\left(r^2 + z^2\right)^{3/2}} r\,dr\,d\phi$$

$$= 9 \times 10^5 \times 15 \frac{dr\,d\phi}{\left(r^2 + z^2\right)^{3/2}}$$

$$E = 135 \times 10^5 \int_2^5 \int_0^{\pi} \frac{dr\,d\phi}{\left(r^2 + z^2\right)^{3/2}}$$

$$= 135\pi \times 10^5 \int_2^5 \frac{dr}{(r^2 + z^2)^{3/2}}$$

$$= 270\pi \times 10^5 \left[\frac{r}{225\sqrt{r^2 + 225}} \right]_2^5$$

$$= \frac{270\pi}{225} \times 10^5 \left[\frac{r}{\sqrt{r^2 + 225}} \right]_2^5 = \left[\frac{5}{\sqrt{250}} - \frac{2}{\sqrt{229}} \right]$$

$$= 3.768 \times 10^5 (0.316 - 0.1322)$$

$$= 3.768 \times 0.1838 \times 10^5$$

$$= 69.25 \text{ kV/m} \quad \textbf{Ans.}$$

It is to be noted here that while considering various charge configurations, the charge is not "on a conductor". When it is given that charge is distributed in the form of a disc, it does not mean a disc shaped conductor with charge on the surface. Rather, it is to be imagined that these charges somehow are suspended in space and are fixed in the specified configuration.

2.7 ELECTRIC FIELD INTENSITY DUE TO AN ELECTRIC DIPOLE

The concept of an electric dipole is quite important in electromagnetic field theory as they form the basis for the behaviour of dielectric materials in electric field and also it justifies the application of method of images for some typical situations in *EM* theory. Two equal and opposite charges separated by a distance d which is much smaller than the distance to the point P at which we are interested to obtain electric field intensity or electric potential. Refer to Fig. 2.4 which shows an electric dipole with

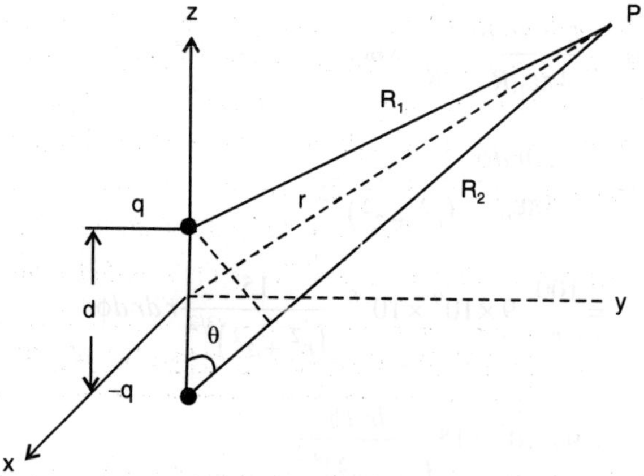

Fig. 2.4 An Electric Dipole

charges q and $-q$ separated by a distance d. In Fig. 2.4 point P is shown not far from the charges. However, since we assume that $r >> d$, the radial vectors R_1, R_2 and r will be more or less parallel lines and if θ is the angle as shown then $R_2 - R_1 = d \cos\theta$ and the product $q \cdot d$ is known as dipole moment.

Since moment is a vector quantity, it's magnitude is given by $q \cdot d$ and the direction is given by the line joining two point charges and from negative charge to the positive charge *i.e.* in Fig. 2.4 it is given by the arrow along positive z-axis $\bar{a}_z$. The distant point is defined here by spherical coordinates, r, θ and ϕ where $\phi = 90°$ here in view of the azimuth symmetry. The coordinates of the positive and negative point charges are (0, 0, $d/2$) and (0, 0, $-d/2$) respectively.

We know there are two ways to find out electric field intensity at a point due to certain charge configuration one, find out the E due to different charges at that point and then the resultant E is the vector sum of these electric field intensities.

Second method is to find potential due to different charges at that point and then the resultant potential is the scalar sum of these potential. Then $E = -\nabla V$ *i.e.* E equals the negative of the gradient of V as evaluated. We prefer second method as it is much simpler to evaluate V as compared to E by the first method.

From Fig. 2.4, potential V at point P due to q and $-q$ is $V = \dfrac{q}{4\pi\varepsilon_0}\left(\dfrac{1}{R_1} - \dfrac{1}{R_2}\right) = \dfrac{q}{4\pi\varepsilon_0}\dfrac{R_2 - R_1}{R_1 R_2}$

It is to be noted that the plane $z = 0$ midway between the two point charges is the loci of point, for which $R_1 \simeq R_2$ and therefore at zero potential as all points are at infinity. Hence $R_1R_2 = r^2$. However, the approximation is used for the term $(R_2 - R_1)$ in the numerator. Also from the Fig. 2.4 we have already obtained $R_2 - R_1 = d\cos\theta$. Hence, substituting we have

$$V = \frac{q}{4\pi\varepsilon_0} \cdot \frac{d\cos\theta}{r^2} = \frac{qd\cos\theta}{4\pi\varepsilon_0 r^2}$$

Here again we see that for $z = 0$ plane $\theta = 90°$ and $V = 0$

We know that in spherical coordinate the expression for gradient of scalar V is given as

$$E = -\nabla V = -\left(\frac{\partial V}{\partial r}\bar{a}_r + \frac{1}{r}\frac{\partial V}{\partial \theta}\bar{a}_\theta + \frac{1}{r\sin\theta}\frac{\partial V}{\partial \phi}\bar{a}_\phi\right)$$

Hence we obtain
$$E = -\left(-\frac{d\cos\theta}{2\pi\varepsilon_0 r^3}\bar{a}_r - \frac{qd}{4\pi\varepsilon_0 r^3}\sin\theta\bar{a}_\theta\right)$$

$$= \frac{qd}{4\pi\varepsilon_0 r^3}(2\cos\theta\bar{a}_r + \sin\theta\bar{a}_\theta)$$

Here qd is known as electric dipole moment. One can really appreciate how simple it is to find out electric field intensity using the relation $E = -\nabla V$ *i.e.* we calculate potential at that point and negative of the ∇V is the electric field intensity.

We now make use of the dipole moment to evaluate potential fields. Dipole moment = $qd = p$. The units of p are Coulomb – m.

Since $d \cdot \bar{a}_r = d\cos\theta$, we then have

$$V = \frac{p \cdot \bar{a}_r}{4\pi\varepsilon_0 r^2}. \text{ Now } \bar{a}_r = \frac{\overline{r - r'}}{|r - r'|} \text{ and } |r - r'|^2 = r^2$$

Hence
$$V = \frac{1}{4\pi\varepsilon_0 |r - r'|^2} \cdot p \cdot \frac{r - r'}{|r - r'|}$$

where $\bar{r}$ is the distance of the field point P and $\overline{r'}$ determines the dipole centre. The equation for V above is independent of any coordinate system.

It is to be noted that for certain dipole moment, the potential at a point is inversely proportional to square of the distance and electric field is inversely proportional to cube of the distance from the dipole. However, these quantities are inversely proportional to the distance and inversely proportional to square of the distance from the point charge respectively. It is on expected lines as dipole charges appear to be closer together at greater distances and these together act like a single point charge of zero coulomb. If we have more than one dipole *i.e.* we have multipoles symmetrically placed the electric fields are inversely proportional to higher powers of *r*.

Example 2.23. A dipole moment $p = 9\bar{a}_z$ n.C-m is located at the origin in free space. Determine (*a*) V at $P(r = 6, \theta = 30°$ and $\phi = 0)$ (*b*) E at P.

Solution. (*a*) Using the relation for $V = \dfrac{qd\cos\theta}{4\pi\varepsilon_0 r^2}$

Here $qd = p = 9\times10^{-9}$ C-m, $\cos\theta = \cos 30° = \dfrac{\sqrt{3}}{2}$, $r = 6$ m

Substituting these values $V = \dfrac{9\times10^{-9}\cdot\sqrt{3}/2}{4\pi\cdot\dfrac{1}{36\pi}\times10^{-9}\times36}$

$$= \frac{81\sqrt{3}}{72} = 1.948 \text{ volts} \quad \textbf{Ans.}$$

(*b*) Expression for electric field

$$E = \frac{qd}{4\pi\varepsilon_0 r^3}\left(2\cos\theta\bar{a}_r + \sin\theta\bar{a}_\theta\right)$$

$$= \frac{9\times10^{-9}}{4\pi\times\dfrac{1}{36\pi}\times10^{-9}\cdot216}\left(2\cos30°\bar{a}_r + \sin30° a_\theta\right)$$

$$= \frac{81}{216}\left(2\times\frac{\sqrt{3}}{2}\bar{a}_r + 0.5\bar{a}_\theta\right)$$

$$= 0.375\left(\sqrt{3}\bar{a}_r + 0.5\bar{a}_\theta\right) = 0.649\bar{a}_r + 0.1875\bar{a}_\theta \quad \textbf{Ans.}$$

Example 2.24. An electric dipole located at the origin in free space has a moment $\left(3\bar{a}_x - 2\bar{a}_y + \bar{a}_z\right)$ n C-m. Determine V at $P(2, 3, 4)$ and V at $r = 2.5$, $\theta = 30°$, $\phi = 60°$.

Solution. (*a*) Using the general expression for V.

$$V = \frac{1}{4\pi\varepsilon_0}\cdot\frac{1}{|r - r'|^2}\cdot p\cdot\frac{r - r'}{|r - r'|}$$

Here $$p = (3\bar{a}_x - 2\bar{a}_y + \bar{a}_z),\ r - r' = 2\bar{a}_x + 3\bar{a}_y + 4\bar{a}_z$$

and $$|r - r'| = \sqrt{4+9+16} = \sqrt{29}$$

Hence $$p \cdot \frac{r - r'}{|r - r'|} = (3\bar{a}_x - 2\bar{a}_y + \bar{a}_z) \cdot (2\bar{a}_x + 3\bar{a}_y + 4\bar{a}_z) \cdot \frac{1}{\sqrt{29}}$$

$$= \frac{4}{\sqrt{29}}$$

$$V = \frac{10^{-9}}{4\pi \cdot \frac{1}{36\pi} \times 10^{-9} \times 29} \cdot \frac{4}{\sqrt{29}}$$

$$= \frac{36}{29\sqrt{29}} = 0.23 \text{ volts. } \textbf{Ans.}$$

(*b*) For the second part

$$V = \frac{\bar{p} \cdot \bar{a}_r}{4\pi\varepsilon_0 r^2} = (3a_x - 2a_y + a_z) \cdot a_r$$

$$= \frac{3\sin\theta\cos\phi - 2\sin\theta\sin\phi + \cos\theta}{4\pi \times \frac{1}{36\pi} \times 10^{-9} \times 2.5^2}$$

$$= \frac{9[3\sin 30°\cos 40° - 2\sin 30°\sin 40° + \cos 30°]}{(2.5)^2}$$

$$= 1.44\left(3 \times \frac{1}{2} \times 0.766 - 2 \times \frac{1}{2} \times 0.643 + 0.866\right)$$

$$= 1.44(1.149 - 0.643 + 0.866)$$

$$= 1.44 \times 1.37 = 1.9756 \text{ volt. } \textbf{Ans.}$$

PROBLEMS

P 2.1 Point charges of 100 nC each are located at *A*(1, 0, 0), *B*(–1, 0, 0), *C*(0, 1, 0) and *D*(0, –1, 0) in free space. Determine the net force on the charge at *A*. **Ans.** $43\bar{a}_x$ μN

P 2.2 Determine electric field intensity *E* at the origin due to a point charge of 32.2 nC located at (–4, 3, 2) m in cartesian coordinates.

Ans. $10 \cdot \dfrac{4\bar{a}_x - 3\bar{a}_y - 2\bar{a}_z}{\sqrt{29}}$ V/m

P 2.3 Develop an expression for electric field intensity due to a charge uniformly distributed over an infinite plane with surface charge density as ρ_s C/m^2.

Ans. $E = \dfrac{\rho_s}{2\varepsilon_0}\bar{a}_n$

P 2.4 A circular disc with radius a lies in the plane $z = 0$ and it is charged with surface charge density $\rho_s = \rho_0 \sin^2\phi$. Determine electric field E at P (0, ϕ, h).

Ans. $E = \dfrac{\rho_0 h}{4\varepsilon_0}\left(\dfrac{-1}{\sqrt{a^2+h^2}} + \dfrac{1}{h}\right)\bar{a}_z$

P 2.5 Two point charges q_1 = 20 nC and q_2 = 10 nC are separated from each other by a distance 25 cm in free space. Determine the electric field at a point P that is 15 cm away from q_1 and 20 cms from q_2. **Ans.** $8.31\,\underline{|15.7°}$ kV/m.

P 2.6 Show that the field F_a is not realizable as a static electric field whereas F_b and F_c are realizable.

(a) $F_a = -y\bar{a}_x + x\bar{a}_y$

(b) $F_b = (\rho_l/2\pi\varepsilon_0 r^2)(\cos\phi\bar{a}_r + \sin\phi\bar{a}_\phi)$ Cylindrical coordinate

(c) $F_c = (\sin\theta\bar{a}_x + r\cos\theta\bar{a}_\theta)$

P 2.7 Determine the electric field intensity required to balance the earth's gravitational force on a charge of q Coulomb having a mass m kg. Also determine the value of this electric field intensity if the charge is an electron.

Ans. $-\dfrac{mg}{q}\bar{a}_r$, 55.7×10^{-12} N/C

P 2.8 A drop of oil is suspended in space between two large metal plates that are 10 cms apart and that have a difference of potential of 10^4 volt. Determine the charge on the drop if its mass is 0.1 gm. **Ans.** 9.8 μC

P 2.9 A charge distribution produces the field $E = r^2\bar{a}_r - r^2\sin\theta\bar{a}_\theta + 10\bar{a}_\phi$. Determine the charge density at (2, π, 0). **Ans.** 12 ε_0 C/m^2

P 2.10 Figure P 2.10 shows three point charges in the xy plane having line charges parallel to the z-axis through three points. Show that the electric field at the fourth point is zero.

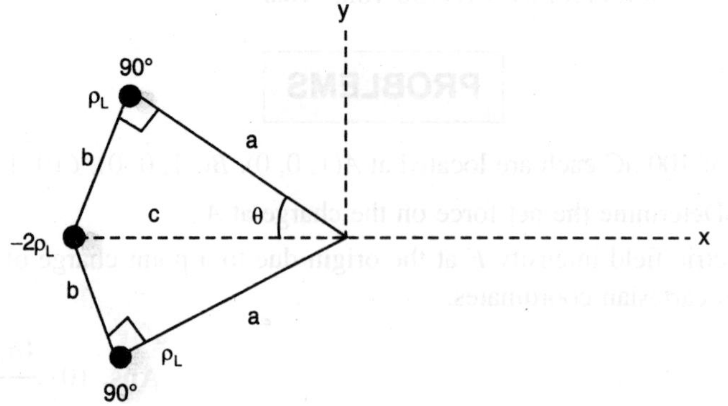

Fig. P 2.10

P 2.11 Three point charges each of mass m and charge Q are suspended by strings of length l from a common point. It is observed that the common point and the points occupied by the three charges form the corners of a tetrahedron. Determine the relation between Q, l, m and the acceleration due to gravity g. **Ans.** $Q^2/\varepsilon_0 l^2 mg = 4\pi/\sqrt{6}$

P 2.12 Three point charges Q, $-2Q$ and Q are located at (0, 0, d), (0, 0, 0) and (0, 0, $-d$) respectively. Determine (a) the electric field intensity at distances large compared to d along the line joining the charges (b) the electric field intensity at distances larger compared to d normal to the line joining the charges.

Ans. (a) $\dfrac{6Qd^2}{4\pi\varepsilon_0 z^4}$ (b) $\dfrac{3Qd^2}{4\pi\varepsilon_0 r^4}$

P 2.13 An infinitely long line charge of uniform linear charge density ρ_L C/m is situated along the z-axis. Determine electric field intensity due to thin line charge using Gauss's law.

Ans. $E = \dfrac{\rho_L}{2\pi\varepsilon_0 r}$

P 2.14 A ring charge of radius R with charge density ρ_L which is a function of ϕ is situated in the xy plane with its centre at the origin. Determine the electric field intensity at a point (0, 0, z) in cartesian coordinates.

Ans.
$$E_x = \frac{-R^2}{4\pi\varepsilon_0 (R^2 + z^2)^{3/2}} \int_0^{2\pi} \rho_L \cos\phi \, d\phi$$

$$E_y = \frac{-R^2}{4\pi\varepsilon_0 (R^2 + z^2)^{3/2}} \int_0^{2\pi} \rho_L \sin\phi \, d\phi$$

$$E_z = \frac{R_z}{4\pi\varepsilon_0 (R^2 + z^2)^{3/2}} \int_0^{2\pi} \rho_L d\phi$$

P 2.15 A volume charge is distributed with uniform density ρ_v C/m^3 throughout an infinitely long cylinder of radius R metre. Determine electric field E for $r \le R$ and $r \ge R$.

Ans. $\dfrac{\rho_v r}{2\varepsilon_0}\bar{a}_r \;\; r \le R, \;\; \dfrac{\rho_v R^2}{2\varepsilon_0 r} \;\; r \ge R$

P 2.16 Identical charges of Q Coulomb are located at the eight corners of a cube with a side l metre. Determine the Coulomb force on each charge. **Ans.** 3.29 $Q^2/4\pi\varepsilon_0 l^2$ N.

P 2.17 The plane $-x + 3y - 6z = 6$ m contains a uniform charge distribution $\rho_s = 0.53$ nC/m^2. Determine E on the side containing the origin.

Ans. $30\left(\dfrac{\bar{a}_x - 3\bar{a}_y + 6\bar{a}_z}{\sqrt{46}}\right)$ V/m.

P 2.18 A circular ring of charge with radius 2 m and charge density 10 nC/m lies in the $z = 0$ plane, with centre at the origin. Determine the point charge Q at the origin which would produce the same electric field E at (0, 0, 5) m. **Ans.** 100.5 nC

P 2.19 Two infinite line charges one having line charge density of ρ_L C/m and the other $-\rho_L$ C/m are separated by a distance D. Determine E at a point P which is at a distance of d m from positive charged line.

Ans. $E = \dfrac{D\rho_L}{2\pi\varepsilon_0 d(D-d)}$

P 2.20 Determine the total charge contained within a cube of side l oriented as shown in Fig. P 2.20 if the electric field within and on the cube is given by

$$E = \frac{1}{l^4}\left(\alpha x^2 yz\bar{a}_x + \beta xy^2 z\bar{a}_y + \gamma xyz^2\bar{a}_z\right)$$

where α, β and γ are constant with units V/m.

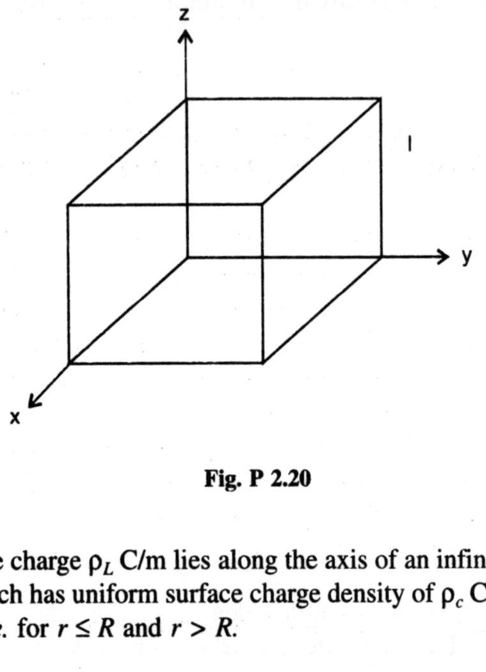

Ans. $Q = \varepsilon_0(\alpha+\beta+\gamma)\cdot l^2/4$

Fig. P 2.20

P 2.21 A uniform line charge ρ_L C/m lies along the axis of an infinitely long cylinder of radius R, the surface of which has uniform surface charge density of ρ_c C/m². Determine the electric field E everywhere *i.e.* for $r \le R$ and $r > R$.

Ans. $E = \dfrac{\rho_L}{2\pi\varepsilon_0 r}\bar{a}_r \quad 0 < r < R$

$= \dfrac{1}{2\pi\varepsilon_0 r}(\rho_L + 2\pi r\rho_s)\bar{a}_r \quad r > R$

P 2.22 A line charge of length $2a$ with line charge density ρ_L C/m is placed along y axis such that $-a \le y \le a$ and a point charge Q is placed along x-axis at a distance R from the y-axis. Determine the force experienced by the charge Q due to line charge.

Ans. $\dfrac{\rho_L Q}{2\pi\varepsilon_0 R\sqrt{\left(1+\dfrac{R}{a}\right)^2}}$ Newtons

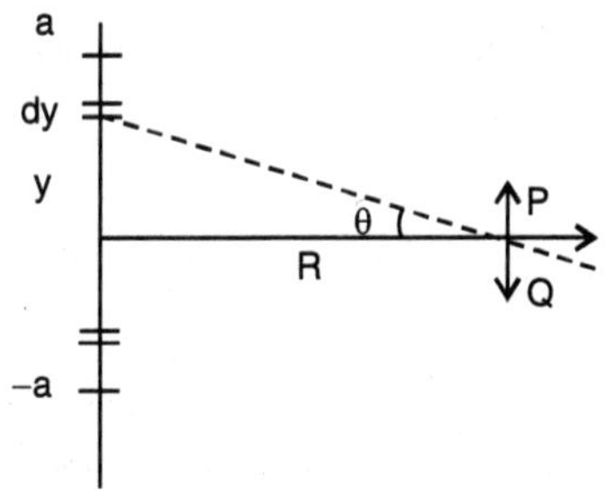

Fig. P 2.22

P 2.23 A charge of 160 nC is distributed uniformly along the z-axis between $z = \pm 4$ m. Determine E at (3, 0, 0) m. **Ans.** $96\bar{a}_x$ V/m

P 2.24 Determine E at P as shown in Fig. P 2.24.

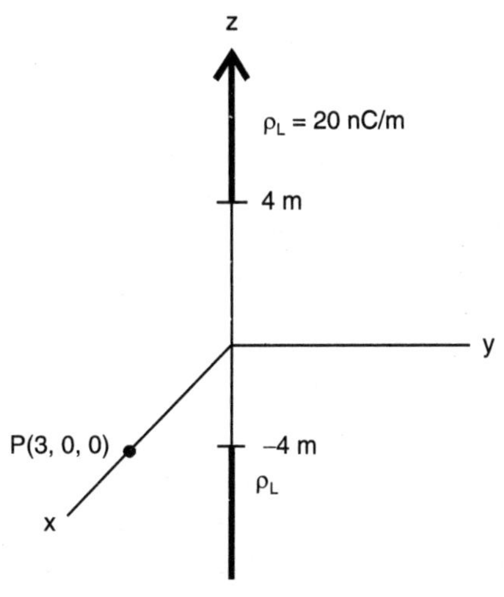

Ans. $24\bar{a}_x$ V/m

Fig. P 2.24

P 2.25 A point charge Q_1 = 25 nC is located at A (4, –2, 7) and a charge Q_2 = 60 nC is at B (–3, 4, –2). (*a*) If $\varepsilon = \varepsilon_0$ determine E at C (1, 2, 3) (*b*) At what point on the y-axis is $E_x = 0$? **Ans.** (*a*) $4.58\bar{a}_x - 0.15\bar{a}_y + 5.51\bar{a}_z$, (*b*) –6.89 or –22.11

Chapter 3

Electric Flux Density and Gauss's Law

3.1 INTRODUCTION

We know that when a positive point charge is placed in space, hypothetical radial lines emanate from this. When a positive test charge is brought in the vicinity of this charge it experiences a force and the direction of movement of the test charge is the direction of hypothetical lines and it is observed that these lines emanate from the charge radially. These lines are termed as streamlines or flux lines. It is to be noted that no physical particle is projected radially outward from the point charge. However, it is very helpful to suggest an electric flux emanates away symmetrically from a point charge and is coincident with the streamlines and to visualise this flux whenever an electric field is present. In this chapter we introduce the concept of electric flux and electric flux density which will be used to solve certain problems with ease which are extremely symmetrical problems.

Charge like mass, is a fundamental property of matter that is not thought of as being derived from other properties as is force or speed. Unlike mass positive and negative charges are common, existing in all atomic system that make up matter. The charge that we are considering is one that has been separated from atomic source, free charge that can be moved about.

In order to understand further the concept of electric flux, Faraday used a pair of concentric metallic spheres, the outer one had two hemisphere which could be clamped firmly together. After dismantling the outer hemisphere he charged the inner sphere with a known quantity of positive charge. The hemispheres were then clamped together around the charge sphere. The outer sphere is then discharged by connecting it momentarily to ground. The outer sphere is separated using tools made of insulating materials so that it does not disturb the induced charge on it. The negative or the induced charge on each hemisphere was measured. It was found that the total charge on the outer sphere was equal in magnitude to the original positive charge placed on the inner sphere and it was found that this came true irrespective of the dielectric material placed in the space between the two spheres. From this experiment he concluded that there was some kind of "displacement" from the inner sphere to the outer sphere which was independent of the medium and we now refer to this flux as displacement flux or simply electric flux. It was also observed that the electric flux is directly proportional to the charge on the inner sphere. The constant of proportionality is dependent on the system of units involved. In SI units the constant of proportionality is unity. If electric flux is denoted by ψ, the total charge on the inner sphere by Q, then $\psi = Q$ and the electric flux ψ is measured in coulombs.

3.2 ELECTRIC FLUX DENSITY

Consider a spherical shell with radius of inner sphere R_1 and that of outer sphere as R_2. Say the inner sphere is charged with Q coulomb charge and hence the negative charge of equal magnitude is induced in the surface of the outer sphere *i.e.* $-Q$ coulomb. As we know flux lines emanate from the positive charge and terminate on to the negative charge. Due to spherical symmetry these electric flux lines emanating from inner sphere terminate on to the outer sphere as shown in Fig. 3.1.

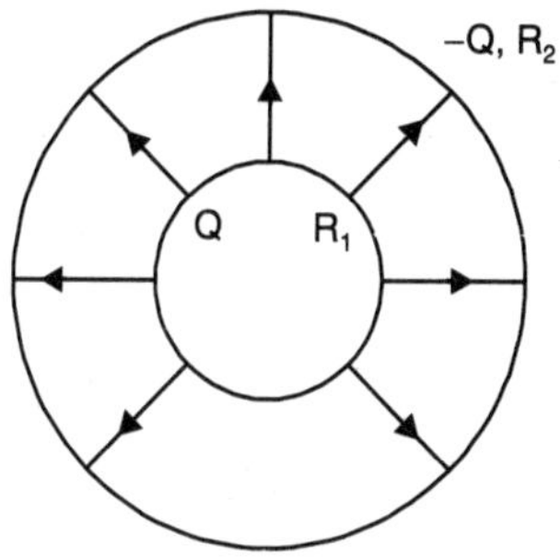

Fig. 3.1 Electric flux between the charged spherical shells

On the inner sphere an electric flux of ψ is produced due to the charge Q on it such that $\psi = Q$ and this flux is uniformly distributed over the spherical surface of area $4\pi R_1^2$. The surface density of this flux is $\psi/4\pi R_1^2$ C/m^2 and is denoted by the letter D which was chosen originally because of the alternate names of displacement flux density or displacement density. Since electric flux density seems more relevant, this term is used invariably. Since D is a vector quantity it has both magnitude and direction. The magnitude is independent of the medium through which flux lines flow and the direction is given by the direction of the flux lines at that point. The magnitude is given by the number of flux lines crossing a surface normal to the lines divided by the surface area.

Referring back to Fig. 3.1 since the flux lines are normal to the surface area, the direction of electric flux density is radial to the surface and the magnitude is given by

$$D = \frac{Q}{4\pi R_1^2}\bar{a}_r \text{ inner sphere} \qquad \text{...(3.1)}$$

$$D = \frac{Q}{4\pi R_2^2}\bar{a}_r \text{ outer sphere} \qquad \text{...(3.2)}$$

and D at any distance r between the spheres is given by

$$D = \frac{Q}{4\pi r^2}\bar{a}_r \qquad R_1 \le r \le R_2 \qquad \text{...(3.3)}$$

If we now tend the inner sphere to a point keeping the charge Q as same, the electric flux density at any distance r from the point charge is given by

$$D = \frac{Q}{4\pi r^2}\bar{a}_r \qquad \text{...(3.4)}$$

as electric flux lines are symmetrically directed outward from the point and pass through an imaginary spherical surface of area $4\pi r^2$.

We know that electric field intensity due to a point charge Q at a distance r is given by (free space)

$$E = \frac{Q}{4\pi\varepsilon_v r^2}\bar{a}_r \quad ...(3.5)$$

Comparing equations for D and E we have

$$D = \varepsilon_0 E \qquad \text{for free space only} \quad ...(3.6)$$

Equation (3.6) is applicable only for free space or vacuum.

However, it is not restricted to point charge only. For a general volume charge distribution in free space.

$$E = \int_v \frac{\rho_v dv}{4\pi\varepsilon_0 r^2}\bar{a}_r \qquad \text{for free space} \quad ...(3.7)$$

Similarly from equation (3.4), we have

$$D = \int_v \frac{\rho_v dv}{4\pi r^2}\bar{a}_r \quad ...(3.8)$$

It is to be noted that in an isotropic medium (properties independent of direction) D and E are in the same direction and $\varepsilon = \varepsilon_0\varepsilon_r$ is a scalar quantity. Here ε_r is the relative permittivity of the medium and ε the permittivity of the medium. However, in an anisotropic medium E and D may be in different direction as ε is a tensor in such media. We will consider isotropic media only here.

Example 3.1. A uniform line charge of 10 nC/m lies along the z-axis. Determine D in the region of the line charge.

Solution. Let us find D at a distance of 5 m *i.e.* $r = 5$ m. The per unit length area through which a charge 10 nC will produce an electric flux is $2\pi r = 10\,\pi$ and the charge is 10 nC, hence the electric flux density is $\frac{10\times10^{-9}}{10\pi} = 0.318\text{ nC/m}^2$ or $0.318\bar{a}_r\text{ nC/m}^2$ and the total flux leasing the 5 m length of the wire is $10 \times 5 = 50$ nC **Ans.**

Example 3.2. A sphere of radius 5 m has a point charge of 10 nC at its centre. Determine the electric flux passing through that part of the sphere between ±60° latitude and ±30° longitude. It is to be noted that latitudes are measured north and south of the equator but θ is measured from the north pole.

Solution. When θ is changed by $d\theta$ and ϕ is changed by $d\phi$ to generate a differential surface on the sphere we obtain differential length as $rd\theta$ and $r\sin\theta d\phi$ and hence the differential surface is $ds = r^2\sin\theta d\theta d\phi$ and hence the solid angle subtended by this differential surface at the centre is $dW = \frac{ds}{r^2} = \sin\theta\, d\theta\, d\phi$. Since the spherical surface of a sphere subtends an angle of $\frac{4\pi r^2}{r^2} = 4\pi$ steradian, the flux per steradian is $\frac{Q}{4\pi}$ and the solid angle subtended by the surface generated by latitudes and longitude is

$$\int_{-\pi/6}^{\pi/6}\int_{30}^{150} \sin\theta d\theta d\phi = \frac{2\pi}{6}(-\cos\theta)_{30}^{150} = \frac{\pi}{3}(-\cos 150° + \cos 30°)$$

$$=\frac{\pi}{3}\cdot 2\cos 30° = \frac{\pi}{3}\sqrt{3} \text{ steradian}$$

Hence flux through this solid angle is $\dfrac{Q}{4\pi}\cdot\dfrac{\pi\sqrt{3}}{3} = \dfrac{Q}{4\sqrt{3}}$

$$= \frac{10\times 10^{-9}}{4\sqrt{3}} = 1.4433 \text{ nC } \textbf{Ans.}$$

Example 3.3. A sphere of radius R = 40 cm has a charge density variation of $\rho = \rho_0 \dfrac{r}{R}$ where ρ_0 = 270 pC/m. Determine the total charge in the sphere.

Solution. The spherical volume of thickness dr is $4\pi \cdot r^2\, dr$. The charge in the differential volume $4\pi r^2 dr \cdot \rho_0 \dfrac{r}{R}$. Hence total charge in the sphere.

$$\frac{4\pi\rho_v}{R}\int_0^R r^3 dr = \frac{4\pi\times 270\times 10^{-12}}{0.4}\int_0^{0.4} r^3 dr$$

$$= 10\pi\times 270\times 10^{-12}\cdot\frac{r^4}{4} = \frac{10\pi\times 270\times 10^{-12}}{4}\times 256\times 10^{-4}$$

$$= 270\pi\times 10^{-11}\times 64\times 10^{-4}$$

$$= 2.7\pi\times 6.4\times 10^{-12} = 54.26 \text{ pC}.$$

Example 3.4. Determine the net flux that crosses surface S which encloses a charge distribution in the form of a plane circular disc of radius 5 m with a charge density $\rho_s = \dfrac{\sin^2\phi}{4r}$ C/m².

Solution. The flux that crosses the surface which encloses the circular disc with certain charge distribution. The differential charge on the disc is $dr \cdot r d\phi\ \rho_s$. Hence total charge is

$$\oint_s \rho_s dr\, r\, d\phi = \int_0^{2\pi}\int_0^5 \frac{\sin^2\phi}{4r}\cdot r\, dr\, d\phi$$

$$= \int_0^{2\pi}\int_0^5 \frac{\sin^2\phi}{4} dr\, d\phi = \int_0^{2\pi}\frac{5}{4}\sin^2\phi\, d\phi = \frac{5}{8}\int_0^{2\pi}(1-\cos 2\phi)\, d\phi$$

$$= \frac{5}{8}\cdot\pi\text{C } \textbf{Ans.}$$

Example 3.5. Charge in the form of a plane sheet with ρ_s = 50 μC/m² is located at z = –0.5 m. In addition to this, a line charge with ρ_L = –5 μC/m lies along y-axis. Determine the net flux that crosses the surface of a cube 3 m on an edge and is centred at the origin.

Solution. Since a part of the surface charge lies within the cube, the flux contribution due to surface charge is 9 × 50 = 450 μC and as the line charge along the *y*-axis, part of this charge is also enclosed by the cube and hence flux contribution by the line charge is 3 × (–5) = –15 μC.

Hence net flux crossing the cube is 450 –15 = 435 μC **Ans.**

Example 3.6. A point charge Q = 50 nC is located at the origin. Determine the electric flux density D at (2, 3, 4) m.

Solution. Since the flux density is required at (2, 3, 4), it is a Cartesian coordinate system and the unit vector from the origin to the point under consideration is

$$\frac{2\bar{a}_x + 3\bar{a}_y + 4\bar{a}_z}{\sqrt{29}}$$

The magnitude of D is $\dfrac{Q}{4\pi r^2} = \dfrac{50\times10^{-9}}{4\pi\times29}$

Hence $D = \dfrac{50\times10^{-9}}{4\pi\times29}\cdot\bar{a}_r = 0.137\times10^{-9}\dfrac{2\bar{a}_x + 3\bar{a}_y + 4\bar{a}_z}{\sqrt{29}}\,\text{C/m}^2$

or the magnitude is 139 pC/m^2 **Ans.**

Example 3.7. If the electric flux density $D = 500\,e^{-0.2x}\,\bar{a}_x$, determine the flux crossing surfaces of area 1m^2 normal to the *x*-axis and located at x = 1 m, x = 5 m and x = 8 m.

Solution. (*i*) The value of D at a distance of 1m is

$$D = 500\,e^{-0.2} = 0.818 \times 500 = 410\ \text{C/m}^2$$

Hence flux crossing 1 m^2 area = 410 × 1 = 410 C **Ans.**

(*ii*) The value of D at a distance of 5 m

$$D = 500\,e^{-1.0} = 500 \times 0.3678 = 184\ \text{C/m}^2$$

Hence flux crossing 1 m^2 area = 184 C **Ans.**

(*iii*) The value of D at a distance of 8 m

$$D = 500\,e^{-1.6} = 101\ \text{C/m}^2$$

Hence flux crossing 1 m^2 area is 101 C. **Ans.**

Example 3.8. A sphere with radius 6 m contains an electric charge with volume density ρ_v = 1.25 (5 – 2r) μC/m^3. Determine the electric flux density 120 m away from the centre of the sphere.

Solution. Total charge within the sphere $\int \rho_v dv$. Since the density is a function of r only $dv = 4\pi r^2 dr$.

Hence total charge enclosed by the sphere.

$$\int_0^6 1.25(5-2r)\cdot 4\pi r^2 dr$$

$$Q = 5\pi\int_0^6 (5-2r)\cdot r^2 dr \times 10^{-6}$$

$$= 5\pi\left[\frac{5r^3}{3} - \frac{2r^4}{4}\right]_0^6 = 5\pi\left[\frac{5}{3}r^3 - \frac{r^4}{2}\right]_0^6$$

$$= 5\pi\left[\frac{5\times 216}{3} - \frac{1296}{2}\right] = 5\pi[360 - 648]$$

$$Q = -5\pi \times 288 \times 10^{-6}\,\text{C}$$

Assuming the charge to remain at the centre, the electric flux density

$$D = -\frac{5\pi\times 288\times 10^{-6}}{4\pi\times 120^2} = -0.025\times 10^{-6}$$

$$= 25\ \text{nC/m}^2\ \textbf{Ans.}$$

3.3 GAUSS'S LAW

Taking the results of Faraday's experiment, consisting of concentric sphere one step further it can be said that the electric flux passing through any imaginary spherical surface lying between the two conducting spheres is equal to the charge enclosed within that imaginary surface. This enclosed charge may be distributed with the outer surface of the inner conductor or a point charge located at the centre of the imaginary sphere or the inner conductor could be of any geometry, only requirement is the conductor should be charged and it could be enclosed by an imaginary surface. However, the magnitude of the flux density will be different at different points of the outer sphere as compared to uniform distribution but Q coulomb of charge on any inner conductor will induce $-Q$ charge on the surrounding sphere. This generalisation of Faraday's experiment led to the following statement which is known as Gauss's law.

The electric flux passing through any closed surface is equal to the charge enclosed by that surface. Gauss provided proof in mathematical form for this statement which is given as follows:

Consider a cluster of point charges enclosed in any shape of surface as shown in Fig. 3.2.

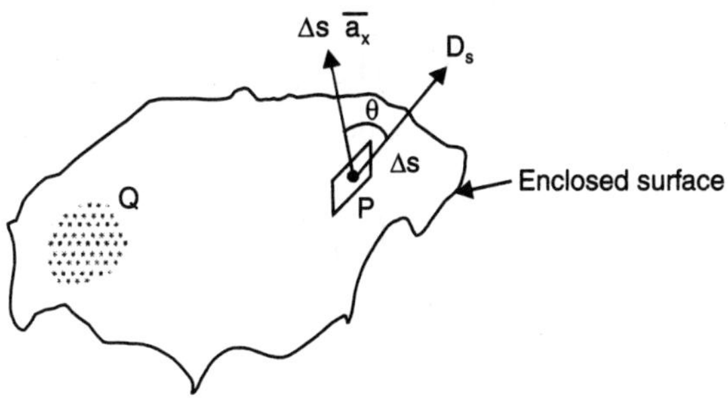

Fig. 3.2 The electric flux density D_s at P due to charge Q and flux passing through Δs is $D_s\,\Delta s$

If the total charge is Q, whatever be the shape of the surface that encloses this charge, the electric flux will be of Q coulomb. This is because one coulomb of electric flux is produced by one coulomb of charge. However, since the shape of the enclosing surface is symmetrical with respect to charge, the electric flux density vector D_s will have different values at different points of surface *i.e.* D_s will in general have different magnitude and direction at different point of the surface.

Let us concentrate on the flux passing through the differential surface element ΔS. Since it is a differential surface element the two coordinates which constitute the surface tend to zero and hence the differential surface would tend to a plane whatever be the curvature of the actual surface. To describe this differential surface we need to know both its magnitude and its orientation in space which means that the differential surface is a vector quantity. As regards magnitude one could take product of differentials of two coordinate e.g. $dx. dy$, $rd\,\phi\,dr$ or $r\sin\theta\,d\theta\,rd\phi$ and the direction associated with ΔS is the direction of the normal to that plane which is tangent to the surface at the point in question e.g. point P in this case. However, there are two normals at this point one outside the surface and other inside it. We are considering here only the one that is outward as we are interested in the flux passing through (outward) the surface.

Consider point P in a differential surface Δs and Ds be the flux density at P which makes angle θ with the normal to the surface Δs as shown in Fig. 3.2, the flux crossing Δs is product of the normal components of Ds and Δs i.e.

$\Delta\psi$ = flux crossing $\Delta S = D_s \bar{a}_n \Delta S = D_s \cos\theta\, \Delta S = \bar{D}_s \cdot \overline{\Delta S}$ *i.e.* the flux passing through ΔS is the dot product of vector $\bar{D}_s$ and $\overline{\Delta S}$. Hence, the total flux passing through the closed surface is the sum of the differential fluxes crossing each of Δs *i.e.*

$$\psi = \oint d\psi = \oint D_s \cdot d\bar{s}$$

Thus the resultant integral is a closed surface integral which is indicated by a small circle on the integral sign and since ds is the differential surface area, it is a double integral and the limits on the two integrals will be such that it represents the closed surface. Such a surface is usually known as Gaussian surface. Therefore, mathematical formulation of Gauss's law is given as

$$\psi = \oint \bar{D} \cdot d\bar{s} = Q = \text{charge enclosed}$$

If the surface encloses more than one point charges, then

$$Q = \sum_{n=1}^{N} Q_n$$

However if the charge enclosed is a line charge with linear charge density as ρ_L C/m.

$$Q = \int \rho_L dl$$

If it were a surface charge with surface charge density ρ_s C/m^2, irrespective of whether the surface is closed or open

$$Q = \int \rho_s d\bar{s}$$

Similarly, for volume charge distribution if the volume charge density is ρ_v C/m^3

$$Q = \int_v \rho_v dv$$

Even though any of the above forms can be used to write mathematical expression for Gauss's law, the last one is usually used *i.e.*

$$\oint \bar{D} \cdot d\bar{s} = \int_v \rho_v dv \qquad ...(3.9)$$

This mathematical expression means that the total electric flux through any closed surface is equal to the charge enclosed. Let us apply Gauss's law to a point charge Q which is placed at the origin of a spherical system of coordinate. The electric field intensity at any distance or radius r is

$$E = \frac{Q}{4\pi\varepsilon_0 r^2}\bar{a}_r \qquad \text{...(3.10)}$$

or

$$\varepsilon_0 \bar{E} = \frac{Q}{4\pi r^2}\bar{a}_r = \bar{D}$$

If R is the radius of the sphere, the electric flux density at the surface of the sphere is

$$\bar{D}_s = \frac{Q}{4\pi r^2}\bar{a}_r \qquad \text{...(3.11)}$$

In spherical coordinates a differential surface area is

$$ds = r^2 \sin\theta d\theta\, d\phi = R^2 \sin\theta d\theta d\phi$$

Hence

$$d\bar{s} = R^2 \sin\theta d\theta d\phi \bar{a}_r$$

or

$$D_s \cdot ds = \frac{QR^2}{4\pi R^2}\sin\theta d\theta\, d\phi\, \bar{a}_r \cdot \bar{a}_r = \frac{Q}{4\pi}\sin\theta d\theta\, d\phi$$

Spherical surface can be generated by rotating R for $\theta = 0$ to 2π and $\phi = 0$ to π
Hence the closed surface integral is

$$\frac{Q}{4\pi}\int_0^{2\pi}\int_0^{\pi}\sin\theta d\theta\, d\phi = \frac{Q}{4\pi}\int_0^{2\pi}[-\cos\theta]_0^{\pi}\, d\phi = \frac{Q}{4\pi}\int_0^{2\pi} 2d\phi = Q$$

which shows that Q coulombs of electric flux crosses the surface which is the expected result as the enclosed charge is Q coulomb.

Next we apply Gauss's law to determine electric flux density D_s if the charge distribution is known. This is a situation where the unknown is an integrand. The solution is possible if we are able to choose the charge enclosing surface which satisfies the following conditions.

1. D_s is either normal or tangential to the closed surface chosen everywhere so that $\bar{D}_s \cdot d\bar{s}$ is either $D_s \cdot ds$ or zero respectively.
2. If $D_s \cdot ds$ is not zero for certain portion of the closed surface for that portion D_s is constant.

From this it is clear that the dot product can be replaced with product of the scalars D_s and ds and D_s can be brought outside the integral sign and the remaining integral can be obtained over that portion of the closed surface for which D_s crosses normally and this is the area of this section of the surface. Therefore, in order to use Gauss's law it is important to know symmetrical surface enclosing the charge distribution. We know that for a point charge the symmetrical surface is a sphere as the flux lines move radially outward from the point charge.

Let us find out electric flux density at any distance r from a point charge located at the origin using Gauss's law. Since for the point charge the surface is normal to the flux if it is spherical with its centre at the origin where the charge is located. Here, in this case D_s is normal to the surface and hence according to Gauss' law.

$$Q = \oint \bar{D}_s \cdot d\bar{s} = \oint D_s \cdot d\bar{s} = D_s \oint ds$$

$$Q = D_s \int_0^{2\pi}\int_0^{\pi} r^2 \sin\theta\, d\theta\, d\phi = D_s \cdot 4\pi r^2$$

Hence $$D_s = \frac{Q}{4\pi r^2}$$

Since D_s is directed radially outward and r may have any value

$$\bar{D}_s = \frac{Q}{4\pi r^2}\bar{a}_r, \qquad \text{...(3.12)}$$

It is to be noted that application of Gauss's law depends upon symmetry of the surface enclosing the charge and if we can not show that symmetry exists then we can not use Gauss's law to obtain solution to a problem.

Next we apply Gauss's law to obtain electric flux density due to a line charge placed along z-axis. So first we have to find out a surface that is symmetrical and encloses the line charge and the flux should leave the surface normally. So a right circular cylinder with radius ρ is the suitable surface as D_ρ is everywhere normal. The periphery of the cylinder is obtained by rotating ρ through 2π radians and the length of the cylinder by say L and hence using Gauss's law

$$Q = \oint D_s \cdot d\bar{s} = D_s \oint_{\text{side}} ds + 0 \oint_{\text{top}} ds + 0 \oint_{\text{bottom}} ds$$

$$= D_s \int_0^{L}\int_0^{2\pi} \rho\, d\phi\, dz = D_s \cdot 2\pi\rho L$$

Hence $$D_s = D_\rho = \frac{Q}{2\pi\rho L}$$

If ρ_L is the line charge density in coulomb/m, then

$$Q = \rho_L L$$

or $$D_\rho = \frac{\rho_L L}{2\pi\rho L} = \frac{\rho_L}{2\pi\rho} \qquad \text{...(3.13)}$$

Hence Electric field intensity $$E_\rho = \frac{\rho_L}{2\pi\varepsilon_0\rho} \qquad \text{...(3.14)}$$

We derived expression for E in chapter 2 due to an infinite line charge. If we compare the computational effort required we find that the computational effort is very small using Gauss's law as compared to the one in Chapter 2.

Next we consider the case of a co-axial cable with radius of inner conductor 'a' and outer conductor 'b' and we assume that the length of the cable quite large as compared to its radii. As far as evaluation of D is concerned, this case is identical to that of a line charge and is an example which is difficult to solve using coulomb's law. Let ρ_s be the surface charge density on the outer surface of the inner conductor and we consider L length of the co-axial cable. We consider Gaussian surface of radius r such that $a < r < b$ and length L and this surface is a right circular cylinder. The Gaussian surface area is $2\pi rL$ which encloses a charge $Q = 2\pi rL\rho_s$ as the flux density D_s will have only radial component due to symmetry of the surface with respect to the charge distribution. The charge enclosed by the Gaussian surface is the charge on the outer surface of the inner conductor *i.e.* $Q = 2\pi aL\, \rho_S = D_s \cdot \text{area} = D_s\, 2\pi rL$.

or $$2\pi aL\, \rho_s = Ds \cdot 2\pi rL$$

or $$D_s r = a\rho_s \text{ or } \bar{D}_s = \frac{a\rho_s}{r}\bar{a}_r \qquad a < r < b \qquad \text{...(3.15)}$$

If we express the charge per unit length of the inner conductor then

$$\rho_L = 2\pi a\rho_s \text{ or } \rho_s = \frac{\rho_L}{2\pi a}$$

Hence $$D_s = \frac{a\rho_s}{r}\bar{a}_r = \frac{a}{r}\cdot\frac{\rho_L}{2\pi a}\bar{a}_r = \frac{\rho_L}{2\pi r}\bar{a}_r \qquad \text{...(3.16)}$$

The solution has the same form as for an infinite line charge.

Since all radial flux lines emanating from the outer surface of inner conductor (as it is positive charge) terminate on to the inner surface of the outer conductor (–ve charge), if we take a Gaussian surface inside the inner conductor, since the charge enclosed by the Gaussian surface is zero, the electric flux density and hence electric field intensity inside is zero. Similarly if we take Gaussian surface enclosing the co-axial cable, the total charge enclosed by the Gaussian surface is zero (positive charge on the outer surface of the inner conductor equals the negative charge on the inner surface of the outer conductor of the cable) hence electric flux density and electric field intensity will be zero. It can therefore be said that the outer conductor provides shield to the electric flux *i.e.* it does not allow the electric flux inside the co-axial cable to move out.

Therefore, $$D_s = 0 \qquad r < a$$
$$= 0 \qquad r > b$$

This result is valid for co-axial cables open at both the ends provided length L of the cable is much greater than the radius b of the outer conductor.

Example 3.9. Two infinite line charges are having line charge density of ρ_L C/m and the other $-\rho_L$ C/m are separated by a distance a. Determine electric field intensity at a point P, at a distance b from ρ_L and a distance $(a - b)$ from $-\rho_L$ using Gauss law.

Solution. The Gaussian surface will be a right circular cylinder with radius b for ρ_L line charge or $(a - b)$ for $-\rho_L$ charge. The surface area per unit length will be $2\pi b$ and $2\pi(a - b)$ for ρ_L and $-\rho_L$ respectively. Let D_1 and D_2 be the electric flux densities due to ρ_L and $-\rho_L$ respectively and since the direction of the two are same the net will be the sum of two densities.

Since $$D_1 = \frac{\rho_L}{2\pi b} \text{ and } D_2 = \frac{\rho_L}{2\pi(a-b)}$$

Therefore, $$D = D_1 + D_2 = \frac{\rho_L}{2\pi}\left[\frac{1}{b} + \frac{1}{a-b}\right] = \frac{\rho_L a}{2\pi b(a-b)}$$

Hence $$E = \frac{D}{\varepsilon_0} = \frac{\rho_L a}{2\pi\varepsilon_0 b(a-b)} \quad \textbf{Ans.}$$

Example 3.10. Determine the total outward electric flux from the cube $0 < x, y, z < 1$ m containing a volume charge density $\rho_v = 16\, xyz\ \mu C/m^3$.

Solution. Using Gauss's law

$$\psi = Q = \int \rho_v dv = 16\int_0^1\int_0^1\int_0^1 xyz\,dx\,dy\,dz$$

$$= \left[16 \cdot \frac{x^2}{2} \cdot \frac{y^2}{2} \cdot \frac{z^2}{2}\right]_0^1 = 16 \cdot \frac{1}{2} \cdot \frac{1}{2} \cdot \frac{1}{2} = 2\,\mu\text{C} \quad \textbf{Ans.}$$

Example 3.11. A co-axial cable of 50 cms length and inner radius of 1 mm and an outer radius of 5 mm has the space between the conductors filled with air. The total charge on the inner conductor is 50 nC. Determine the charge density on each conductor and flux density D on the surfaces of the conductor.

Solution. The surface charge density on the inner conductor.

$$= \frac{Q}{\text{surface area}} = \frac{50 \times 10^{-9}}{2\pi \cdot r \cdot L} = \frac{50 \times 10^{-9}}{2\pi \cdot 10^{-3} \times 0.5} = 15.92 \times 10^{-6}\ \text{C/m}^2$$

The negative charge density on the inner surface of the outer conductor.

$$\frac{50 \times 10^{-9}}{2\pi(5 \times 10^{-3}) \times 0.5} = -3.18\ \mu\text{C/m}^2$$

The internal field density $D = \dfrac{a\rho_s}{r} = \dfrac{10^{-3} \times 15.9 \times 10^{-6}}{r}$

$$= \frac{15.9}{r}\,\text{nC/m}^2$$

for $1 < r < 5$ **Ans.**

Example 3.12. A point charge of 0.5 μC is located at $r = 0$ and uniform surface charge densities are located as follows: 4 mC/m² at $r = 1$ cm and –1.2 mC/m² at $r = 1.8$ cm. Determine D at (*i*) $r = 0.5$ cm (*ii*) $r = 1.5$ cm (*iii*) 2.5 cm.

Solution. Since the Gaussian spherical surface with radius 0.5 cm encloses the point charge.

(*i*) $$D = \frac{Q}{4\pi r^2}\bar{a}_r = \frac{0.5 \times 10^{-6}}{4\pi \times (0.5)^2 \times 10^{-4}} = 0.1592 \times 10^{-2}\ \text{C/m}^2$$

$$= 1592\ \text{C/m}^2 \quad \textbf{Ans.}$$

(*ii*) At $r = 1.5$ cm, the Gaussian surface encloses point charge and surface charge 4 mC/m².

$$D_1 = \frac{0.5 \times 10^{-6}}{4\pi(1.5)^2 \times 10^{-4}} = \frac{0.5 \times 10^{-6}}{4\pi \times 2.25 \times 10^{-4}} = 0.0177 \times 10^{-2} = 177\,\mu\text{C/m}^2$$

The charge $Q = 4\pi r^2 \rho_s = 4\pi r^2 \times 4 \times 10^{-3}$

$$= 4\pi \times 1 \times 10^{-4} \times 4 \times 10^{-3}$$

$$D_2 = \frac{4\pi\times10^{-7}\times4}{4\pi\times2.25\times10^{-4}} = \frac{4\times10^{-7}}{2.25\times10^{-4}} = 1.778\times10^{-3} = 1778\times10^{-6}$$

$$D = D_1 + D_2 = (177+1778)\times10^{-6} = 1955\ \mu\text{C/m}^2 \ \textbf{Ans.}$$

$$(iii)\ D_1 = \frac{0.5\times10^{-6}}{4\pi\times6.25\times10^{-4}} = 0.00637\times10^{-2} = 63.7\ \mu\text{C/m}^2$$

$$D_2 = \frac{4\pi\times10^{-7}\times4}{4\pi\times6.25\times10^{-4}} = 0.64\times10^{-6}\ \text{C/m}^2 = 640\ \mu\text{C/m}^2$$

$$D_3 = \frac{4\pi\times48^2\times10^{-4}\times1.2\times10^{-3}}{4\pi\times6.25\times10^{-4}} = 0.622\times10^{-3}\ \text{C/m}^2 = -622\ \mu\text{C/m}^2$$

Hence $D = 63.7 + 640 - 622$

$= 81.7\ \mu\text{C/m}^2$ **Ans.**

Example 3.13. A volume charge distribution is given by $\rho_v = 6\,r$ C/m³ in spherical coordinates. Find out D using Gauss's law.

Solution. According to Gauss's law

$$\oint D\cdot d\bar{s} = \oint_v \rho_v dv$$

$$D\cdot4\pi r^2 = \int 6r\cdot4\pi r^2 dr = 24\pi\frac{r^4}{4}$$

$$D = \frac{6\pi r^4}{4\pi r^2} = \frac{3}{2}r^2\bar{a}_r \ \textbf{Ans.}$$

Example 3.14. A charge distribution in cylindrical coordinates is given by $\rho = 5\,re^{-2r}$ C/m³. Determine D using Gauss's law.

Solution. Since the charge density is a function of r and is independent of ϕ and z, the electric flux density is radial. The Gaussian surface is taken as a right circular cylinder. The flux passing through the plane surfaces is zero as the flux lines are radial. Therefore, according to Gauss's law

$$Q = \int D\cdot ds = \int \rho_v dv = \int_0^L\int_0^{2\pi}\int_0^r 5re^{-2r}r\,dr\,d\phi\,dz = 5L\int_0^{2\pi}\int_0^r r^2dre^{-2r}d\phi$$

$$10\pi L\int_0^r r^2\cdot e^{-2r}dr = 10\pi L\left[-r^2\cdot\frac{e^{-2r}}{2}+\int 2r\cdot\frac{e^{-2r}}{2}\right] = 10\pi L\left[-r^2\cdot\frac{e^{-2r}}{2}-\frac{re^{-2r}}{2}+\frac{1}{2}\int_0^r e^{-2r}dr\right]$$

$$5\pi L\left[-r^2e^{-2r}-re^{-2r}-\frac{1}{4}e^{-2r}+\frac{1}{4}\right] = 2\pi rLD \text{ or } D = \frac{2.5}{r}\left[\frac{1}{2}-e^{-2r}\left(r^2+r+\frac{1}{2}\right)\right]\bar{a}_r\ \text{C/m}^2 \ \textbf{Ans.}$$

Example 3.15. The volume in a cylindrical coordinate between $r = 3$ cm and $r = 5$ cm contains a constant volume charge density of ρ_v C/m³. Using Gauss's law determine D in all regions.

Solution. The various regions are $r < 3$ cm $3 \le r \le 5$ and $r > 5$ cm.

Since no charge is enclosed by the Gaussian surface for $r < 3$ cm $D = 0$.

In the second region, let the radius be r, such that $3 \le r \le 5$, hence the volume charge enclosed by this surface

$$\pi\left(r^2 - 9\right) \cdot L \cdot \rho_v$$

Hence
$$D = \frac{\pi\left(r^2-9\right)L \cdot \rho_v}{2\pi r L} = \frac{\left(r^2-9\right)\rho_v}{2r}\bar{a}_r \text{ C/m}^2 \textbf{ Ans.}$$

In the third region $r > 5$, the charge enclosed by the Gaussian surface

$$\pi(25-9)L\rho_v C$$

Hence $D = \dfrac{16\pi L\rho_v}{2\pi r L} = \dfrac{8\rho_v}{r}$ C/m^2 **Ans.**

Example 3.16. Electric flux density in cylindrical coordinates is given by $D = 2r\cos\phi\, a_\phi - \dfrac{\sin\phi}{3r} a_z$. Determine flux passing through the portion of the $z = 0$ plane defined by $r \le a,\ 0 \le \phi \le \pi/2$. Repeat for $\dfrac{3\pi}{2} \le \phi \le 2\pi$. Assume flux is positive in the $\bar{a}_z$ direction.

Solution. (*i*) Since the flux is along a_z, therefore the component along $\bar{a}_\phi$ is to be ignored for finding flux through $z = 0$ plane. Hence $\bar{a}_\phi$ component of D is ignored

$$\int_0^a \int_0^{\pi/2} -\frac{\sin\phi}{3r} \cdot r dr d\phi = -\frac{a}{3}\int_0^{\pi/2} \sin\phi\, d\phi = -\frac{a}{3}\left[\cos\phi\right]_0^{\pi/2} = -\frac{a}{3} \textbf{ Ans.}$$

$$(ii)\quad -\frac{a}{3}\int_{3\pi/2}^{2\pi} \sin\phi\, d\phi = -\frac{a}{3}\left[-\cos\phi\right]_{3\pi/2}^{2\pi} = \frac{a}{3} \textbf{ Ans.}$$

Example 3.17. If electric flux density D is given as $\bar{D} = 8xyz^4\bar{a}_x + 4x^2z^4\bar{a}_y + 16x^2yz^3\bar{a}_z$ pC/m^2. Determine the total electric flux passing through the rectangular surface $z = 2,\ 0 < x < 2,\ 1 < y < 3$ in the $\bar{a}_z$ direction. Also determine approximate value of charge contained in an incremental sphere located at $P(2, -1, 3)$ and having a volume of 10^{-12} m^3 and E at P.

Solution. (*i*) Since the flux required is that which passes through $z = 2$ and since $\bar{a}_x \cdot \bar{a}_z = 0$ and $\bar{a}_y \cdot \bar{a}_z = 0$, the flux crossing will be due to the component of D normal to $\bar{a}_z$.

Since z is fixed and is $z = 2$

$$\psi = \int D \cdot ds = 16\int_0^2\int_1^3 8 \cdot x^2 y\, dx\, dy = 16 \cdot \frac{x^3}{3} \cdot \frac{y^2}{2} = 16\frac{8}{3}\frac{1}{2}(9-1)\cdot 8$$

$$\psi = 16 \cdot \frac{8 \times 8 \times 8}{3 \times 2} = 1365 \text{ pC} \textbf{ Ans.}$$

(*ii*) Next we evaluate partial derivatives of the three components of *D*.

$$\frac{\partial D_x}{\partial x} = 8yz^4 \quad \frac{\partial D_y}{\partial y} = 0 \quad \frac{\partial D_z}{\partial z} = 48x^2yz^2$$

Charge enclosed in differential volume Δv

$$= \left(\frac{\partial D_x}{\partial x} + \frac{\partial D_y}{\partial y} + \frac{\partial D_z}{\partial z} \right) \Delta v$$

$$= 8(-1) \cdot 81 + 0 + 48 \cdot 4 \cdot (-1)9$$

$$= -641 + 0 - 1728 = 2369 \times 10^{-12}$$

Hence charge enclosed in Δv *i.e.* 10^{-12} m^2,

$$2369 \times 10^{-12} \times 10^{-12} = 2.369 \times 10^{-21} \text{ C } \textbf{Ans.}$$

Next we can find out *E* at the point *P*(2, –1, 3)

$$E = E_x\bar{a}_x + E_y\bar{a}_y + E_z\bar{a}_z$$

$$E_x = \frac{D_x}{\varepsilon_0} = \frac{8 \times 2(-1) \times 81}{8.854 \times 10^{-12}} = -146.4$$

$$E_y = \frac{D_y}{\varepsilon_0} = \frac{4 \times 4 \times 81}{8.854 \times 10^{-12}} = 146.4$$

$$E_z = \frac{D_z}{\varepsilon_0} = \frac{16 \times 4 \times (-1) \times 27}{8.854 \times 10^{-12}} = -195.166$$

Hence $$E = -146.4\bar{a}_x + 146.4\bar{a}_y - 195.166\bar{a}_z \text{ V/m } \textbf{Ans.}$$

3.4 DIVERGENCE THEOREM

We know from Chapter 1 that divergence of a vector *D* is given as

$$\text{div}\, D = \frac{\int D \cdot ds}{\underset{\Delta v \to 0}{\text{Lt}} \Delta v} \qquad \ldots(3.17)$$

and if we make use of this relation to a differential volume in rectangular coordinates, we obtain

$$\text{div}\, D = \frac{\partial D_x}{\partial x} + \frac{\partial D_y}{\partial y} + \frac{\partial D_z}{\partial z} \qquad \ldots(3.18)$$

and from Gauss Law we know that $\oint D \cdot ds = Q$ and if the surface chosen is a differential surface the charge enclosed is also differential charge and hence $\underset{\Delta v \to 0}{\text{Lt}} \frac{\Delta Q}{\Delta v} = \rho_v$ at that point and hence

$$\text{div } D = \rho_v \qquad \ldots(3.19)$$

Again according to Gauss's law

$$Q = \int_S D \cdot ds = \oint_v \rho_v \, dv$$

Now since $\operatorname{div} D = \rho_v$ or $\nabla \cdot D = \rho_v$, substituting for ρ_v in the above equation we have

$$\oint D \cdot ds = \oint \nabla \cdot D \, dv \qquad \ldots(3.20)$$

This is known as divergence theorem which may be stated in words as

The integral of the normal component of any vector field over a closed surface is equal to the integral of the divergence of this vector field throughout the volume enclosed by the closed surface.

It is to be noted that divergence theorem applies to any vector field not necessarily D as we have taken here.

The importance of the divergence theorem lies in the fact that it relates a triple integration throughout some volume to a double integration over the surface of that volume. We will take up an example to illustrate the application of Divergence theorem.

Example 3.18. Evaluate both sides of the divergence theorem for the electric flux densi.y $D = 3xy\bar{a}_x + x^2\bar{a}_y$ C/m^2 and the rectangular parallelopiped formed by the planes $x = 0$ and 2, $y = 0$ and 3 and $z = 0$ and 4.

Solution. From the expression for D, it is clear that since a_z component is about that means D is parallel to the surfaces at $z = 0$ and 4. For the remaining four surfaces, we have $D_x = 3xy$, $D_y = x^2$, $D_z = 0$

$$\oint D \cdot ds = \int_0^4 \int_0^3 D_{x=0} (-dy\, dz)\bar{a}_x + \int_0^4 \int_0^3 (D)_{x=1} \, dy\, dz\, \bar{a}_x$$

$$+ \int_0^4 \int_0^2 (D)_{y=0} (-dx\, dz)\bar{a}_y + \int_0^4 \int_0^2 (D)_{y=2} dx\, dz\, \bar{a}_y$$

$$= \int_0^4 \int_0^3 (D_x)_{x=1} \, dy\, dz = \int_0^4 \int_0^3 6y\, dy\, dz$$

$$= \int_0^4 \int_0^3 6y\, dy\, dz$$

$$= \int_0^4 \left(\frac{6y^2}{2}\right)_0^3 dz = \int_0^4 27\, dz$$

$$= 27 \times 4 = 108 \text{ coulomb.}$$

as $\qquad D_x\big|_{x=0} = 0$ and $D_y\big|_{y=0} = 0 \;\; D_y\big|_{y=3} = 0$

Now calculating the R.H.S. of the divergence expression

$$\nabla \cdot D = \frac{\partial}{\partial x}(3xy) + \frac{\partial}{\partial y}(x^2) = 3y + 0$$

the volume integral becomes

$$\int \nabla \cdot D\, dv = \int_0^4\int_0^3\int_0^2 3y\, dx\, dy\, dz = \int_0^4\int_0^3 6y dy\, dz$$

$$= \int_0^4 27\, dz = 108.$$

Here we find the computational effect is smaller as compared to using the left hand side of the expression.

3.5 MAXWELL'S FIRST EQUATION (ELECTROSTATIC FIELDS)

Rewriting equation (3.19) we have

$$\text{div } D = \rho_v \qquad \text{...(3.21)}$$

and this is the first of Maxwell's four equation as they apply to electrostatics and steady magnetic field. This equation states that the electric flux per unit volume $\frac{\Delta Q}{\Delta v}$ leaving an infinitesimal volume is equal to the volume charge density ρ_v at that point. This equation is also known as the point form of Gauss's law. Gauss's law thus relates the flux leaving any closed surface to the charge enclosed by the same surface. Maxwell's first equation also holds good on a per unit volume basis for an infinite small volume as this differential volume tends to zero *i.e.* at a point. It is to be noted that since divergence of a vector can be expressed as the sum of three partial derivatives, Maxwell's first equation is also described as the differential equation form of Gauss's law and hence conversely Gauss's law is recognised as the integral form of Maxwell's first equation.

As a specific example let us consider a point charge Q located at the origin and we find out divergence of D in the region of the point charge. Now

$D = \frac{Q}{4\pi r^2}\bar{a}_r$ and we make use of the expression of divergence in spherical coordinate system which is reproduced here

$$\text{div} D = \frac{1}{r^2}\frac{\partial}{\partial r}\left(r^2 D_r\right) + \frac{1}{r\sin\theta}\frac{\partial}{\partial\theta}\left(\sin\theta D_\theta\right) + \frac{1}{r\sin\theta}\frac{\partial D_\phi}{\partial\phi} \qquad \text{...(3.22)}$$

Now the expression for D above shows that D_θ and D_ϕ components are zero and hence

$$\text{div} D = \frac{1}{r^2}\frac{d}{dr}\left(r^2 D_r\right) = \frac{1}{r^2}\cdot\frac{d}{dr}\left(r^2\cdot\frac{Q}{4\pi r^2}\right) - 0 \text{ for } r \neq 0$$

Thus $\rho_v = 0$ everywhere except at the origin. It is true also as the charge is located at a point only and hence charge density $\rho_v = 0$ everywhere except at the origin where it will be infinity.

It has already been mentioned that divergence operation is not limited to electric flux density, it can be applied to any vector field.

Example 3.19. Determine the volume charge density if the electric flux density D is given as follows:

(*i*) $D = \frac{6xy}{z}\bar{a}_x + \frac{3x^2}{z}\bar{a}_y + \frac{2x^2y}{z^2}\bar{a}_z$

(*ii*) $D = z\sin\phi\bar{a}_\rho + z\cos\phi\bar{a}_\phi + \rho\sin\phi\bar{a}_z$

(*iii*) $D = \sin\theta\sin\phi\bar{a}_r + \cos\theta\sin\phi\bar{a}_\theta + \cos\phi\bar{a}_\phi$

Solution. (*i*) In Cartesian coordinates the divergence of a vector D is given as

$$\text{div}\, D = \frac{\partial D_x}{\partial x} + \frac{\partial D_y}{\partial y} + \frac{\partial D_z}{\partial z}$$

Here $$D_x = \frac{6xy}{z},\ D_y = \frac{3x^2}{z},\ D_z = \frac{2x^2y}{z^2}$$

Hence $$\frac{\partial D_x}{\partial x} = \frac{6y}{z},\ \frac{\partial D_y}{\partial y} = 0,\ \frac{\partial D_z}{\partial z} = -2x^2y\cdot z^{-3}(-3) = +6\frac{x^2y}{z^3}$$

Hence $\frac{6y}{z} + \frac{6x^2y}{z^3} = \frac{6y}{z^3}\left(z^2 + x^2\right)$ **Ans.**

(*ii*) This is related to cylindrical coordinate system where

$$\text{div}\, D = \frac{1}{\rho}\frac{\partial}{\partial\rho}\left(\rho D_\rho\right) + \frac{1}{\rho}\frac{\partial D_\phi}{\partial\phi} + \frac{\partial D_z}{\partial z}$$

and $$D_\rho = z\sin\phi \quad D_\phi = z\cos\phi \quad D_z = \rho\sin\phi$$

$$\frac{1}{\rho}\frac{\partial}{\partial\rho}(\rho z\sin\phi) + \frac{1}{\rho}\cdot\frac{\partial}{\partial\phi}z\cos\phi + 0$$

$$= \frac{1}{\rho}\cdot z\sin\phi - \frac{z\sin\phi}{\rho} + 0 = 0 \quad \textbf{Ans.}$$

(*iii*) $$\text{div}\, D = \frac{1}{r^2}\frac{\partial}{\partial r}\left(r^2 D_r\right) + \frac{1}{r\sin\theta}\frac{\partial}{\partial\theta}\left(\sin\theta D_\theta\right) + \frac{1}{r\sin\theta}\frac{\partial D_\phi}{\partial\phi}$$

$$D_r = \sin\theta\sin\phi,\ D_\theta = \cos\theta\sin\phi,\ D_\phi = \cos\phi$$

$$\frac{1}{r^2}\cdot\frac{\partial}{\partial r}\left(r^2\sin\theta\sin\phi\right) + \frac{1}{r\sin\theta}\cdot\frac{\partial}{\partial\theta}\cdot\sin\theta\cos\theta\sin\phi + \frac{1}{r\sin\theta}\frac{\partial\cos\phi}{\partial\phi}$$

$$= \frac{2r}{r^2}\sin\theta\sin\phi + \frac{1}{r\sin\theta}\cos 2\theta\sin\phi - \frac{\sin\phi}{r\sin\theta}$$

$$= \frac{\sin\phi}{r}\left[2\sin\theta + \frac{\cos 2\theta}{\sin\theta} - \frac{1}{\sin\theta}\right] = \frac{\sin\phi}{r}\left[\frac{2\sin^2\theta + \cos 2\theta - 1}{\sin\theta}\right] = 0 \quad \textbf{Ans.}$$

PROBLEMS

P 3.1 The electric field density in cartesian coordinates is given by $D = 4xy\bar{a}_x + 2(x^2 + z^2)\bar{a}_y + 4yz\bar{a}_z$. Determine the total charge enclosed in the rectangular parallelopiped $0 < x < 2$, $0 < y < 3$, $0 < z < 5$m. **Ans.** 360 C

P 3.2 A volume charge density $\rho_v = \frac{k}{r}$ ($r \neq 0$ and k = constant) exists within a sphere of radius a. Determine the magnitude of a point charge placed at the origin which will produce the same electric field at $r > 0$. **Ans.** $Q = 2\pi ka^2$

P 3.3 A continuous, spherically symmetric distribution of electric flux density is given by

$$D = \begin{cases} \frac{5}{4} r^2 \bar{a}_r & r \leq 2 \\ \frac{20}{r^2} \cdot \bar{a}_r & r > 2 \end{cases}$$

Determine the charge density in each region. **Ans.** $5r$, $r < 2$ and 0, $r > 2$

P 3.4 Show that $\bar{D}$ and $\bar{E}$ are zero at all points in the plane of a uniformly charged circular ring which are inside the ring. Use Gauss's law.

P 3.5 The electric field intensity E in air is given by

$$E = \rho_0 \left[\rho\bar{a}_\rho + \bar{a}_\phi + z\bar{a}_z\right] \text{V/m}$$

(*i*) Determine the charge density (*ii*) Verify Gauss's law in integral form by taking the volume $\rho \leq 1, 0 \leq \phi \leq \pi, 0 \leq z \leq 1$. **Ans.** $\rho_v = 3\varepsilon_0\rho_0, Q = 3\varepsilon_0\rho_0\,\pi/2$

P 3.6 Determine the total charge enclosed in a cube defined by the six planes for which $0 \leq x \leq 3$, $0 \leq y \leq 3, 2 \leq z \leq 5$ if $D = x^2\bar{a}_x + 2y\bar{a}_y + 3z^3\bar{a}_z$ nC/m^2. Solve by evaluating both sides of the expression $\oint \nabla \cdot D\,dv = \oint D \cdot ds$. **Ans.** 3294 C

P 3.7 If the electric flux density D in cylindrical coordinates is given by $D = \frac{\bar{a}_r}{r^2} + \sin\phi\bar{a}_\phi + \frac{a_z}{\sqrt[3]{z}}$ C/m^2, determine the charge density at 1 m, 15°, 3m. **Ans.** –0.11 C/m^3

P 3.8 In cylindrical coordinates let $\rho_v = 0$ for $\rho < 1$ mm $\rho_v = 2 \sin 2000\,\pi\rho$ nC/m^3 for 1 mm $< \rho < 1.5$ mm and $\rho_v = 0$ for $\rho > 1.5$ mm. Determine D everywhere.

Ans. $D = 0$ $\rho < 1$ mm

$$D = \frac{10^{-15}}{2\pi^2\rho}\left[(\sin 2000\,\pi\rho) + 2\pi\left(1 - 10^3\rho\cos 2000\,\pi\rho\right)\right] \text{C/m}^2 \quad 1 < \rho < 1.5 \text{ mm.}$$

$$D = \frac{2.5 \times 10^{-15}}{\pi\rho}\text{C/m}^2 \quad \rho > 1.5 \text{ mm.}$$

P 3.9 Spherical surfaces at $r = 2$, 4 and 6 m carry uniform surface charge densities of 20 C/m² and ρ_{S_0} respectively. (*i*) Determine D at $r = 1$, 3 and 5 m (*ii*) Determine ρ_{S_0} such that $D = 0$ at $r = 7$ m.

Ans. (*i*) $D = 0$ $r < 2$, $D = 8.9 \times 10^{-9}$ C/m² $r = 3$

$D = 6.4 \times 10^{-10}$ C/m² $r = 5$ (*ii*) $\rho_{S_0} = -\frac{4}{9} \times 10^{-9}$ C/m²

P 3.10 Determine the electric flux density due to the following volume charge distributions in spherical coordinates using Gauss's law.

$$(i)\ \rho = \begin{cases} 0 & 0 < r < a \\ \rho_0 & a < r < b \\ 0 & b < r < \infty \end{cases} \qquad (ii)\ \rho = \begin{cases} \rho_0 \frac{r}{a} & 0 < r < a \\ 0 & a < r < \infty \end{cases}$$

$$(iii)\ \rho = \begin{cases} \rho_0\left(1 - \frac{r^2}{a^2}\right) & 0 < r < a \\ 0 & a < r < \infty \end{cases}$$

where ρ_0 is a constant

Ans. (*i*) 0 for $r < a$, $\frac{\rho_0}{3r^2}\left(r^3 - a^3\right)\bar{a}_r$ for $a < r < b$, $\frac{\rho_0}{3r^2}\left(b^3 - a^3\right)\bar{a}_r$ for $r > b$

(*ii*) $\frac{\rho_0 r^2}{4a}\bar{a}_r$ for $r < a$, $\frac{\rho_0 a^3}{4r^2}$ for $r > a$

(*iii*) $\frac{\rho_0\left(5a^2r^3 - 3r^3\right)}{15a^2r^2}\bar{a}_r$ for $r < a$, $\frac{2\rho_0 a^3}{15r^2}\bar{a}_r$ for $r > a$.

P 3.11 A 100 μC point charge is located at the origin of a cartesian coordinate system. Determine the electric flux passing through the square $|x| < 1$ m, $|y| < 1$ m, $z = 2$ m. **Ans.** 16.67 μC

P 3.12 Determine the total outward flux from the spherical shell $1 \le r \le 2$ m, if the charge density in the volume is $\rho_v = \frac{5\cos^2\phi}{r^4}$ C/m³. **Ans.** 5 πC

P 3.13 Show that the electric flux density given by $D = r\sin\phi\bar{a}_r + 2r\cos\phi\bar{a}_\phi + 2z^2\bar{a}_z$ can be realised by a charge density increasing linearly with z.

P 3.14 Show that the volume charge density corresponding to $D = 10\sin^2\phi\bar{a}_r + r\bar{a}_\phi + \frac{10z}{r}\cos^2\phi a_z$ is independent of ϕ and z.

P 3.15 The electric flux density within the cylindrical volume bounded by $r = 2$ m, $z = 0$ and $z = 5$ m is given by $D = 30e^{-r}\,\bar{a}_r - 2z\bar{a}_z$ C/m². Determine the total outward flux crossing the surface of the cylinder. **Ans.** 129.44C

P 3.16 A point charge $Q = 3$ nC, is located at the origin of a cartesian coordinate system. Determine the electric flux that crosses the portion of the $z = 2$ m plane for which $-4 \leq x < 4$ m and $-4 \leq y \leq 4$ m. **Ans.** 0.5 nC

P 3.17 A uniform charge density of 2 C/m^3 exists in the volume $2 \leq x \leq 4$m (Cartesian co-ordinates). Determine D in all regions. **Ans.** $-2\bar{a}_x$ C/m^2, $2(x-3)\bar{a}_x$ C/m^2, $2\bar{a}_x$ C/m^2

P 3.18 A point charge $Q = 2000$ pC is at the origin of spherical coordinates. A concentric spherical distribution of charge at $r = 1$ m has a charge density $\rho_s = 40\pi$ pC/m^2. Determine the surface charge density on a concentric shell at $r = 2$ m which would result in $D = 0$ for $r > 2$ m. **Ans.** -71.2 pC/m^2.

P 3.19 Suppose $D = 5r^2\bar{a}_r$ mC/m^2 for $r \leq 0.08$ m and $D = 0.205\ a_r/r^2$ μC/m^2 for $r \geq 0.08$. Determine (*i*) ρ_v for $r = 0.06$ (*ii*) ρ_v for $r = 0.1$ m. Determine surface charge density which could be located at $r = 0.08$ m to cause $D = 0$ for $r \geq 0.08$ m. **Ans.** (*i*) 1.2 mC/m^3 (*ii*) zero (*iii*) -32 μ C/m^2

P 3.20 A spherical surface of radius 3 mm is centred at $P(4, 1, 5)$ in free space. Let $D = x\bar{a}_x$. Determine the net electric flux leaving the spherical surface. **Ans.** 113 nC

P 3.21 Within the cylindrical region $r \leq 4$ m, the electric flux density

$$D = \frac{5}{2r}\left[\frac{1}{2} - e^{-2r}\left(r^2 + r + \frac{1}{2}\right)\right]\bar{a}_r \text{ C/m}^2$$. Determine the charge density at $r = 2$ m.

Ans. $10\ e^{-4}$ C/m^3.

Chapter 4

The Electrostatic Field Energy and Potential

4.1 INTRODUCTION

In Chapter 2 we studied Coulomb's law and its application in finding out the electric field intensity when there are different charge distributions. We studied Gauss' law also for finding out the electric field intensity E when the charge distribution is symmetrical and it was possible to obtain Gaussian surface which enclosed the symmetrical charge distribution and since the area of the Gaussian surface was easy to calculate, the need for integration did not arise and the electric flux density D could be calculated without much effort and hence electric field intensity due to symmetrical charge distribution could be obtained as $E = \frac{D}{\varepsilon}$. However, if we try to obtain with slightly more difficult charge distribution viz. two unlike point charges separated by a small distance (an electric dipole), we find it rather impossible to choose a suitable Gaussian surface to obtain solution to the problem. Under such circumstances Coulomb's law comes to our rescue. Coulomb's law is more powerful as compared to Gauss' law as it can be applied to any system of charge distribution. However, the application of Coulomb's law requires lot of computational effort and is quite involved. The reason behind this is the fact that electric field intensity which is a vector field needs to be obtained directly from the charge distribution. In a three dimensional space three different components of electric field intensity are required and for each component one integration operation is required and for the solution of the complete problem three integration operations are required. The resolution of the electric field intensity vector into three components usually adds to the complexity of the integral.

The third possibility for obtaining electric field intensity could be one where we could think of a scalar function which can be obtained by a single integration (as we don't have three components of scalar function) and then determine electric field intensity E from this scalar function with a simple operation on the scalar function such as differentiation. Fortunately such a function exists and is known as the potential scalar or potential field. This scalar function is known to us and it not only simplifies evaluation of electric field intensity but it gives clear physical interpretation of electric field.

4.2 WORK DONE IN MOVING A POINT CHARGE IN ELECTRIC FIELD

We know that electric field intensity at any point in an electric field is the force per unit charge at that point. We also know that E has both magnitude and direction. If we move a positive test charge in the direction of the electric field, then work is done on the charge by the electric field and if it is a differential charge

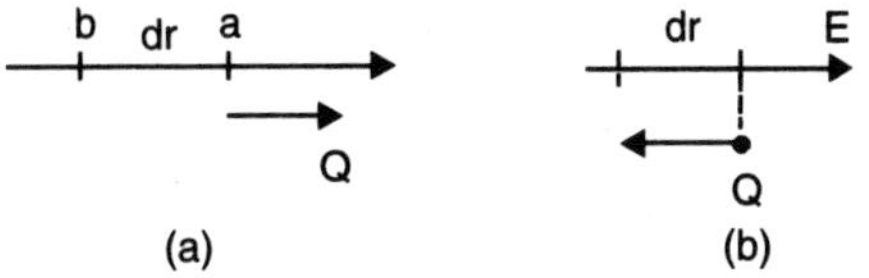

Fig. 4.1

ΔQ, positive, the differential work done is $dW = -\Delta Q\bar{E}dr$. Figure 4.1(a) and is equivalent to gravitation field of earth which performs work on a mass M as it is moved from a higher elevation to a lower one. Here the charge is moved from higher potential to a lower potential. However, if we move the test charge against the direction of electric field, we have to exert a force equal and opposite to that exerted by the field and this requires us to spend energy or do work and thus the work done on the charge movement is from an external source and is taken as positive i.e.

$$dW = \Delta QEdr = \Delta Q\bar{E}\cdot dr \qquad \ldots(4.1)$$

This positive work is equivalent to lifting a mass M against the gravitation field *i.e.* the mass M is taken to a higher elevation. Similarly in electric field the charge is moved from a lower potential to a higher potential.

The differential work in equation (4.1) can be zero under several conditions viz when $\Delta Q = 0$ or $E = 0$ or $dr = 0$ and the most important condition is when E and dr are orthogonal to each other as the dot product of the field $\bar{E}$ and radial displacement dr is zero. With regard to gravitation field it is equivalent to sliding a mass M around with constant velocity on a frictionless surface which does not require any effort. Or it is equivalent to applying a force on a mass perpendicular to the direction of motion. The mass does not move and hence work done is zero. Taking equation (4.1) further, the work required to move the charge a finite distance which requires integration over the distance moved and we have

$$W = Q\int_{\text{initial}}^{\text{final}} \bar{E}\cdot d\bar{r} \qquad \ldots(4.2)$$

Here the path must be specified before the integral can be evaluated. It is to be noted that the charge is at rest at both its initial and final positions.

Equation (4.2) is known as line integral. Here we have some electric field, uniform or non-uniform and when we take a charge Q in this field from initial point say b to a point final a, the path between a and b could be a straight line joining the two points or it could be any curvilinear path, the work done on the charge is same.

The differential lengths in different coordinate system have been derived in Chapter 1 and are reproduced here

$$\bar{d}_r = dx\bar{a}_x + dy\bar{a}_y + dz\bar{a}_z \text{ (cartesian)}$$

$$= d\rho\bar{a}_\rho + \rho d\phi\bar{a}_\phi + dz\bar{a}_z \text{ (cylindrical)}$$

$$= dr\bar{a}_r + rd\theta a_\theta + r\sin\theta d\phi\bar{a}_\phi \text{ (spherical)} \qquad \ldots(4.3)$$

Let us now find out work done in moving a charge Q in an electric field set up by an infinite line charge placed along z-axis. We know that electric field intensity is given as

$$E = E_r \bar{a}_r = \frac{\rho_L}{2\pi\varepsilon_0 r}\bar{a}_r \quad \text{...(4.4)}$$

If we now take a positive charge Q about a circular path with centre at the origin, and move along the circle the work done is zero as the electric field is always normal to the circular path

$$W = -Q\int_0^{2\pi} \frac{\rho_L}{2\pi\varepsilon_0}\frac{\bar{a}_r}{r}\cdot r\cdot d\phi\cdot \bar{a}_\phi \quad \text{...(4.5)}$$

Since $\bar{a}_r \cdot d\phi = 0$

Hence $W = 0$ *i.e.* the line integral around a circular path is zero. However if we move the charge Q along the direction of electric field from a point a to b such that $a < b$, then the work done is

$$W = -Q\int_a^b \frac{\rho_L}{2\pi\varepsilon_0 r}dr = -Q\frac{\rho_L}{2\pi\varepsilon_0}\ln\frac{b}{a} \quad \text{...(4.6)}$$

Since here $b > a$ the $\ln\frac{b}{a}$ is positive and hence the work done on the charge is negative which means that external agency which is moving the charge receives energy.

However if we move the charge from b to a we have

$$W = -Q\int_b^a \frac{\rho_L}{2\pi\varepsilon_0}\cdot\frac{dr}{r} = -\frac{Q\rho_L}{2\pi\varepsilon_0}\ln\frac{a}{b} = Q\frac{\rho_L}{2\pi\varepsilon_0}\ln\frac{b}{a} \quad \text{...(4.7)}$$

Since here we are moving in opposition to the direction of electric field the work done is positive *i.e.* the energy level of the charge is increased as it has been moved to a higher potential level.

In the circular path we have found that

$$W = -Q\oint E\cdot dr = 0 \text{ or } \oint E\cdot dr = 0 \quad \text{...(4.8)}$$

Equation (4.8) is valid for static field and for time varying fields it requires certain correction, this will be discussed in a later chapter.

Any field that satisfies equation (4.8) *i.e.* the line integral of the field over a closed line path is zero is said to be conservative field. The property conservative arises from the fact that no work is done or that energy is conserved around a closed path. The gravitational field is also conservative where if we take a mass M from a particular point and move some distance and come back to the same point, the work done is zero as work is the product of force applied and the displacement of the mass. Since here even though we have moved some distance but displacement is zero hence the work done is zero.

4.3 POTENTIAL AND POTENTIAL DIFFERENCE

Using the expression for work done by an external agency in moving charge Q from one point to another in an electric field we define the term known as potential difference.

$$W = -Q\int_{\text{initial}}^{\text{final}} \bar{E}\cdot d\bar{r}$$

Just as we define electric field intensity at a point as the force per unit charge at that point, we now define potential difference as the work done by an external agency to move a unit positive charge from one point to another in an electric field *i.e.*

$$\frac{W}{Q} = -\int_{\text{initial}}^{\text{final}} E \cdot dr = V \qquad \ldots(4.9)$$

If we find potential difference as V_{AB} which signifies the potential difference between points A and B and is the work done in moving a unit charge from B (last letter) to A (first letter). Thus in determining V_{AB}, B is the initial point and A is the final point. This convention is used as B is often taken at infinity (as potential at that point is zero) and A as a point close or near to the charge. Potential difference is measured in terms of joules per coulomb or it is expressed in volts which is the common unit. Hence potential difference between points A and B is

$$V_{AB} = -\int_{B}^{A} E \cdot d\bar{r} \text{ volts}$$

and V_{AB} is positive if work is done in carrying the charge from point B to A. From equation (4.7) we have

$$V = \frac{W}{Q} = \frac{\rho_L}{2\pi\varepsilon_0} \ln \frac{b}{a} \qquad \ldots(4.10)$$

when a point charge Q is moved from $r = b$ to $r = a$.

Next we find out potential difference between two points r_a and r_b when the field is due to a point charge. We know for point charge $E = \dfrac{Q}{4\pi\varepsilon_0 r^2}\bar{a}_r$ and differential length is $dr\,\bar{a}_r$.

Hence

$$V_{ab} = -\int_{b}^{a} E \cdot dr = -\int_{r_b}^{r_a} \frac{Q}{4\pi\varepsilon_0 r^2}\bar{a}_r \cdot dr \cdot \bar{a}_r$$

$$= -\frac{Q}{4\pi\varepsilon_0} \cdot \left[\frac{1}{r}\right]_{r_b}^{r_a} = -\frac{Q}{4\pi\varepsilon_0}\left(\frac{1}{r_a} - \frac{1}{r_b}\right) \qquad \ldots(4.11)$$

Since $r_b > r_a$ as the charge is moved from a point almost at infinity *i.e.* r_b to r_a which is closer to the charge and V_{ab} is positive. Here charge is being moved from a larger distance to a smaller distance from charge, which gives rise to electric field and hence V_{ab} is positive which indicates that work is done by an external agency to bring the positive charge from r_b to r_a. Since the two charges are identical *i.e.* positive, therefore there will be force of repulsion between the two charges and hence an external agency will have to work against the force of repulsion to bring the test charge closer to each other.

Next we define what is known as potential at a point or absolute potential at a point. If in equation (4.11) we let $b \to \infty$ the equation becomes

$$V_a = \frac{Q}{4\pi\varepsilon_0 r_a} \qquad \ldots(4.12)$$

Therefore, for any general point r from the point charge Q the potential or absolute potential is given as

$$V = \frac{Q}{4\pi\varepsilon_0 r} \qquad \text{...(4.13)}$$

Here we have taken infinity as the zero potential or zero reference point.

Equation 4.13 represents the potential field of a point charge. Since potential is a scalar quantity, it does not involve any unit vector.

Next we define an equipotential surface as a surface which consists of all the points which have same value of potential. On such a surface, no work is done when we move a unit charge around on an equipotential surface as no potential difference between any two point of the surface exists. The equipotential surfaces in the potential field of a point charge are spheres with centre at the location of the point charge. However, maximum amount of work per unit distance is performed moving normal or perpendicular to an equipotential surface in the direction of the electric field.

4.3.1 Superposition of Potentials

The electric scalar potential due to a point charge at a point is a linear function of charge. If there are N charges in space, the electric scalar potential due to all the charges at a point is the scalar or algebraic addition of potential due to individual charges at that point. This is known as principle of superposition as applied to electric potential.

$$V_p = \frac{1}{4\pi\varepsilon_0} \sum_{n=1}^{N} \frac{Q_n}{r_n} \text{ volts} \qquad \text{...(4.14)}$$

If there is a line charge distribution with ρ_L as the line charge density, the potential at point P due to this charge is given as

$$V_p = \frac{1}{4\pi\varepsilon_0} \int \frac{\rho_L}{r} dr \text{ volts} \qquad \text{...(4.15)}$$

The integration is carried out over the entire length of the line charge.

However, if it is a surface charge distribution, the potential at any point P is (surface charge density being ρ_s).

$$V_p = \frac{1}{4\pi\varepsilon_0} \oint_s \frac{\rho_s}{r} ds \text{ volts} \qquad \text{...(4.16)}$$

where ds is a differential surface of the total surface area. The integration (double) is to be carried out over the complete surface.

Again if the charge distribution is volumetric with ρ_v as the volume c ensity in C/m^3, the electric potential at point P is given as

$$V_p = \frac{1}{4\pi\varepsilon_0} \oint_v \frac{\rho_v dv}{r} \text{ volts} \qquad \text{...(4.17)}$$

where dv is the differential volume in a complete volume enclosing the charge. The integ ion (triple) is to be carried out over the volume containing the charge.

However, if the charge distribution contains all combinations of point, line, surface and volume charge distribution and potential V at some point due to all these charge distributions are say, V_P, P_L, V_s and V_v respectively, then

$$V = V_P + V_L + V_S + V_v \qquad \text{...(4.18)}$$

i.e. if all charge distributions are present simultaneously the total electric potential at any point P due to all these distributions is, by the principle of superposition, the algebraic sum of the individual components of potentials. Thus in, general

$$V = \frac{1}{4\pi\varepsilon_0}\left[\sum_{1}^{N}\frac{Q_n}{r_n} + \int\frac{\rho_L}{r}dr + \iint\frac{\rho_s}{r}ds + \iiint\frac{\rho_v}{r}dv\right] \qquad \text{...(4.19)}$$

Example 4.1. Determine the potential difference due to a point charge $Q = 222$ pC between the two points at a distance of 400 mm and 100 mm respectively from the point charge.

Solution. The system is shown in Fig. E 4.1.

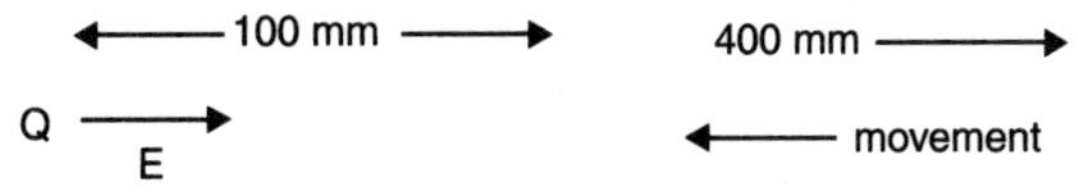

Fig. E 4.1

$$V = -\frac{Q}{4\pi\varepsilon_0}\int_{400}^{100}\frac{dr}{r^2} = +\frac{Q}{4\pi\varepsilon_0}\cdot\frac{1}{r}$$

Here –ve sign is given as we are moving the test charge in a direction opposite to that of the electric field.

$$V = \frac{222\times10^{-12}}{4\pi\times\frac{1}{36\pi}\times10^{-9}}\left[\frac{1}{0.1} - \frac{1}{0.4}\right] = 222\times9\times10^{-3}(10-2.5)$$

$$= 14.98 \text{ volts}$$

In order to obtain absolute potential at a distance 0.1 m from the charge, we move the test charge from infinity to a point at 0.1 m from the charge.

Hence
$$V_{0.1} = -\frac{Q}{4\pi\varepsilon_0}\int_{\infty}^{0.1}\frac{dr}{r^2} = \frac{Q}{4\pi\varepsilon_0}\left[\frac{1}{r}\right]_{\infty}^{0.1} = \frac{Q}{4\pi\varepsilon_0}\times0.1$$

$$= \frac{222\times10^{-12}}{4\pi\times\frac{1}{36\pi}\times10^{-9}}[10] = 222\times9\times10^{-3}\times10$$

$$= 19.98 \text{ volts.}$$

Similarly absolute potential at a distance of 0.4 m, we have

$$222\times9\times10^{-3}\times2.5 = 4.995 \text{ volts. } \textbf{Ans.}$$

Example 4.2. A circular ring 2.5 m in diameter has a charge density of 8 μC/m. Determine voltage at the centre of the ring.

Solution. Consider the differential element $rd\theta$ which could

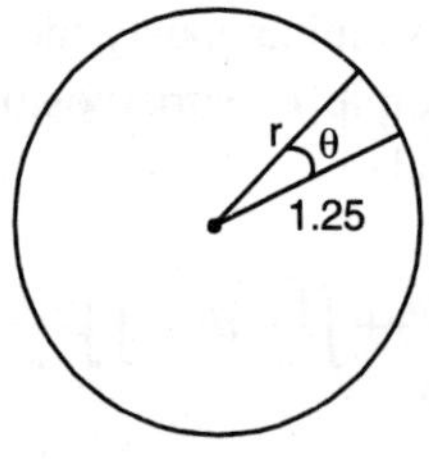

Fig. E 4.2

be considered as a point charge as $d\theta \to 0$, the potential

$$dV = \frac{\rho_L r d\theta}{4\pi\varepsilon_0 r}. \quad \text{Hence } V = \frac{\rho_L}{4\pi\varepsilon_0}\int_0^{2\pi} d\theta = \frac{\rho_L}{4\pi\varepsilon_0} \times 2\pi$$

$$V = 8 \times 10^{-6} \times 9 \times 10^{9} \times 2\pi = 452.16 \text{ kV} \quad \textbf{Ans.}$$

Example 4.3. A circular disc with radius 2 m has uniformly charge distribution of $\frac{50}{3}$ nC. Determine the potential at a point on the axis of the disc, 2 m from the disc. Compare this potential with the one which would result if the total charge were a point charge at the centre of the disc.

Solution. Considering a differential surface $ds = rd\theta \cdot dr$

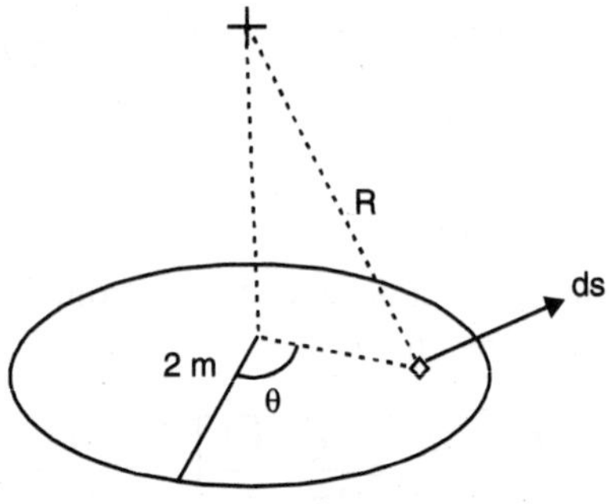

Fig. E 4.3

$$\rho_s = \frac{50}{3} \times 10^{-9} \cdot \frac{1}{\pi \cdot 4}$$

$$= 1.327 \text{ nC/m}^2$$

Considering this charge as a point as $d\theta \to 0$ and $dr \to 0$, the potential at a distance R is given by

$$dV = \frac{\rho_s ds}{4\pi\varepsilon_0 R} = \frac{\rho_s}{4\pi\varepsilon_0} \cdot \frac{rdrd\theta}{\sqrt{4+r^2}} = \frac{9\rho_s}{10^{-9}} \cdot \int_0^{2\pi}\int_0^{2} \frac{rdrd\phi}{\sqrt{4\pi r^2}}$$

$$= 1.327 \times 9 \int_0^{2\pi}\int_0^{2} \frac{rdrd\phi}{\sqrt{4+r^2}}$$

Let $4 + r^2 = z^2$

$2rdr = 2zdz$

$rdr = zdz$

$$= 1.327 \times 9 \int_0^{2\pi} \int_2^{2\sqrt{2}} \frac{z dz d\phi}{z}$$

$$= 1.327 \times 9 \left(2\sqrt{2} - 2\right) \cdot 2\pi = 75(1.414) = 106$$

If we consider point charge at the centre

$$V = \frac{\frac{50}{3} \times 10^{-9}}{4\pi \times \frac{1}{36\pi} \times 10^{-9} \times 2} = \frac{50}{3} \times \frac{9}{2} = 75 \text{ volts} \quad \textbf{Ans.}$$

Example 4.4. Electric field intensity $E = \left(\frac{x}{2} + 2y\right) \cdot \bar{a}_x + 2x\bar{a}_y$ V/m. Determine the work done in moving a point charge –20 μC (*i*) from the origin to (4, 0, 0) m and (*ii*) from (4,0,0) to (4, 2, 0) m.

Solution. From the coordinates it is clear that the charge is first moved along $\bar{a}_x$ and in the second case along $\bar{a}_y$

(*i*) $dW = -QE \cdot dr = 20 \times 10^{-9} \int_0^4 \left(\frac{x}{2} + 2y\right) \bar{a}_x \cdot dx \bar{a}_x$

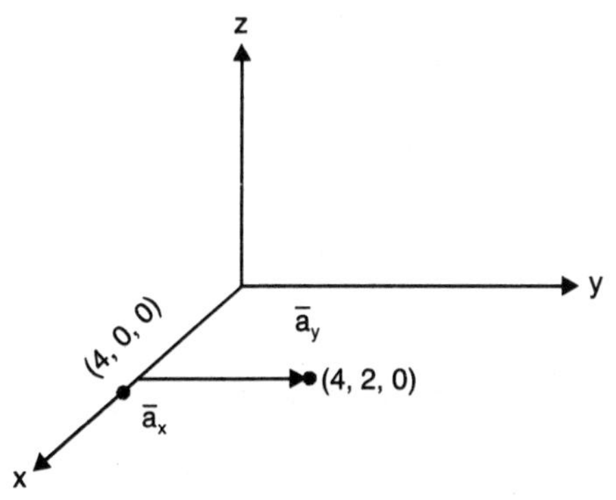

Fig. E 4.4

$$= 20 \times 10^{-9} \left[\frac{x^2}{4}\right]_0^4 = 80 \text{ μJ. Since } y = 0 \text{ at this point}$$

(*ii*) In this case the differential length $dr = dy\bar{a}_y$

Hence $dW = -QE \cdot dr = -QEdy\bar{a}_y$

$$= 20 \times 10^{-9} \int_0^2 2x\, dy = 20 \times 10^{-6} \cdot 2x[y]_0^2$$

$$= 40 \times 10^{-6} \times 4 \times 2 = 320 \text{ μJ} \quad \textbf{Ans.}$$

Example 4.5. Determine the work done in moving a point charge 10 μC from the origin to $\left(2m, \frac{\pi}{4}, \frac{\pi}{2}\right)$ point in spherical coordinate in the field $E = 5e^{-r/y}\bar{a}_r + \frac{10}{r\sin\theta}\bar{a}_\phi$ V/m.

Solution. Differential length in spherical coordinates is given as

$$d\bar{R} = dr\bar{a}_r + rd\theta\bar{a}_\theta + r\sin\theta d\phi\bar{a}_\phi$$

$$dW = -RE\cdot d\bar{R} = -Q\left(5e^{-r/4}\bar{a}_r + \frac{10}{\sin\theta}\bar{a}_\phi\right)\cdot\left(dr\bar{a}_r + rd\theta\bar{a}_\theta + r\sin\theta d\phi\bar{a}_\phi\right)$$

$$= -\left(10\times10^{-6}\right)\cdot 5e^{-r/4}dr - 10\times10^{-6}\times\frac{10}{\sin\theta}r\sin\theta d\phi$$

$$W = -\left(10\times10^{-6}\right)5\int_0^2 e^{-r/4}dr - 100\times10^{-6}\int_0^{\pi/2} rd\phi$$

$$= -50\times10^{-6}\left[\frac{e^{-r/4}}{-1/4}\right]_0^2 - 100\times10^{-6}\times2\times\pi/2$$

$$= 200\times10^{-6}[0.606-1] - 314\times10^{-6}$$

$$= 200\times10^{-6}\times0.394 - 314\times10^{-6}$$

$$= 78\times10^{-6} - 314\times10^{-6}$$

$$= -236\ \mu\text{J}$$

Since it is negative quantity which means the field does 236 μJ of work on the moving charge.

Example 4.6. The electric field intensity due to a line charge is given by $E = \frac{\rho_L}{2\pi\varepsilon_0 r}\bar{a}_r$. Determine V_{AB} where A is (2m, π/2, 0) and B is (4m, π, 5m) $\rho_L = (10^{-9}/2)$ C/m.

Solution. $V_{AB} = V_B - V_A = -\int_4^2 \frac{\rho_L}{2\pi\varepsilon_0 r}dr = \frac{\rho_L}{2\pi\varepsilon_0}\ln\frac{4}{2}$

$$= \frac{10^{-9}}{2}\cdot\frac{1}{2\pi\times\frac{1}{36\pi}\times10^{-9}}\ln\frac{4}{2} = 9\ln\frac{4}{2} = 9\times0.693$$

= 6.24 volts **Ans.**

Example 4.7. Four point charges are located at the corners of a square of length 2 L on each side symmetrically located relative to the origin as shown here. Derive an expression for the potential V at a point x along the x-axis.

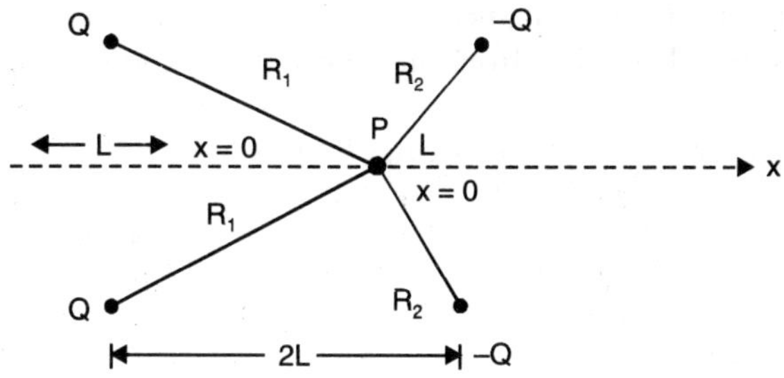

Fig. E 4.7

Solution. Potential at point P due to the four charges is

$$V(x) = \frac{1}{4\pi\varepsilon_0}\left(\frac{Q}{R_1} + \frac{Q}{R_1} - \frac{Q}{R_2} - \frac{Q}{R_2}\right)$$

$$= \frac{2Q}{4\pi\varepsilon_0}\left(\frac{1}{R_1} - \frac{1}{R_2}\right)$$

$$= \frac{Q}{2\pi\varepsilon_0}\left(\frac{1}{R_1} - \frac{1}{R_2}\right)$$

Now from the Fig. $R_1^2 = (L+x)^2 + (L)^2$

or $$R_1 = \sqrt{(L+x)^2 + L^2}$$

and $$R_2^2 = L^2 + (L-x)^2 \text{ or } R_2 = \sqrt{L^2 + (L-x)^2}$$

Hence $$V(x) = \frac{Q}{2\pi\varepsilon_0}\left[\frac{1}{\sqrt{(L+x)^2 + L^2}} - \frac{1}{\sqrt{L^2 + (L-x)^2}}\right]$$ **Ans.**

Example 4.8. An infinitely long cylindrical surface has radius R and has uniform surface charge density ρ_s C/m^2. Determine potential difference between two points at radial distance of r_a and r_b from the axis of the cylindrical surface where $r_b > r_a > R$.

Solution. The charge per unit length is $2\pi R \cdot 1 \cdot \rho_s$

Using symmetry and Gauss' law $D = \dfrac{2\pi R\rho_s}{2\pi r} = \dfrac{R\rho_s}{r}$

and hence $$E = \frac{R\rho_s}{r\varepsilon_0}$$

The potential difference $V_b - V_a = -\displaystyle\int_{r_a}^{r_b} \frac{R\rho_s}{\varepsilon_0 r} dr = \frac{R\rho_s}{\varepsilon_0} \ln\frac{r_a}{r_b}$ volts **Ans.**

Example 4.9. A circular loop 6 m in diameter carries a uniform charge of 2 μC/m. Determine the potential at a point on its axis 4 m away from the centre of the loop.

Solution. The distance of every point in the circular loop is $\sqrt{3^2+4^2}=5$ m and since the charge due to the $rd\theta$ is $rd\theta \cdot 2\times 10^{-6}$ C. Hence the potential at a distance 5 m is

$$dV = \frac{2\times 10^{-6}\cdot 3d\theta}{4\pi\varepsilon_0 \cdot 5} \text{ or } V = \frac{6\times 10^{-6}\times 9}{5}\cdot 2\pi\times 10^9 = 67.8 \text{ kV} \quad \textbf{Ans.}$$

Example 4.10. A slab of charge has a uniform charge density ρ. At $x = x_1$, $V = V_1$ and $E_x = E_1$. There is no variation of E_x in the y and z directions. Determine V_x within the slab

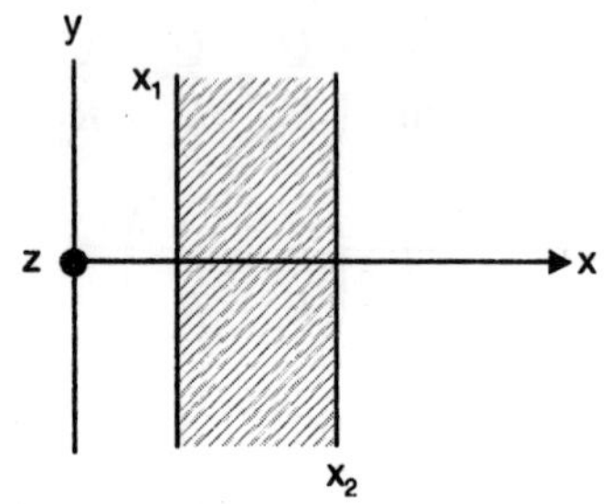

Fig. E 4.10

Solution. Since $\nabla\cdot D = \rho$ or $\nabla\cdot E = \frac{\rho}{\varepsilon_0}$

or $$\frac{dE}{dx} = \frac{\rho}{\varepsilon_0} \quad \text{or} \quad \int dE = \int \frac{\rho}{\varepsilon_0}dx = \frac{\rho}{\varepsilon_0}x + c$$

or $$E_x = \frac{\rho}{\varepsilon_0}x + c$$

At $$x = x_1 \quad E = E_1$$

Hence $$E_1 = \frac{\rho}{\varepsilon_0}x_1 + c \quad \text{or} \quad c = E_1 - \frac{\rho}{\varepsilon_0}x_1$$

or $$E_x = \frac{\rho}{\varepsilon_0}x + E_1 - \frac{\rho}{\varepsilon_0}x_1 = \frac{\rho}{\varepsilon_0}(x - x_1) + E_1$$

Now for $x_1 < x < x_2$

$$\frac{dV}{dx} = -E_x = -\frac{\rho}{\varepsilon_0}(x - x_1) - E_1$$

$$V = -\int_{x_1}^{x}\left\{\frac{P}{\varepsilon_0}(x - x_1) - E_1\right\}dx$$

$$= -\frac{\rho}{2\varepsilon_0}(x - x_1)^2 - E_1(x - x_1) + V_1 \quad \textbf{Ans.}$$

Example 4.11. Determine the work done in moving a point charge $Q = 3\ \mu C$ from (4m, π, 0) to (2m, π/2, 2m) cylindrical coordinates in the field $E = \frac{10^5}{r}\bar{a}_r + 10^5 z a_z$ V/m.

Solution. Work done $W = QV = -3 \times 10^{-6} \int E \cdot dl$

$$dl = dr\bar{a}_r + rd\phi\bar{a}_\phi + a_z\bar{a}_z$$

Hence
$$E \cdot dl = \left(\frac{10^5}{r}\bar{a}_r + 10^5 \bar{z}\right) \cdot \left(dr\bar{a}_r + rd\phi\bar{a}_\phi + dz\bar{a}_z\right)$$

$$= \frac{10^5}{r}dr + 10^5 z\,dz$$

$$W = -3 \times 10^{-6}\left[\int_4^2 \frac{10^5}{r}dr + \int_0^2 10^5 z\,dz\right]$$

$$= -3 \times 10^{-1}\left[\ln\frac{1}{2} + \left(\frac{z^2}{2}\right)_0^2\right]$$

$$= -0.3[-0.693 + 2] = -0.3 \times 1.307$$

$$= -0.392 \text{ J}$$ **Ans.**

4.4 POTENTIAL GRADIENT

We define gradient of a scalar quantity at a point as ratio of maximum change in the scalar quantity over a differential charge in the space. We find in respect of electric field the increase in scalar over a differential distance *dr* when moving in the direction of increasing potential is say *dV*, then

$$E = -\frac{dV}{dr} \text{ V/m} \qquad \text{...(4.20)}$$

Here we have assigned negative sign as the increase in potential is witnessed when we move in the opposite direction of the electric field. In order to find out gradient at a point we have to look in all direction and gradient will be in that direction where the change in the scalar quantity is maximum and hence gradient of a scalar quantity is a vector quantity. Since it is a vector quantity it may have components in all the three directions *i.e.*

$dr = dx\bar{a}_x + dy\bar{a}_y + dz\bar{a}_z$ in cartesian coordinates and hence the above equation for *E* can be written in cartesian coordinates as

$$E = -\left(\frac{\partial}{\partial x}\bar{a}_x + \frac{\partial}{\partial y}\bar{a}_y + \frac{\partial}{\partial z}\bar{a}_z\right)V \qquad \text{...(4.21)}$$

$$= -\nabla V \qquad \text{...(4.22)}$$

where
$$\text{del}\,\nabla = \left(\frac{\partial}{\partial x}\bar{a}_x + \frac{\partial}{\partial y}\bar{a}_y + \frac{\partial}{\partial z}\bar{a}_z\right) \qquad \text{...(4.23)}$$

If, however, we move in a direction normal to the direction of the gradient we should expect zero increase in the potential as is clear from the expression

$$\Delta V = -E \cdot dr \qquad \text{...(4.24)}$$

Since here neither E is zero nor dr is zero, and if $\Delta V = 0$ the only possibility is dr is normal to $\overline{E}$. Hence

$$\Delta V = -E \cdot dr = 0 \qquad \text{...(4.25)}$$

From this it is clear that the magnitude of $\overline{E}$ is given by the maximum space rate of change of V and the direction of $\overline{E}$ is normal to the equipotential surface, of course, in the direction of decreasing potential.

Equation 4.21 represent gradient of V and in equation 4.23 ∇ is known as del operator in cartesian coordinates.

Electric field E is normally represented by

$$E = -\text{grad } V = -\nabla V$$

From the above discussion following observations can be made regarding potential gradient.

(*i*) The magnitude of the electric field intensity is given by the maximum value of rate of change of potential with distance.

(*ii*) The maximum value is obtained when the direction of the distance increment is opposite to E or the direction of E is opposite to the direction in which potential V increase most rapidly.

The expressions for gradient may be expressed in other coordinate systems also which are given as under

$$\nabla V = \frac{\partial V}{\partial r}\overline{a}_r + \frac{1}{r}\frac{\partial V}{\partial \phi}\overline{a}_\phi + \frac{\partial V}{\partial z}\overline{a}_z \text{ ...cylindrical} \qquad \text{...(4.26)}$$

$$\nabla V = \frac{\partial V}{\partial r}\overline{a}_r + \frac{1}{r}\frac{\partial V}{\partial \theta}\overline{a}_\theta + \frac{1}{r\sin\theta}\cdot\frac{\partial V}{\partial \phi}\overline{a}_\phi \text{ ...spherical} \qquad \text{...(4.27)}$$

It is to be noted that the denominator of such term has change in length in the respective unit vector direction viz in spherical coordinates when ϕ is changed to $\phi + d\phi$, the change in length is $r \sin \theta d\phi$ and hence the component along $\overline{a}_\phi$ in spherical coordinates is

$\dfrac{1}{r\sin\theta}\cdot\dfrac{\partial V}{\partial \phi}$. Only difference is that in gradient it is partial derivative and hence $r \sin \theta\ \partial\phi$ rather than $r \sin \theta\ d\phi$.

4.5 ENERGY DENSITY IN THE ELECTROSTATIC FIELD

Consider the work required to assemble charge by charge with three charges Q_1, Q_2, Q_3 in a space which is charge free. This means if we bring a point charge Q_1 from infinity to some point in a charge free region the work done is zero as the electric field present is zero. Next we bring Q_2 in the region where Q_1 is already present. Due to the field of Q_1, some work will be required to bring Q_2 from infinity to some point near Q_1 and similarly if we bring Q_3, then we have to put in work to overcome the potentials due to Q_1 and Q_2. Hence total work done when we bring three charges from infinity to some points in the initially charge free space is

$$\begin{aligned} W &= W_1 + W_2 + W_3 \\ &= 0 + Q_2V_{21} + Q_3V_{31} + Q_3V_{32} \end{aligned} \qquad \text{...(4.28)}$$

Here potential V_{21} is the potential at point 2 where the charge Q_2 is placed, due to the charge Q_1 at position 1. Here W is the energy stored in the electric field due to the charge distribution (Here three point charges).

Now, if we bring the three charges in free space in the reverse order *i.e.* Q_3, Q_2 and Q_1, the energy in the electric field in space will be

$$W = W_3 + W_2 + W_1$$

$$= 0 + Q_2V_{23} + Q_1(V_{13} + V_{12}) \qquad ...(4.29)$$

Adding the two expression as given above have

$$2W = Q_1(V_{12} + V_{13}) + Q_2(V_{21} + V_{23}) + Q_3(V_{31} + V_{32}) \qquad ...(4.30)$$

The terms $Q_1(V_{12} + V_{13})$ is the work done against fields due to Q_2 and Q_3, the only other charges in the region. Hence $V_{12} + V_{13} = V_1$, the potential at point 1 where the charge Q_1 is located. Similarly for the other charges, we thus have

$$2W = Q_1V_1 + Q_2V_2 + Q_3V_3$$

or

$$W = \frac{1}{2}\left(Q_1V_1 + Q_2V_2 + Q_3V_3\right) \qquad ...(4.31)$$

However, if we have N charges in the free space (originally)

$$W = \frac{1}{2}\sum_{n=1}^{N} Q_nV_n \qquad ...(4.32)$$

If we are finding out expression for the energy stored in a region of continuous charge distribution, each charge is replaced by $\rho_v\ dv$ and the summation becomes an integral

$$W = \frac{1}{2}\int_v \rho_v V\,dv \qquad ...(4.33)$$

Equation 4.33 is a general expression for energy stored in all types of charge distribution e.g. point, line, surface or volume charge distribution as various charge distribution could be considered to be continuous distributions of volume charge density over very small or infinitesimally small regions. We now try to write equation 4.33 in terms of $\bar{E}$ and $\bar{D}$. ρ_v can be replaced by $\rho_v = \bar{\nabla}\cdot\bar{D}$ and we make use of a vector identity which is true for any scalar function V and any vector function $\bar{D}$.

$$\nabla\cdot(VD) = V(\nabla\cdot D) + D\cdot(\nabla V) \qquad ...(4.34)$$

Using these relation, equation 4.33 becomes

$$W = \frac{1}{2}\int_v (\nabla\cdot D)V\,dv$$

Now replacing $(\nabla\cdot D)V = \nabla\cdot(VD) - D\cdot(\nabla V)$ we have

$$W = \frac{1}{2}\int_v \left[\nabla\cdot(V\bar{D}) - D\cdot(\nabla V)\right]dv$$

The first-volume integral in the above equation can be converted into a surface integral where the closed surface surrounds the volume considered. This volume should contain each and every charge and no charge is left outside this volume *i.e.* we could consider volume $\rightarrow \infty$. Thus

$$W = \frac{1}{2}\oint_S (V \cdot D)\,ds - \frac{1}{2}\int D(\nabla V)\,dv \qquad \ldots(4.35)$$

When we look at the volume charge distribution from a distance r where $r \to \infty$, these charges look like point charges and here $V \propto \frac{1}{r}$ and $D \propto \frac{1}{r^2}$ *i.e.* the integrand $(VD) \propto \frac{1}{r^3}$ whereas the surface looks like a part of a sphere and hence $ds \propto r^2$ therefore as $r \to 0$, the first integral reduces to zero and we have

$$W = -\frac{1}{2}\int D \cdot (\nabla V)\,dv \qquad \ldots(4.36)$$

Now $$E = -\nabla V$$

Hence $$W = \frac{1}{2}\int D \cdot E\,dv = \frac{1}{2}\int \varepsilon_0 E^2 dv = \frac{1}{2}\int \frac{D^2}{\varepsilon_0} dv \text{ Joules} \qquad \ldots(4.37)$$

If we look at equation 4.33 we find that

$$\frac{1}{2}\int \rho_v V dv = \frac{1}{2} Q \cdot V \qquad \ldots(4.38)$$

where Q is the charge in the volume and V is the potential due to the charges and we find that this is the expression for energy stored in a capacitor where Q is the charge on one of the plates of the capacitor and V is the potential between the two plates of the capacitor. It is to be noted that the energy of an electric field or a charge distribution is stored in the field itself. To determine energy density in the electric field we differentiate expression 4.37 w.r. to differential volume and we have

$$dW = \frac{1}{2} D \cdot E\,dv \text{ or } \frac{dW}{dv} = \frac{1}{2} D \cdot E \text{ J/m}^3 \qquad \ldots(4.39)$$

From this expression also it is clear that electric energy is stored in the electric field.

Example 4.12. The electric field between two concentric cylindrical conductors at r = 0.01 m and r = 0.05 m is given by $E = (10^5/r)\, a_r$. Determine the energy stored in a 0.6 m length. Assume free space and fringing to be neglected.

Solution. The energy stored is given by

$$W_E = \frac{1}{2}\int_v \varepsilon_0 E^2 dv = \frac{\varepsilon_0}{2}\int_h^{h+0.6}\int_0^{2\pi}\int_{0.01}^{0.05} \frac{10^{10}}{r^2} r\,dr\,d\phi\,dz$$

$$= 10^{10}\frac{\varepsilon_0}{2}\int_h^{h+0.6}\int_0^{2\pi}\int_{0.01}^{0.05} \frac{dr}{r} d\phi\,dz$$

$$= \frac{10^{10}\varepsilon_0}{2}\ln\frac{5}{1} \cdot 2\pi \times 0.6$$

$$= \frac{8.854 \times 10^{-12} \times 10^{10}}{2} \times 2\pi \times 0.6 \times 1.609$$

= 0.2683 J **Ans.**

Example 4.13. Determine the energy stored in free space for the region 2 mm < r < 3 mm, $0 < \theta < \frac{\pi}{2}$, $0 < \phi < \frac{\pi}{2}$ given the potential field $V = \frac{200}{r}$ V.

Solution. $$E = -\nabla V = -\frac{\partial V}{\partial r}\bar{a}_r = -\frac{\partial}{\partial r}\left(\frac{200}{r}\right) = -\frac{200}{r^2}$$

$$W_E = -\frac{1}{2}\int \varepsilon_0 E^2 dv = \frac{\varepsilon_0}{2}\int_{2\times10^{-3}}^{3\times10^{-3}}\int_0^{\pi/2}\int_0^{\pi/2}\frac{4\times10^4}{r^4}r^2\sin\theta dr\, d\theta\, d\phi$$

$$= 4\times10^4 - \frac{\varepsilon_0}{2}\int_{2\times10^{-3}}^{3\times10^{-3}}\int_0^{\pi/2}\int_0^{\pi/2}\frac{1}{r^2}dr\sin\theta\, d\theta\, d\phi$$

$$= 2\times10^4\varepsilon_0\left[\frac{1}{3}-\frac{1}{2}\right]\times10^{-3}[\cos\theta]_0^{\pi/2}\cdot\frac{\pi}{2}$$

$$= 2\times8.854\times10^{-8}\times\frac{1}{6}\times10^{-3}\times1\times\frac{\pi}{2}$$

$= 46.33$ pJ **Ans.**

Example 4.14. Determine the energy stored in a system of four identical point charges of $Q = 5$ nC at the corners of a square 1 m on a side. Determine the energy stored in the system when only two charges at opposite corners of the square are present.

Solution. The potential of the point at location 1.

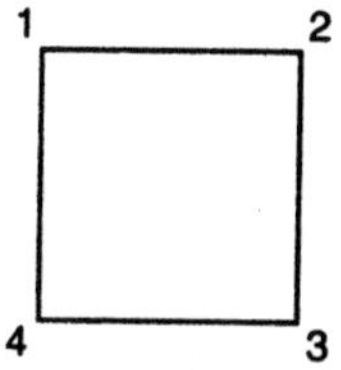

Fig. E 4.14

$$V_1 = \frac{Q_2}{4\pi\varepsilon_0 r_{12}} + \frac{Q_3}{4\pi\varepsilon_0 r_{13}} + \frac{Q_4}{4\pi\varepsilon_0 r_{14}}$$

$$= 5\times10^{-9}\times9\times10^9\left(\frac{1}{r_{12}}+\frac{1}{r_{13}}+\frac{1}{r_{14}}\right)$$

$$= 45\left(\frac{1}{1}+\frac{1}{\sqrt{2}}+\frac{1}{1}\right) = 45\times2.707 = 121.8 \text{ volts.}$$

Since $2W = Q_1V_1 + Q_2V_2 + Q_3V_3 + Q_4V_4 = 4\,Q_1V_1$ as

$Q_1 = Q_2 = Q_3 = Q_4$ and $V_1 = V_2 = V_3 = V_4$ as it is a symmetrical system

Hence $W = 2\,Q_1V_1 = 2\times5\times10^{-9}\times121.8 = 1210$ J

If only two charges are placed

$$2W = Q_1V_1 = Q_1\frac{Q_1}{4\pi\varepsilon_0 r_1} = \frac{5\times10^{-9}\times5\times10^{-9}\times9\times10^9}{\sqrt{2}}$$

= 159 nJ **Ans.**

Example 4.15. Determine the distance at which the absolute potential due to a 120 nC point charge equals 1*k* when (*i*) in vacuum and (*ii*) in a medium with $\varepsilon_r = 4$.

Solution. (*i*) Absolute potential at a point is given by $V = \dfrac{Q}{4\pi\varepsilon_0 r}$ where *r* is the distance from the point charge.

$$1000 = \frac{120\times10^{-9}}{4\pi\varepsilon_0 r} = \frac{120\times 9}{r}$$

or $$r = 1.08 \text{ m}$$

(*ii*) However if the $\varepsilon_r = 4$ for the medium then

$$1000 = \frac{120\times10^{-9}\times 9\times10^{9}}{4\times r} = \frac{120\times 9}{4r}$$

or $$r = 0.27 \text{ m} \quad \textbf{Ans.}$$

Example 4.16. An $\bar{E}$ field is given by $\bar{E} = 4x\bar{a}_x + 2\bar{a}_y$ V/m. Determine the work required to move a unit positive charge along the curve $xy = 4$ from (2, 2) to (4, 1).

Solution. $V = -\int E\cdot dl = (4x\bar{a}_x + 2\bar{a}_y)\cdot(dx\bar{a}_x + dy\bar{a}_y) = -\int(4x dx + 2dy)$

Now $$xy = 4 \text{ or } y = \frac{4}{x} \text{ or } dy = -\frac{4}{x^2}dx$$

Hence $$V = -\int_2^4\left(4x - \frac{8}{x^2}\right)dx = -\left(2x^2 + \frac{8}{x}\right)_2^4$$

$$= -(34 - 12) = -22 \text{ J} \quad \textbf{Ans.}$$

that is, the field does 22 J of work on the charge.

Example 4.17. The value of *E* at *r* ($r = 2$, $\phi = 40°$, $z = 3$) is given as $E = 100\,\bar{a}_z - 200\,\bar{a}_\phi + 300\,\bar{a}_z$ V/m. Determine the incremental work required to move a 40 μC charge a distance of 6 μm (*i*) in the direction of $\bar{a}_r$ (*ii*) in the direction of $\bar{a}_\phi$ (*iii*) in the direction of $\bar{a}_z$.

Solution. (*i*) Since we have to find work done along $\bar{a}_r$, and since $\bar{a}_\phi$ and $\bar{a}_z$ are normal to $\bar{a}_r$, these do not come to into the picture.

Hence $$V = -\int_2^{2+6\times10^{-6}} 100\bar{a}_r\cdot dr\cdot\bar{a}_r = -100\times6\times10^{-6}$$

Hence $$W = QV = -600\times10^{-6}\times40\times10^{-6} = -24000\times10^{-12} = -24 \text{ nJ} \quad \textbf{Ans.}$$

(*ii*) Similarly here

$$V = -\int_{40°}^{40+d\phi} -200\,d\phi = 200\times3\times10^{-6} = 600\times10^{-6}$$

Since r is 2 m, the angle subtended by the differential length 6 μC will be $\dfrac{6\times10^{-6}}{2} = 3\times10^{-6}$ radian

For 40°, the radians are $40\times\dfrac{\pi}{180} = 0.698$

$$W = QV = 40\times10^{-6}\times600\times10^{-6} = 24 \text{ nJ} \quad \textbf{Ans.}$$

(*iii*) In the direction of $\bar{a}_z$

Again $$V = -\int_{3}^{3+6\times10^{-6}} 300\, dz = -300\times6\times10^{-6}$$

$$\text{Work done} = QV = -40\times10^{-6}\times300\times6\times10^{-6} = 72 \text{ nJ} \quad \textbf{Ans.}$$

Example 4.18. The potential scalar field in cylindrical coordinates is given by $V = \dfrac{100}{z^2+1} r\cos\phi$ volt and point P at $r = 3$ m, $\phi = 60°$, $z = 2$ m, determine values at P for (*i*) voltage (*ii*) $\bar{E}$ (*iii*) E (*iv*) ρ_v in free space.

Solution.

(*i*) Voltage $$V = \frac{100}{2^2+1} r\cos\theta = \frac{100}{4+1}\times3\times\cos60°$$

$$= 60\cos60° = 30 \text{ volts}$$

(*ii*) $$\bar{E} = -\nabla V = \frac{\partial V}{\partial r}\bar{a}_r + \frac{1}{r}\frac{\partial V}{\partial\phi}\bar{a}_\phi + \frac{\partial V}{\partial z}\bar{a}_z$$

$$= -\frac{\partial}{\partial r}\frac{100}{2^2+1}\cdot r\cos\phi = -\frac{100}{z^2+1}\cos\phi = -20\times0.5 = -10\bar{a}_r$$

$$= -\frac{1}{r}\frac{\partial}{\partial\phi}\frac{100}{z^2+1} r\cos\phi = \frac{100}{z^2+1}\sin\phi$$

$$= 20\sin60° = 17.3 a_\phi$$

Similarly $$\frac{\partial V}{\partial z} = \frac{\partial}{\partial z}\frac{100}{z^2+1} r\cos\phi = 100 r\cos\phi \frac{\partial}{\partial z}\frac{1}{z^2+1}$$

$$= 100\times3\times\frac{1}{2}\times\frac{4}{25} = 24\bar{a}_z$$

Hence $$\bar{E} = -10\bar{a}_r + 17.3\bar{a}_\phi + 24\bar{a}_z$$

$$E = \sqrt{10^2 + 17.3^2 + 24^2} = \sqrt{100 + 299 + 576} = 31.2 \text{ V/m}$$

Now $$D = \varepsilon_0\bar{E} = \varepsilon_0\left(-10\bar{a}_r + 17.3\bar{a}_\phi + 24\bar{a}_z\right)$$

$$\nabla \cdot D = \rho_v = \varepsilon_0 \nabla \cdot E = \rho_v$$

$$= \varepsilon_0 \left[\frac{1}{r}\frac{\partial (rD_r)}{\partial r} + \frac{1}{r}\frac{\partial D_\phi}{\partial \phi} + \frac{\partial D_z}{\partial z} \right]$$

$$= \varepsilon_0 \left[\frac{1}{r}\frac{\partial}{\partial r} r(-10) + \frac{1}{r}\frac{\partial\, 17.3}{\partial \phi} + 0 \right]$$

$$= -8.854 \times 10^{-12} \times \frac{10}{3} + 0 + 0 = -29.5 \text{ pC/m}^3$$

Example 4.19. The scalar potential function is given as $V = 2x + 4y$ volt in free space. Determine the energy stored per unit volume centered at the origin.

Solution. Energy $W = \frac{1}{2}\varepsilon_0 E^2 \text{ J/m}^3$

Now
$$E = -\nabla V = -\left(\frac{\partial V}{\partial x}\bar{a}_x + \frac{\partial V}{\partial y}\bar{a}_y + \frac{\partial V}{\partial z}\bar{a}_z \right)$$

$$= -2\bar{a}_x - 4\bar{a}_y$$

The magnitude of $E = \sqrt{2^2 + 4^2} = \sqrt{20}$

Since the electric field is of constant magnitude, the energy density is $\frac{1}{2}\varepsilon_0 E^2 \text{ J/m}^3$

$$\frac{1}{2} \times 8.854 \times 10^{-12} \times 20 = 88.54 \text{ pJ/m}^3 \quad \textbf{Ans.}$$

Example 4.20. A volume charge is distributed throughout a sphere of radius R and continued at the origin with uniform density ρ_v C/m³. Determine the potential field of the volume charge distribution.

Solution. (*i*) Potential for $r < R$

Consider a radius $r < R$ and the charge enclosed is $\frac{4}{3}\pi r^3 \rho_v$ which could be considered as point charge centred at the origin and hence the potential at r is $\frac{Q}{4\pi\varepsilon_0 r} = \frac{4}{3}\pi r^3 \rho_v \cdot \frac{1}{4\pi\varepsilon_0 r} = \frac{\rho_v r^2}{3\varepsilon_0}$

$$\text{For } r > R \;\; Q = \frac{4}{3}\pi R^3 \rho_v \;\; \text{Hence } \frac{4}{3}\pi R^3 \rho_v \frac{1}{4\pi \times \frac{1}{36\pi} \times 10^{-9}}$$

$$\frac{4}{3}\pi R^3 \rho_v \cdot \frac{1}{4\pi\varepsilon_0 r} = \frac{\rho_v R^3}{3\varepsilon_0 r} \quad \text{for } r > R$$

Example 4.21. A circular ring of radius 2 m has 40 nC of charge uniformly distributed. Determine the potential at a point on the axis 5 m from the plane of the ring. Compare the result if 40 nC were a point charge located at the centre of the ring.

Solution. Here $\rho_L = \dfrac{40 \times 10^{-9}}{2\pi \times 2} = \dfrac{10^{-8}}{\pi}$ C/m.

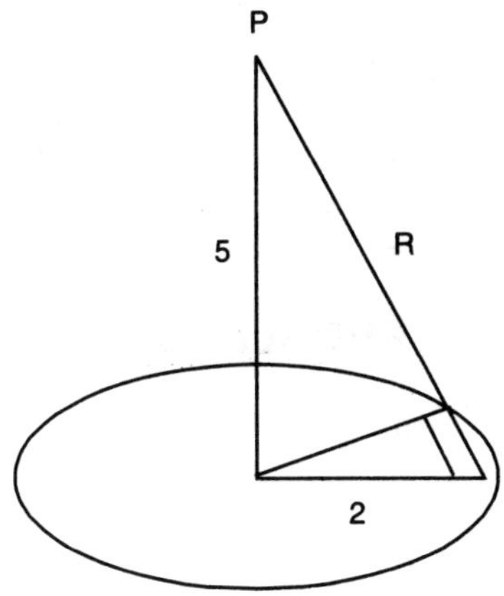

Fig. E 4.21

The charge on the differential length $= 2d\phi \cdot \rho_L = \dfrac{2 \times 10^{-8}}{\pi} d\phi$

Hence differential potential at point P

$$dV = \int \frac{\rho_L dl}{4\pi\varepsilon_0 \sqrt{29}} = \frac{2 \times 10^{-8}}{\pi} \int_0^{2\pi} d\phi \cdot \frac{1}{4\pi\varepsilon_0 \sqrt{29}}$$

$$= \frac{2 \times 10^{-8}}{\pi\sqrt{29}} \times 9 \times 10^9 \cdot 2\pi = \frac{4 \times 9 \times 10}{\sqrt{29}} = 66.85 \text{ volts}$$

However, if the total charge of 40 nC is located at the centre the potential due to point charge at point P is

$$\frac{40 \times 10^{-9}}{4\pi \times \dfrac{1}{36\pi} \times 10^{-9} \times 5} = \frac{40 \times 9}{5} = 72 \text{ volts. } \textbf{Ans.}$$

QUESTIONS

Q. 4.1 Discuss briefly and compare some methods of determining electric field intensity.

Q. 4.2 Show that the curl of a static electric field is zero.

Q. 4.3 Derive an expression for work done on a positive point charge in an electric field and explain the significance of negative and positive work in such a system.

Q. 4.4 Explain clearly conservative property of electric fields compare it with gravitational fields.

Q. 4.5 Define and differentiate between potential, potential difference and absolute potential.

Q. 4.6 Explain what do you mean by superposition of potentials. Derive an expression for potential when all charge distribution systems are present *i.e.* point, line, surface and volume charge distribution.

Q. 4.7 Explain what do you mean by gradient of a scalar quantity. Derive an expression for potential gradient for an electric field using cartesian coordinates.

Q. 4.8 Derive an expression for energy stored in electric field due to point charges, starting with first principles. Explain clearly where this energy is stored.

Q. 4.9 Derive expressions for energy density in an electric field using D and E.

PROBLEMS

P 4.1 A charge of 30 nC is situated at the point (0, 2, 0) m and a charge of –30 nC at the point (0, –2, 0) m. Determine electric field intensity E at points (*i*) (4, 2, 4) m (*ii*) (4, 0, 4) m and (*iii*) (4, –2, 4) m.

Ans. (*i*) $\left(2.71\bar{a}_x - 3.25\bar{a}_y + 2.71\bar{a}_z\right)$ (*ii*) $-5\bar{a}_y$ V/m (*iii*) $9.21\bar{a}_x - 3.25\bar{a}_y + 9.21\bar{a}_z$

P 4.2 If the scalar potential due to certain electric field is given by $V = \phi Z^{1/2}/r$. Determine the electric field intensity.

Ans. $\phi\dfrac{z^{1/2}}{r^2}\bar{a}_r - z^{1/2}r^{-2}\bar{a}_\phi - \dfrac{1}{2}\phi\dfrac{z^{-1/2}}{r}$ V/m

P 4.3 Determine if the following fields are realisable as static electric field.

(*i*) $E = -y\bar{a}_x + x\bar{a}_y$ (*ii*) $E = \dfrac{\rho_L}{2\pi\varepsilon_0 r^2}\left(\cos\phi\bar{a}_r + \sin\phi\bar{a}_\phi\right)$ Cylindrical coordinates

(*iii*) $E = \sin\theta\bar{a}_r + \cos\theta\bar{a}_\theta$ Spherical coordinates.

Ans. (*i*) No (*ii*) Yes (*iii*) Yes.

P 4.4 Charge is uniformly distributed along a straight line of length $2L$. Show that for two external points near the midpoint, such that r_1 and r_2 are small compared to the length, the potential V_{12} is the same as for an infinite length of charge.

Ans. $V_{12} = \dfrac{\rho_L}{2\pi\varepsilon_0}\ln\dfrac{r_2}{r_1}$

P 4.5 If the electric flux density within a sphere of radius R is given by $D = \dfrac{5}{r^2}\bar{a}_r - r^2\phi\sin\theta\bar{a}_\phi$ C/m^2, determine the volume charge density in the sphere. **Ans.** $-r$ C/m^3

P 4.6 An infinitely long cylinder of radius a is filled with a charge of uniform density ρ. If the potential on the surface of the cylinder is V_0, determine the potential within the cylinder.

Ans. $V = V_0 + \dfrac{\rho}{4\varepsilon_0}\left(a^2 - r^2\right)$

P 4.7 A point charge of –4 C is situated at the origin and another point charge of 6 nC is located at the point (3, 2, 1) m. Determine V at the point (1, 1, 1) m. **Ans.** 3.3 volts

P 4.8 A uniform line charge of density 1 nC/m is arranged in the form of a square 6 m on a side with its centre at the origin. Determine the potential at (0, 0, 5) m. **Ans.** 35.6

P 4.9 A scalar potential field in free space is expressed as $v = 20/(xyz)$ V (*i*) Determine the total energy stored within the cube $1 < x, y, z < 2$.

P 4.10 Two uniform line charges 8 nC/m each are located at $x = 1$, $z = 2$ and at $x = -1$, $y = 2$ in free space. If the potential at the origin is 100 V, determine V at $P(4, 1, 3)$ **Ans.** –68.4 volts

P 4.11 Determine the amount of electric charge associated with the scalar potential $V = xy^2 z$ in air in the volume $0 \le x, y, z < 1$. **Ans.** $Q = -\varepsilon_0/2$

P 4.12 If the electric field $E = -\dfrac{16}{r^2}\bar{a}_r$ V/m in spherical coordinates, determine the potential of point P (2m, π, $\pi/2$) with respect to Q (4m, 0, π). **Ans.** –4 volt

P 4.13 A uniform sheet of charge $\rho_s = \dfrac{1}{6\pi}\ \text{nC/m}^2$ is at $x = 0$ and a second sheet $\rho_s = -\dfrac{1}{6\pi}\ \text{nC/m}^2$ is at $x = 10$ m. Determine V_{AB}, V_{BC} and V_{AC} for A (10 m, 0, 0), B (4 m, 0, 0) and C (0, 0, 0). **Ans.** –36 V, –24 V, –60 V

P 4.14 Determine the electric field of a long line charge at a radial distance where the potential is 24 V higher than at a radial distance $r_1 = 3$ m where $|E| = 4$V/m. **Ans.** 29.5 V/m

P 4.15 Two thin conducting half planes at $\phi = 0$ and $\phi = \pi/6$ are insulated from each other along the z-axis. Assume that the scalar potential function for $0 \le \phi \le \pi/6$ is $V = \left(\dfrac{-60\phi}{\pi}\right)$ volt, determine the energy stored between the half planes for $0.1 \le r \le 0.6$ m and $0 \le z \le 1$ m. Assume free space.

Ans. 1.51 nJ

P 4.16 A line charge of finite length is situated along the z-axis between $z = -z_0$ and $z = z_0$. Assuming the charge density ρ_2 to be an arbitrary function of z, show that the potential at any point in the xy plane at a distance r from the origin is given by

$$V = \frac{1}{4\pi\varepsilon_0}\int_{-z_0}^{z_0}\frac{\rho_L dz}{\sqrt{r^2+z^2}}$$

Evaluate the integral for the following charge distributions:
(*i*) $\rho_L = \rho_{L0}$ a constant $-z_0 < z < z_0$
(*ii*) $\rho_L = |z|$ $\quad -z_0 < z < z_0$
(*iii*) $\rho_L = +z$ $\quad -z_0 < z < z_0$

Ans. (*i*) $\dfrac{\rho_{L_0}}{4\pi\varepsilon_0}\ln\dfrac{\sqrt{r^2+z_0^2}+z_0}{\sqrt{r^2+z_0^2}-z_0}$, (*ii*) $\dfrac{1}{2\pi\varepsilon_0}\left(\sqrt{r^2+z^2}-r\right)$, (*iii*) 0

P 4.17 The scalar potential for a system is given as $V = 3x^2 + 4y^2$ volt. Determine the energy stored in the volume described by $0 \le x \le 1$ m, $0 \le y \le 1$ m and $0 \le z \le 1$. **Ans.** 147 pJ

P 4.18 If electric flux density due to certain charge distribution is given by $D = \frac{\bar{a}_r}{r^2} + \sin\phi\,\bar{a}_\phi + \frac{\bar{a}_z}{3\sqrt{z}}\ \text{C/m}^2$. Determine the charge density at the point (1 m, 15°, 3 m). **Ans.** –0.11 C/m³

P 4.19 A uniform surface charge density of 5 nC/m² is present at the $z = 0$ plane, a uniform line charge density of 8 nC/m be located at $x = 0$, $z = 4$ and a point charge of 2 μC is present at $P(2, 0, 0)$. If $V = 0$ at $P'(0, 0, 5)$, determine V at $P''(1, 2, 3)$. **Ans.** 1.98 kV

P 4.20 Three identical point charges of 4 pC each are located at the corners of an equilateral triangle 0.5 mm on a side in free space. Determine the work to be done to move one charge to a point equidistant from the other two and on the line joining them. **Ans.** 576 pJ

P 4.21 A cube of 2 m side is centred at the origin with the edges parallel to the axes. Within the cube $D = \frac{10x^3}{3}\bar{a}_x\ \text{C/m}^3$. Determine the total charge enclosed by the cube. **Ans.** 80/3C

P 4.22 The scalar potential field V is given as $V = 2x^2y - 5z$. Determine the potential V, electric field intensity E, the electric flux density D and the volume charge density ρ_v at a point $P(-4, 3, 6)$.

Ans. $66\,\text{V},\ 48\bar{a}_x - 32\bar{a}_y + 5\bar{a}_z\ \text{V/m}$

$\rho_v = -106\,\text{pC/m}^3$

P 4.23 Determine the charge density function corresponding to the field.

$$D = \begin{cases} \rho_0 \dfrac{r^2 - a^2}{2r}\bar{a}_r & a < r < b \\ \rho_0 \dfrac{b^2 - a^2}{2r}\bar{a}_r & b < r \\ 0 & r < a \end{cases}$$

Ans. $\rho = \begin{cases} \rho_0 & a < r < b \\ 0 & b < r \\ 0 & r < a \end{cases}$

P 4.24 Electric field intensity E is given as $E = -5\bar{e}^{r/a}\bar{a}_r$ in cylinderical coordinates. Determine the energy stored in the volume described by $r \le 2a$ and $0 \le z \le 5a$. **Ans.** $7.89 \times 10^{-10}\,a^3$

Chapter 5

Current and Conductors

5.1 INTRODUCTION

The concept of charge is based on atomic theory. An atom has positive charges (protons) in its nucleus and an equal number of electrons (negative charges) surround the nucleus making the atom neutral. Removal of an electron leaves the atom positive charged and addition of an electron makes the atom negatively charged. The basic unit of charge is the charge on an electron. The MKS unit of charge is Coulomb. An electron has a charge of 1.062×10^{-9}C.

5.2 ELECTRIC CURRENT

When a test charge q is introduced in an electric field E as shown in Fig. 5.1, it experiences a

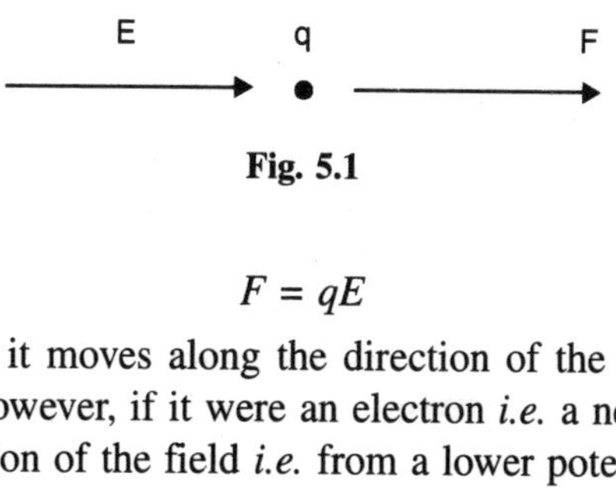

Fig. 5.1

force F given by

$$F = qE \qquad ...(5.1)$$

Since the charge is positive it moves along the direction of the electric field *i.e.* from a higher potential to a lower potential. However, if it were an electron *i.e.* a negative charge, the electron will move in opposition to the direction of the field *i.e.* from a lower potential to a higher potential.

If the charge in Fig. 5.1 is free to move, it will receive an acceleration which according to Newtons second law of motion is

$$F = ma \qquad ...(5.2)$$

or

$$a = \frac{F}{m} \text{ m/sec}^2$$

where m is the mass of the charged particle in kg. From this it is clear that if $\bar{E}$ were uniform and if there were no restrictions in the path of movement of the charge, the particle's velocity $v = at$ will increase indefinitely with time t. However, in solid, liquid or gaseous medium the particle collides repeatedly with other particles thereby losing some of its energy and causing random changes in its direction of motion. If E is constant and the medium is homogeneous, the net effect of the collisions is to restrain the charged particle to a constant average velocity called the drift velocity v_d. This drift velocity has the same direction as the electric field and is related to it by a constant called the mobility μ_m. Thus

$$v_d = \mu_m E \text{ m/sec. Drift velocity} \qquad \text{...(5.3)}$$

where μ_m is mobility in m²/volt/sec and E = V/m. Naturally good conductors will have higher mobility as compared to bad conductors. Typical values for mobility are 0.0012 for aluminium, 0.0032 for copper and 0.0056 for silver.

If a medium of uniform cross section S contains many free-to-move charged particles with volume charge density ρ_v, then these moving charges will form a current I in Coulombs per second passing given reference point or a cross section and is given as

$$I = \rho_v v_d S \text{ Ampere} \qquad \text{...(5.4)}$$

An electric current is defined as the time rate of flow of charge through a certain section. Its unit is ampere. A current is said to be of one ampere when a charge of 1 Coulomb flows through a section per second *i.e.* $i = \dfrac{dq}{dt}$.

If charge q is expressed in Coulombs and time in second 1 amp flow of current through a section is equivalent to approximately 6.24×10^{18} electrons per second through the section.

For good conductors a drift velocity of a few centimeters is good enough to produce sufficient heat to cause the wire to melt if the heat is not removed quickly by conduction or radiation.

From equation 5.4,

$$J = \frac{I}{S} = \rho_v v_d \qquad \text{...(5.5)}$$

Substituting equation (5.3) in equation 5.5, we have

$$J = \rho_v \mu_m E \qquad \text{...(5.6)}$$

For a free electron ρ_v is negative and hence

$$J = -\rho_v \mu_e E \qquad \text{...(5.6a)}$$

The total charge density ρ_v is zero because equal positive and negative charge is present in the neutral material. The negative value of ρ_v and the minus sign lead to a current density J which is in the same direction as E, the electric field intensity. Since ρ_v represents the volume charge density and μ_e the mobility of electrons in a certain material the product of the two gives the conductivity of the material. Hence J and E for a metallic conductor are related through conductivity of the material as

$$J = \sigma E \qquad \text{...(5.6b)}$$

where σ is the conductivity of the conductor and is measured in Siemens. Typical values of σ for aluminium is 3.82×10^7 Siemen, for silver 6.17×10^7 and for copper 5.80×10^7 s. It is to be noted that conductivity of several conductors is constant over wide ranges of current density and electric field intensity. Ohm's law and the metallic conductors are also known as isotropic or having the same properties in every direction. A material which is not isotropic is known as anisotropic.

The conductivity of a metal is a function of temperature. The conductivity decreases with increase in temperature. For some metals the conductivity increases abruptly to a high value at a temperature of a few Kelvin. This property is termed as superconductivity. Copper and silver are not superconductors, though aluminium is for temperature below 1.14 K.

From equation (5.6*a*) and (5.6*b*), we have

$$\sigma = -\rho_v \mu_e \qquad ...(5.6c)$$

It is to be noted that the electric charges flow from a lower potential to a higher potential and these contribute to electronic current, whereas the conventional current is considered to flow from higher potential to lower potential.

5.3 CURRENT DENSITY

It is to be noted that we are normally interested in finding out a quantity at a point rather than within some large region and we should find the concept of current density measured in amperes for square metre more useful. Current density is a vector quantity represented by J.

If ΔI is the incremental current crossing an incremental surface Δs normal to the current density it is given as

$$\Delta I = J_N \Delta s \qquad ...(5.6d)$$

However, if the current density is not normal to the surface Δs, then

$$\Delta I = \bar{J} \cdot \Delta s \qquad ...(5.7)$$

Therefore, the total current passing through a surface s is given as

$$I = \oint_s J \cdot ds = JS \qquad ...(5.8)$$

Equation 5.8 is also known as Ohm's law in pointform.

From equation (5.4) we have

$$J = \frac{I}{S} = \rho_v v_d \qquad ...(5.9)$$

From equation (5.9) it is clear that charge in motion constitutes a current known as convection current and J or $\rho_v v_d$ is the convection current density. The convection current density is related linearly to volume charge density as well as to velocity. The current density is expressed in amperes per square metre.

5.4 CONTINUITY OF CURRENT

The law of conservation of charge states that charges can neither be created nor destroyed, even though equal number of positive and negative charges may be simultaneously created. These may be obtained by separation from their atoms or these may be lost by recombination of positive and negative charges to make neutral atoms. The continuity equation is possible only if we consider any region enclosed by a closed surface. From equation (5.8), the current through the closed surface is

$$I = \oint_s J \cdot \overline{ds}$$

and this outward flow of positive charge is balanced by a decrease of positive charge or perhaps an increase of negative charge within the closed surface. If the charge inside the closed surface is denoted by Q, then the rate of decrease (due to outward flow of positive charge from the closed surface) is $-\frac{dQ}{dt}$ and the principle of conservation of charge requires

$$I = \oint_s J \cdot d\bar{s} = -\frac{dQ}{dt} \qquad ...(5.10)$$

The presence or absence of a negative sign depends on what current and charge we consider. In circuit theory, the conventional current flows from a higher to lower potential. The current I indicated in equation (5.10) is an outward flowing current. Equation (5.10) is the integral form of the continuity equation and the differential or point form is obtained by using the divergence theorem to change the surface integral into a volume integral.

$$\oint_s J \cdot d\bar{s} = \int_v (\nabla \cdot J)\, dv \qquad \text{...(5.11)}$$

If ρ_v is the volume charge density, then the charge enclosed by the surface is $\int_v \rho_v\, dv$ and hence the current I is given as

$$I = -\frac{d}{dt}\oint_v \rho_v dv = \oint_v (\nabla \cdot J)\, dv \qquad \text{...(5.12)}$$

If we keep the surface constant, the derivative becomes partial derivative and hence

$$\oint_v (\nabla \cdot J)\, dv = \oint_v -\frac{\partial \rho_v}{\partial t} dv \qquad \text{...(5.13)}$$

Since the expression holds good for any volume, however small, it holds good for an incremental volume

$$(\nabla \cdot J)\Delta v = -\frac{\partial \rho_v}{\partial t}\Delta v$$

or

$$(\nabla \cdot J) = \frac{-\partial \rho_v}{\partial t} \qquad \text{...(5.14)}$$

This is known as point form of continuity equation. This equation indicates that the current or charge per sec, diverging from a small volume per unit volume is equal to the time rate of decrease of charge per unit volume at every point.

When an electric field is applied to a conductor; valence electrons are free to move and hence static conditions do not exist. However, these charges do not form the net charge as each conduction electron is balanced by a proton in the nucleus such that there is zero net charge in every Δv of the material. Suppose, however, that through a temporary imbalance a region within a solid conductor has a net charge density ρ_0 at a time $t = 0$. Then, since $J = \sigma E = \frac{\sigma}{\varepsilon} D$.

$$\nabla \cdot \frac{\sigma}{\varepsilon} D = -\frac{\partial \rho}{\partial t} \qquad \text{...(5.15)}$$

Since σ and ε for a homogeneous material are constant therefore, the above equation becomes

$$\frac{\sigma}{\varepsilon}\nabla \cdot D = -\frac{\partial \rho}{\partial t}$$

$$\frac{\sigma}{\varepsilon}\rho = -\frac{\partial \rho}{\partial t}$$

or $$\frac{\partial \rho}{\partial t} + \frac{\sigma}{\varepsilon}\rho = 0$$

Hence $$\rho = \rho_0 e^{-(\sigma/\varepsilon)t} \qquad \text{...(5.16)}$$

Thus ρ decays exponentially with a time constant of $\tau = \frac{\varepsilon}{\sigma}$. This is also known as relaxation time. At $t = \tau$, ρ decays to 36.8% of it's initial value. For conductor τ is extremely small on the order of 10^{-19}. This confirms that free charge can not remain within a conductor and rather is distributed evenly over the conductor surface.

Example 5.1. Determine the relaxation time for silver given that $\sigma = 6.17 \times 10^7$ S/m. If charge of density ρ_0 is placed within a silver block, find the charge density ρ after one and also after four time constants.

Solution. Since $\varepsilon = \varepsilon_0$ as $\varepsilon_r = 1$ for silver

$$\tau = \frac{\varepsilon_0}{\sigma} = \frac{8.854 \times 10^{-12}}{6.17 \times 10^7} = 1.435 \times 10^{-19}$$

Therefore at $t = \tau \quad \rho = \rho_0 e^{-1} = 0.368\ \rho_0$ C/m³ **Ans.**

at $t = 4\tau \quad \rho = \rho_0 e^{-4} = 0.0183\ \rho_0$ C/m³ **Ans.**

Example 5.2. The current density $J = 10\rho^2 z\bar{a}_\rho - 4\rho\cos^2\phi\bar{a}_\phi$ mA/m². Determine (*i*) current density at point $P(\rho = 3, \phi = 30°, z = 2)$

(*ii*) the total current flowing outward through the circular band $\rho = 3, 0 \le \phi < 2\pi, 2 < z < 2.8$.

Solution. (*i*) Since $J = 10\rho^2 z\bar{a}_\rho - 4\rho\cos^2\phi\bar{a}_\phi$, its value at P is

$$J = 10 \times 3^2 \times z\bar{a}_\rho - 4 \times 3 \times \frac{3}{4}\bar{a}_\phi = 180\bar{a}_\rho - 9\bar{a}_\phi \quad \textbf{Ans.}$$

(*ii*) At ρ = 3 $$J = 90z\bar{a}_\rho - 4\rho\cos^2\phi\bar{a}_\phi = 90z\bar{a}_\rho - 12\cos^2\phi\bar{a}_\phi$$

$$dI = J \cdot ds$$

$$= 90z\bar{a}_\rho \cdot \rho\, d\phi\, dz\,\bar{a}_\rho - 12\cos^2\phi\bar{a}_\phi\, d\rho\, dz\,\bar{a}_\theta$$

Since $d\rho$ is zero $$dI = 90z\rho\, d\phi\, dz$$

$$I = 270\int_0^{2\pi}\int_2^{2.8} z\, d\phi\, dz$$

$$= 270 \times 2\pi\left[\frac{z^2}{2}\right]_2^{2.8}$$

$$= 540\pi\left(\frac{2.8^2 - 2^2}{2}\right)$$

$$= 540\pi \times \frac{3.84}{2} \text{ mA}$$

$$= 3.255 \text{ A} \quad \textbf{Ans.}$$

Example 5.3. The current density in cylindrical coordinates is given as $J = -10^6 z^{1.5} \bar{a}_z \text{ A/m}^2$ in the region $0 \le \rho \le 25$ μm and for $\rho \ge 25$ μm J = 0. Determine (*i*) the total current crossing the surface z = 0.1 m in the $\bar{a}_z$ direction (*ii*) If the charge velocity is 2×10^6 m/s at z = 0.1 m, find ρ_v (*iii*) If the volume charge density at z = 0.15 m is –2000 C/m³, find the charge velocity.

Solution. (*i*) $I = J \cdot ds = -10^6 \times 0.1^{1.5} \times \pi \times 25^2 \times 10^{-12} = -62.06 \mu\text{A}$ **Ans.**

(*ii*) $\rho_v v = J$ or $\rho_v = \dfrac{J}{v} = \dfrac{-10^6 \times 0.1^{1.5}}{2 \times 10^6} = 15.8 \text{ mc/m}^3$ **Ans.**

(*iii*) $v = \dfrac{J}{\rho_v} = \dfrac{-10^6 \times 0.15^{1.5}}{2000} = 29 \text{ m/sec}.$ **Ans.**

5.5 CONDUCTORS, SEMICONDUCTORS AND INSULATORS

As there are as many energy bands in a solid as there are energy levels in the parent atom, the energy band diagrams for most of the materials would look quite complicated. However, we are mostly interested in the upper bands of energy levels mainly the upper two called the conduction band and the valence band. The valence band contains energies of the same level as those of valence electrons. The valence electrons are attached to individual atoms and, therefore, not free to move about. If electric field at sufficient strength is applied, electrons in the valence band will jump to the conduction band crossing the forbidden energy band (energy gap E_g) as shown in Fig. 5.2. The size of E_g (the forbidden band) is a prime factor in determining whether a solid is a conductor, an insulator or a semiconductor.

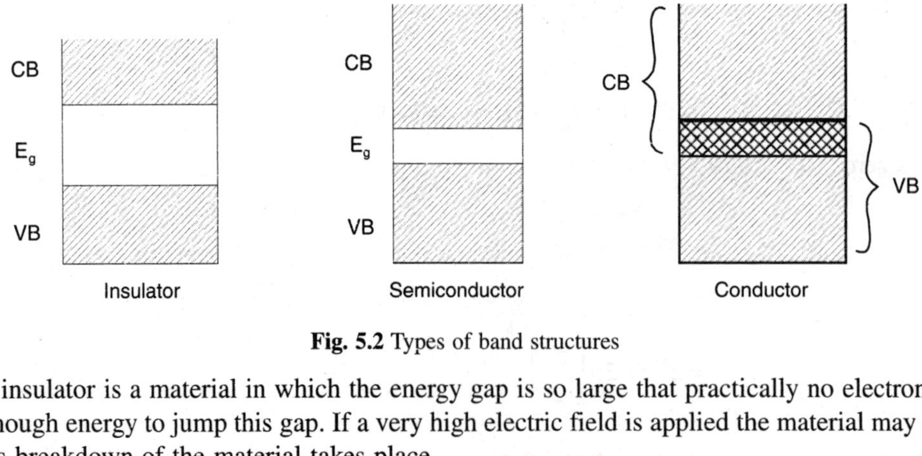

Fig. 5.2 Types of band structures

An insulator is a material in which the energy gap is so large that practically no electrons can be given enough energy to jump this gap. If a very high electric field is applied the material may puncture and thus breakdown of the material takes place.

A semiconductor is a solid with an energy gap small enough for electrons to cross rather easily from the valence band to the conduction band. At room temperature sufficient energy is available for a few valence electrons to bridge the energy gap to the conduction band, thus the material sustains some electric current.

Conductors contain a large number of electrons in the conduction band at room temperature. No energy gap or forbidden band exists between the valence band and the conduction band. In fact the valence band and the conduction bands overlap and a small value of electric field applied will lead to flow of electrons and the electric current.

In ordinary metallic conductors such as aluminium or copper the highest energy electrons are readily detached from their atoms by an applied electric field and are free to migrate. However, the atoms remain fixed in the conductor lattice, so only the electrons have the mobility. Therefore, the conductivity for conductor has only one term as given by

$$\sigma = \rho_e \mu_e \text{ Siemen/m} \qquad \ldots(5.17)$$

However, in semiconductors, the normal conduction by electrons is supplemented by another charge carrier called a hole which represents a vacant space in the lattice structure of the semiconductor. These vacant spaces or holes left by the electrons can migrate from atom to atom in a semiconductor so that the hole tends to behave like a positively charged ion but its mobility is more like an electron.

Thus the conductivity of a semiconductor is given by

$$\sigma = \rho_e \mu_e + \rho_h \mu_h \text{ Siemen/m} \qquad \ldots(5.18)$$

Germanium and silicon are typical semiconductors. In their intrinsic or pure form, the electrons and vacancies *i.e.* the electron hole pair have only a short life-time, disappearing as electrons and holes recombine. However, new pairs are continually being formed and so some are always present. In case of semiconductors, increase as temperature increases the pair formation *i.e.* increase the density of electron-hole pairs and this increases the conductivity contrary to conductor where the conductivity decreases with increase in temperature.

If small amount of certain impurities are added to the semiconductor either during crystal growth or by diffusion, the carrier density and conductivity can be greatly enhanced. Impurities such as boron introduce more holes and are called acceptors forming *p*-type semiconductor with holes predominating. However, if phosphorous is added, it provides more electrons and are called donors forming *n*-type semiconductors in which electrons constitute most of the carriers, holes being in minority. The process of introducing impurities is called doping. It is observed that if donor concentrations of less than one part per million can increase the conductivity by a factor of nearly a million.

In plasma, gases and liquids, charges of both signs are usually present and free to migrate. Assuming that all negative particles are of the same kind and all positive ions are of the same kind, the conductivity will have two terms as follows

$$\sigma = \rho_- \mu_- + \rho_+ \mu_+ \text{ S/m} \qquad \ldots(5.19)$$

The first term represents the contribution to the conductivity from negatively charged particles moving opposite to the field $\bar{E}$ whereas the second term represents the contribution from positively charged particles moving with E.

5.6 CONDUCTOR PROPERTIES AND BOUNDARY CONDITIONS

Suppose we have a conductor placed in free space and we allow a large number of electron in the interior of the conductor. The electric fields set up by these electrons are not counteracted by any positive charges and the electrons, therefore, begin to accelerate away from each other. This process continues until a number of electrons equal to the no. of injected have reached the surface. The electrons can not move in space on its own as no external electric field exists in space. No charge remains inside the conductor and hence the charge density inside the conductor is zero and a surface charge density is present on the exterior surface. This is an important characteristic of a good conductor. Since charge resides on the surface of the conductor, for static electric fields, the electric field intensity inside a conductor is zero. Physically also it is true as, if an electric field were present, the conduction electrons would move and produce a current which results in a non-static condition. Therefore, for electrostatic, no charge and no electric field may exist at any point within a conducting material. We next study, in such a situation, the electric fields external to the conductor due to the charge on the surface of the conductor.

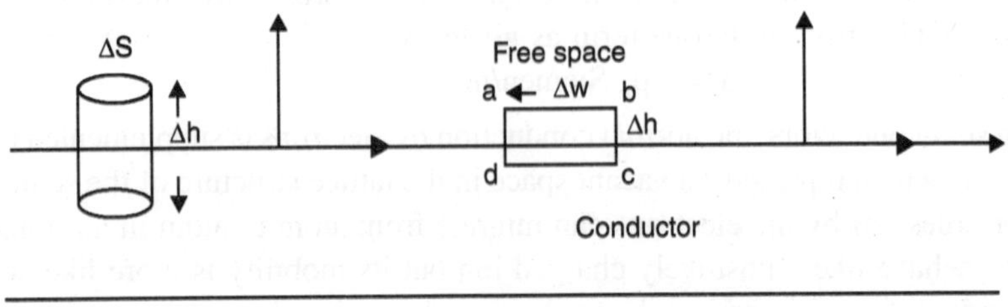

Fig. 5.3 Boundary conditions conductor—Free space

If the external electric field intensity is decomposed into two components one tangential and another normal to the surface of conductor since the surface charges are static no potential difference exists and hence electric field intensity tangent to the surface is zero and so also the electric field density is zero.

Normal component of electric flux density can be obtained by applying Gauss' law along the differential surface of boundary between the conductor and the free space as shown in Fig. 5.3. Since the charge in the lower $\frac{\Delta h}{2}$ depth inside the conductor is zero as no charge resides inside the conductor. Therefore, the flux must leave the surface normally. We, therefore, can say that according to Gauss' law the surface charge will give rise to electric flux that will leave the surface normally and the electric flux density in Coulombs per square metre leaving the surface normally is equal to the surface charge density in Coulombs per square metre or $D_n = \rho_s$.

We may prove these relations $E_{t_1} = E_{t_2} = 0$ and $D_{n_1} = D_{n_2} = \rho_s$ mathematically as follows using the relation.

$$\oint E \cdot dl = 0 \text{ and } \oint D \cdot ds = Q \text{ respectively.}$$

Applying the line integral equation to the rectangular boundary surface in Fig. 5.3, we have

$$E_t \Delta w - E_{n_b} \cdot \frac{\Delta h}{2} + E_{n_a} \frac{\Delta h}{2} = 0 \qquad \text{...(5.20)}$$

as $\Delta h \to 0$, even if $E_{n_b} \neq E_{n_a}$, the two cancel each other and $E_t \Delta W = 0$

therefore, $E_t = 0$

Similarly, for the normal component, we integrate over the three surfaces of the differential cylinder as shown in Fig. 5.3.

$$\int_{top} + \int_{bottom} + \int_{sides} = Q \qquad \text{...(5.21)}$$

Since the bottom surface lies inside the conductor where both E and D are zero. For the side integral even though flux densities at the two surfaces will be different; as the depth $\frac{\Delta h}{2} \to 0$ the integral also tends to zero.

So we are left with only one integral corresponding to top surface of the differential cylinder

Hence $$D_n \cdot \Delta S = Q = \rho_s \, \Delta\rho$$

or $$D_n = \rho_s \qquad \ldots(5.22)$$

From the foregoing analysis following observations can be made in regard to electric field intensity and electric flux density for a conductor to free space boundary condition in electrostatic fields.

1. The static electric field intensity inside a conductor is zero.
2. The static electric field intensity at the surface of a conductor is everywhere directed normal to that surface. Similarly the electric flux density is also normal.
3. The conductor surface is an equipotential surface.

5.6.1 Boundary Conditions: Conducting Media

Consider boundary between two conductors of different conductivity σ_1 and σ_2, the normal components of the current density are continuous. (See Fig. 5.4)

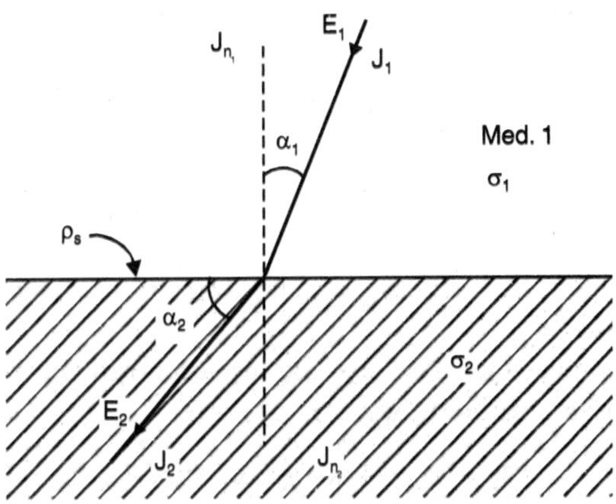

Fig. 5.4 Boundary condition—conducting media

Hence $$J_{n_1} = J_{n_2}$$

We also have $E_{t_1} = E_{t_2}$ since $J_{t_1} = \sigma_1 E_{t_1}$ and $J_{t_2} = \sigma_2 E_{t_2}$

Hence $$\frac{J_{t_1}}{\sigma_1} = \frac{J_{t_2}}{\sigma_2} \quad \text{or} \quad \frac{J_{t_1}}{\sigma_1 J_{n_1}} = \frac{J_{t_2}}{\sigma_2 J_{n_2}}$$

$$\text{Since } \frac{J_{t_1}}{J_{n_1}} = \tan\alpha_1 \quad \text{Hence } \frac{\tan\alpha_1}{\sigma_1} = \frac{\tan\alpha_2}{\sigma_2}$$

or $$\frac{\tan \alpha_1}{\tan \alpha_2} = \frac{\sigma_1}{\sigma_2} \quad \text{...(5.23)}$$

where α_1 and α_2 are as shown in Fig. 5.4.

5.6.2 Boundary Conditions: Dielectric Media

Consider two dielectric media with relative permittivities ε_1 and ε_2 as shown in Fig. 5.5.

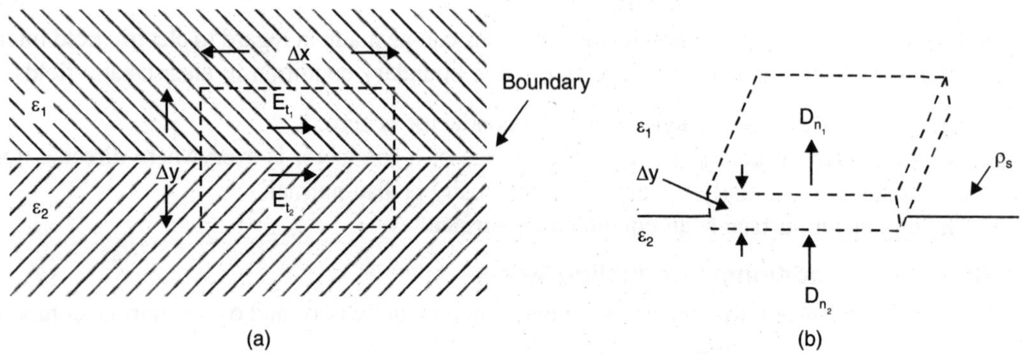

Fig. 5.5 Boundary condition— Dielectric media

We know that line integral of the static electric field around a closed path is zero. Thus referring to Fig. 5.5(*a*) as $\Delta y \to 0$ so as to consider the boundary condition

$$\oint E \cdot dl = E_{t_1} \Delta x - E_{t_2} \Delta x = 0 \quad \text{...(5.24)}$$

or $$E_{t_1} = E_{t_2}$$

i.e. the tangential component at the boundary is same for both the dielectrics even though then have different relative permittivities.

Referring to Fig. 5.5(*b*) and using Gauss' law, the surface integral of the electric flux density D over a closed path equals the charge enclosed. Again as $\Delta y \to 0$

$$\oint_s D \cdot ds = D_{n_1} s - D_{n_2} s = \rho_s s \quad \text{...(5.25)}$$

or $$D_{n_1} - D_{n_2} = \rho_s$$

Thus at the boundary of two dielectric media, the electric flux density is discontinuous if the boundary has surface charge density ρ_s C/m^2. However, the tangential component of the electric field intensity is continuous. If the surface charge density is zero, the electric flux density is also continuous at the boundary conditions whereas the normal component of electric field intensity will be discontinuous as even though $D_{n_1} = D_{n_2}$, t_1 and t_2 being unequal $E_{n_2} \neq E_{n_1}$.

Example 5.4. Consider a charge free boundary between two different dielectrics and electric field E_1 is incident obliquely at the boundary from medium 1 to 2 at angle 26.6° as shown in Fig. E 5.4. If E_1 = 20 V/m at P. Determine (*i*) α_2 (*ii*) E_2 and D_2 at P.

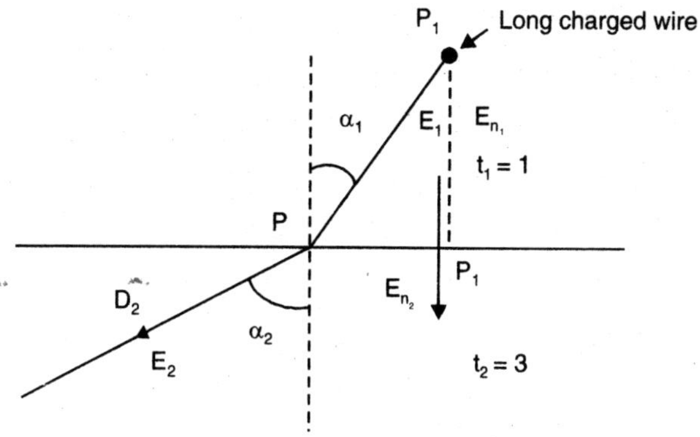

Fig. E 5.4

Solution. (*i*) The tangential component of E_1 along the boundary

$$E_{t_1} = E_1 \sin\alpha_1 \text{ and } E_{t_2} = E_2 \sin\alpha_2$$

Since at the boundary the tangential components are equal

$$E_1 \sin\alpha_1 = E_2 \sin\alpha_2 \qquad \text{...(5.26)}$$

Also, since it is a charge free boundary $D_{n_1} = D_{n_2}$

$$D_{n_1} = D_1 \cos\alpha_1 \text{ and } D_{n_2} = D_2 \cos\alpha_2$$

Hence $$D_1 \cos\alpha_1 = D_2 \cos\alpha_2 \qquad \text{...(5.27)}$$

Now $$D_1 = \varepsilon_0 E_1 \text{ and } D_2 = 3\varepsilon_0 E_2$$

Now $$\frac{E_1}{D_1}\tan\alpha_1 = \frac{E_2}{D_2}\tan\alpha_2$$

$$\frac{1}{\varepsilon_0}\tan\alpha_1 = \frac{1}{3\varepsilon_0}\tan\alpha_2$$

or $$\tan\alpha_1 = \frac{\tan\alpha_2}{3} \text{ or } \tan\alpha_2 = 3\tan\alpha_1 = 3\times\frac{1}{2} = 1.5$$

Hence $$\alpha_2 = 56.30° \textbf{ Ans.}$$

(*ii*) $E_2 \text{ at } P = E_1 \dfrac{\sin\alpha_1}{\sin\alpha_2} = 20\cdot\dfrac{\sin 26\cdot 6°}{\sin 56.30°} = 20\times\dfrac{0.448}{0.832} = 10.77 \text{ V/m}$

$$D_2 \text{ at } P = D_1\frac{\cos\alpha_1}{\cos\alpha_2} = \varepsilon_0 E_1 \frac{\cos\alpha_1}{\cos\alpha_2} = 8.854\times10^{-12}\times20\times\frac{\cos 26.6°}{\cos 5.63°}$$

$$= 177\times10^{-12}\times\frac{0.897}{0.555} = 286 \text{ pC/m}^2 \textbf{ Ans.}$$

5.7 METHOD OF IMAGES

If we have, some charge a point charge or a line charge, in space at a height of h from the surface of the earth which is infinite in extent and is held at zero potential, and we are asked to find out electric field

Line charge

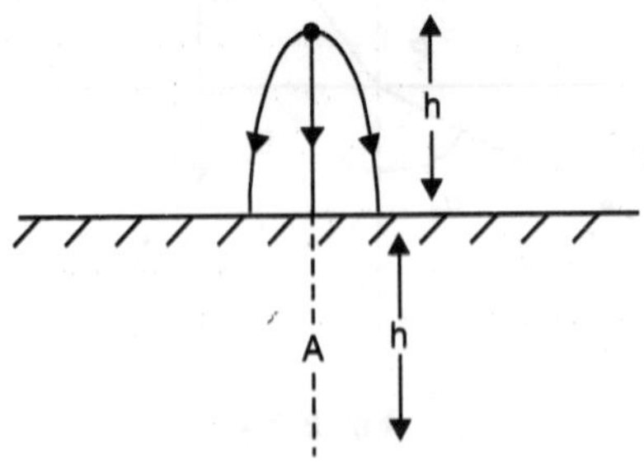

Fig. 5.6 Method of images

intensity or electric potential at a point we make use of what is known as method of images. If the infinite plane were not there the electric flux lines would not have terminated at the surface of the earth and the potential of these lines is forced to zero. Now this kind of situation can be simulated by assuming exactly identical charge configuration with opposite polarity and at the same depth as is the height of original charge above the earth. This can be seen very easily that since the two charges are equidistant from the zero potential surface of infinite extent, one could consider an imaginary very thin sheet of metal of infinite extent in the middle of the two charges which will simulate earth at zero potential and now we may remove the equipotential surface and analysis in terms of evaluation of electric field intensity or electric potential any where above or below the equipotential (zero potential) surface. The second charge below the original charge is known as the image of the first charge.

Since both electric field intensity or potential distribution due to charges are a linear function of charge and the net effect of the charge distribution can be superimposed to get the total result. Thus any charge distribution above an infinite ground plane may be replaced by an arrangement composed of the given charge distribution, its' image and no conducting plane (earth surface). With this method of images, therefore it becomes very simple to evaluate potential field of the new system as it does not contain the conduction plane with its' unknown surface charge distribution.

Example 5.5. The following data are obtained by measurements on two materials A and B

	Electron mobility	Hole mobility in m^2/volt-sec
A	6	0.2
B	0.3	0.1

At room temperature 300 kelvin resistivity is found to be for A 2.5×10^{-4} and for B 2.5×10^{-3} ohm–m respectively. Assuming both the materials to be intrinsic, determine their intrinsic carrier densities (ρ_v) at 300 K.

Solution. Since conductivity $\sigma = \rho_e(\mu_e + \mu_n)$

or
$$\rho_e = \frac{\sigma}{e(\mu_s + \mu_n)} = \frac{0.4 \times 10^4}{1.6 \times 10^{-19} \times 6.2}$$

$$= 4.032 \times 10^{21} \textbf{ Ans.}$$

For $$B,\ \rho_v = \frac{0.4 \times 10^3}{1.6 \times 10^{-14} \times 0.4} = 0.625 \times 10^{22} \text{ per cm}^3.$$

Example 5.6. Determine the resistance of a circular section copper wire 16 gauge which has a diameter of 1.291 mm and 1 km long. If the current carrying capacity is 10 A dc, determine the current density, electric field intensity, drift velocity and volume charge density of free electrons.

Solution. The resistance of 1 km length of wire

$$\frac{1000}{5.8 \times 10^7 \times 1.308 \times 10^{-6}} = 13.18\Omega$$

$$J = \text{The current density} = \frac{I}{A} = \frac{10}{1.308 \times 10^{-6}} = 7.65 \text{ A/mm}^2$$

The voltage drop $IR = 10 \times 13.18 = 131.8$ **Ans.**

Therefore, $$E = \frac{V}{L} = \frac{131.8}{1000} = 0.131 \text{ V/m} \textbf{ Ans.}$$

$$v_d = \mu_e E = 0.004 \times 0.131 = 0.000524 \text{ m/sec} \textbf{ Ans.}$$

$$\rho_v = \frac{J}{v_d} = \frac{7.65 \times 10^6}{5.24 \times 10^{-4}} = 1.46 \times 10^{10} \text{ C/m}^3 \textbf{ Ans.}$$

Example 5.7. If 10 V is applied between the ends of a 1 mm diameter copper wire 100 m long, determine (*i*) current (*ii*) electric field intensity E (*iii*) electron velocity assume $T = 290$ K

$$\mu_e = 0.004 \text{ m}^2 \text{v}^{-1}\text{s}^{-1} \text{ and } \rho_v = 1.4 \times 10^{10} \text{ C/m}^3.$$

Solution. The conductivity of the copper $\sigma = \rho_v\, \mu_e$.

$$\sigma = 0.004 \times 1.4 \times 10^{10} = 5.6 \times 10^7 \text{ S/m}.$$

$$R = \frac{l}{\sigma a} = \frac{100}{5.6 \times 10^7 \times \frac{\pi}{4} \times 10^{-6}} = 2.27\Omega$$

The current is $\dfrac{10}{2.27} = 4.4\text{ A}$

$$E = \frac{V}{L} = \frac{10}{100} = 0.1 \text{ V/m}$$

$$v_d = \mu_E = 0.004 \times 0.1 = 4 \times 10^{-4} \text{ m/sec} \textbf{ Ans.}$$

Example 5.8. Determine the image of a charge Q_1 placed at a distance of R_1 from the centre of a spherical conductor maintained at zero potential. Also determine electric field intensity at any point in the space.

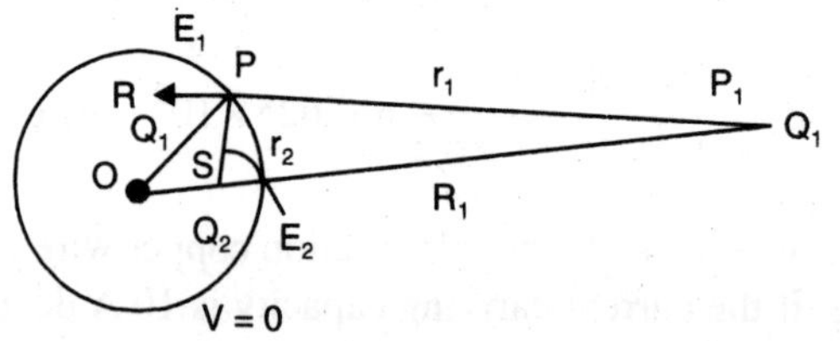

Fig. E 5.8

Solution. This is a situation where we have to find out image of Q_1 into the spherical conductor. Suppose the image of the conductor is at point S and magnitude $-Q_2 \neq -Q_1$ as the sphere is not at a plane of infinite extent. Since the sphere is at zero potential, the potential at P due to the two charges Q_1 and $-Q_2$ should be zero.

Hence
$$V_p = \frac{Q_1}{4\pi\varepsilon_0 r_1} - \frac{Q_2}{4\pi\varepsilon_0 r_2} = 0$$

or
$$\frac{Q_1}{r_1} = \frac{Q_2}{r_2} \quad \text{or} \quad Q_2 = Q_1 \cdot \frac{r_2}{r_1}$$

If we know r_2, the image charge Q_2 can be obtained. To find out r_2, we make use of the above relation that the relation must be valid for any choice of P on the surface of the sphere. This requires that the ratio $\frac{r_2}{r_1}$ must be constant. This condition is met if triangle OSP and OPP$_1$ are similar. If R is the radius of the sphere $OP = R$

$$\frac{OP}{OP_1} = \frac{PS}{PQ_1} = \frac{r_2}{r_1} = \frac{R}{R_1}$$

or hence
$$Q_2 = Q_1 \frac{r_2}{r_1} = Q_1 \cdot \frac{R}{R_1}$$

By using the magnitude and location of these charges, the original charge Q_1 and its image Q_2, we can determine the electric field intensity at any point outside the surface of the spherical conductor using the relation

$$E_1 = \frac{Q_1}{4\pi\varepsilon_0 r_1^2}\bar{a}_{r_1} \text{ and } E_2 = \frac{Q_2}{4\pi\varepsilon_0 r_2^2}\bar{a}_{r_2}$$

The vector sum of the two will give the resultant electric field intensity at that point.

It is to be noted that if the potential of the sphere were V and not zero, then to find out the location of the image charge and its magnitude, we proceed as follows. Since the surface of the sphere is equipotential at V volts, the image charge is to be located at the centre of the sphere with opposite polarity and thus, the magnitude of the image charge can be obtained.

Example 5.9. Determine the electric field intensity and current density corresponding to drift velocity of 6×10^{-4} m/sec in a silver conductor. Assume $\sigma = 61.7$ MS/m and $\mu = 5.6 \times 10^{-3}$ m^2/v.s. for silver.

Solution. $v_d = \mu E \text{ or } E = \dfrac{v_d}{\mu} = \dfrac{6\times10^{-4}}{5.6\times10^{-3}} = 1.07\times10^{-1} \text{ V/m}$

$$J = \sigma E = 61.7\times10^{6}\times1.07\times10^{-1} = 6.61\times10^{6} \text{ A/m}^2 \textbf{ Ans.}$$

Example 5.10. Determine the current in a circular wire if the current density is $J = 20\left(1-e^{-500r}\right)\bar{a}_z$ A/m^2. The radius of the wire is 4 mm.

Solution. Since $dI = J\cdot ds = 20\left(1-e^{-500r}\right)\bar{a}_z \cdot r\,dr\,d\phi\,\bar{a}_z$

$$= 20\left(1-e^{-500r}\right) r\,dr\,d\phi$$

$$I = 20\int_0^{2\pi}\int_0^{0.004}\left(1-e^{-500r}\right) r\,dr\,d\phi$$

$$= 40\pi\int_0^{0.004}\left(r - re^{-500r}\right)dr$$

$$= 40\pi\left[\left(\frac{r^2}{2}\right)_0^{0.004} - \left\{\frac{e^{-500r}}{(-500)^2}(-500r-1)\right\}_0^{0.004}\right]$$

$$= 40\pi\left[8\times10^{-6} - \frac{1}{500^2}\left\{e^{-2}(-2-1) - e^{-0}(-1)\right\}\right]$$

$$= 40\pi\left[8\times10^{-6} - \frac{e^{-2}}{500^2}\{(-3)+1\}\right]$$

$$= 40\pi\left[8\times10^{-6} - \frac{e^{-2}}{500^2}(-3+1)\right]$$

$$= 40\pi\left[8\times10^{-6} - \frac{0.135}{25\times10^{4}}\times2\right]$$

$$= 40\pi\left[8\times10^{-6} - 0.010\times10^{-4}\right]$$

$$= 40\pi\left[8\times10^{-6} - 1\times10^{-6}\right]$$

$$= 40\pi\times7\times10^{-6}$$

$$= 0.879 \text{ mA } \textbf{Ans.}$$

Example 5.11. Determine the current crossing the portion of the x = 0 plane defined by $-\dfrac{\pi}{4} \le y \le \dfrac{\pi}{4}$ m and $-0.01 \le z \le 0.01$ m if J = 100 cos 2y a_x.

Solution. Since $dI = J \cdot ds$

$$dI = 100\cos 2y\bar{a}_x \cdot dy\,dz\bar{a}_x = 100\cos y\,dy\,dz$$

$$I = \int_{-0.01}^{0.01}\int_{-\pi/4}^{\pi/4} 100\cos 2y\,dy\,dz$$

$$= \int_{-\pi/4}^{\pi/4} [z100\cos 2y\,dy]_{-0.01}^{0.01}$$

$$= 0.02\int_{-\pi/4}^{\pi/4} 100\cos 2y dy = \frac{2}{2}[\sin 2y]_{-\pi/4}^{\pi/4}$$

$= 2A$ **Ans.**

Example 5.12. The scalar potential field is given by V = 500 $(x^2 - y^2)$ and a point $P(2, -1, 3)$ is assumed to lie on a conductor to free space boundary. Determine the equation of the conductor surface and also, the potential at point *P*, the electric field intensity, electric flux density *D* and surface charge density ρ_s at *P*.

Solution. The potential at point *P* is given as

$$V_P = 500[2^2 - (-1)^2] = 500 \times 3 = 1500 \text{ volts}$$

Since the conductor is an equipotential surface, the potential at the entire surface of the conductor is same *i.e.* 1500 volts. If the conductor is solid, the electric field intensity inside the conductor is zero *i.e.* $E = 0$ inside. The equation of the conductor surface is

$$1500 = 500\,(x^2 - y^2) \text{ or } x^2 - y^2 = 3$$

which is an equation to a hyperbola.

Assume that the conductor lies to the right and above the equipotential surface at *P* whereas the free space is down and to the left of the conductor surface.

We find *E* by using the gradient operation.

$$E = -500\nabla\left(x^2 - y^2\right) = -1000x\bar{a}_x + 1000y\bar{a}_y$$

At point *P* $\qquad x = 2,\ y = -1$

Hence $\qquad E_p = -2000\bar{a}_x - 1000\bar{a}_y \text{ V/m}$

Since $D = \varepsilon_0 E$ we have

$$D_p = -17.708\bar{a}_x - 8.854\bar{a}_y \text{ nC/m}^2$$

Since, the D_p is negative along $\bar{a}_x$ and $\bar{a}_y$, this shows that the field is directed downward and to the left at *P*, it is normal to the equipotential surface. Therefore,

$$D_n = \left|D_p\right| = \sqrt{313.57 + 78.39} = 19.8 \text{ nC/m}^2$$

Thus, the surface charge density at *P* is 19.8 nC/m^2 **Ans.**

Example 5.13. The electric field intensity at a point on a conductor surface is given by $0.2\bar{a}_x - 0.4\bar{a}_y - 0.3\bar{a}_z$ V/m. Determine the surface charge density at the point.

Solution. Since $\rho_s = D = \varepsilon_0 |E|$

$$\rho_s = 8.854 \times 10^{-12} \sqrt{0.04 + 0.16 + 0.09} = 8.854 \times 10^{-12} \times 0.5385$$

$$= 4.768 \text{ pC/m}^2 \quad \textbf{Ans.}$$

Example 5.14. A conductor of uniform cross section and 100 m long has a voltage drop of 1.5 V and a current density of 4.5×10^5 A/m^2. Determine the conductivity of the material in the conductor.

Solution. The electric field intensity $E = \dfrac{V}{l} = \dfrac{1.5}{100}$

$$= 1.5 \times 10^{-2} \text{ V/m.}$$

Since $J = \sigma E$

$$4.5 \times 10^5 \times 1.5 \times 10^{-2} = 6.75 \times 10^3 \text{ S/m} \quad \textbf{Ans.}$$

Example 5.15. A long copper conductor has a circular cross section of diameter 4 mm and carries a current of 20 A. Each second, what percent of the conductor electrons must leave (to be replaced by others) a 100 mm length. Avogadro's number is $N = 6.02 \times 10^{26}$ atoms/kmol. Specific gravity of copper is 8.96 and atomic weight is 63.54. Assume one conduction electron per atom.

Solution. No. of atoms per unit volume

$$\frac{6.02 \times 10^{26}}{63.54} \times 8.96 \times 10^3 = 8.49 \times 10^{28} \text{ electron/m}^3$$

The volume of the wire $= \dfrac{\pi}{4} \times \left(4 \times 10^{-3}\right)^2 \times 0.1 = 1.256 \times 10^{-6} \text{ m}^3$

Hence no. of electrons $= 1.256 \times 10^{-6} \times 8.49 \times 10^{28} = 10.66 \times 10^{22}$

Since charge on an electron is 1.6×10^{-19} C, the electron required for 20 A or 20 C/sec

$$= \frac{20}{1.6 \times 10^{-19}} = 12.5 \times 10^{19} \text{ electron}$$

Hence percent of electrons leaving the 100 mm length is

$$\frac{12.5 \times 10^{19}}{10.66 \times 10^{22}} \times 100 = 0.1172\% \text{ per sec. } \textbf{Ans.}$$

Example 5.16. Determine the current that would result if all the conduction electrons in a 2 cm cube of aluminium passed a specified point 2 sec. Density of aluminium is 2.7×10^3 kg/m^3 and atomic weight is 26.98 kg/k.mol. Assume one conduction electron per atom.

Solution. No. of electron/m^3 $= \dfrac{6.02 \times 10^{26}}{2.7 \times 10^3} \times 2.7 \times 10^3 = 6.02 \times 10^{28}$

Total no. of electron in 2 cm cube $= 6.02 \times 10^{28} \times 8 \times 10^{-6}$

$$= 48.16 \times 10^{22}$$

Hence total charge = $48.16 \times 10^{22} \times 1.6 \times 10^{-19} = 77 \times 10^3$ C

Hence the current $= \frac{Q}{t} = \frac{77 \times 10^3}{2} = 38.5$ kA **Ans.**

Example 5.17. Determine the conductivity of n-type germanium at 300 K, assuming one donor atom in each 10^8 atoms. The density and atomic weight of Ge are 5.32×10^3 kg/m^3 and 72.6 kg/kmol respectively. Assume intrinsic concentration n_i for Ge at 300 K to be 2.5×10^{19}/m^3.

Solution. The carriers in the n-type semiconductor material are electrons. 1k mol of a substance has 6.02×10^{26} atoms, the carrier density is given by

$$N_c = \frac{6.02 \times 10^{26}}{72.6} \times 5.32 \times 10^3 \times \frac{1}{10^8}$$

$$= 4.41 \times 10^{20} \text{ electrons/m}^3$$

According to mass-action law $n_i^2 = N_e N_h$

$$\text{Hence } N_b = \frac{(2.5 \times 10^{19})^2}{4.41 \times 10^{20}} = 1.417 \times 10^{18} \text{ moles/m}^3$$

Since $N_e >> N_h$, the conductivity will be controlled by the donated electrons. The mobility of electrons at 300 K is 0.38 m^2/v.s. Hence conductivity σ is given as

$$\sigma = N_e e \mu_e = 4.41 \times 10^{20} \times 1.6 \times 10^{-19} \times 0.38 = 2.681 \text{ S/m } \textbf{Ans.}$$

Example 5.18. Determine the image charges for a point charge Q midway between two infinite, grounded conducting plates as shown in Fig. E 5.18

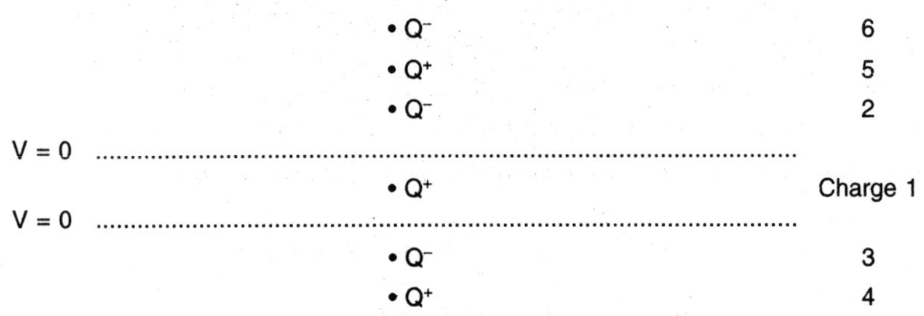

Fig. E 5.18

Solution. The image of charge 1(Q^+) through the copper plate gives charge 2(Q^-) such that the upper plate is at zero potential. However with charge 1 and 2, the potential of the lower plate is no longer zero. Hence we image charge 1 and 2 across the lower plate to produce charges 3 and 4. Now the potential along the position of lower plate is zero but the potential of the upper plate is not zero, so we image across this plate to produce charges 5 and 6 and so on. This, therefore, leads to an infinite set of image charges having alternating signs and separation equal to the plate separation. The situation is similar to one when an object is placed between two parallel plane mirrors. One would see an infinite number of images.

Example 5.19. Determine the capacitance per unit length of a single conductor (infinitely long) placed at a height of h above the ground using method of images.

Solution. The image of the linear charge above the ground h will be below the zero potential ground at a depth of h units.

Since the charges are linear, the electric field intensity due to two charges at point P is

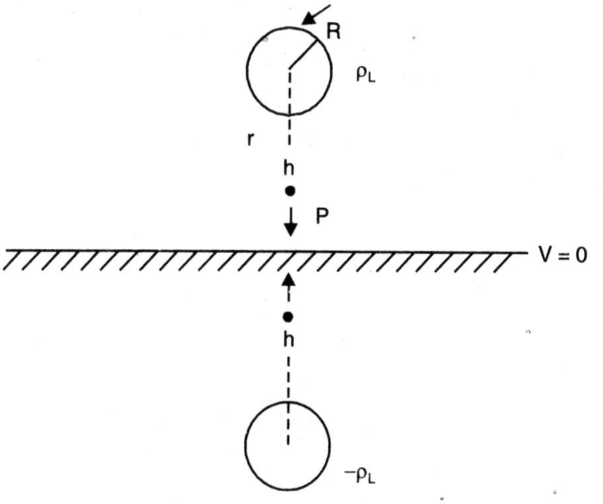

Fig. E 5.19

$$E = \frac{\rho_L}{2\pi\varepsilon_0 r} + \frac{\rho_L}{2\pi\varepsilon_0 (2h-r)}$$

Hence potential difference between the two conductors is

$$V = \int E \cdot dl = \frac{\rho_L}{2\pi\varepsilon_0} \int_R^{2\pi-R} \left(\frac{1}{r} + \frac{1}{2h-r}\right) dr$$

$$= \frac{\rho_L}{2\pi\varepsilon_0} \left[\ln r - \ln(2h-r)\right]_R^{2h-R}$$

$$= \frac{\rho_L}{2\pi\varepsilon_0} \left[\ln(2h-R) - \ln R - \ln(2h-2h+R) + \ln R\right]$$

$$= \frac{\rho_L}{2\pi\varepsilon_0} \left[\ln(2h-R) - \ln R\right]$$

$$V = \frac{\rho_L}{2\pi\varepsilon_0} \ln \frac{2h-R}{R}$$

$$C = \frac{\rho_L}{V} = \frac{2\pi\varepsilon_0}{\ln(2h-R)/R} \text{ F/m} \quad \textbf{Ans.}$$

Example 5.20. Two slabs of dielectric materials have a common interface in the xy plane of rectangular coordinates. The electric field intensity in slab 2 is given by

$$E_1 = 3\bar{a}_x + 4\bar{a}_y - 5\bar{a}_z \text{ V/m for } z \geq 0$$

$\varepsilon_2 = 4$ is given determine E_1. Also determine D_1 in material 1 which has $\varepsilon_1 = 2$

Solution. Since the slabs are placed in the *xy* plane, therefore components of *E* along $\bar{a}_x$ and $\bar{a}_y$ are same at the interface, however the normal component along $\bar{a}_z$ will have different values in the two dielectric *i.e.* along $\bar{a}_z$. Also we know that at the interface the normal components are related as follows

$$\varepsilon_1 E_{1n} = \varepsilon_2 E_{2n} \quad \text{or} \quad E_{1n} = \frac{\varepsilon_2}{\varepsilon_1} E_{2n}$$

$$E_1 = 3\bar{a}_x + 4\bar{a}_y - 5 \times 2\bar{a}_z = 3\bar{a}_x + 4\bar{a}_y - 10\bar{a}_z \text{ V/m}$$

and since $\quad D_1 = \varepsilon_0 \varepsilon_1 E_1$

Hence $\quad D_1 = \varepsilon_0 \left(6\bar{a}_x + 8\bar{a}_y - 20\bar{a}_z\right) \text{ C/m}^2$ **Ans.**

PROBLEMS

P 5.1 Explain what you mean by drift velocity when referred to a conducting media. Derive an expression for drift velocity.

P 5.2 Differentiate between electronic current and conventional electric current. Derive expression for current density in terms of electric field intensity and the parameters of the material of the conductor.

P 5.3 Explain and derive point form of continuity equation for electric currents. Also write down the integral form of the continuity equation for currents.

P 5.4 Derive expressions for conductivity of a (*i*) Conductor (*ii*) Semiconductor (*iii*) Plasma or gases.

P 5.5 Explain and differentiate between conductors, semiconductors and insulators in terms of energy band theory.

P 5.6 Explain with neat diagrams and analytical expressions how, the electric field intensities and electric flux densities in the two media are related at the boundary of the two media when the media are (*i*) Conductor-free space (*ii*) Dielectric and free space (*iii*) Dielectric and dielectric with different relative permittivities.

P 5.7 Explain clearly "Method of Images" when referred to electrostatic fields. Derive an expression for capacitance per unit length of an infinite long conductor placed on poles at the height of '*h*' from the ground. The radius of the cylindrical conductor is '*r*'.

P 5.8 Determine the number of conduction electrons in a 1-metre cube of copper if $\sigma = 58 \times 10^6$ S/m and mobility $\mu = 3.2 \times 10^{-3}$ m^2/V.S. Atomic weight of copper 63.54 and the density is 8.96×10^3 kg/m^3. **Ans.** 1.13×10^{29}

P 5.9 Determine the magnitude of the current density in a sample of sliver for which $\sigma = 6.17 \times 10^7$ S/m and $\mu_e = 5.6 \times 10^{-3}$ m^2/v.s. if (*i*) the drift velocity is 1.5×10^{-6} m/sec. (*ii*) the electric field intensity is 1 mV/m. **Ans.** 16.5 kA/m^2, 61.7 kA/m^2

P 5.10 Refer to Fig. P 5.10 where medium 1 is carbon and medium 2 is copper. If α_1 for *J* in carbon is 10°, determine α_2 for *J* in copper. σ for carbon is 3×10^4 S/m, σ(copper) $= 5.7 \times 10^7$ S/m.

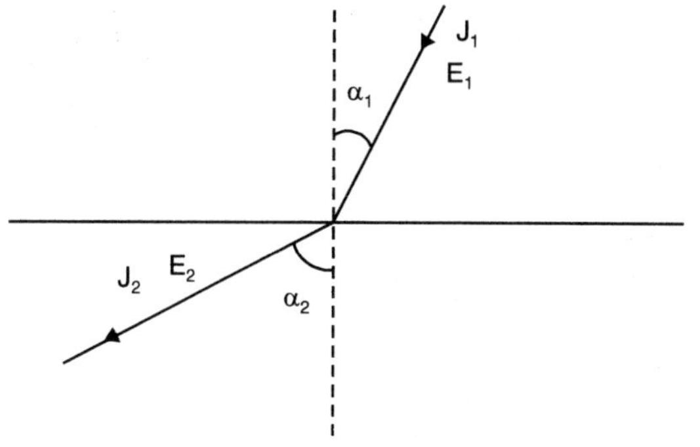

Fig. P 5.10

Ans. 89.8°

P 5.11 A boundary is formed between medium 1 a perfect conductor and medium 2 a perfect dielectric. Determine the conditions on the normal and tangential component of D and E at the interface of the media. Assume surface charge density ρ_s on the interface.

Ans. $E_1 = D_1 = 0,\ E_{t_2} = 0,\ E_{n_2} = \dfrac{\rho_s}{\varepsilon_0},\ D_{t_2} = 0,\ D_{n_2} = \rho_s$

P 5.12 In a cylindrical conductor of radius 2 mm, the current density J varies with the distance from the axis according to $J = 10^3\, e^{-400r}\ \text{A/m}^2$
Determine the current through the conductor.

Ans. 7.51 mA

P 5.13 A point charge Q is located at a distance h above an infinite conducting plate. Using method of images, determine the electric flux density normal to the plane and hence show that the surface charge density on the plane is $\rho_s = -\dfrac{Qh}{2\pi r^3}$ where r is the distance from the charge Q to the point on the plane.

P 5.14 The electric current density J is given as

$$J = 1000 \sin\theta \bar{a}_r\ \text{A/m}^2$$

in spherical coordinate, determine the current crossing the spherical shell r = 0.02 m.

Ans. 3.95 A

P 5.15 Determine the concentration of holes N_h in p-type germanium where $\sigma = 10^4$ S/m and the hole mobility is μ_h = 0.18 m²/VS.

Ans. 3.47×10^{23}

P 5.16 At time $t = 0$, excess free charge of volume density ρ_0 is injected into the interior of a conductor. Determine the time for the charge density to decrease to $\dfrac{1}{e}$ of its initial value. Assume uniform conductor properties.

Ans. $\tau = \varepsilon/\sigma$

P 5.17 Two perfectly conducting cylindrical surfaces of length l are located at $\rho = 3$ and $\rho = 5$ cms. The total current passing radially outward through the medium between the cylinders is 3 A d.c.

(*i*) Determine the voltage and resistance between the cylinders, and $\bar{E}$ in the region between the cylinders, if a conducting material having $\sigma = 0.05$ S/m is present for $3 < \rho < 5$ cm.

Ans. $E = \dfrac{9.55}{\rho_a l}\bar{a}_p$ V/m, $V = \dfrac{4.88}{L}$ volts, $R = \dfrac{1.63}{l}\,\Omega$

(where l is the length of the cylinder)

P 5.18 The current density J in a conductor is

$$J = \left(\frac{100}{r^2}\cos\theta\right)\bar{a}_\theta$$

in spherical coordinates. Determine the current crossing the conical strip $\theta = \dfrac{\pi}{4}$, $0.001 \le r \le 0.08$ m. **Ans.** 13.8 kA

Chapter 6

Dielectric Materials and Capacitance

6.1 INTRODUCTION

The basic difference between conductors and dielectric materials is that in dielectric materials there are no free charges which can be transferred within them to produce conduction current. In fact, all charges are confined to molecular or lattice structures by Coulomb's forces. An external field has the effect of displacing the charges slightly leading to the formation of what are known as electric dipoles. This formation (magnitude) depends upon the relative permittivities of the dielectric materials. The charge displacement results into an energy storage mechanism that becomes useful in the construction of capacitors. Further, the response of the dielectric to time-varying fields, particularly electromagnetic waves is extremely important in order to understand many physical phenomenon and to the development of useful devices which will be considered in the following chapters.

The charges within the dielectric are known as bound charges in contrast to free charges in a conductor or a conducting material.

6.2 THE CHARACTERISTICS OF DIELECTRIC MATERIALS

Whether the dielectric materials are in solid or liquid or gaseous form, these have one common characteristic of their ability to store electric energy. This storage takes place by means of a drift in the relative positions of the internal bound positive and negative charges against the normal molecular and atomic force.

A simplified theory of dipole formation can be obtained by treating an atom of the dielectric as two superimposed positive and negative charge regions as shown in Fig. 6.1(*a*).

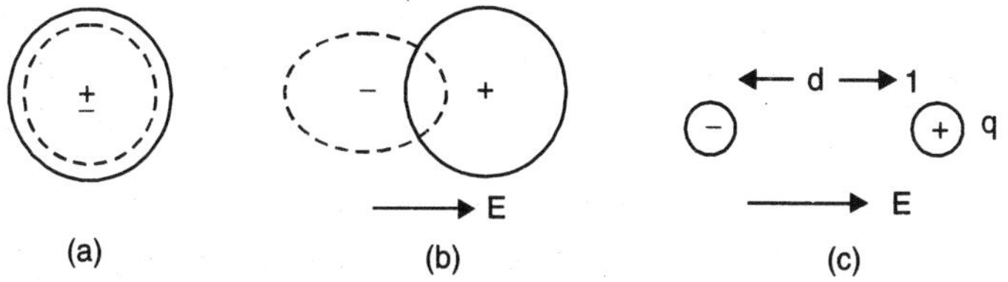

Fig. 6.1

When an electric field E is applied to a dielectric material a force qE is exerted on the positive charge which moves in the direction of the electric field E whereas a force $-qE$ is experienced by the negative charge which moves in a direction opposite to the applied field E. These opposite forces cause the two charges to move apart a distance d to a new equilibrium position where their attractive interaction balances the applied electric force. This separation of ⊕ and ⊖ electric poles creates an electric dipole whose dipole moment is defined as $p = qd$, its magnitude equals the charge times the separation distance d and its vector direction is from the negative to the positive charge as shown in Fig. 6.1(c). For most materials, the charge regions will return to their original superimposed position when the applied field is removed. The displacement against a restraining force (electric field E) is analogous to lifting a weight or stretching a spring and represents potential energy. The source of the energy is the electric field E (external field), the motion of the shifting charges resulting in a transient current through a battery which is producing the field.

In a dielectric material (polar molecules), normally the dipoles are oriented in a random way, throughout the interior of the material and the action of the external field E is to align these molecules to some extent in the same direction. A sufficiently strong field may even produce an additional displacement between the positive and negative charges.

If there are N dipoles per unit volume and we consider a differential volume Δv, then these are $N\Delta v$ dipoles and the total dipole moment is obtained by the vector sum.

$$p_{\text{total}} = \sum_{i=1}^{N\Delta v} P_i \qquad \text{...(6.1)}$$

If all the dipoles are aligned in the same general direction p_{total} may have a significant value. However, a random orientation may lead p_{total} to zero value. We may define the polarization P as the dipole moment per unit volume with units of coulombs per square metre.

$$P = \underset{\Delta v \to 0}{\text{Lt}} \frac{1}{\Delta v} \sum_{i=1}^{N\Delta v} P_i \qquad \text{...(6.2)}$$

Polarization is the ability of the dielectric material to produce electric dipoles.

This suggests a smooth and continuous distribution of electric dipole moments throughout the volume, which, of course is not true. In the macroscopic view, however, polarization P can account for the increase in the electric flux density, the equation being

$$D = \varepsilon_0 E + P \qquad \text{...(6.3)}$$

This equation permits P and E to have different directions, as they do in certain crystalline dielectrics. In an isotropic linear material P and E are parallel at each point which is expressed by

$$P = x_e\, \varepsilon_0 E \text{ (isotropic material)} \qquad \text{...(6.4)}$$

Where the electric susceptibility x_e is a dimensionless constant. Substituting equation (6.4) in equation (6.3) we have

$$D = \varepsilon_0(1 + x_e)E = \varepsilon_0\, \varepsilon_r E \qquad \text{...(6.5)}$$

where $\varepsilon_r = 1 + x_e$ which is also a pure number. Since $D = \varepsilon E$, $\varepsilon = \varepsilon_0 \varepsilon_r$ or $\varepsilon_r = \dfrac{\varepsilon}{\varepsilon_0}$ ...(6.6)

where ε_r is called the relative permittivity of the dielectric material.

Anisotropic material cannot be described in terms of a simple susceptibility or permittivity parameter. Instead we find that each component of D may be a function of every component of E_y and E, and $D = \varepsilon E$ becomes a matrix equation and E becomes a tensor quantity.

It is to be noted from equation (6.5) that ε and ε_0 relate $\bar{D}$ and $\bar{E}$, the two vector electric fields. This involves a change of magnitude and a change of units. The different units imply a relation between the two different physical aspects of the fields, as a force generator (E) and as a flux emanation (D).

Example 6.1. Determine the magnitudes of D and P for a dielectric material in which E = 200 kV/m and x_e = 5.0.

Solution. The relative permittivity of the dielectric material $\varepsilon_r = 1 + x_e = 1 + 5 = 6$

$$D = \varepsilon_0\, \varepsilon_r\, E = 8.854 \times 10^{-12} \times 6 \times 200 \times 10^3$$
$$= 10.624 \times 10^{-12} \times 10^6 = 10.624\ \mu\text{C/m}^2$$
$$P = x_e\, \varepsilon_0\, E = 5 \times 8.854 \times 10^{-12} \times 200 \times 10^3 = 8.854\ \mu\text{C/m}^2 \quad \textbf{Ans.}$$

Example 6.2. A slab of dielectric material has a relative permittivity of 4 and has a uniform electric flux density D of 10 nC/m^2. If the dielectric loss is zero determine (*i*) E (*ii*) P (*iii*) the average number of dipoles per cubic metre if average dipole moment is 10^{-28}C-m.

Solution. (*i*) $E = \dfrac{D}{\varepsilon_0 \varepsilon_2} = \dfrac{10 \times 10^{-9}}{8.854 \times 10^{-12} \times 4} = 282\ \text{V/m}.$ **Ans.**

(*ii*)
$$P = x_e\, \varepsilon_0\, E = 3 \times 8.854 \times 10^{-12} \times 282$$
$$= 7.49\ \text{nC/m}^2 \quad \textbf{Ans.}$$

(*iii*) Average number of dipoles/m$^3 = \dfrac{\text{dipole moment per unit volume}}{\text{average dipole moment}}$.

$$= \frac{7.49 \times 10^{-9}}{10^{-28}} = 7.49 \times 10^{19}/\text{m}^3 \quad \textbf{Ans.}$$

Example 6.3. Determine the polarization P in a dielectric material with $\varepsilon_r = 3.5$ if $D = 3 \times 10^{-7}\, \bar{a}\, \text{C/m}^2$. Assume the material to be homogeneous and isotropic.

Solution. Since the material is homogeneous and isotropic $P = x_e\, \varepsilon_0\, E$
and since $D = \varepsilon_0\, \varepsilon_r\, E$ and $x_e = \varepsilon_r - 1$

$$P = (\varepsilon_r - 1)\varepsilon_0 \frac{D}{\varepsilon_0 \varepsilon_r} = \frac{\varepsilon_r - 1}{\varepsilon_r} D$$
$$= \frac{3.5 - 1}{3.5} \times 3 \times 10^{-7}\, \bar{a}$$
$$= \frac{2.5}{3.5} \times 3 \times 10^{-7}\, \bar{a} = 2.143 \times 10^{-7}\, \bar{a}\, \text{C/m}^2 \quad \textbf{Ans.}$$

Example 6.4. Determine the value of E in a dielectric material for which the electric susceptibility is 3.8 and $P = 2.5 \times 10^{-7}\, \bar{a}\, \text{C/m}^2$. Assume the dielectric to be isotropic.

Solution. Since the material is isotropic P and E are in the same direction.

For isotropic material

$$P = x_e\, \varepsilon_0\, E$$

Hence $$E = \frac{P}{x_e \varepsilon_0} = \frac{2.5 \times 10^{-7}}{3.8 \times 8.854 \times 10^{-12}} = 0.0743 \times 10^5 \text{ V/m}$$

$= 7.43 \times 10^3$ V/m **Ans.**

6.3 BOUNDARY CONDITIONS FOR PERFECT DIELECTRIC MATERIALS

Consider the interface between two dielectrics having permittivities ε_1 and ε_2 in regions 1 and 2 respectively as shown in Fig. 6.2. We first consider the tangential components.

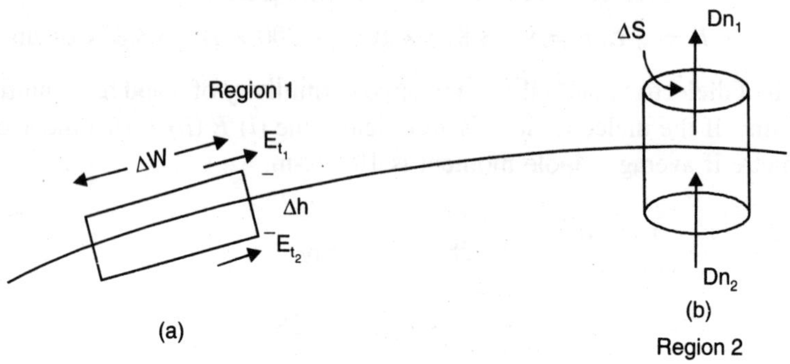

Fig. 6.2 Boundary between the perfect dielectric materials

We know that line integral of the static electric field around a closed path is zero $\therefore \oint E \cdot dl = 0$.

Using this expression we have from Fig. 6.2(*a*)

$$E_{t_1} \Delta W - E_{t_2} \Delta W = 0$$

or $$E_{t_1} = E_{t_2} \quad \ldots(6.7)$$

The differential contribution to the line integral by the normal component of E along the length Δh tends to zero as we approach the interface as then $\Delta h \to 0$.

Equation (6.7) shows that the potential difference between any two points on the boundary that are separated by a differential distance ΔW is same immediately above or below the boundary.

If the tangential component of electric field intensity is continuous at the boundary, then the tangential component of electric flux density D is discontinuous at the boundary as

$$\frac{D_{t_1}}{\varepsilon_1} = E_{t_1} = E_{t_2} = \frac{D_{t_2}}{\varepsilon_2}$$

or $$D_{t_1} = \varepsilon_1 E_{t_1} \text{ and } D_{t_2} = \varepsilon_2 E_{t_2}$$

Since $t_1 \neq t_2$ $$D_{t_1} \neq D_{t_2} \text{ and } \frac{D_{t_1}}{D_{t_2}} = \frac{\varepsilon_1}{\varepsilon_2} \quad \ldots(6.8)$$

Next we consider the normal components of E and D at the boundary by applying Gauss' law to the differential cylinder as shown in Fig. 6.2(*b*). The normal component of electric flux density will leave the electric flux from the top differential surface and will enter the lower differential surface ΔS. Hence the net flux enclosed by the surfaces is

$$D_{n_1}\Delta S - D_{n_2}\Delta S = \rho_s \Delta S \qquad \ldots(6.9)$$

where ρ_s is the surface charge density on the interface of the two dielectrics.

From equation (6.9) we have

$$D_{n_1} - D_{n_2} = \rho_s$$

which shows that the normal component of electric flux is discontinuous at the boundary. It is to be noted that ρ_s is not due to bound charges of the dielectric as we are taking the polarization of the dielectric into account by using different relative permittivities other than unity *i.e.* instead of considering bound charges in free space we are using an increased permittivity. Since no free charge is available in perfect dielectrics. The free charge under ρ_s is the charge deliberately placed on the interface between the two dielectrics. However, if no charge is placed on the interface of the two dielectrics $\rho_s = 0$ and hence

$$D_{n_1} = D_{n_2}$$

i.e. in absence of free charge on the surface between the two dielectrics, the normal component of the electric flux densities is continuous. Therefore,

$$\varepsilon_1 E_{n_1} = \varepsilon_2 E_{n_2}$$

and normal components of electric flux densities is discontinuous.

Next we consider D and E when they are incident at the boundary at angles other than zero (tangential) and 90° (normal) say D_1 and E_1 make an angle θ_1 with a normal to the surface as shown in Fig. 6.3.

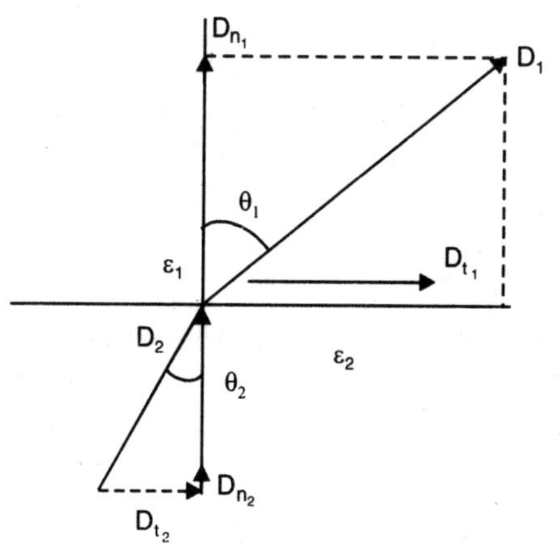

Fig. 6.3 Refraction of *D* at a dielectric interface

Since no interface charge is present the normal components of D are continuous. Hence

$$D_{n_1} = D_1 \cos\theta_1 = D_2 \cos\theta_2 = D_{n_2} \qquad \ldots(6.10)$$

The ratio of tangential components is given by equation 6.8 as

$$\frac{D_{t_1}}{D_{t_2}} = \frac{D_1 \sin\theta_1}{D_2 \sin\theta_2} = \frac{\varepsilon_1}{\varepsilon_2}$$

or

$$\varepsilon_2 D_1 \sin\theta_1 = \varepsilon_1 D_2 \sin\theta_2 \qquad \ldots(6.11)$$

Using equation (6.10) and (6.11) we have

$$\frac{\varepsilon_2 D_1 \sin\theta_1}{D_1 \cos\theta_1} = \frac{\varepsilon_1 D_2 \sin\theta_2}{D_2 \cos\theta_2}$$

or

$$\varepsilon_2 \tan\theta_1 = \varepsilon_1 \tan\theta_2$$

or

$$\frac{\tan\theta_1}{\tan\theta_2} = \frac{\varepsilon_1}{\varepsilon_2} \qquad ...(6.12)$$

In Fig. 6.3 we have assumed $\varepsilon_1 > \varepsilon_2$ and hence $\theta_1 > \theta_2$. Since we are considering isotropic media $D = \varepsilon E$, hence the direction of E and D are same *i.e.* E_1 is along D_1 and E_2 is along D_2.

Using equations (6.10) and (6.11) magnitudes of D_2 and E_2 can be obtained in terms of D_1, E_1 and θ_1

$$\text{magnitude of } D_2 = D_1\sqrt{\cos^2\theta_1 + \left(\frac{\varepsilon_2}{\varepsilon_1}\right)^2 \sin^2\theta_1} \qquad ...(6.13)$$

and

$$\text{magnitude of } E_2 = E_1\sqrt{\sin^2\theta_1 + \left(\frac{\varepsilon_1}{\varepsilon_2}\right)^2 \cos^2\theta_1} \qquad ...(6.14)$$

From the above analysis following observations can be made:

(*i*) D is larger in the region of larger permittivity unless $\theta_1 = \theta_2 = 0$ where the magnitude remains unchanged.

(*ii*) E is larger in the region of smaller permittivity unless $\theta_1 = \theta_2 = 90°$ where its magnitude is uncharged.

(*iii*) If we know the field (E, D and θ) on one side of the interface between two dielectrics, we can find out the field on the other side using the relation derived above.

Example 6.5. The electric field intensity in medium 1 is given by $E_1 = 3\bar{a}_x - 4\bar{a}_y + 5\bar{a}_z$ V/m at the charge free dielectric interface as shown in Fig. E 6.5. Determine D_2 and the angles θ_1 and θ_2.

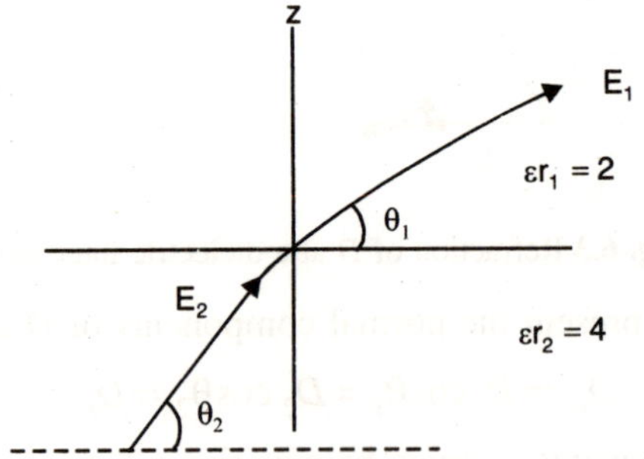

Fig. E 6.5

Solution. The interface is $z = 0$ constant. The x, y components of electric field E_1 are tangent to x = constant or y = constant plane and the z–components are normal. Since the tangential components of E and normal components of D are continuous at the boundary,

$$E_1 = 3\bar{a}_x - 4\bar{a}_y + 5\bar{a}_z$$

$$E_2 = 3\bar{a}_x - 4\bar{a}_y + Ez_2\bar{a}_z = 3\bar{a}_x - 4\bar{a}_y + 2.5\bar{a}_z$$

$$D_1 = \varepsilon_0\varepsilon_1 E_1 = 6\varepsilon_0\bar{a}_x - 8\varepsilon_0\bar{a}_y + 10\varepsilon_0\bar{a}_z$$

$$D_2 = D_{x_2}\bar{a}_x + D_{y_2}\bar{a}_y + 10\varepsilon_0\bar{a}_z$$

Now $$D_2 = \varepsilon_0\varepsilon_{r_2}E_2$$

$$D_{x_2}\bar{a}_x + D_{y_2}\bar{a}_y + 10\varepsilon_0\bar{a}_z = 3\varepsilon_0\varepsilon_{r_2}\bar{a}_x - 4\varepsilon_0\varepsilon_{r_2}\bar{a}_y + \varepsilon_0\varepsilon_{r_2}E_{z_2}\bar{a}_z$$

Comparing the coefficients

$$D_{x_2} = 12\varepsilon_0,\ D_{y_2} = -16\varepsilon_0,\ \ \varepsilon_0 \cdot 4 \cdot E_{z_2} = 10\varepsilon_0,\ E_{z_2} = 2.5$$

$$D_2 = \varepsilon_0\left[12\bar{a}_x - 16\bar{a}_y + 10\bar{a}_z\right]$$

Component of E_1 along z axis is $E_1 \cdot \bar{a}_z = |E|\cos(90 - \theta_1)$

$$5 = \sqrt{50}\cos(90 - \theta_1)$$

$$\cos(90 - \theta_1) = \frac{5}{\sqrt{50}} = 0.707$$

$$90 - \theta_1 = 45° \text{ or } \theta_1 = 45° \textbf{ Ans.}$$

Similarly $$E_2 \cdot \bar{a}_z = |E_2|\cos(90 - \theta_2)$$

$$2.5 = \sqrt{31.25}\sin\theta_2$$

$$\sin\theta_2 = 0.447 \text{ or } \theta_2 = 26.55° \textbf{ Ans.}$$

Example 6.6. The electric flux density in medium 1 with ε_1 = 3.2 is given by $D_1 = -30\bar{a}_x + 40\bar{a}_y + 50\bar{a}_z\ \text{nC/m}^2$. The $\varepsilon_2 = 2$. Determine *(i)* D_n, *(ii)* $\bar{D}_{t_1}$, *(iii)* D_{t_1}, *(iv)* D_1, *(v)* θ_1 *(vi)* P_1, *(vii)* D_{n_2}, *(viii)* Dt_2, *(ix)* $\bar{D}_2$, *(x)* P_2, *(xi)* θ_2.

Solution. Since the slabs are lying in the xy plane, the tangent components of D_1 are along $\bar{a}_x$ and $\bar{a}_y$ whereas the normal component is along $\bar{a}_z$.

(i) $D_{n_1} = 50\ \text{nC/m}^2$ along $\bar{a}_z$

(ii) $\bar{D}_{t_1}$ *i.e.* the tangent components of D_1 and along $\bar{a}_x$ and $\bar{a}_y$

Hence $\bar{D}_{t_1} = -30\bar{a}_x + 40\bar{a}_y\ \text{nC/m}^2$

(iii) $|D_{t_1}|$ *i.e.* the magnitude of D_{t_1} is $\sqrt{30^2 + 40^2} = 50\ \text{nC/m}^2$

(iv) $|D_1|$ *i.e.* the magnitude of D_1 is $\sqrt{30^2 + 40^2 + 50^2} = 70.7\ \text{nC/m}^2$

(*v*) $D_1 \cos\theta_1 = 50$ or $\cos\theta_1 = \dfrac{50}{70.7}$ or $\theta_1 = 45°$

(*vi*) $P_1 = x_1\varepsilon_0 E_1 = x_1\varepsilon_0 \dfrac{D_1}{\varepsilon_0\varepsilon_r} = \dfrac{x_1}{\varepsilon_r} D_1 = \dfrac{2.2}{3.2} D_1 = 0.6875\ D_1$

$$= -20.6\bar{a}_x + 27.5\bar{a}_y + 34.4\bar{a}_z \quad \textbf{Ans.}$$

As mentioned in the text earlier, if quantities in one dielectric material are known, the quantities in the second dielectric can be obtained using various relations given in the last section. From equation (6.10) $D_{n_1} = D_{n_2}$

(*vii*) Hence $\bar{D}_{n_2} = 50\bar{a}_z\ \text{nC/m}^2$

(*viii*) From (6.11)

$$\frac{D_{t_1}}{D_{t_2}} = \frac{\varepsilon_1}{\varepsilon_2} \quad \text{or} \quad D_{t_2} = D_{t_1} \cdot \frac{\varepsilon_2}{\varepsilon_1}$$

$$D_{t_1} = -30\bar{a}_x + 40\bar{a}_y$$

Hence $$D_{t_2} = \frac{2}{3.2}\left(-30\bar{a}_x + 40\bar{a}_y\right)$$

$$= -18.75\bar{a}_x + 25\bar{a}_y \quad \textbf{Ans.}$$

(*ix*) $\bar{D}_2 = -18.75\bar{a}_x + 25\bar{a}_y + 50\bar{a}_z$ **Ans.**

(*x*) $P_2 = x_2\varepsilon_0 E_2 = x_2\varepsilon_0 \dfrac{D_2}{\varepsilon_0\varepsilon_2} = \dfrac{x_2}{\varepsilon_2} D_2$

$$P_2 = \left(-18.75\bar{a}_x + 25\bar{a}_y + 50\bar{a}_z\right) \cdot \frac{1}{z}$$

$$= -9.38\bar{a}_x + 12.5\bar{a}_y + 25\bar{a}_z \quad \textbf{Ans.}$$

(*xi*) $\dfrac{\tan\theta_1}{\tan\theta_2} = \dfrac{\varepsilon_1}{\varepsilon_2} = \dfrac{3.2}{2} = 1.6$

Since $\theta_1 = 45°$ $\tan\theta_1 = 1$

$1.6\tan\theta_2 = 1$ or $\tan\theta_2 = 0.625$

or $\theta_2 = 32°$ **Ans.**

6.4 CAPACITANCE

Whenever two conducting bodies, could be in the form of two circular conductors, two conducting parallel plates or two concentric spheres, are separated by free space or a dielectric material, they constitute a capacitance between them. When a voltage V is applied between these conducting bodies a charge Q is established on the surface of one conductor and there is induced charge $-Q$ on the other conductor. By definition, the ratio of the absolute value of charge on one plate of the conductor to the absolute value of voltage is the capacitance of the system. In general terms we determine Q by the surface integral over the positive conductor and V is the work done in moving a unit positive charge from the negative to the positive surface *i.e.* against the direction of the electric field *i.e.*

$$C = \frac{\oint_S \varepsilon E \cdot ds}{-\oint E \cdot dl} \qquad \ldots(6.15)$$

It is to be noted that the capacitance is independent of the potential and the total charge as their ratio is constant. Consider a parallel plate capacitor shown in Fig. 6.4.

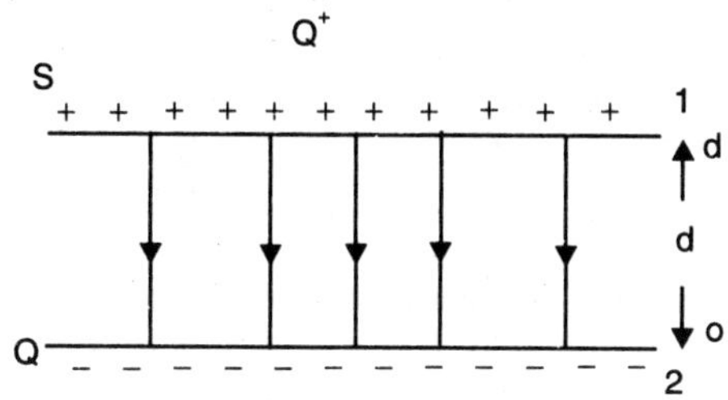

Fig. 6.4 Parallel plate capacitor

If we place charge Q^+ on plate 1 and Q^- on plate 2 which creates a flux field as shown. Flux line leave the +ve plate normally and terminate on plate 2 again normally. The D and E fields are thus set up. If we double the charges on each plate D and E would also be doubled and therefore, the potential difference would also be doubled and the ratio Q/V thus remains constant. This shows that the capacitance is a function only of the physical dimensions of the system of conductors and of the permittivity of the dielectric medium.

Capacitance is measured in terms of farads (F) where a farad is defined as one coulomb per volts. Common values of capacitance are in terms of μF or nF or pF as F is a very large value for the actual physical systems. We can apply this definition of capacitance to two conductor system in which the conductors are identical, infinite parallel plane with separation d. However, if we consider the distance d between the plates very small as compared to the length and width of the parallel plates, this could be approximated to an infinite parallel planes as the flux fringing effect at the corners of the parallel plates will be negligible and could be ignored for all practical purposes.

Let S be the surface area of each plate and Q be the charge on each plate, the electric flux density between the plates is

$D = \dfrac{Q}{S} = \rho_s$ and electric field intensity is $E = \dfrac{D}{\varepsilon} = \dfrac{\rho_s}{\varepsilon}$ and since it is uniform between the plates

$$V = E \cdot d = \frac{\rho_s}{\varepsilon} d = \frac{Q}{S\varepsilon} d$$

Hence
$$\frac{Q}{V} = \frac{S}{d} \varepsilon = C \text{ F} \qquad \ldots(6.16)$$

It can be seen that capacitance is independent of the voltage and the charge and depends only on the dimensions of the conductors and the permittivity of the medium between the plates.

6.4.1 Multiple Dielectric Capacitors

Consider the two dielectrics present in a capacitor with the interface parallel to D and E as shown in Fig. 6.5, the equivalent capacitance can be obtained by treating the arrangement on two capacitors in parallel. The voltage across each capacitance and the thickness of the dielectrics are same, hence $E_1 = E_2 = \frac{V}{d}$. Also these are tangential fields at the interface, they must be equal.

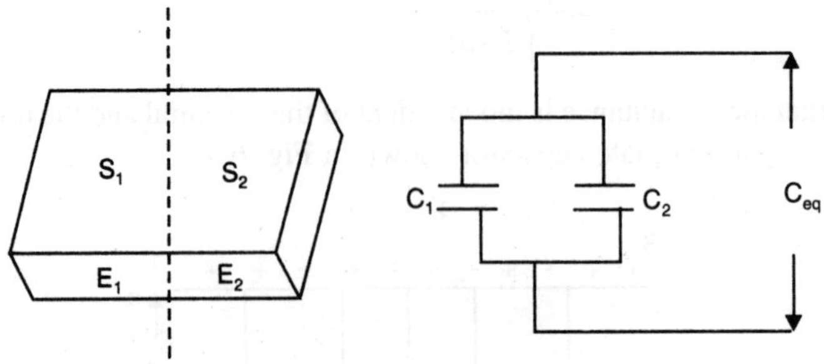

Fig. 6.5 Two dielectrics with interface || to D

Since $$C_1 = \frac{\varepsilon_1 S_1}{d} \text{ and } C_2 = \frac{\varepsilon_2 S_2}{d}$$

$$C = C_1 + C_2 = \frac{\varepsilon_1 S_1 + \varepsilon_2 S_2}{d} \quad \text{...(6.17)}$$

Next we consider two dielectrics are present such that the interface is normal to D and E as shown in Fig. 6.6

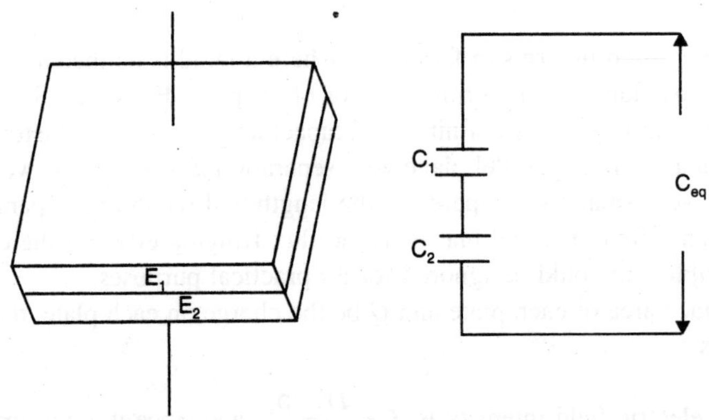

Fig. 6.6 Two dielectrics with interface normal to D

Since $$D_{n_1} = D_{n_2}$$

$$\varepsilon_1 E_1 = \varepsilon_2 E_2$$

The electric field intensities will be different *i.e.* $E_1 \neq E_2$ but E_1 and E_2 are uniform within their respective dielectrics.

If d_1 is thickness of material 1 and d_2 that of material 2.

Total voltage across the plates $V = E_1 d_1 + E_2 d_2$.

Now $$E_2 = E_1 \cdot \frac{\varepsilon_1}{\varepsilon_2} \text{ hence } V = E_1 d_1 + E_1 \frac{\varepsilon_1}{\varepsilon_2} d_2$$

or $$E_1 = \frac{V}{d_1 + \frac{\varepsilon_1}{\varepsilon_2} d_2} \qquad ...(6.18)$$

Since $$D_1 = \varepsilon_1 E_1 = \frac{V}{d_1 + \frac{\varepsilon_1}{\varepsilon_2} d_2} \cdot \varepsilon_1 = \frac{V}{\frac{d_1}{\varepsilon_1} + \frac{\varepsilon_1}{\varepsilon_1 \varepsilon_2} d_2} = \frac{V}{\frac{d_1}{\varepsilon_1} + \frac{d_2}{\varepsilon_2}}$$

Now $$Q = D_1 S = \frac{V_s \cdot S}{\frac{d_1}{\varepsilon_1} + \frac{d_2}{\varepsilon_2}}$$

or $$\frac{Q}{V} = C = \frac{1}{\frac{d_1}{\varepsilon_1 S} + \frac{d_2}{S \varepsilon_2}} = \frac{1}{\frac{1}{C_1} + \frac{1}{C_2}}$$

or $$\frac{1}{C} = \frac{1}{C_1} + \frac{1}{C_2} \text{ where } C_1 = \varepsilon_1 \frac{S}{d_1},\ C_2 = \varepsilon_2 \frac{S}{d_2}$$

Next we consider a situation when a thin conducting plate is placed on the interface between the two dielectrics. Here the flux lines emanating from the upper positive charge will terminate on the upper side of the thin sheet and induce negative charge on the sheet whereas the other side of the sheet will have positive charges. These charges will give rise to electric field emanating from it and terminating on to the lower plate. The capacitance thus remains unchanged, provided of course that the added conductor sheet is of negligible thickness. However, if the conducting sheet inserted is thick, it will increase the capacitance if the separation of the outer plates is kept constant and this is an example of a more general theorem which states that the replacement of any portion of the dielectric by a conducting body will cause increase in the capacitance.

Next we consider capacitance of a coaxial cable with inner radius a and outer b and length L. Let ρ_L be the charge per unit length. The electric field intensity within the dielectric is $E = \dfrac{\rho_L}{2\pi\varepsilon_0 r}$

Hence potential difference $-\int E \cdot dr$

$$V = -\int_b^a \frac{\rho_L}{2\pi\varepsilon r} dr = -\frac{\rho_L}{2\pi\varepsilon} \ln\frac{a}{b} = \frac{\rho_L}{2\pi\varepsilon} \ln\frac{b}{a}$$

Hence capacitance per unit length $= \dfrac{\rho_L}{V} = \dfrac{2\pi\varepsilon}{\ln\frac{b}{a}}$ F/m. ...(6.19)

Next we consider capacitance of two concentric spheres where the space between the two spheres is occupied by dielectric of permittivity ε. The radius of inner sphere is a and that of outer b units.

We charge the inner sphere with Q charges.

The electric field intensity anywhere between b and a is given as $E = \dfrac{Q}{4\pi\varepsilon_0 r^2}$

Hence
$$V = -\int_b^a \frac{Q}{4\pi\varepsilon_0 r^2}\,dr = +\left[\frac{Q}{4\pi\varepsilon_0}\cdot\frac{1}{r}\right]_b^a$$

$$= \frac{Q}{4\pi\varepsilon_0}\cdot\left(\frac{1}{a}-\frac{1}{b}\right).$$

or
$$C = \frac{Q}{V} = \frac{4\pi\varepsilon_0}{\dfrac{1}{a}-\dfrac{1}{b}}\,F \qquad \ldots(6.20)$$

However, for an isolated sphere $b \to \infty$ and hence
$$c = 4\pi\,\varepsilon\,a\,F \qquad \ldots(6.21)$$

6.4.2 Two Infinite Lines of Charge

Consider the two infinite line charges as shown in Fig. 6.7. The line charge density of one is ρ_L coulomb/metre and that of the other is $-\rho_L$ C/metre and say they are separated by a distance $2s$. Let *'O'* be the origin and the centre point between the charges. Since the charges are of opposite polarity a plane passing through *'O'* will be the neutral plane and, therefore, *'O'* is taken as the origin for voltage calculation. It is required to find out the potential of point $P(x, y)$ with respect to *'O'* (the neutral point

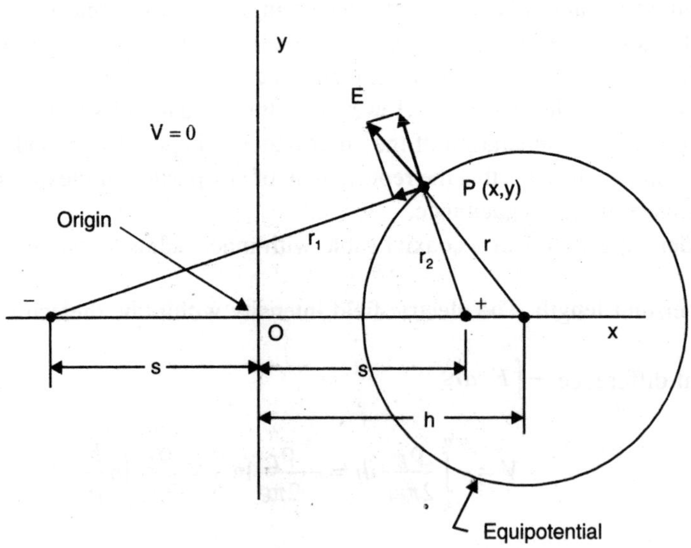

Fig. 6.7 Two infinite lines of charges

or zero potential point) due to the two infinite line charges. The potential at P due to positive linear charge is given as

$$V_+ = \frac{\rho_L}{2\pi\varepsilon_0} \ln \frac{s}{r_2} \qquad \text{...(6.22)}$$

and the potential due to negative charge

$$V_- = -\frac{\rho_L}{2\pi\varepsilon_0} \ln \frac{s}{r_1} \qquad \text{...(6.23)}$$

Total potential at P is $V = V_+ + V_-$

$$V = \frac{\rho_L}{2\pi\varepsilon_0} \ln \frac{r_1}{r_2} \qquad \text{...(6.24)}$$

Now it is required to find out the locus of the equipotential lines.
To find out that, equation (6.24) is written in the form

$$\ln \frac{r_1}{r_2} = \frac{2\pi\varepsilon_0 V}{\rho_L}$$

or

$$\frac{r_1}{r_2} = \exp\left(2\pi\varepsilon_0 V / \rho_L\right) \qquad \text{...(6.25)}$$

Now for a particular voltage V at P, the term on the right side of equation (6.25) is constant.

$$\therefore \qquad \frac{r_1}{r_2} = K \qquad \text{...(6.26)}$$

From Fig. 6.7

$$r_1 = \sqrt{(s+x)^2 + y^2}$$

and

$$r_2 = \sqrt{(s-x)^2 + y^2}$$

Substituting these relations in equation (6.26)

$$\frac{\sqrt{(s+x)^2 + y^2}}{\sqrt{(s-x)^2 + y^2}} = K$$

Squaring both the sides and rearranging, we get

$$x^2 - 2xs\frac{K^2+1}{K^2-1} + s^2 + y^2 = 0 \qquad \text{...(6.27)}$$

For completing the square add on both the sides of equation (6.27), the term $s^2 \dfrac{\left(K^2+1\right)^2}{\left(K^2-1\right)^2}$; we get

$$\left(x - \frac{K^2+1}{K^2-1}s\right)^2 + y^2 = \left(\frac{2Ks}{K^2-1}\right)^2 \qquad \text{...(6.28)}$$

which represents an equation to a circle

$$(x - h)^2 + (y - g)^2 = r^2 \qquad \text{...(6.29)}$$

where

$$r = \frac{2Ks}{K^2 - 1} \qquad \text{...(6.30)}$$

and centre at

$$x = h = \frac{K^2 + 1}{K^2 - 1}s,\ y = g = 0 \qquad \text{...(6.31)}$$

An equipotential line corresponding to voltage V at P is drawn in Fig. 6.7. For higher potentials, K increases and it can be seen from the equation of the equipotential lines, r decreases *i.e.* r approaches zero and h approaches 's' so that the equipotentials are smaller circles with their centres more nearly at the line of charge.

6.4.3 Capacitance of a 1–ϕ Transmission Line

Before an expression for the capacitance is derived we define the following terms:

Capacitor: It is an electrical device which consists of two conductors separated by a dielectric medium and is used for storing electrostatic energy.

Capacitance: The capacitance of a capacitor is the ratio of the charge on one of its conductors to the potential difference between the conductors.

We make use of this definition of capacitance and other results derived previously in this chapter for finding out the capacitance of the transmission lines.

Consider a 1–ϕ transmission line as shown in Fig. 6.8. Let a fixed potential V be applied between the conductors so that the charge per unit length of each conductor is ρ_L coulomb per metre. The length of the line is very large as conductor is very small as compared to the distance of separation. It is to be noted that the charge ρ_L coulomb/metre is distributed on the surface of the conductor which is non-uniformly distributed over the surface such that it has higher density on the adjacent sides of the conductors. This charge distribution can be considered as a line charge as in the previous section. The surface of the conductor represents an equipotential surface with circular cross section and radius equal to r. So the objective will be to find out the equivalent line charge distribution for a system of two conductors with operating voltage V, distance of separation h and radius of the equipotential surface r. This equivalent charge distribution, as can be seen from Fig. 6.8 and the results of the previous section, will be a line charge ρ_L coulomb/metre separated by a distance s where s can be obtained from any of the equations (6.30) or (6.31).

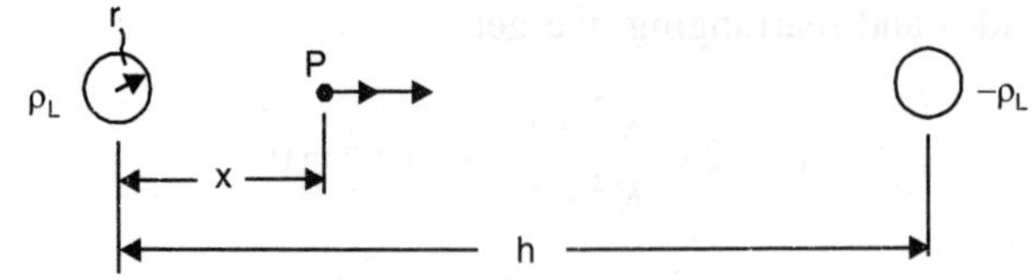

Fig. 6.8 1–ϕ transmission line

Refer to section 6.4.2 V is a function of K and is given by

$$V = \frac{\rho_L}{\pi \varepsilon_0} \ln K$$

From this $$C = \frac{\rho_L}{V} = \frac{\pi\varepsilon_0}{\ln K} \text{ F/metre}$$

The value of K can be obtained from equations (6.30) and (6.31) by eliminating s from these equations.

$$r = \frac{2Ks}{K^2 - 1}$$

$$h = \frac{K^2 + 1}{K^2 - 1} s$$

or $$\frac{h}{r} = \frac{K^2 + 1}{2K}$$

or $$K^2 - 2K\frac{h}{r} + 1 = 0$$

or $$K = \frac{2\frac{h}{r} \pm \sqrt{\frac{4h^2}{r^2} - 4}}{2}$$

$$= \frac{h}{r} \pm \sqrt{h^2/r^2 - 1} \quad \text{...(6.32)}$$

Since the capacitance of a given system is constant only one of the two values of K is to be used.

Since $\frac{h}{r} >> 1$.

$$K \simeq \frac{h}{r} + \frac{h}{r} \quad \text{...(6.33)}$$

Since $K \neq 0$ only positive sign is taken into account. Therefore

$$C = \frac{\pi\varepsilon_0}{\frac{h}{r} + \sqrt{\frac{h^2}{r^2} - 1}} \text{ F/metre} \quad \text{...(6.34)}$$

The expression for capacitance obtained above is very accurate. However, it could be assumed that the charge is uniformly distributed which is not very far from the actual condition for power system problems where $\frac{h}{r} >> 1$. The derivation is much more simplified and is as follows (refer to Fig. 6.8).

Since the charge is assumed to be uniformly distributed over the surface of the conductor, this could be considered as concentrated along the axis of conductor. The electric field intensity at point P due to ρ_L is

$$E_+ = \frac{\rho_L}{2\pi\varepsilon_0 x} \quad \text{...(6.35)}$$

and is directed along $\bar{a}_x$. Similarly electric field intensity at P due to $-\rho_L$

$$E_- = \frac{\rho_L}{2\pi\varepsilon_0 (h-x)} \quad \text{...(6.36)}$$

along $\bar{a}_x$ again, as this time the force experienced by a unit positive charge at P will be towards the negative charge (force of attraction).

Total electric field intensity at P

$$E = \frac{\rho_L}{2\pi\varepsilon_0}\left[\frac{1}{x} + \frac{1}{h-x}\right] \quad \text{...(6.37)}$$

The potential difference between the conductors

$$V = -\int_{h-r}^{r} E \cdot dx$$

$$= \frac{\rho_L}{2\pi\varepsilon_0}\int_{r}^{h-r}\left(\frac{1}{x} + \frac{1}{h-x}\right)dx$$

$$= \frac{\rho_L}{2\pi\varepsilon_0}\left[\ln x + \ln(h-x)\right]_r^{h-r}$$

$$= \frac{\rho_L}{2\pi\varepsilon_0}\left[\ln(h-r) - \ln r - \ln\{h-(h-r)\} + \ln(h-r)\right]$$

$$= \frac{\rho_L}{2\pi\varepsilon_0} 2\ln\frac{h-r}{r}$$

$$= \frac{\rho_L}{\pi\varepsilon_0}\ln\frac{h-r}{r} \quad \text{...(6.38)}$$

Since $h >> r$, $h - r \simeq h$.

$$\therefore \quad V = \frac{\rho_L}{\pi\varepsilon_0}\ln\frac{h}{r}$$

or

$$C = \frac{\rho_L}{V} = \frac{\pi\varepsilon_0}{\ln h/r} \text{ F/metre} \quad \text{...(6.39)}$$

Equation (6.39) corresponds to the expression for capacitance of a single phase transmission line.

Example 6.7. Determine the relative permittivity of the dielectric material present in a parallel plate capacitor if (*i*) $s = 0.12$ m^2, $d = 100$ μm, $v = 15$V and the capacitor contains 1.2 μJ of energy. (*ii*) the stored energy density is 120 J/m^3, $v = 220$ V and $d = 50$ μm (*iii*) $E = 220$ kVm^{-1} $\rho_s = 25$ μC/m^{-2} and $d = 125$ μm.

Solution. (*i*) Stored energy $= \frac{1}{2}CV^2 = \frac{1}{2}C \times 15^2 = 1.2 \times 10^{-6}$

or
$$C = \frac{2.4 \times 10^{-6}}{225} = \varepsilon_0 \varepsilon_r \frac{A}{d} = \frac{8.854 \times 10^{-12} \times 0.12 \times \varepsilon_r}{100 \times 10^{-6}}$$

$$\varepsilon_r = \frac{2.4 \times 10^{-6} \times 100 \times 10^{-6}}{8.854 \times 10^{-12} \times 0.12 \times 225} = 1.003$$

(*ii*) Energy density $= \frac{1}{2}\varepsilon E^2 = 120 \text{ J/m}^3$

$$\Rightarrow \frac{1}{2} \times 8.854 \times 10^{-12} \varepsilon_r \cdot \left(\frac{220}{50 \times 10^{-6}}\right)^2 = 120$$

$$\Rightarrow \frac{1}{2} \times 8.854 \times 10^{-12} \varepsilon_r \cdot 19.36 \times 10^{12} = 120$$

$$\varepsilon_r = \frac{240}{8.854 \times 19.36} = 1.4$$

(*iii*) $E = \frac{\rho_s}{\varepsilon} = 220 \times 10^3 = \frac{25 \times 10^{-6}}{8.854 \times 10^{-12} \times \varepsilon_r}$

$$\varepsilon_r = \frac{25 \times 10^{-6}}{8.854 \times 220 \times 10^{-9}} = 12.8 \textbf{ Ans.}$$

Example 6.8. Determine the capacitance of a parallel plate capacitor containing two dielectrics where E and D parallel to the interface, $\varepsilon r_1 = 2$ and $\varepsilon r_2 = 4$ each comprising half of the volume. Assume $S = 2 \text{ m}^2$ and $d = 10^{-3}$ m.

Solution. Since the interface is parallel to E and D and these occupy equal volumes $s_1 = s_2 = 1 \text{ m}^2$

$$C_1 = \frac{\varepsilon_0 \varepsilon_r S}{d} = \frac{8.854 \times 10^{-12} \times 2 \times 1}{10^{-3}} = 17.7 \text{ nF}, \ C_2 = 35.4 \text{ nF}$$

Hence $C = C_1 + C_2 = 17.5 + 35.4 = 52.9$ nF **Ans.**

Example 6.9. If in example 6.8 the dielectrics are so placed that the interface is normal to E and D, determine the equivalent capacitance.

Solution. Since the two dielectrics are so placed that the interface is normal to E and D, then equivalent capacitance $C = \frac{C_1 C_2}{C_1 + C_2}$ as the two capacitances are connected in series.

$$C_1 = \frac{8.854 \times 10^{-12} \times 2 \times 1}{0.5 \times 10^{-3}} = 35.4 \text{ nF}, C_2 = 70.8 \text{ nF}$$

Hence $$C = \frac{34.5 \times 70.8}{35.4 + 70.8} = 23.6 \text{nF } \textbf{Ans.}$$

Example 6.10. A pair of 500 mm long concentric cylindrical conductors of radii 50 and 100 mm is filled with a dielectric $\varepsilon = 10\ \varepsilon_0$. A voltage is applied between the conductors to establish an electric field $E = \left(\frac{10^5}{r}\right) a_r \text{ Vm}^{-1}$ between the cylinders. Calculate the energy stored.

Solution. The differential volume of the dielectric where the energy is stored in cylindrical coordinates $dr\ rd\phi\ dz$. The energy density is given as $\frac{1}{2}\varepsilon E^2\ \text{J/m}^3$

$$\text{Hence total energy stored} = W = \frac{1}{2}\oint_v \varepsilon E^2\, dv$$

$$= \frac{1}{2}\int \varepsilon \cdot \left(\frac{10^5}{r}\right)^2 \cdot r\,dr\,d\phi\,dz$$

$$= \frac{1}{2}\int_{z=0}^{0.5}\int_{\phi=0}^{2\pi}\int_{r=0.05}^{0.1} \frac{10^{10}}{r^2} r\,dr\,d\phi\,dz$$

$$= \frac{10 \times 10^{-9}}{36\pi \times 2} \times 0.5 \cdot 2\pi \times 10^{10} \ln \frac{0.1}{0.05}$$

$$= 1.39 \ln 2 = 0.963 \text{ J } \textbf{Ans.}$$

Example 6.11. In a cylindrical capacitance with internal radius a and outer radius b having length L metres has two dielectrics where each dielectric occupies one half volume. The dielectrics interface is parallel to D and E. Determine the capacitance. The permittivities are ε_1 and ε_2.

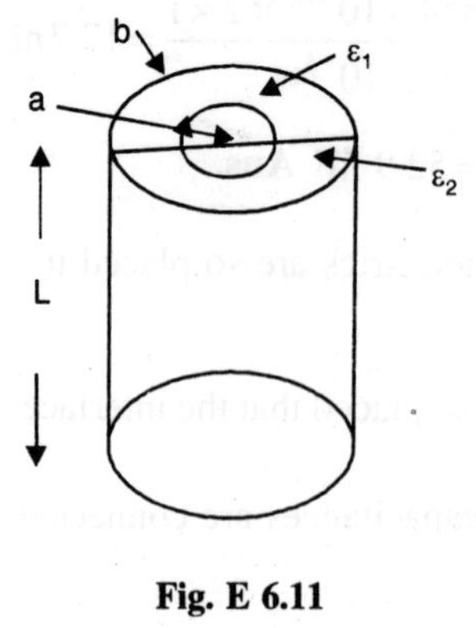

Fig. E 6.11

Solution. Since the dielectric interface is parallel to D and E, the configuration can be considered as two capacitors in parallel. Each capacitor carries half as much charge as a full cylinder. Let ρ_s be the charge per unit area. The surface charge on the inner conductor $\pi a \cdot 1 \cdot \rho_s = \pi \rho_s a$ per unit length.

D at a distance $r < b$ is $D = \dfrac{\pi \rho_s a}{\pi r}\bar{a}_r = \dfrac{\rho_s a}{r}\bar{a}_r$

$E = \dfrac{D}{\varepsilon} = \dfrac{\rho_s a}{\varepsilon r}\bar{a}_r$ and the voltage difference between conductors

$$V = -\int E\,dl = -\int_b^a \frac{\rho_s a}{\varepsilon r}dr = \frac{\rho_s a}{\varepsilon}\ln\frac{b}{a}$$

Hence $\dfrac{\rho_s \pi a \varepsilon}{\rho_s \ln b/a} = C = \dfrac{\pi \varepsilon_0 \varepsilon_r a}{\ln b/a}$ for half the surface

Hence for permittivity ε_1 we have $C_1 = \dfrac{\pi \varepsilon_0 \varepsilon_1 L}{\ln b/a}$

and $C_2 = \dfrac{\pi \varepsilon_0 \varepsilon_2 L}{\ln b/a}$. Hence $C = C_1 + C_2 = \dfrac{\pi \varepsilon_0 L}{\ln b/a}(\varepsilon_1 + \varepsilon_2)$ **Ans.**

Example 6.12. A parallel plate capacitor has two dielectric slabs between the plates with the interface normal to E and D. The thickness of each slab is 1 cm, area of plates 100 cm^2. Neglect fringing effects. Determine E_1, E_2, V_1, V_2, D_1 and D_2 and the capacitance of the capacitor. Voltage applied is 10 volts.

$$\varepsilon_1 = 2 \text{ and } \varepsilon_2 = 3$$

Solution.

Since $\quad D_1 = D_2$ or $\varepsilon_0\,\varepsilon_1\,E_1 = \varepsilon_0\,\varepsilon_2\,E_2$

or $\quad 2E_1 = 3E_2$ or $E_1 = \dfrac{3}{2}E_2$

Now $\quad E_1 d + E_2 d = 10 = \dfrac{3}{2}E_2 d + E_2 d = 10$

or $\quad \dfrac{5}{2}E_2 d = 10$ or $E_2 = 4\,\text{V cm}^{-1}$ and $E_1 = 6\ \text{V cm}^{-1}$

Hence $\quad V_1 = E_1 d = 6 \times 1 = 6$ volts $V_2 = 4$ volts

$$E_1 = 600\ \text{V/m and } E_2 = 400\ \text{V/m}$$

$$D_1 = 8.854 \times 10^{-12} \times 2 \times 600 = 10.6\ \text{nCm}^{-2} = D_2$$

$$C_1 = \frac{\varepsilon_0 \varepsilon_1 A}{d} = \frac{8.854 \times 2 \times 100 \times 10^{-4} \times 10^{-12}}{10^{-2}} = 17.7\ \text{pF}$$

$$C_2 = \frac{8.854 \times 3 \times 100 \times 10^{-4}}{10^{-2}} = 26.56 \text{ pF}$$

$$C = \frac{C_1 C_2}{C_1 + C_2} = \frac{17.7 \times 26.56}{44.26} = 10.6 \text{ pF} \textbf{ Ans.}$$

Example 6.13. Determine the voltage across each dielectric in the capacitor shown below when the applied voltage is 300 V.

Solution. Capacitance of the slab with $\varepsilon_r = 5$ is

$$C_1 = \frac{\varepsilon_0 \varepsilon_r A}{d} = \frac{\varepsilon_0 \cdot 5 \cdot 1}{10^{-3}} = 5000\,\varepsilon_0$$

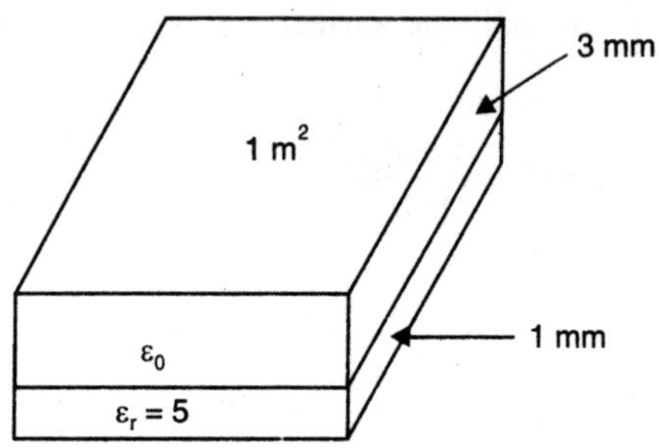

Fig. E 6.13

$$C_2 = \frac{\varepsilon_0 \cdot 1 \cdot 1}{3 \times 10^{-3}} = \frac{1000\,\varepsilon_0}{3}$$

Since the interface is normal to D and E, the two capacitors are in series.

Hence
$$C = \frac{C_1 C_2}{C_1 + C_2} = \frac{5000 \times 1000/3\varepsilon_0}{5000 + \frac{1000}{3}} = \frac{500 \times 10^3}{16000}\varepsilon_0$$

$$\frac{5000}{16} \times 8.854 \times 10^{-12} = 2.77 \text{ nF} \textbf{ Ans.}$$

The ratio $C_1 : C_2 :: 15 : 1$

and the voltage distribution is in inverse proportions to the capacitance as higher the capacitance the smaller voltage drop, so that the product $VC = q$ is same.

Hence voltage across $\varepsilon_r = 5$ relative permittivity material

$$V_1 = 300 \times \frac{1}{16} = 18.75 \text{ volts}$$

$$V_2 = 300 \times \frac{15}{16} = 281.25 \text{ volts} \textbf{ Ans.}$$

Example 6.14. Determine the capacitance per unit length of the cable shown in Fig. E 6.14.

Solution. It is a case of co-axial cylinders with two layers of permittivities with ε_1 and ε_2, then these form two capacitors in series.

For a co-axial cable capacitance per unit length is given as

$$\frac{2\pi\varepsilon_0\varepsilon_r}{\ln r_2/r_1}$$

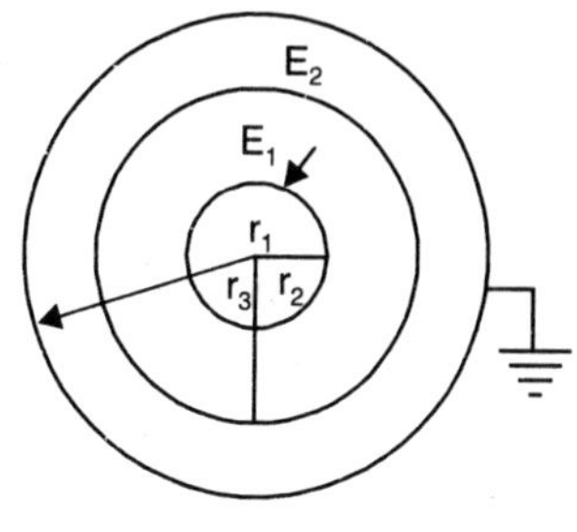

Fig. E 6.14

Hence the two capacitances per unit length are

$$C_1 = \frac{2\pi\varepsilon_0\varepsilon_1}{\ln\frac{r_2}{r_1}},\quad C_2 = \frac{2\pi\varepsilon_0\varepsilon_2}{\ln\frac{r_3}{r_2}}$$

Hence $$C = \frac{C_1C_2}{C_1+C_2} = \frac{2\pi\varepsilon_1\varepsilon_2\varepsilon_0}{\varepsilon_2\ln\frac{r_2}{r_1}+\varepsilon_1\ln\frac{r_3}{r_2}} \quad \textbf{Ans.}$$

Example 6.15. For the system of example 6.14 assume conductor at potential V_0 and the outer shield grounded, determine maximum electric field intensity in each dielectric assuming V_0 = 1.5 kV, $\varepsilon_{r_1} = 4.5$, $\varepsilon_{r_2} = 3$, r_1 = 10 mm, r_2 = 20 mm and r_3 = 40 mm.

Solution. $$C_1 = \frac{2\pi\times4.5\times10^{-9}}{\ln 2\times36\pi} = 0.36\text{ nFm}^{-1},\quad C_2 = \frac{2\pi\times3\times10^{-9}}{\ln 36\pi} = 0.24\text{ nFm}^{-1}.$$

Since the two capacitances are connected in series $\frac{V_1}{V_2} = \frac{C_2}{C_1} = \frac{0.24}{0.36} = \frac{2}{3}$

$$V_1 = \frac{2}{3}V_2 \text{ and } V_1+V_2 = 1500\text{ V or } \frac{2}{3}V_2+V_2 = 1500 = \frac{5}{3}V_2$$

or V_2 = 900 volts and V_1 = 600 volts, $q = c_1\,v_1 = 0.36\times10^{-9}\times600 = 216\text{ nCm}^{-1}$

We know that the maximum E_r occurs at the inner surface of either dielectric and is given after the inner dielectric

$E_{r\max} = \frac{q}{2\pi\varepsilon_r}$. Hence at $r = r_1$

$$E_{r_1\max} = \frac{216\times36\pi\times10^{-9}}{2\pi\times10^{-9}\times4.5\times0.01} = 86.4\text{ kVm}^{-1}$$

$$E_{r_2\max} = \frac{216\times36\pi\times10^{-9}}{2\pi\times10^{-9}\times3\times0.02} = 64.8\text{ kVm}^{-1} \quad \textbf{Ans.}$$

6.5 CURRENT AND ELECTRIC FLUX ANALOGIES

A useful analogy between electric current density in conducting media and electric flux density in dielectric medium exists. For direct currents in conducting medium, analogy can be easily demonstrated using ohm's law and gradient relationship. Ohm's law in conducting medium is given as

$$J = \sigma E_\sigma$$

$$E_\sigma = -\nabla V_\sigma$$

whereas in dielectric medium, electric flux density and electric field intensity are related as

$$D_\varepsilon = \varepsilon E_\varepsilon$$

$$E_\varepsilon = -\nabla V_\varepsilon$$

The subscripts σ and ε indicate quantities for the conducting medium and dielectric medium respectively.

It is evident that the current density J and flux density D, E_σ and E_c, σ and ε, the potentials V_σ and V_ε form analogous pairs.

Suppose we want to find out field within a co-axial cable or capacitor. The analogous to this will be to take two co-axial copper cylinders and fill the region between them with some electrolytic solution. The conductivity of the electrolytic solution is analogous to the permittivity of the dielectric material in a co-axial capacitor. We now apply a potential difference between the co-axial cylinders and we use a probe to determine the potential at any intermediate point in the electrolytic solution or we can move the probe to determine equipotential point in the electrolyte. Therefore any condition in a co-axial cable can be simulated using an electrolytic tank and therefore an electrolytic tank is analogous to a co-axial capacitor. The greatest advantage of this method is that this can be used for three dimensional problems also.

We make use of analysis to determine the capacitance from such measurements. The total current leading the more positive conductor is

$$I = \oint_S J_\sigma \cdot ds = \sigma \oint_S E_\sigma \cdot ds$$

where s represents the integral to cover entire conductor surface. Also potential difference is given by the negative of the line integral of E_σ from less to the more positive plate

$$V_{\sigma 0} = -\int E_\sigma \cdot dl$$

and the total resistance is, therefore, given as

$$R = \frac{V_{\sigma 0}}{I} = \frac{-\oint E \cdot dl}{\sigma \oint_s E_\sigma \cdot ds}$$

The capacitance is the ratio of total charge Q to the potential difference

$$C = \frac{Q}{V_{\varepsilon 0}} = \frac{\varepsilon \oint_s E_\varepsilon \cdot ds}{-\int E_\varepsilon \cdot dl}$$

Using analogies $V_{\varepsilon 0} = V_{\sigma 0}$ and $E_\varepsilon = E_\sigma$

$$RC = \frac{V_{\sigma 0}}{I} \cdot \frac{Q}{V_{\sigma 0}} = \frac{Q}{I} = \frac{It}{I} = t = \frac{\varepsilon}{\sigma} \qquad \text{...(6.40)}$$

RC is the time constant of the capacitive circuit and $\frac{\varepsilon}{\sigma}$ is also known as relaxation time of a dielectric material. Therefore, knowing the conductivity of the electrolyte and permittivity of the dielectric, the capacitance can be obtained by a simple resistance measurement. Another application of the above equation is to evaluate resistance of a given two conductor configuration, knowing its capacitance or vice versa. It is to be noted that equation (6.40) holds good only for linear, homogeneous and isotropic media.

PROBLEMS

P 6.1 Explain the characteristics of a dielectric when it is subjected to an electric field.

P 6.2 What is meant by polarization of a dielectric material? Derive relation between *D, E* and *P.*

P 6.3 Define susceptibility of a dielectric material and explain its physical significance.

P 6.4 Explain with mathematical analysis the behavior of Electric field intensity and Electric flux density in terms of their tangential and normal components at the interface of two perfect dielectrics.

P 6.5 Comment on the statement: *D* is larger in the region of larger permittivity whereas *E* is larger in the region of smaller permittivity.

P 6.6 Show that the capacitance of any conducting system is independent of the charge on the conducting surface and voltage across the conducting medium.

P 6.7 Derive an expression for the capacitance of a parallel plate capacitor consisting of two different dielectric with their interface parallel to *D* and *E.*

P 6.8 Derive expression for capacitance when their interface is normal to *D* and *E.*

P 6.9 What happens to the capacitance of a parallel plate with two dielectric in between as in *P* 6.8 and a thick conducting plate is placed on the interface keeping the distance of separation between the outer plates fixed.

P 6.10 Explain analogies between electric current in a conducting medium and electric flux in a dielectric medium.

P 6.11 Region 1 is defined by $x < 0$ is free space whereas region 2 $x > 0$ is a dielectric material for which $\varepsilon r_2 = 2.4$ as shown in Fig. P 6.11

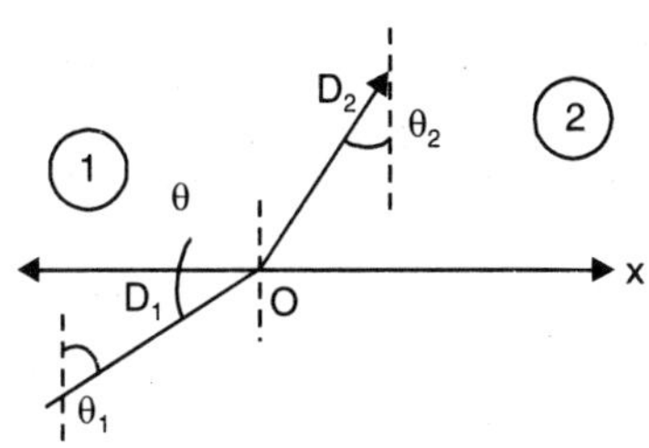

Fig. P 6.11

The electric flux density in medium 1 is given by

$$D_1 = 3\bar{a}_x - 4\bar{a}_y + 6\bar{a}_z \text{ cm}^{-2}$$

Determine E_2 and the angles θ_1 and θ_2.

Ans. $E_2 = \frac{1.25}{\varepsilon_0}\bar{a}_x - \frac{4}{\varepsilon_0}\bar{a}_y + \frac{6}{\varepsilon_0}\bar{a}_z$

$\theta_1 = 22.6°$ $\theta_2 = 9.83°$

P 6.12 The $y - z$ plane is the boundary between two dielectrics of relative permittivity $\varepsilon_r = 3$ and $\varepsilon_r = 9$. For $x < 0$, $E = 5\bar{a}_x + \bar{a}_y \text{ Vm}^{-1}$, determine the magnitude and direction of D for $x > 0$.

Ans. $15\varepsilon_0\bar{a}_x + 9\varepsilon_0\bar{a}_y$

P 6.13 For the E-fields shown at the interface of two dielectrics relate the anyles θ_1 and θ_2.

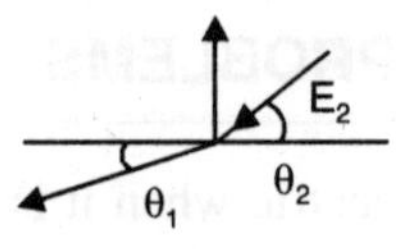

Ans. $\frac{\tan\theta_1}{\tan\theta_2} = \frac{\varepsilon_2}{\varepsilon_1}$

Fig. P 6.13

P 6.14 A parallel plate capacitor has plates 0.15 mm apart a plate area of 1000 cm^2 and a dielectric with $\varepsilon_r = 3$. Determine the electric flux density, electric field intensity and the voltage between the plates if the capacitor has a charge of 0.5 μC.

Ans. 5 μCm^{-2}, 188.5 kVm^{-1}, 28.3 V

P 6.15 A co-axial conductor has radii $a = 0.8$ mm and $b = 3$ mm and dielectric with kVm^{-1} $\varepsilon_r = 2.56$. If $P = (2/\rho)\bar{a}_\rho \text{ nCm}^{-2}$ in the dielectric, determine (*i*) D and E as a function of ρ (*ii*) V_{ab} and x_e.

Ans. $D = (3.28/\rho)\bar{a}_\rho \text{ nCm}^{-2}$, $E = \left(\frac{144.9}{\rho}\right)a_\rho \text{ Vm}^{-1}$.

$V_{ab} = 192$ V, $x_e = 1.56$

P 6.16 A parallel plate capacitor has 2 × 3 cms plates spaced by 5 mm. Determine the capacitance, (*i*) if the medium is air (*ii*) if $\varepsilon_r = 12$. Neglect fringing.

Ans. (*i*) 1.06 pF (*ii*) 12.7 pF

P 6.17 A single core cable has an inner diameter of 5 cms and a core diameter of 1.5 cm. It's paper dielectric has a maximum dielectric stress of 60 kVcm^{-1}. Calculate the maximum permissible voltage.

Ans. 54.2 kV

P 6.18 Determine the maximum working voltage of a single core lead sheathed cable having a conductor 1 cm dia and sheath of inside dia 5 cm. Two insulating materials with permittivities and maximum stresses 4, 2.5 and 60 kVcm^{-1} and 50 kVcm^{-1} respectively are used.

Ans. 65 kV

P 6.19 Two perfect dielectrics have relative permittivities $\varepsilon_{r_1} = 2$ and $\varepsilon_{r_2} = 8$. The planar interface between them is the surface $x - y + 2z = 5$. The origin lies in region 1. If $E_1 = 20\bar{a}_x - 10\bar{a}_y + 50\bar{a}_z \text{ Vm}^{-1}$. Find E_2.

Ans. $125\bar{a}_x + 175\bar{a}_y \text{ Vm}^{-1}$

P 6.20 In the free space region $x < 0$ the electric field intensity is $E_1 = 3\bar{a}_x + 5\bar{a}_y - 3\bar{a}_z \text{ Vm}^{-1}$. The region $x > 0$ is a dielectric for which $\varepsilon_r = 3.6$. Determine angle θ_2 that the field in the dielectric makes with the $x = 0$ plane.

Ans. $\theta_2 = 8.13°$

P 6.21 A dielectric sphere of permittivity ε and radius R is centred at the origin of a spherical coordinate system and polarized radially such that $\rho = kr\bar{a}_r$ where k is a constant. Determine the electric potential at the centre of the sphere.

Ans. $V = -\dfrac{kR^2}{2\varepsilon}$

P 6.22 A 2 cm diameter conductor is suspended in air with its axis 5 cm from a conducting plane. The potential of the cylinder is 100 V and that of the plane is zero. Determine the surface charge density on the (*i*) cylinder at a point nearest the plane (*ii*) plane at a point nearest the cylinder.

Ans. (*i*) 473 nCm^{-2} (*ii*) -15.8 nCm^{-2}

P 6.23 The electric field intensity in a mass of porcelain with $\varepsilon_r = 6$ in air is 1 kVcm^{-1}. At the inner surface of the porcelain the field makes an angle 45° to the normal and emerge into the air. Determine the angle of emergence of the external field and it's magnitude.

Ans. 9.5° with interface or 80.5° with normal, 4.3 kVcm^{-1}.

Chapter 7

Poisson's and Laplace's Equations

7.1 INTRODUCTION

We have learnt to find out electric field intensity at a point due to certain charge distribution or a potential field by using Coulomb's law. With this law we obtain the vector sum of the fields from all point charges that comprise the distribution. However if the charge distribution is symmetrical, Gauss's law is used to advantage as it provides the solution very fast. A still simpler method to obtain electric field intensity indirectly is to obtain the potential field associated with a charge distribution and then taking the negative of the gradient of the result to obtain electric field intensity *E*.

In this chapter we develop certain mathematical techniques for the solution of such problems in a more direct way. Poisson's and Laplace's equations are fundamental to science and engineering wherever fields exists. These equations enable us to find potential fields within regions bounded by known potentials or charge densities. Problems involving one to three dimensions can be solved either analytically or numerically depending upon the boundary conditions. Laplace's and Poisson's equations when compared with other methods are probably the most widely used as many problems in engineering practice involve devices which applied potential differences which are known and where constant potentials occur on the boundaries. Once the potential functions are known, electric field intensity can be obtained by taking negative of the gradient.

The solution of a boundary value problem can be obtained if it is set up in a coordinate system in which the boundaries can be specified in a simpler manner. For example, if a problem involved a rectangular object which can be handled more readily using rectangular coordinates, a cylindrical object by cylindrical coordinates and a spherical object by spherical coordinates etc. unfortunately in many cases formal mathematical techniques cannot be used for solution on the boundaries in many practical situation cannot be expressed in any coordinate system and under such situations we resort to other methods using graphical, experimental, analog-computer and numerical techniques.

7.2 DERIVATION OF POISSON'S AND LAPLACE'S EQUATIONS

In order to derive Poisson's equation we make use of point form of Gauss's law

$$\nabla \cdot D = \rho \qquad \ldots(7.1)$$

Since $D = \varepsilon E$ and $E = -\nabla V$

$$\nabla \cdot \varepsilon E = \rho$$

or $$\varepsilon \nabla \cdot E = \rho$$

or $$-(\varepsilon \nabla \cdot \nabla V) = \rho$$

or $$\nabla \cdot \nabla V = -\rho/\varepsilon \qquad \text{...(7.2)}$$

Equation (7.2) holds good for a homogeneous region where ε is constant and this equation is Poisson's equation.

However, the double ∇ operator must be suitably expressed is say at least rectangular coordinates (as this is the simplest of the three coordinates) before the equation can be made use of. We know that in rectangular coordinates

$$\nabla \cdot A = \frac{\partial A_x}{\partial x} + \frac{\partial A_y}{\partial y} + \frac{\partial A_z}{\partial z}$$

where A_x, A_y and A_z are the components of vector A along $\bar{a}_x$, $\bar{a}_y$ and $\bar{a}_z$ respectively.

Also gradient of V in rectangular coordinates is given by

$$\nabla V = \frac{\partial V}{\partial x}\bar{a}_x + \frac{\partial V}{\partial y}\bar{a}_y + \frac{\partial V}{\partial z}\bar{a}_z$$

and therefore,

$$\nabla \cdot \nabla V = \frac{\partial}{\partial x}\left(\frac{\partial V}{\partial x}\right) + \frac{\partial}{\partial y}\left(\frac{\partial V}{\partial y}\right) + \frac{\partial}{\partial z}\left(\frac{\partial V}{\partial z}\right)$$

$$= \frac{\partial^2 V}{\partial x^2} + \frac{\partial^2 V}{\partial y^2} + \frac{\partial^2 V}{\partial z^2} \qquad \text{...(7.3)}$$

Hence, equation (7.2) is rewritten as

$$\nabla \cdot \nabla \cdot V = \nabla^2 V = \frac{\partial^2 V}{\partial x^2} + \frac{\partial^2 V}{\partial y^2} + \frac{\partial^2 V}{\partial z^2} = -\frac{\rho_v}{\varepsilon} \qquad \text{...(7.4)}$$

Equation (7.4) is Poisson's equation in rectangular coordinates. Suppose in a particular situation $\rho_v = 0$ *i.e.* the volume charge density is zero, yet we allow point charges, line charges and surface charge density to exist at singular locations as sources of the electric field, then

$$\nabla^2 V = 0 \qquad \text{...(7.5)}$$

Equation (7.5) is known as Laplace's equation. The ∇^2 operator is known as Laplacian of V.

Again in rectangular coordinates Laplace's equation is given as

$$\frac{\partial^2 V}{\partial x^2} + \frac{\partial^2 V}{\partial y^2} + \frac{\partial^2 V}{\partial z^2} = 0 \qquad \text{...(7.6)}$$

To obtain Laplace's equation in cylindrical and spherical coordinates, we can use expressions for the divergence and gradient (∇ · ∇V) already obtained in Chapter 1. Making use of these two operation Laplace's equation in the respective coordinate system is given as follows:

$$\nabla^2 V = \frac{1}{\rho}\frac{\partial}{\partial \rho}\left(\rho \frac{\partial V}{\partial \rho}\right) + \frac{1}{\rho^2}\left(\frac{\partial^2 V}{\partial \phi^2}\right) + \frac{\partial^2 V}{\partial z^2} \text{ (cylindrical)} \qquad \text{...(7.7)}$$

and in spherical coordinate system is

$$\nabla^2 V = \frac{1}{r^2}\frac{\partial}{\partial r}\left(r^2 \frac{\partial V}{\partial r}\right) + \frac{1}{r^2 \sin\theta}\frac{\partial}{\partial \theta}\left(\sin\theta \frac{\partial V}{\partial \theta}\right) + \frac{1}{r^2 \sin^2\theta}\frac{\partial^2 V}{\partial \phi^2} \quad \text{...(7.8)}$$

Since, Poisson's equation or Laplace's equations are second order partial differential equation, a complete solution is possible only if we have two boundary conditions relating to the system for which the solution is sought. The solution so obtained if satisfies both the Laplace's equation/Poisson's equation and the boundary conditions. Thus this is the only correct solution. If somehow we try to solve to Laplace's equation using two different techniques the answers must be same and should not be different. We try to prove in the next article that the two answers are identical.

7.3 UNIQUENESS THEOREM

Suppose solutions of the Laplaces' equation after taking into consideration the boundary conditions are V_1 and V_2 where V_1 and V_2 are general functions of the coordinates used. Therefore

$$\nabla^2 V_1 = 0$$

and $$\nabla^2 V_2 = 0$$

and hence $$\nabla^2(V_1 - V_2) = 0 \quad \text{...(7.9)}$$

Each solution must also satisfy the boundary conditions. Let the boundary potential be V_b, then the value of V_1 on the boundary V_{1b} and the value of V_2 on the boundary V_{2b} and both V_{1b} and V_{2b} must be identical to V_b. Hence

$$V_{1b} = V_{2b} = V_b$$

or $$V_{1b} - V_{2b} = 0$$

Further, we make use of the vector identity

$$\nabla \cdot (V\bar{D}) = V(\nabla \cdot \bar{D}) + \bar{D}(\nabla \cdot V) \quad \text{...(7.9a)}$$

which holds good for any scalar field V and any vector $\bar{D}$. Now using $V = V_1 - V_2$ and the vector as $\nabla(V_1 - V_2)$ we have

$$\nabla \cdot [(V_1 - V_2)\, \nabla (V_1 - V_2)] = (V_1 - V_2)[\nabla \cdot \nabla (V_1 - V_2) + \nabla (V_1 - V_2) \cdot \nabla (V_1 - V_2)] \quad \text{...(7.10)}$$

we next integrate throughout the volume enclosed by the boundary surfaces specified

$$\underbrace{\int_v \nabla \cdot \left[(V_1 - V_2)\nabla(V_1 - V_2)\right] dV}_{I} = \underbrace{\int_v (V_1 - V_2)\left[\nabla \cdot \nabla(V_1 - V_2)\right] dV}_{II}$$

$$+ \underbrace{\int_v \left[\nabla(V_1 - V_2)\right]^2 dV}_{III} \quad \text{...(7.11)}$$

Using Divergence theorem to replace the volume integral on the left hand side of the equation with the closed surface integral over the surface surrounding the volume. This surface consists of the boundaries already specified where $V_{1b} = V_{2b}$ and, therefore,

$$\int_v \nabla \cdot \left[(V_1 - V_2)\nabla(V_1 - V_2)\right] dV = \oint_s \left[(V_{1b} - V_{2b})\nabla(V_{1b} - V_{2b})\right] \cdot dS = 0 \quad \text{...(7.12)}$$

In equation (7.11) one part of II is

$$\nabla \cdot \nabla (V_1 - V_2) \text{ or } \nabla^2 (V_1 - V_2) \quad \text{...(7.13)}$$

which according to (7.9) is zero. Hence III part on the right hand side of equation (7.11) is

$$\int_v \left[\nabla\left(V_1 - V_2\right)\right]^2 dV = 0 \qquad \ldots(7.14)$$

Now there are two possibilities for this integral to be zero. First possibility is that the integrand is zero and the other possibility is that the integrand is positive in some region and negative in others and thus the net is zero. Here the first possibility holds good because $[\nabla\ (V_1 - V_2)]^2$ cannot be negative. Therefore,

$$\left[\nabla\left(V_1 - V_2\right)\right]^2 = 0$$

or

$$\nabla\ (V_1 - V_2) = 0 \qquad \ldots(7.15)$$

Finally if the gradient of a scalar is zero, the scalar is a constant *i.e.* $V_1 - V_2 = C$

The constant can easily be evaluated by considering a point on the boundary. Here $V_1 - V_2 = V_{1b} - V_{2b} = 0$ and therefore we see that the constant C is zero and hence

$V_1 = V_2$ giving two identical solutions.

It is to be noted that the uniqueness theorem applies to Poisson's equation also. If $\nabla^2 V_1 = -\rho_v/\varepsilon$ and $\nabla^2 V_2 = -\rho_v/\varepsilon$.

Hence $\nabla^2(V_1 - V_2) = 0$ as in case of Laplace equation. Boundary conditions also require that $V_{1b} - V_{2b} = 0$ and hence the proof is identical to the one for Laplace equation. This shows that if we use two different methods for the solution of a system expressed by Laplace's equation or Poisson's equation, the answer must be same.

Example 7.1. Determine the numerical values for V and ρ_v at point P in free space if (*i*) $V = \dfrac{5yz}{x^2+1}$ at $P(1, 2, 4)$ (*ii*) $V = 6\rho^2 \cos 2\phi$ at $P\left(\rho = 3,\ \phi = \dfrac{\pi}{3},\ z = 2\right)$ (*iii*) $V = \dfrac{2\cos\phi}{r^2}$ at $P(r = 0.6, \theta = 30°, \phi = 45°)$

Solution. (*i*) Since $V = \dfrac{5yz}{x^2+1}$ Now at $P(1, 2, 4)$

$$V = \frac{5\times 2\times 4}{1^2+1} = 20 \text{ volts}$$

To determine ρ_v we make use of Poisson's equation

$$\frac{\partial^2 V}{\partial x^2} + \frac{\partial^2 V}{\partial y^2} + \frac{\partial^2 V}{\partial z^2} = -\rho_v/\varepsilon_0$$

$$\frac{dV}{dx} = \frac{-5yz\cdot 2x}{\left(x^2+1\right)^2} = \frac{-10xyz}{\left(x^2+1\right)^2}$$

$$\frac{d^2V}{dx^2} = \frac{\left(x^2+1\right)(-10yz) - (-10xyz)\cdot 2\left(x^2+1\right)2x}{\left(x^2+1\right)^4}$$

$$= -\frac{10yz(x^2+1)+10xyz(x^2+1)2x}{16\times 8.854\times 10^{-12}}$$

$$= \frac{-10\times 2\times 4\times 2+10\times 2\times 4\times 2\times 2}{16}\times 8.854\times 10^{-12}$$

$$= \frac{(-160+320)}{16}8.854\times 10^{-12} = -88.54\,\text{pC/m}^2 \quad \textbf{Ans.}$$

(*ii*) $V = 6\rho^2\cos 2\phi = 6\times 9\cos\frac{2\pi}{3} = 27$ volts.

$$\frac{1}{\rho}\frac{\partial}{\partial\rho}\left(\rho\frac{\partial V}{\partial\rho}\right)+\frac{1}{\rho^2}\left(\frac{\partial^2 V}{\partial^2\phi}\right)$$

$$\frac{1}{\rho}\cdot\frac{\partial}{\partial\rho}(\rho\cdot 12\rho\cos 2\phi) = \frac{4}{\rho^2}(6\rho^2\cos 2\phi)$$

$$\frac{\cos 2\phi}{\rho}\frac{\partial}{\partial\rho}(12\rho^2) - 24\cos 2\phi$$

$$\cos 2\phi\cdot 24 - 24\cos 2\phi = 0.\ \rho_v = 0 \quad \textbf{Ans.}$$

(*iii*) $V = \dfrac{2\cos\phi}{r^2} \qquad r = 0.5,\ \phi = 60°$

$$= \frac{2\times 1/2}{(1/2)^2} = 4 \text{ volts.}$$

Now using spherical coordinate as V is not a soln. of θ

$$\nabla^2 V = \frac{1}{r^2}\frac{\partial}{\partial r}\left(r^2\frac{\partial V}{\partial r}\right)+\frac{1}{r^2\sin^2\theta}\frac{\partial^2 V}{\partial\phi^2}$$

$$-\frac{1}{r^2}\frac{\partial}{\partial r}\left(r^2\frac{4\cos\theta}{r}\right)+\frac{1}{r^2\sin^2\theta}\cdot\frac{\partial}{\partial\phi}\frac{(-2\sin\phi)}{r^2}$$

$$-\frac{1}{r^2}\cdot 4\cos\phi - \frac{1}{r^4}\cdot\frac{1}{\sin^2\theta}2\cos\phi$$

$$16\cos\phi - 16\times\frac{1}{2}\times 2\cos\phi = 0 \quad \textbf{Ans.}$$

Hence the surface charge density is zero.

Example 7.2. Determine the potential function for the region between the parallel circular disc placed in the *xy* plane.

Solution. Since the circular discs are placed in the *xy* plane *V* is not a function of *x* and *y* and hence

$$\frac{d^2V}{dz^2} = 0 \text{ or } \frac{dV}{dz} = A \text{ or } V = Az + B \textbf{ Ans.}$$

However, if the plates were placed in the *xz* plane the potential solution will be $V = Ay + B$.

Example 7.3. Two parallel conducting plates are placed in free space and are located at $y = 0$ and $y = 0.02$ m and the zero voltage reference is at $y = 0.01$ m. If $D = 250\,\bar{a}_y\ \text{nC/m}^2$ between the conductors, determine the conductor voltage.

Solution. Since the plates are placed parallel to *xz* plane *i.e.* the *y*-axis is normal to the plate and hence

$$\frac{d^2V}{dy^2} = 0 \text{ and } \frac{dV}{dy} = A$$

$$V = Ay + B$$

$$E = -\nabla V = -\frac{\partial V}{\partial y} = -A\bar{a}_y$$

$$D = \varepsilon_0 E \text{ or } E = \frac{D}{\varepsilon_0} = \frac{250 \times 10^{-9}}{8.854 \times 10^{-12}} = -A\bar{a}_y$$

$$A = -2.82 \times 10^4 \text{ V/m}$$

Now at $\quad y = 0.01$ m $\quad V = 0$

$$0 = -2.82 \times 10^4 \times 0.01 + B \text{ or } B = 2.82 \times 10^2 \text{ volts}$$

Hence $\quad V = Ay + B = -2.82 \times 10^4 y + 2.82 \times 10^2$

At $y = 0$ $\quad V = 2.82 \times 10^2 = 282$ volts

and $Ay = 0.02$ $\quad V = -2.82 \times 10^4 \times 0.02 + 282$

$$= -5.64 \times 10^2 + 282 = -282 \text{ volts } \textbf{Ans.}$$

Example 7.4. The spherically symmetric potential distribution within a hydrogen atom is given by

$$V(r) = \frac{A\rho^{-\alpha r}}{r}\left(1 + \frac{\alpha r}{2}\right)$$

where α and *A* are positive constants. Determine the spherical charge distribution required to produce this potential. Origin to be excluded from the calculation.

Solution. Since the potential function does not depend on θ and ϕ, hence

$$\frac{1}{r^2}\frac{\partial}{\partial r}\left(r^2\frac{\partial V}{\partial r}\right) = -\frac{\rho}{\varepsilon_0} \quad r > 0$$

In order to obtain ρ we substitute the value of *V* in this expression

$$\frac{1}{r^2}\frac{\partial}{\partial r}\left[r^2 \cdot \frac{\partial}{\partial r}\left\{\frac{Ae^{-\alpha r}}{r}\left(1 + \frac{\alpha r}{2}\right)\right\}\right]$$

Now $$\frac{\partial}{\partial r}\left(\frac{Ae^{-\alpha r}}{r}+\frac{Ae^{-\alpha r}}{r}\cdot\frac{\alpha r}{2}\right)=\frac{\partial}{\partial r}\left(\frac{Ae^{-\alpha r}}{r}+\frac{A\alpha e^{-\alpha r}}{2}\right)$$

$$\frac{\alpha Ar\cdot e^{-\alpha r}-Ae^{-\alpha r}}{r^2}-\frac{A\alpha^2 e^{-\alpha r}}{2}$$

$$\frac{A}{r^2}\frac{\partial}{\partial r}\left[r^2\left(\frac{\alpha e^{-\alpha r}}{r}-\frac{e^{-\alpha r}}{r^2}-\frac{\alpha^2 e^{-\alpha r}}{2}\right)\right]$$

$$=\frac{A}{r^2}\left(\alpha e^{-\alpha r}+\alpha^2 re^{-\alpha r}-\alpha e^{-\alpha r}+\frac{\alpha^3}{2}r^2e^{-\alpha r}-\alpha^2 re^{-\alpha r}\right)$$

$$=\frac{A}{2}\alpha^3 e^{-\alpha r}=-\frac{\rho}{\varepsilon_0}$$

$$\text{or } \rho=-\frac{1}{2}\varepsilon_0 A\alpha^3 e^{-\alpha r} \quad \textbf{Ans.}$$

Example 7.5. Determine the potential function and the electric field intensity for the region between two concentric right circular cylinders where the inner cylinder with $r = 1$ mm $V = 0$ and outer cylinder with radius 25 mm $V = 175$ volts. Neglect fringing.

Solution. Suppose the axis of the cylinder lies along z-axis. Therefore, the potential is independent of ϕ and z and the Laplace's equation reduces to (in cylindrical coordinates)

$$\frac{1}{r}\frac{d}{dr}\left(r\frac{dV}{dr}\right)=0.$$

Integrating once we have

$$r\frac{dV}{dr}=A \text{ or } dV=\frac{A}{r}dr$$

or $$V = A \ln r + B$$

Using boundary condition's *i.e.* at $r = 1$ mm $V = 0$ and at $r = 25$ mm $V = 175$, we have

$$0=A\ln\frac{1}{100}+B \qquad 175 = A \ln 0.025 + B$$

or $$175 = A(\ln 0.025 - \ln 0.001)$$

$$= A(-3.69 + 6.907)$$

or $$A = 54.4$$

and $$175 = 54.4 \ln 0.025 + B$$

$$= -200.67 + B$$

or $$B = 375.67$$

Therefore, $$V = 54.4 \ln r + 375.67 \quad \textbf{Ans.}$$

Hence $$E=-\nabla V=-\frac{54.4}{r}\bar{a}_r \text{ V/m} \quad \textbf{Ans.}$$

7.4 PRODUCT SOLUTION OF LAPLACE'S EQUATION

Let us consider a potential which is a function of x and y alone so that $\dfrac{\partial^2 V}{\partial x^2}+\dfrac{\partial^2 V}{\partial y^2}=0$ is Laplace equation in two variables. Next we assume that the potential can be expressed as the product of a function of x alone and a function of y alone. Since Laplace's equation is linear and the sum of any two solutions is also a solution *i.e.* we could treat $V = x + y$ as the sum of two solutions $V_1 = x$ and $V_2 = y$ where each of these latter potentials is now a product solution.

Representing the function of x by X and the function of y by Y, we have

$$V = XY \qquad \text{...(7.16)}$$

which when substituted into the two variable partial differential equation given above gives

$$Y\frac{\partial^2 X}{\partial x^2}+X\frac{\partial^2 Y}{\partial y^2}=0 \qquad \text{...(7.17)}$$

Since Y does not involve x and X does not involve y ordinary derivatives (total differential) may be used and we have

$$Y\frac{d^2 X}{dx^2}+X\frac{d^2 Y}{dy^2}=0 \qquad \text{...(7.18)}$$

Equation (7.18) may be solved by separating the variables through division by XY, thus giving

$$\frac{1}{X}\frac{d^2 X}{dx^2}+\frac{1}{Y}\frac{d^2 Y}{dy^2}=0$$

or

$$\frac{1}{X}\frac{d^2 X}{dx^2}=-\frac{1}{Y}\frac{d^2 Y}{dy^2} \qquad \text{...(7.19)}$$

Since $\dfrac{1}{X}\dfrac{d^2 X}{dx^2}$ does not have a term in y and $\dfrac{1}{Y}\dfrac{d^2 Y}{dy^2}$ does not have a term in x and since the two terms are equal then $\dfrac{1}{X}\dfrac{d^2 X}{dx^2}$ cannot be a function of x either and similarly the term $\dfrac{1}{Y}\dfrac{d^2 Y}{dy^2}$ cannot be a function of y. This means that each of these terms must be a constant. For convenience let us call this constant as a^2 and hence let

$$\frac{1}{X}\frac{d^2 X}{dx^2}=a^2 \qquad \text{...(7.20)}$$

and

$$\frac{1}{Y}\frac{d^2 Y}{dy^2}=-a^2 \qquad \text{...(7.21)}$$

The constant a^2 is called the separation constant as its use results in separating one equation into two equations. Equation (7.20) can be rewritten as

$$\frac{d^2 X}{dx^2}=a^2 X$$

The general solution for X for a given 'a' is

$$X = A_1 e^{ax} + A_2 e^{-ax} \quad \text{...(7.22)}$$

or equivalently

$$X = A_3 \cosh x + A_4 \sinh x \quad \text{...(7.23)}$$

Similarly for equation (7.21) where a^2 is negative *i.e.* $(\pm ja)^2$

$$Y = B_1 e^{jay} + B_2 e^{-jay} \quad \text{...(7.24)}$$

or equivalently

$$Y = B_3 \cos ay + B_4 \sin ay \quad \text{...(7.25)}$$

Therefore, the potential function in the variables x and y can be written as

$$V = (A_1\, e^{ax} + A_2\, e^{-ax})\,(B_1\, e^{jay} + B_2\, e^{-jay})$$

or

$$V = (A_3 \cosh ax + A_4 \sinh ax)\,(B_3 \cos ay + B_4 \sin ay)$$

As mentioned earlier that as Laplace's equation is a linear homogeneous equation, a sum of products of the above form each product corresponding to a different value of a— is also a solution.

Example 7.6. In spherical coordinates $V = 0$ for $r = 0.10$ m and $V = 150\ V$ for $r = 2.5$ m. Assume free space between these spherical shells, use Laplace's equation to find E and D.

Solution. Since V is not a function of ϕ and θ, Laplace's equation reduces to

$$\frac{1}{r^2}\frac{d}{dr}\left(r^2\frac{dV}{dr}\right) = 0$$

or

$$\frac{d}{dr}\left(r^2\frac{dV}{dr}\right) = 0$$

On integration we have

$$r^2\frac{dV}{dr} = A$$

or

$$dV = \frac{A\,dr}{r^2}$$

$$V = -\frac{A}{r} + B$$

At $r = 0.1$ $V = 0$ and $r = 2.5$ m $V = 150$ V

$$0 = -\frac{A}{0.1} + B \quad \text{or} \quad B = 10A$$

Another boundary condition $150 = -\dfrac{A}{2.5} + B$

since $B = 10A$ $\quad 150 = -\dfrac{A}{2.5} + 10A = \dfrac{24}{2.5}A$

or

$$A = \frac{150 \times 2.5}{24} = 15.625$$

and hence $B = 156.25$ volts

Hence $$V = -\frac{15.625}{r} + 156.25$$

$$E = -\nabla V = +\frac{dV}{dr}\bar{a}_r = -\frac{15.6}{r^2}\bar{a}_r \text{ V/m}$$

$$D = \varepsilon_0 E = \frac{8.854 \times 10^{-12} \times 15.6}{r^2}\bar{a}_r = \frac{138}{r^2} \times 10^{-12}\ \bar{a}_r \text{ C/m}^2 \text{ } \textbf{Ans.}$$

Example 7.7. Consider a parallel plate capacitor. The plates are infinite in extent and are separated by a distance x_1. The left hand plate is located at $x = 0$ and its potential is zero, whereas the right hand plate is at $x = x_1$ and has a potential V_1. Use Laplace's equation to find the potential distribution between the plates.

Solution. Since potential is not a function of y or z, the Laplace's equation is in one dimension and is given as

$$\frac{d^2V}{dx^2} = 0 \quad \text{or} \quad \frac{dV}{dx} = C_1 \quad \text{or} \quad V = C_1 x + C_2$$

To determine C_1 and C_2 we make use of boundary condition at $x = 0$ $V = 0$ and at $x = x_1$ $V = V_1$

$$0 = C \cdot 0 + C_2 \quad \text{or} \quad C_2 = 0$$

$$V_1 = C_1 x_1 + C_2 = 0 \text{ or } C_1 = \frac{V_1}{x_1}$$

Hence $V = \frac{V_1}{x_1}x$ is a solution of Laplace's equation that satisfies the boundary conditions and gives the potential variation between the capacitor plates.

Example 7.8. A volume charge density for a system is approximated by

$$\rho_V = 2\rho_0 \operatorname{sec h}\frac{x}{a}\tanh\frac{x}{a}.$$

Using Poisson's equation determine potential function for the system. Assume $E \to 0$ as $x \to \infty$ and at $x = 0$ $V = 0$.

Solution. Since volume charge density is a function of x only Poisson's equation will be with one variable x

$$\frac{d^2V}{dx^2} = -\frac{\rho_V}{\varepsilon} = -\frac{2\rho_{v_0}}{\varepsilon}\operatorname{sec h}\frac{x}{a}\tanh\frac{x}{a}$$

On integrating once, we have

$$\frac{dV}{dx} = \frac{2\rho_0}{\varepsilon} a \operatorname{sec h}\frac{x}{a} + C_1$$

Now $$E = -\frac{dV}{dx} = -\frac{2\rho_0 a}{\varepsilon} \cdot \frac{2}{e^{x/a} + e^{-x/a}} + C_1$$

as $$x \to \infty,\ E \to 0$$

$$0 = 0 + C_1 \text{ as } e^{x/a} \to \infty \text{ as } x \to \infty$$

or $$C_1 = 0$$

Hence $$\frac{dv}{dx} = -\frac{2\rho_0 a}{\varepsilon}\operatorname{sech}\frac{x}{a}$$

Integrating once again we have

$$V = \frac{4\rho_v a^2}{\varepsilon}\tan^{-1} e^{x/a} + C_2$$

Now at $x = 0$ $V = 0$

$$0 = \frac{4\rho_v a^2}{\varepsilon}\tan^{-1} e^0 + C_2$$

$$= \frac{4\rho_v a^2}{\varepsilon}\cdot\frac{\pi}{4} + C_2$$

or $$C_2 = -\frac{4\rho_v a^2}{\varepsilon}\frac{\pi}{4}$$

Hence $$V = \frac{4\rho_v a^2}{\varepsilon}\left[\tan^{-1} e^{x/a} - \frac{\pi}{4}\right] \quad \textbf{Ans.}$$

Example 7.9. Consider 7.7 for the parallel plate capacitor but now suppose that a uniformly distributed cloud of electrons exists between the plates. Determine the potential function between the plates.

Solution. Since free charges exist in the region between the plates Poisson's equation can be applied

$$\nabla^2 V = -\frac{\rho}{\varepsilon_0}$$

$$\frac{d^2V}{dx^2} = -\frac{\rho}{\varepsilon_0}$$

or $$\frac{dV}{dx} = -\frac{\rho}{\varepsilon_0}x + A$$

$$V = -\frac{\rho}{\varepsilon_0}\frac{x^2}{2} + Ax + B$$

At $$x = 0,\ V = 0$$

$$x = l,\ \ V = V_0,\ \ 0 = B$$

$$V_0 = \frac{-\rho}{2\varepsilon_0}l^2 + Al$$

or
$$A = \frac{V_0}{l} + \frac{\rho}{2\varepsilon_0} l$$

Hence the complete solution

$$V = -\frac{\rho}{2\varepsilon_0} x^2 + \left(\frac{V_0}{l} + \frac{\rho l}{2\varepsilon_0}\right) x$$

Hence the potential function V is given as

$$V = \left(\frac{V_0}{l} + \frac{\rho l}{2\varepsilon_0}\right) x - \frac{\rho}{2\varepsilon_0} x^2 \quad \textbf{Ans.}$$

PROBLEMS

P 7.1 Starting from first principle derive Poisson's equation and Laplace's equation.

P 7.2 For certain system the potential field $V = (A\rho^4 + B\rho^{-4}) \sin 4\phi$, show that $\nabla^2 V = 0$ (*i*) Select A and B so that $V = 100$ volts and $|E| = 500$ V/m at $P(\rho = 1, \phi = 22.5, z = 2)$

Ans. $A = 112.5$ $B = -12.5$ or $A = -12.5$ $B = 112.5$

P 7.3 For the spherical conducting shells filled with two dielectrics as shown in Fig. P 7.3, determine the potential distribution in the region $c < r < a$.

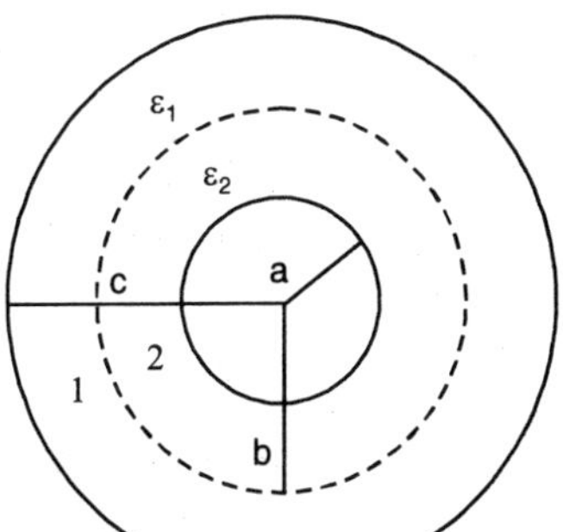

Fig. P 7.3

Ans. $$V_1 = \left[\frac{\frac{1}{a} - \frac{1}{r}}{\left(\frac{\varepsilon_2}{\varepsilon_1}\right)\left(\frac{1}{c} - \frac{1}{b}\right) + \left(\frac{1}{b} - \frac{1}{a}\right)} + 1\right] V_0,$$

$b < r < a$

$$V_2 = \frac{\frac{1}{c} - \frac{1}{r}}{\left(\frac{1}{c} - \frac{1}{b}\right) + \frac{\varepsilon_1}{\varepsilon_2}\left(1 + b - \frac{1}{a}\right)} \cdot V_0,$$

$c < r < b$

P 7.4 In spherical coordinates $V = -25$ V on a conductor at $r = 2$ cm and $V = 150$ V at $r = 35$ cm. The space between the conductors is a dielectric for which $\varepsilon_r = 3.12$. Determine the surface charge densities on the conductors.

Ans. $\rho_s = -256$ nC/m^2 at $r = 2$ cm
$= 0.837$ nC/m^2 at $r = 35$ cm

P 7.5 In cylindrical coordinates $\rho = 111/r$ pC/m^2. Assume $V = 0$ at $r = 1$m and $V = 50$ volts at $r = 3$m due to this charge configuration. Determine expression for E.

Ans. $\left(12.5 - \frac{68.3}{r}\right) \bar{a}_r$ V/m

P 7.6 Show that the rectangular components of the E field satisfy Laplace's equation in a charge free region.

P 7.7 Show that the function $f(r, \phi) = \dfrac{\cos\phi}{r}$ satisfies Laplace's equation in cylindrical coordinates.

P 7.8 In free space let $\rho_v = 200\ \varepsilon_0/r^{2.4}$. Use Poisson's equation to find $V(r)$ if it is assumed that $r^2 E_r \to 0$ as $r \to 0$ and also that $V \to 0$ as $r \to \infty$. **Ans.** $833.3\ r^4$ volts

P 7.9 The cross section of a coaxial cable is shown in Fig. P 7.9. For the potential of the electrode shown, determine the E field within the dielectric.

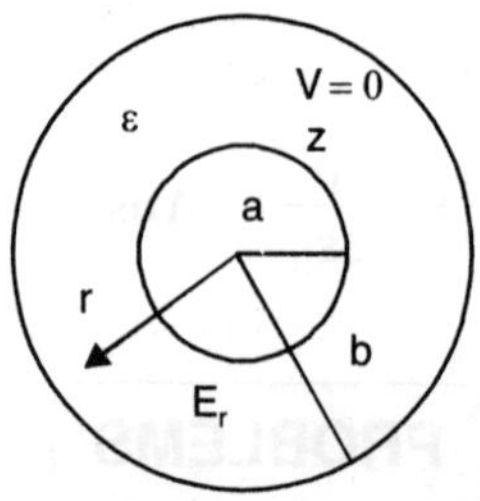

Ans. $E_r = \dfrac{-V_0}{\ln b/a} \cdot \dfrac{1}{r},\ a < r < b$

Fig. P 7.9

P 7.10 In Cartesian coordinates a plane at $z = 3$ cm is the voltage reference. Determine the voltage and the charge density on the conductor $z = 0$ if $E = 6.67 \times 10^3\ \bar{a}_z$ V/m for $z > 0$ and the region contains a dielectric for which $\varepsilon_r = 4.5$. **Ans.** 200 V, 266 nc/m^2

P 7.11 State and explain Uniqueness Theorem.

P 7.12 Determine E in spherical coordinates from Poisson's equation assuming a uniform charge density ρ.

Ans. $\left(\dfrac{\rho r}{3\varepsilon} - \dfrac{A}{r^2}\right) a_r$ V/m

Chapter 8

The Steady Magnetic Fields

8.1 INTRODUCTION

After studying electrostatics which is due to static charge distribution, we now study magnetic fields which are due to certain electric current distribution. The effect of this field on other current distribution will be discussed here assuming free space as we did in electrostatics where we later on considered the presence of dielectric materials. The relation between the source (charge distribution) in electrostatics is much more simpler as to the magnetic field and its source (current distribution). Just as we developed electrostatic field by assuming certain laws which were proved subsequently, we assume in magnetic fields also certain laws and we will prove them in subsequent chapter/s. Usually the source of steady magnetic field could be a permanent magnet or an electric field varying linearity with time or a direct current. We will of course not consider permanent magnet as source of steady magnetic fields and the time varying fields will be considered in subsequent discussion.

8.2 BIOT-SAVART LAW

We consider this law where the magnetic field is produced by a differential dc element in free space. Here a differential dc current element means a vanishing small section of a current carrying filamentary conductor, where a filamentary conductor is the limiting case of a cylindrical conductor of circular cross section as the radius approaches zero.

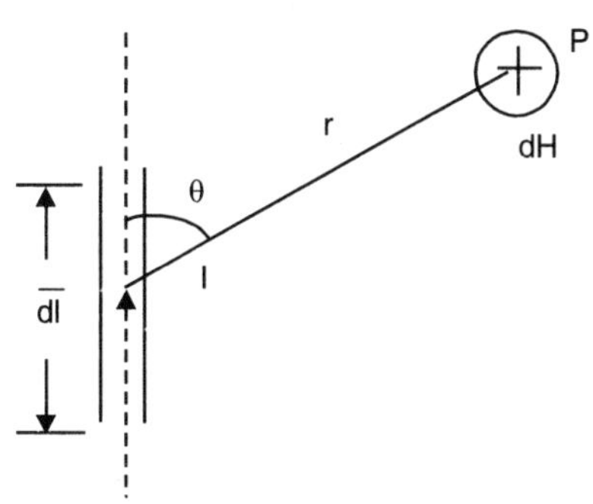

Fig. 8.1 Magnetic field intensity dH due to a current $Id\bar{l}$ at P

A current carrying conductor produces a magnetic field which is in the form of closed circular loops around the conductor. The relation of the magnetic field direction to the current direction can be easily remembered by means of the right hand thumb rule. With the thumb pointing in the direction of the curser the fingers of the right hand encircling the conductor, point in the direction of the magnetic field.

Consider a differential current $Id\bar{l}$ as shown in Fig. 8.1.

Then according to Biot-Savart's law the magnitude of magnetic field intensity dH produced due to the current element. $Id\bar{l}$ at point P is proportional to the product of the current, the magnitude of the differential length $d\bar{l}$ and sine of the angle (θ) lying between the filament axis to the point P at which the field is required, also the magnitude of the magnetic field intensity is inversely proportional to the square of the distance from the differential element to the point $P(r)$.

Biot-Savart's law can be described in the mathematical terms as

$$\overline{dH} = \frac{Id\bar{l} \times \bar{a}_r}{4\pi r^2} \quad \ldots(8.1)$$

and the magnitude of

$$dH = \frac{Idl \sin\theta}{4\pi r^2} \quad \ldots(8.2)$$

The units of H are clearly A/m. The Biot-Savart's law is sometimes referred to as Ampere's law for the current element. However, we shall retain the former name (Biot-Savart's) because of possible confusion with Ampere's circuital law which will be discussed in subsequent sections of this chapter.

The direction of the magnetic field intensity is normal to the plane containing the differential element a and the line drawn from the filament to the point P. Of the two possible normals, that one is to be chosen which is in the direction of progress of a right handed screw turned from $d\bar{l}$ through the smaller angle to the line from the filament to P or as given by right hand thumb rule as explained earlier.

Next we consider magnetic field intensity due to an infinitely long current carrying conduct at a point P with R units from the conductor as shown in Fig. 8.2.

From Fig. 8.2.

$$dl \sin\theta = rd\theta \text{ and } R = r \sin\theta$$

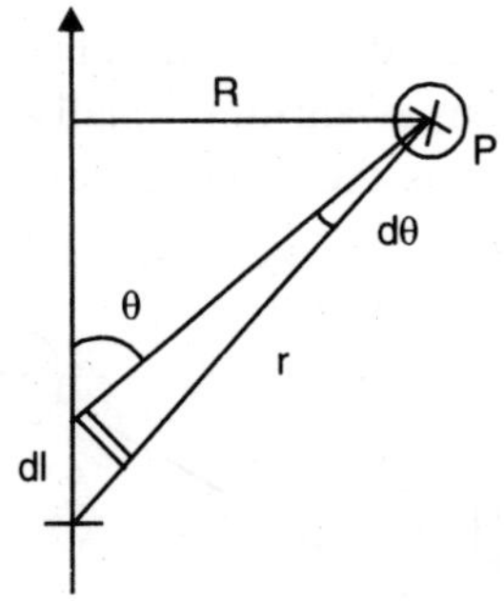

Fig. 8.2 Magnetic field intensity at P due to an infinitely long wire

Substituting this relation in equation 8.2, we have

$$dH = \frac{I}{4\pi}\int \frac{r\,d\theta}{r^2} = \frac{I}{4\pi}\int \frac{d\theta}{r} = \frac{I}{4\pi}\int \frac{\sin\theta}{R} d\theta = \frac{I}{4\pi R}\int_0^{\pi} \sin\theta\, d\theta \quad \ldots(8.3)$$

Here to cover the complete length θ varies from 0 to π

Hence
$$H = \frac{I}{4\pi R}[-\cos\theta]_0^{\pi} = \frac{I}{2\pi R}\ \text{A/m}. \qquad \ldots(8.4)$$

Next we consider the field due to a finite length current element as shown in Fig. 8.3. Using equation (8.3) and changing the limits of integration from α_1 to α_2 we have

$$H = \frac{I}{4\pi R}\int_{\alpha_1}^{\alpha_2} \sin\theta\, d\theta$$

$$= \frac{I}{4\pi R}[-\cos\theta]_{\alpha_1}^{\alpha_2}$$

$$= \frac{I}{4\pi R}\left[\cos\alpha_1 - \cos\alpha_2\right] \qquad \ldots(8.5)$$

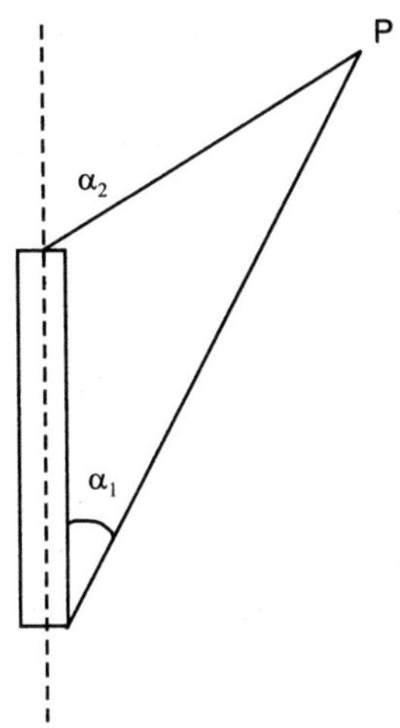

Fig. 8.3 Finite length of current loop

Next we consider magnetic field intensity due to a current loop as shown in Fig. 8.4.

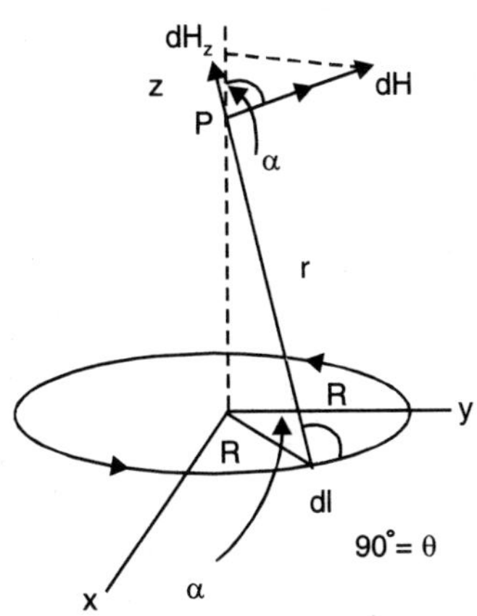

Fig. 8.4 *H* due to current loop

Referring to equation (8.2) we have here $\theta = 90°$,
However to find H at the centre of the loop $z = 0$

Hence
$$H = \oint_0^{2\pi} \frac{Idl}{4\pi R^2} = \int_0^{2\pi} \frac{IRd\phi}{4\pi R^2}$$

$$= \frac{I}{4\pi R}\int_0^{2\pi} d\phi = \frac{2\pi I}{4\pi R} = \frac{I}{2R} \text{ A/m}.$$

(*b*) Next we find magnetic field intensity at P on the z-axis. Here again $\theta = 90°$ and the direction of dH is normal to plane containing the current element and the line joining the current element and the point P. dH is shown in Fig. 8.4. This will have two components one along the z-axis and the other normal to dH_z. This component gets cancelled due to the corresponding element diametrically opposite to the current element under consideration.

Hence
$$dH = \frac{Idl}{4\pi\left(R^2 + z^2\right)}$$

and $dH_z = dH \cos\alpha$ and from Fig. 8.4 $\cos\alpha = \frac{R}{r}$

Hence
$$dH_z = \frac{I \cdot Rd\phi}{4\pi\left(R^2 + z^2\right)} \cdot \frac{R}{r} = \frac{IR^2}{4\pi\left(R^2 + z^2\right)^{3/2}}\int_0^{2\pi} d\phi$$

$$H_z = \frac{IR^2}{4\pi\left(R^2 + z^2\right)^{3/2}} \cdot 2\pi = \frac{IR^2}{2\left(R^2 + z^2\right)^{3/2}} \quad \textbf{Ans.} \qquad \ldots(8.6)$$

However at the centre of the loop $z = 0$, hence

$$H = \frac{IR^2}{2R^3} = \frac{I}{2R}$$

(*c*) To find out H at a height $z \to \infty$
Taking the expression for H_z

$$H_z = \frac{IR^2}{2\left(R^2 + z^2\right)^{3/2}}$$

Let
$$z \to \infty \quad R << Z$$

$$H_z = \frac{IR^2}{2z^3} \quad \textbf{Ans.} \qquad \ldots(8.6a)$$

Thus when $z \to \infty$, the magnetic field intensity H varies inversely as the cube of the axial distance z.

Example 8.1. A long straight wire carries a current $I = 20$ A. Determine the distance from the wire at which magnetic field intensity $H = 1$ A/m.

Solution. We know that for a straight long wire Equation (8.4).

$$H = \frac{I}{2\pi R} \quad \text{or} \quad R = \frac{I}{2\pi H} = \frac{20}{2\pi \times 1} = 3.18 \text{ m} \quad \textbf{Ans.}$$

Example 8.2. A current loop of radius 0.5 m carries certain current to produce magnetic field H = 1mA/m at the centre. (*a*) Determine the current (*b*) Determine the magnetic field H at a distance of 2 m from the loop axis.

Solution. (*a*) H at centre of current loop is given as

$$H = \frac{I}{2R} = 1 \times 10^{-3} = \frac{I}{2 \times 0.5} \quad \text{or } I = 1 \text{ mA} \quad \textbf{Ans.}$$

(*b*) Now H at a distance of 2 m from the centre of the loop and along z-axis.

$$H_z = \frac{IR^2}{2\left(R^2 + Z^2\right)^{3/2}} = \frac{1 \times 10^{-3} \times 0.5^2}{2\left(0.5^2 + 4\right)^{3/2}} = \frac{1 \times 10^{-3} \times 0.25}{2 \times 8.76}$$

$$= 14.27 \ \mu\text{A/m} \quad \textbf{Ans.}$$

8.3 AMPERE'S CIRCUITAL LAW

While solving problem in electrostatics we made use of Coulomb's law to begin with. Later on we found that application of Gauss's law for the solution of the same problem makes it very simple job. The only requirements for the application was that a high degree of symmetry in the system was very essential. Similarly we have solved some problems in magnetic fields using Biot-Savart's law. However, if symmetry in the system exists the same problem can be solved with very little efforts by using Ampere's circuital law.

Ampere's circuital law states that the line integral of H about any closed path is exactly equal to the direct current enclosed by the path

$$\oint H \cdot dl = I \qquad \text{...(8.6}b\text{)}$$

We define positive current as flowing in the direction of advance of a right handed screw turned in the direction in which the closed path is traversed. The application of Gauss's law provides the total charge enclosed by the closed surface, the application of Ampere's circuital law provides the current enclosed by the closed path.

If we try to find out magnetic field intensity due to an infinitely long conductor placed along z-axis with the current flowing along $\bar{a}_z$. Symmetry considerations show that there is no variation of H with z or ϕ. Also we know that the direction of H is normal to the plane containing $\overline{dl}$ and $\bar{R}$ and, therefore, H is in the direction of $\bar{a}_\phi$. Hence the only component of H is H_ϕ and it is a function of ρ only.

We, therefore, select a path to any section of which H is either perpendicular or tangential and along which H is constant. We know that the dot product of two vectors is zero whereas if they are largest to each either then the dot product is simply the product of magnitudes of the two vectors. Regarding the second requirement *i.e.* constancy of H, permits us to remove the magnetic field intensity from the integral sign. The integration required is usually simple and consists of finding the length of that portion of the path to which H is parallel.

For the problem under consideration (long length conductor along z-axis), the path must be a circle of radius ρ and Ampere's circuital law gives

$$\oint \bar{H} \cdot d\bar{l} = \int_0^{2\pi} H_\phi \rho d\phi = H_\phi \rho \oint_0^{2\pi} d\phi = H_\phi \rho \cdot 2\pi = I$$

or $$H_\phi = \frac{I}{2\pi\rho} \qquad \text{...(8.7)}$$

as before using Biot-Savart's law.

We next take another example and solve it using Ampere's circuital law. We consider an infinitely long co-axial transmission line carrying a uniformly distributed total current I in the inner conductor and $-I$ in the outer conductor as shown in Fig. 8.5.

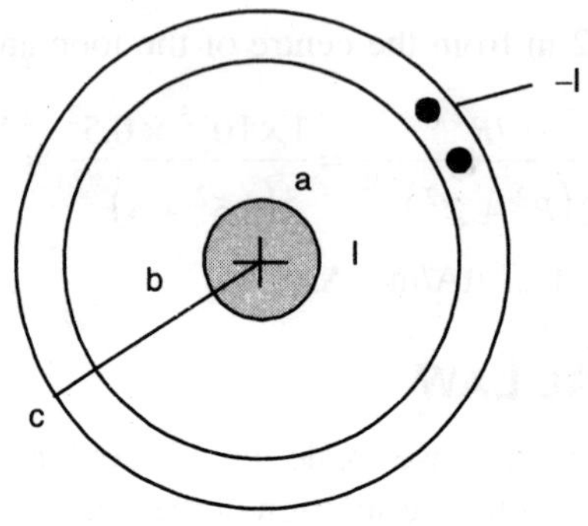

Fig. 8.5 Co-axial line

Consider a circular path within the inner conductor at a distance r such that $r < a$, the current enclosed by this path is $I \cdot \frac{r^2}{a^2}$, and let it magnetic field intensity be H_r, then

$H_r \cdot 2\pi r$ or $H \cdot dl$ and this should equal the current enclosed by the path *i.e.*

$$H_r \cdot 2\pi r = I \frac{r^2}{a^2} \qquad \text{...(8.8)}$$

Hence $$H_r = \frac{I}{2\pi} \cdot \frac{r^2}{a^2 r} = \frac{I}{2\pi a^2} \cdot r \; (r < a) \qquad \text{...(8.9)}$$

and at $r = a$ $$H_a = \frac{I}{2\pi a} \qquad \text{...(8.10)}$$

Next we take a circular path of a radius r such that $a < r < b$ *i.e.* it lies between the inner and outer conductor, the current enclosed is I, hence

$$H_r \cdot 2\pi r = I \text{ or } H_r = \frac{I}{2\pi r} \quad a < r < b$$

However, if the path lies within the entire conductor we have

$$I - I \frac{r^2 - b^2}{c^2 - b^2}$$

Hence $$2\pi r H_r = I - I\frac{r^2 - b^2}{c^2 - b^2} \quad b < r < c$$

$$= I\left[\frac{c^2 - b^2 - r^2 + b^2}{c^2 - b^2}\right] = \frac{c^2 - r^2}{c^2 - b^2}$$

Hence $$H_r = \frac{I}{2\pi r}\cdot\frac{c^2 - r^2}{c^2 - b^2} \quad b < r < c \qquad \text{...(8.11)}$$

when $r = b$ $$H_b = \frac{I}{2\pi b} \qquad \text{...(8.12)}$$

If we now take the path outside the outer conductor since the current enclosed is zero, the magnetic field intensity is zero and $r = c$ also $H_c = 0$. The variation of H as r varies from $r = 0$ to $r = c$ is shown in Fig. 8.6. Making use of the expression for different regions, the variation is given here.

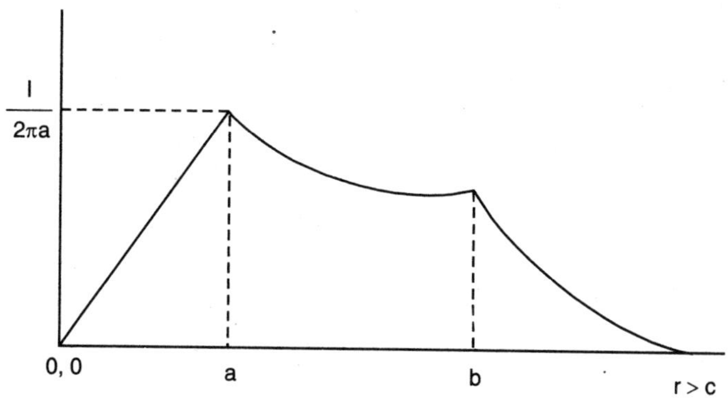

Fig. 8.6 Variation of H as a function of r for a co-axial line

Since the magnetic field outside the co-axial cable is zero, the outer conductor provides magnetic shield for the adjacent circuit.

Magnetic field is a solenoid: Refer to Fig. 8.7 for a solenoid.

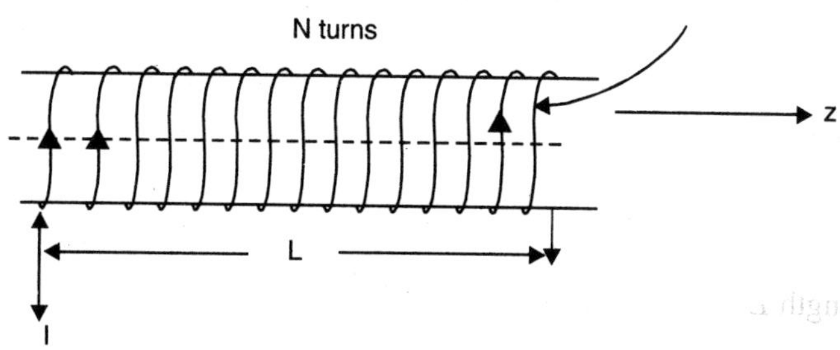

Fig. 8.7 Solenoid with N turns

Let L is the length of the solenoid, N the number of turns and I the current flowing in these turns.

If the N turns are closely wound then a differential current, loop will be $\frac{NI}{L}dz$.

To solve this problem we have assumed that the diameter of the solenoid is large compared with the diameter of the current carrying wire. In this way it is assumed that the solenoid is made up of a stack of current carrying loops as shown in Fig. 8.4. The current carrying loop contains conductor to width $\frac{N}{L}dz$ and the expression for H due to this loop is given by equation (8.6) as

$$dH_z = \frac{IR^2}{2\left(R^2+z^2\right)^{3/2}}\frac{N}{L}\cdot dz$$

The total field at point P is then

$$H_z = \frac{NI}{2L}\int_{z_1}^{z_2}\frac{R^2}{\left(R^2+z^2\right)^{3/2}}dz$$

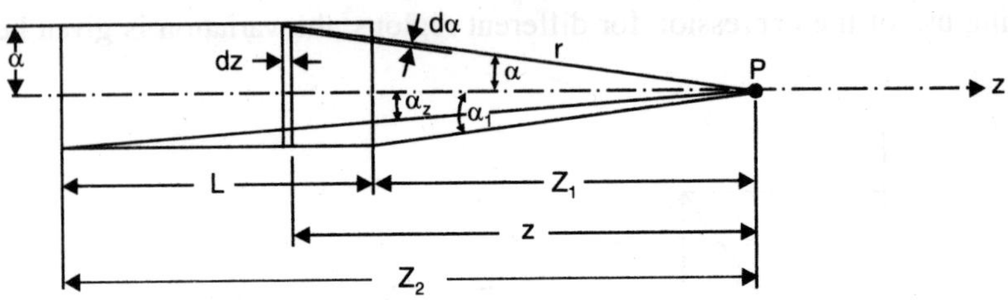

Fig. 8.8 A projection view of a solenoid

From Fig. 8.8 we have

$z = R\cot\alpha$, hence $R^2+z^2 = R^2+R^2\cot^2\alpha = R^2\,\mathrm{cosec}^2\,\alpha$ and $dz = -R\,\mathrm{cosec}^2\,\alpha\,d\alpha$

Substituting these values in expression for H_z above we have

$$H_z = -\frac{NI}{2L}\int_{\alpha_1}^{\alpha_2}\sin\alpha\,d\alpha$$

$$= -\frac{NI}{2L}\left(\cos\alpha_1-\cos\alpha_2\right) \qquad \ldots(8.13)$$

as the desired expression for H for a solenoid. Now two special cases of importance are

(*i*) Point P at the end of a solenoid of finite length. Here $\alpha_1 = 90$ and hence

$$H_z = \frac{NI}{2L}\cos\alpha_2 \text{ where } \cos\alpha_2 = \frac{L}{\sqrt{R^2+L^2}}$$

However, if the length L is large compared with the radius R, this expression reduces to

$$H_z = \frac{NI}{2L} \text{ for } L >> R \qquad \ldots(8.14)$$

(*ii*) If point P is at the centre of the solenoid which is finite in length, then $\alpha_2 = (\pi + \alpha_1)$ then

$$H_z = \frac{NI}{L}\cos\alpha \text{ where } \cos\alpha = \frac{L/2}{\sqrt{R^2+(L/2)^2}}$$

Hence $$H_z = \frac{NI}{L}\frac{L/2}{\sqrt{R^2+(L/2)^2}}$$

If $L >> R$, then

$$H_z = \frac{NI}{L} \quad \text{...(8.15)}$$

Comparing equation (8.15) with (8.14), it is clear that the magnetic field intensity at the ends of the solenoid is half of what it is at the centre of the solenoid. This means that half the flux lines at the centre of the solenoid pass out through the surface of the solenoid before they reach the end of the solenoid.

Magnetic field H in a toroid: Figure 8.9 gives a schematic view of a toroid with N closely wound turns. In case of a solenoid we have learnt that the magnetic field vector is directed along the axis of

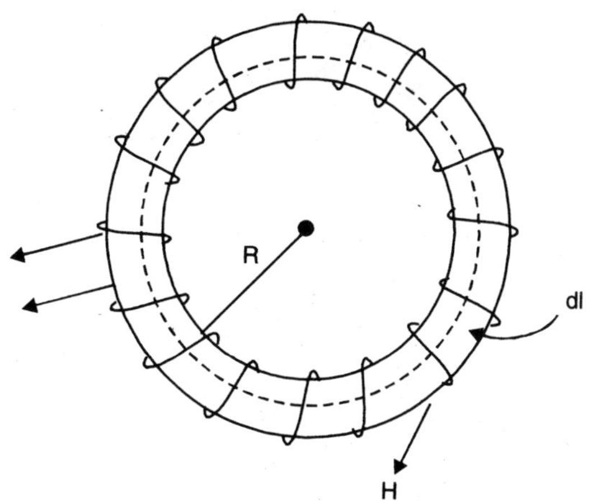

Fig. 8.9 A schematic view of a Toroid with N closely wound turns

the solenoid and the toroid may be considered as a solenoid that has been bent into a circle with the ends joined. Since the axis of the resulting bent solenoid is circular, the H-lines are also circular. As H inside the toroid is a constant at a distance R from the centre and is directed along dl, a direct application of Ampere's law we have

$$\oint H \cdot d\bar{l} = NI \quad \text{or} \quad H\oint dl = NI$$

Now $\oint dl$ for toroid $= 2\pi R$

Hence $$H \cdot 2\pi R = NI.$$

or $$H = \frac{NI}{2\pi R} \quad \text{...(8.16)}$$

From above it is clear that H varies inversely with R so that the value of H is not uniform across a section of the toroid. It is to be noted that if the path of integration lies outside of the toroid, the current enclosed is zero, so that H is zero everywhere outside.

8.4 STOKE'S THEOREM

Consider the surface S of Fig. 8.10 which is broken up into incremental surfaces of area ΔS. From chapter I, the definition of curl to one of these incremental surfaces can be applied and we have

$$(\nabla \times H)_N = \lim_{\Delta S \to 0} \frac{\oint H \cdot dL_{\Delta S}}{\Delta S} \qquad \ldots(8.17)$$

where N subscript indicates the right hand normal to the surface.

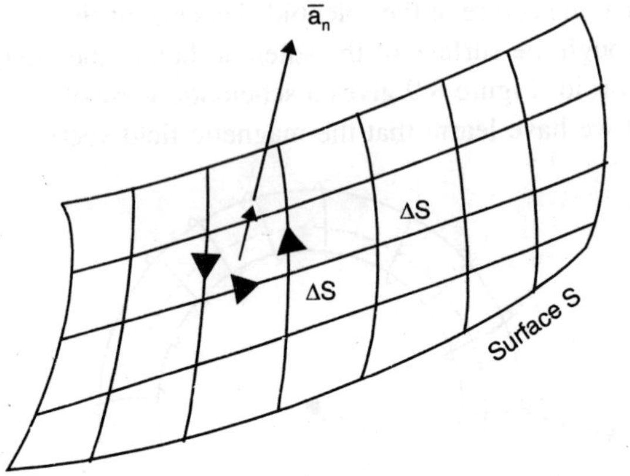

Fig. 8.10 The sum of the closed line integrals about the perimeter of every ΔS is the same as the closed line integral about the perimeter of S because of cancellation on every interior path

The subscript on $dL_{\Delta S}$ indicates that the closed path is the perimeter of an incremental area ΔS. Equation (8.17) may also be written as

$$\frac{\oint H \cdot dL_{\Delta S}}{\Delta S} = (\nabla \times H) \cdot \bar{a}_n$$

or

$$\oint H \cdot dL_{\Delta S} = (\nabla \times H) \cdot \bar{a}_n \Delta S = (\nabla \times H) \cdot \Delta \bar{S} \qquad \ldots(8.18)$$

where $\bar{a}_n$ is a unit vector in the direction of the right hand normal to ΔS (as according to cross product of ∇ and H, the resultant is normal to plane containing ∇ and H).

If we consider circulation (curl) for every ΔS a part of S and sum the results. If we evaluate the closed line integral for each ΔS, some cancellation will occur because every interior wall is covered once in each direction as shown here. The only boundaries on which cancellation cannot occur from the outside boundary, the path, enclosing S. Therefore, we have

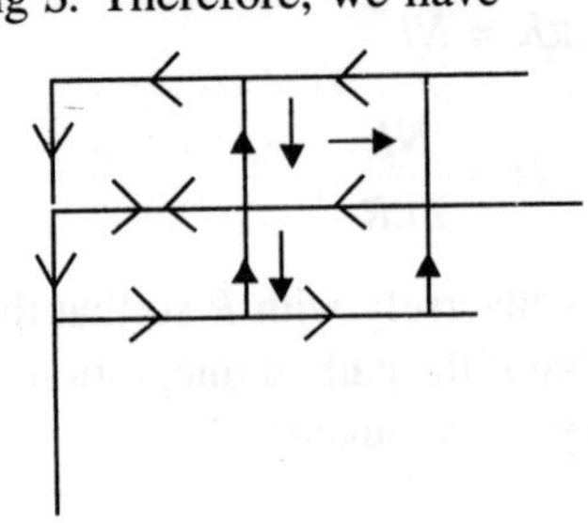

Fig. 8.11

$$\oint \bar{H} \cdot d\bar{L} = \int_s (\nabla \times \bar{H}) \cdot d\bar{s} \qquad \ldots(8.19)$$

where $d\bar{L}$ is taken only on the perimeter of *S*.

Equation (8.19) is an identity, valid for any vector field and is known as Stoke's theorem.

From equation (8.19) it is clear that Stoke's theorem relates a surface integral to a closed line integral. We know that whereas divergence theorem relates a volume integral to a closed surface integral.

We next derive Ampere's law from $\nabla \times H = J$. Dotting both the sides by $d\bar{S}$, integral each side over the same open surface *S* (Fig. 8.10) and apply Stoke's theorem we have

$$\int_s (\nabla \times H) \cdot d\bar{s} = \int J \cdot d\bar{S} = \oint \bar{H} \cdot d\bar{l} \qquad \ldots(8.20)$$

We have used $\nabla \times H = J$ in the above equation. The integral of the current density *J* over the surface *S* is the total current *I* passing through the surface and, therefore

$$\oint \bar{H} \cdot d\bar{l} = I \qquad \ldots(8.21)$$

The expression $\nabla \times H = J$ is known as point form of Ampere's circuital law and expression (8.21) is known as the integral form of Ampere's circuital law. From this it is clear that we can find the integral form of Ampere's circuital law from its point form and vice versa.

Example 8.3. A long straight conductor with radius *a* has magnetic field intensity $H = \frac{Ir}{2\pi r^2}\bar{a}_\phi$ for $r < a$ and $H = \frac{I}{2\pi r}\bar{a}_\phi$ for $r > a$. Determine *J* in both region.

Solution. Within the conductor $r < a$

Since $\nabla \times H = J$

and from the expression for curl of *H*, the component of H_ϕ are to be taken into consideration.

Hence
$$J = \nabla \times H = -\frac{\partial}{\partial z}\left(\frac{Ir}{2\pi a^2}\right)\bar{a}_1 + \frac{1}{r}\left[\frac{\partial}{\partial r} r\, H_\phi\right]\bar{a}_z$$

$$= 0 + \frac{1}{r}\frac{\partial}{\partial r}\left(\frac{r^2 I}{2\pi a^2}\right)\bar{a}_z = \frac{I}{2\pi a^2}\cdot\frac{1}{r}\cdot 2r\cdot\bar{a}_z$$

$$= \frac{I}{\pi a^2}\bar{a}_z$$

which corresponds to a current *I* in the +*z* direction which is distributed uniformly over the cross sectional area πa^2.

For $r > a$

$$J = \nabla \times H = -\frac{\partial}{\partial z}\left(\frac{I}{2\pi r}\right)\bar{a}_r + \frac{1}{r}\frac{\partial}{\partial r}\left(\frac{rI}{2\pi r}\right)\bar{a}_z$$

$$= 0 + \frac{1}{r}\frac{\partial}{\partial r}\left(\frac{I}{2\pi}\right)\bar{a}_z$$

$$= 0 + 0 = 0.$$

Example 8.4. The current density J in cylindrical coordinates for a region $0 < r \leq 0.6$ m, $J = 5e^{-2r}\bar{a}_r$ A/m^2 and $J = 0$ elsewhere. Determine magnetic field intensity H using Ampere's law.

Solution. Since J is along $\bar{a}_z$, the current density is symmetrical about origin and hence circular path can be used in Ampere's law. For $r < 0.6$ m.

$$H_\phi \cdot (2\pi r) = \int_0^{2\pi}\int_0^{r} 5e^{-2r} r\, dr\, d\phi$$

$$= 10\pi \int_0^r re^{-2r} dr$$

$$= 10\pi\left[-\frac{r}{2}e^{-2r} + \int \frac{e^{-2r}}{2} dr\right]_0^r$$

$$= 10\pi\left[-\frac{r}{2}e^{-2r} - \frac{e^{-2r}}{4}\right]_0^r$$

$$(2\pi r H_\phi) = 10\pi\left[-\frac{r}{2}e^{-2r} - \frac{e^{-2r}}{4} + 1\right]$$

$$H_\phi = \frac{2.5\pi}{2\pi r}[1 - e^{-2r} - 2re^{-2r}]$$

$$= \frac{1.25}{r}[1 - e^{-2r} - 2re^{-2r}]\bar{a}_\phi$$

For $r = 0.6$

$$H_\phi = \frac{1.25}{0.6} \times (1 - 0.3 - 1.2 \times 0.3)$$

$$= \frac{1.25}{0.6} \times 0.34 = 0.7 \text{ A/m} \quad \textbf{Ans.}$$

Example 8.5. A circular conductor of radius $r_0 = 1.5$ cm has magnetic field intensity.

$$H = \frac{10^4}{r}\left(\frac{1}{a^2}\sin ar - \frac{r}{a}\cos ar\right)\bar{a}_\phi \text{ A/m}$$

where $a = \pi/2r_0$. Determine the total current in the conductor.

Solution. Using Ampere's law and substituting for $a = \dfrac{\pi}{2r_0}$

We have
$$I = \oint_{r=r_0} H \cdot d\bar{l} = \int_0^{2\pi} \frac{10^4}{r_0}\left(\frac{4r_0^2}{\pi^2}\sin\frac{\pi}{3} - \frac{2r_0^2}{\pi}\cos\frac{\pi}{3}\right) r_0 d\phi$$

$$= 10^4\left(\frac{4\times1.5^2\times10^{-4}}{\pi^2}\cdot\frac{\sqrt{3}}{2} - \frac{2\times1.5^2\times10^{-4}}{\pi}\cdot\frac{1}{2}\right)$$

$$= (0.79 - 0.716) = 0.074 \text{ Amp} \quad \textbf{Ans.}$$

Example 8.6. Determine the magnetic field intensity at the centre of a square loop of side L carrying a current I as shown in Fig. E 8.6.

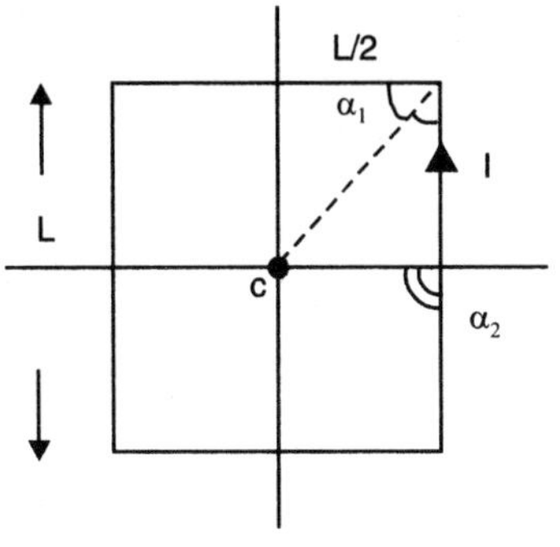

Fig. E 8.6

Solution. Using equation (8.5) we have the field intensity due to $\frac{L}{2}$ length of the wire.

$$H = \frac{I}{4\pi L/2}(\cos\alpha_1 - \cos\alpha_2) = \frac{I}{4\pi\frac{L}{2}}\cdot\frac{1}{\sqrt{2}}$$

and since for all the 8 parts of the wire which gives cumulative effect of intensities we have

$$H = \frac{8I}{4\pi L}\frac{2}{\sqrt{2}} = \frac{8\sqrt{2}\,I}{4\pi L} = \frac{2\sqrt{2}\,I}{\pi L}$$

and the direction will be according to right hand thumb rule along $\bar{a}_z$. Hence

$$H = \frac{2\sqrt{2}}{\pi L} I\,\bar{a}_z \quad \textbf{Ans.}$$

8.5 MAGNETIC FLUX AND MAGNETIC FLUX DENSITY

The magnetic flux density B is defined in free space as $B = \mu_0 H$ where B is in wb/m^2 or Tesla and μ_0 is the permeability in free space and is equal to $\mu_0 = 4\pi \times 10^{-7}$ H/metre. B in magnetic field is analogous to D in electric field and similarly H the magnetic field intensity in magnetic field is analogous to electric field intensity E in electric fields. Suppose ϕ represents magnetic flux in webers and we consider ϕ to be the flux passing through any designated area. Then

$$\phi = \oint \bar{B} \cdot d\bar{S} \text{ Wb} \quad \ldots(8.22)$$

The magnetic flux ϕ in magnetic fields is analogous to electric flux ψ expressed in coulombs and according to Gauss' law the total flux passing through any closed surface is equal to the charge enclosed

$$\psi = \oint \bar{D} \cdot d\bar{S} = Q$$

Here charge Q is the source for the electric flux lines and these lines emanate from a positive charge and terminate on to negative charge.

It is to be noted that no such source exists for the lines of magnetic flux. We know that a conductor carrying direct current I produces flux or H field in the form of concentric circles about the current carrying filament. Since $B = \mu_0 H$, B field is identical to H and hence the magnetic flux lines are closed and do not terminate on a magnetic "charge". For this reason Gauss's law for magnetic field is

$$\oint \bar{B} \cdot d\bar{S} = 0 \quad \ldots(8.23)$$

Using divergence theorem, we have

$$\nabla \cdot \bar{B} = 0 \quad \ldots(8.24)$$

Equation (8.24) is the last of maxwell's four equations applied to static electric and steady magnetic fields. The four equations are

$$\begin{aligned} \nabla \cdot D &= \rho_v \\ \nabla \times E &= 0 \\ \nabla \times H &= J \\ \nabla \cdot B &= 0 \end{aligned} \quad \ldots(8.25)$$

Other useful relations in electric and magnetic fields are (in free space) $D = \varepsilon_0 E$, $B = \mu_0 H$

and $$E = -\nabla V$$

The four equations in equation (8.25) specify the divergence and curl of an electric and magnetic fields. The corresponding set of four equations in integral form which apply to static electric fields and steady magnetic fields is

$$\begin{aligned} \oint_s D \cdot dS &= \int_V \rho_v dv = Q \\ \oint \bar{E} \cdot d\bar{l} &= 0 \\ \oint \bar{H} \cdot d\bar{l} &= \oint J \cdot d\bar{S} = I \\ \oint \bar{B} \cdot d\bar{S} &= 0 \end{aligned} \quad \ldots(8.26)$$

By making use of vector analysis/calculus we can readily obtain one from the other by applying the divergence theorem or Stoke's theorem. With the use of flux and flux density in magnetic fields, we find the flux between the conductors of the co-axial cable with inner radius a and outer conductor radius b. The magnetic field intensity at any radius $a < r < b$

$$H_\phi = \frac{I}{2\pi r} a_\phi \quad a < r < b$$

and, therefore, $$B = \mu_0 H = \frac{\mu_0 I}{2\pi r} a_\phi$$

The magnetic flux contained between the conductors in a length L of co-axial cable

$$\phi = \oint \bar{B} \cdot d\bar{s} = \int_0^L \int_a^b \frac{\mu_0 I}{2\pi r} \bar{a}_\phi \cdot dr\, dz\, \bar{a}_\phi$$

$$= \frac{\mu_0 I}{2\pi} L \int_a^b \frac{dr}{r}$$

$$= \frac{\mu_0 IL}{2\pi} \ln \frac{b}{a}$$

Example 8.7. A solid conductor of circular cross section is made of a homogeneous non magnetic material. If the radius of the conductor is 2 mm and the current flowing is 20 A along z-axis determine (*i*) H_ϕ at r = 1 mm (*ii*) B_ϕ at r = 1.5 mm (*iii*) the total flux per unit length inside the conductor (*iv*) the total flux for r < 1 mm (*v*) the total flux outside the conductor.

Solution. (*i*) The current enclosed by the radius r = 1 mm

$$20 \times \frac{1}{4} = 5\text{A. Hence} \quad H = \frac{I}{2\pi r} = \frac{5000}{2\pi \times 1}$$

$$= 796 \text{ A/m}$$

(*ii*) The current upto r = 1.5 mm

$$20 \times \left(\frac{1.5}{2}\right)^2 = 11.25\,\text{A}$$

Hence $H = \dfrac{11.25 \times 1000}{2\pi \times 1.5}$ $\qquad B = \dfrac{11.25 \times 10^3 \times 4\pi \times 10^{-7}}{2\pi \times 1.5} = 1.5\,\text{mT}$

(*iii*) At any radius r, the current enclosed is

$$H = I \cdot \frac{r^2}{R^2} \cdot \frac{1}{2\pi r} = H$$

$$B = \frac{\mu_0}{2\pi} \frac{Ir}{R^2}$$

Hence $$d\phi = \frac{\mu_0 I}{2\pi} \frac{r}{R^2} \cdot dr \cdot 1$$

$$= \frac{\mu_0 I}{2\pi R^2} \cdot \left[\frac{r^2}{2}\right]_0^R$$

$$\phi = \frac{\mu_0 I}{4\pi} = \frac{4\pi \times 10^{-7} \times 20}{4\pi}$$

$$= 2\mu \text{ Wb} \quad \textbf{Ans.}$$

(*iv*) Total flux for $r < 1$ mm

$$\frac{\mu_0 I}{2\pi R^2}\cdot\left[\frac{r^2}{2}\right]_0^1 = \frac{4\pi\times 10^{-7}\times 1\times 10^{-6}}{2\pi\times 4\times 10^{-6}}$$

$$= 0.5\mu \text{ Wb}.$$

(*v*) Total flux outside the conductor would be infinity as the flux lines encircle from radius $r = 0$ to $r = \infty$. Hence flux lines would be infinity.

Example 8.8. A current filament carrying a current 5A in placed along z-axis. Determine the flux crossing the portion cf the plane $\phi = \frac{\pi}{4}$ defined by $0.01 < r < 1$ m and $0 < z < 2$ m.

Solution. For a straight circular conductor.

$$H = \frac{I}{2\pi r} \quad \text{or} \quad B = \frac{\mu_0 I}{2\pi r}\bar{a}_\phi$$

and $$d\bar{s} = dr\, dz\, \bar{a}_\phi$$

Hence $$d\phi = B\cdot ds = \frac{\mu_0 I}{2\pi\cdot r} dr\, dz$$

and $$\phi = \frac{\mu_0 I}{2\pi}\int_0^2\int_{0.01}^1 \frac{dr\, dz}{r} = \frac{2\mu_0 I}{2\pi}\ln\frac{1}{0.01}$$

$$= \frac{2\times 4\pi\times 10^{-7}\times 5}{2\pi}\times 4.6$$

$$= 20\times 10^{-7}\times 4.6$$

$$= 9.2\ \mu\text{Wb}.$$

8.6 THE SCALAR AND VECTOR MAGNETIC POTENTIALS

In electrostatic while finding out electric field intensity at a certain point due to certain charge distribution, we found it very convenient to find out scalar potential at that point and the negative of the gradient of scalar potential gives us the electric field intensity at that point.

Let us assume that there is a scalar magnetic potential and we designate it as V_m whose negative gradient gives us magnetic field intensity.

$$H = -\nabla V_m \qquad \ldots(8.27)$$

We have chosen V_m so that it does not create any problem with our previous results for the magnetic field. Therefore,

$$\nabla\times H = J = \nabla\times(-\nabla V_m) \qquad \ldots(8.28)$$

We know from our knowledge of vector calculus that the curl of the gradient of any scalar is zero. Therefore, if we want to define H as the gradient of a scalar magnetic potential, the current density J must be zero throughout the region in which the scalar magnetic potential is so defined.

$$H = -\nabla V_m \text{ and } J = 0$$

Since in most of the situations the current carrying conductor occupies relatively much smaller space as compared to the remaining space, the concept of scalar magnetic potential (valid for $J = 0$) is quite useful. The unit of V_m is amperes.

The scalar potential V_m also satisfies Laplace's equation. For free space

$$\nabla \cdot B = \mu_0 \nabla \cdot H = 0$$

and hence

$$\mu_0 \nabla \cdot (-\nabla V_m) = 0$$

or

$$\nabla^2 V_m = 0 \qquad (J = 0) \qquad \ldots(8.29)$$

V_m satisfies Laplace's equation as long as the current density in the space is zero. The main difference between the electric scalar potential and the magnetic scalar potential is that whereas the electric scalar potential is single valued function once a zero reference is defined, there is only one value V associated with each point in space, V_m is not a single valued function of position.

The multivaluedness of V_m as compared to single valuedness of V can be explained as follows. For an electrostatic field we know that

$$\nabla \times E = 0$$

$$\oint E \cdot dl = 0$$

and therefore scalar potential between two point a and b is

$$V_{ab} = -\oint E \cdot dl \qquad \ldots(8.30)$$

We know that in case of electrostatic fields the line integral between any two points is same irrespective of the path traversed. In magnetostatic fields, however,

$$\nabla \times H = 0 \text{ for } J = 0$$

but

$$\oint H \cdot dl = I$$

even if J is zero along the path of integration. Every time we make a round around the current, the result of integration increases by I. If no current I is enclosed by the path, a single valued potential function may be defined.

In general, however,

$$V_{m,\,ab} = -\int_b^a H \cdot dl \text{ for a specified path.} \qquad \ldots(8.31)$$

It is to be noted that whereas electrostatic potential V is a conservative field, the magnetic scalar potential V_m is not a conservative field.

Next we consider vector magnetic potential. This concept is quite useful in studying radiation from antennas and from apertures and radiation leakage from transmission lines, wave guides. The vector magnetic potential can be used for both the situation whether $J = 0$ or $J \neq 0$.

One choice of a vector magnetic potential is indicated by noting that

$$\overline{\nabla} \cdot \overline{B} = 0$$

and also we know that divergence of the case of any vector field is zero. Let

$$\overline{B} = \nabla \times \overline{A} \qquad \ldots(8.32)$$

where A represents a vector magnetic potential and this automatically satisfies the condition that the magnetic flux density shall have zero divergence. The H field is

$$H = \frac{1}{\mu_0} \nabla \times A$$

and
$$\nabla \times H = J = \frac{1}{\mu_0} \nabla \times \nabla \times A \quad \text{...(8.33)}$$

The curl of the curl of a vector ($\nabla \times \nabla \times A$) is not zero. If for a specific case $\bar{A}$ is known, the curl operation may be used twice to obtain the current density J.

Equation (8.32) provides a useful definition of the vector magnetic potential $\bar{A}$. Since the curl operation implies differentiation with respect to length, the units of A are webers per metre.

Using Biot-Savart's law, A may be defined as

$$A = \oint \frac{\mu_0 I dL}{4\pi R} \quad \text{...(8.34)}$$

The significance of the various terms used in equation (8.34) is the same as in Biot-Savart's law *i.e.* a direct current I flows through a filamentary conductor of which any differential element IdL is at a distance R from the point at which $\bar{A}$ is to be obtained. If we compare expression for electrostatic scalar potential with that equation (8.34), there is a definite similarity between vector magnetic potential and the electric scalar potential. Expression for scalar electric potential is given as

$$V = \int \frac{\rho_L dL}{4\pi\varepsilon_0 R}$$

In each expression integral is along a line source in one case line charge and in the other line current, each integrand is inversely proportional to the distance from the source to the point of interest and each involves characteristic of the media *i.e.* μ_0 and ε_0.

We now write equation (8.34) in its differential form

$$dA = \frac{\mu_0 I\, dL}{4\pi R} \quad \text{...(8.35)}$$

Let us consider a differential current element placed along z-axis as shown in Fig. 8.12. Here $dL = dz\bar{a}_z$. We use cylindrical coordinates to find $d\bar{A}$ at the point (r, ϕ, z).

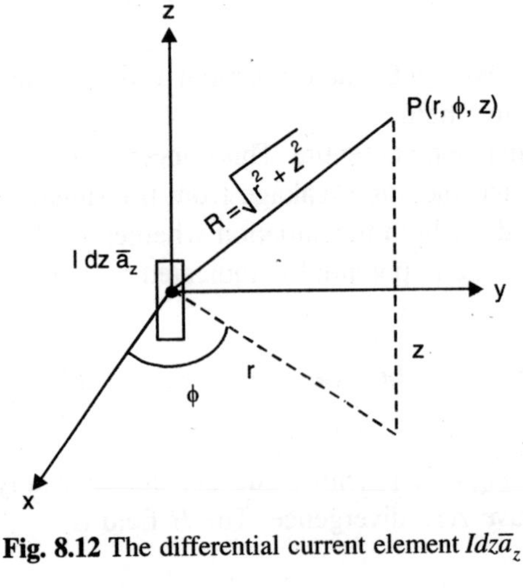

Fig. 8.12 The differential current element $Idz\bar{a}_z$

The differential vector magnetic potential at P is given as

$$dA = \frac{\mu_0 I dz \bar{a}_z}{4\pi\sqrt{r^2 + z^2}} \quad \ldots(8.36)$$

or

$$dA_z = \frac{\mu_0 I\, dz}{4\pi\sqrt{r^2 + z^2}}. \text{ However, } dA_r = dA_\phi = 0.$$

In order to find magnetic field intensity using the vector magnetic potential, we have

$$dH = \frac{1}{\mu_0}\nabla \times d\bar{A} = \frac{1}{\mu_0}\left(-\frac{\partial A_z}{\partial r}\right)\bar{a}_\phi$$

Now

$$\nabla \times dA = \left(-\frac{\partial}{\partial r} dA_z\right)a_\phi$$

$$= -\frac{\partial}{\partial r}\frac{\mu_0 I\, dz}{4\pi\sqrt{r^2 + z^2}}$$

$$= \frac{\mu_0 I dz \cdot r}{4\pi\left(r^2 + z^2\right)^{3/2}}$$

Hence

$$dH = \frac{1}{\mu_0}\cdot\frac{\mu_0 I\, dz\, r}{4\pi\left(r^2 + z^2\right)^{3/2}}\bar{a}_\phi$$

$$= \frac{Idz}{4\pi}\cdot\frac{r}{\left(r^2 + z^2\right)^{3/2}}\bar{a}_\phi \quad \ldots(8.37)$$

We have obtained this expression for dH using the vector magnetic potential, in fact we get the same expression using Biot-Savart law.

This shows that expression for dA as suggested in equation (8.35) is correct.

We can obtain expression for vector magnetic potential for a current source which is distributed *e.g.* it is in the form of a sheet with current density K in A/m. The differential current element $Id\bar{L} = \bar{K}d\bar{s}$.

If it is a volume current distribution with volume current density as J, we have

$$IdL = Jdv$$

With these differential current elements the expressions for A are given as

$$A = \int_s \frac{\mu_0 kds}{4\pi R} \quad \ldots(8.38)$$

and

$$A = \int \frac{\mu_0 Jdv}{4\pi R} \quad \ldots(8.39)$$

Equations (8.35), (8.38) and (8.39) express vector magnetic potential for different current distributions. Comparing these expressions with those of electrostatic field we find that just as in case of electrostatic field the reference for electric potential is at infinity. Here also for vector magnetic potential the reference for A is at infinity as no finite current element can produce any contribution to magnetic field as $R \to \infty$.

Next we make use of equation (8.33) as an alternative approach to understand vector magnetic potential.

$$\mu J = \nabla \times \mu H = \nabla \times (\nabla \times A)$$
$$= \nabla (\nabla \cdot A) - (\nabla \cdot \nabla) A \qquad \text{...(8.40)}$$

From equation we know that $\nabla \times A = B$. We don't yet know what $\nabla \cdot A = ?$. Let us choose $\nabla \cdot A = 0$.

In fact we will prove later that even for time varying fields $\nabla \cdot A = 0$. If we make this assumption then equation (8.40) reduces to

$$\nabla^2 A = -\mu J \qquad \text{...(8.41)}$$

which shows that vector A satisfies the vector form of Poisson's equation. Again equation (8.41) indicates that each component of A satisfies the scalar Poisson's equation with corresponding scalar component of J as it's source, just as the scalar electric potential V satisfies the Poisson equation with the scalar charge density ρ as its source. Here we have three scalar component equations.

$$A_x = \frac{\mu}{4\pi} \int \frac{J_x}{r} dv$$

$$A_y = \frac{\mu}{4\pi} \int \frac{J_y}{r} dv \qquad \text{...(8.42)}$$

$$A_z = \frac{\mu}{4\pi} \int \frac{J_z}{r} dv$$

Since the solution of equation (8.41) must be the vector sum of the individual component solutions, it follows that

$$\bar{A} = \frac{\mu}{4\pi} \int \frac{J}{r} dv \qquad \text{...(8.43)}$$

It is to be noted that each element of $\bar{A}$ is parallel to the corresponding element $\bar{I}$ or $\bar{J}$.

Example 8.9. Calculate the vector potential in the central plane of a straight current carrying conductor segment in air.

Solution. Since I is along z-axis then all contributions to A are parallel to $\bar{a}_z$. The vector integration, therefore, reduces to a scalar integration of the current element $Idl = Idz$. From Fig. E 8.9, we have

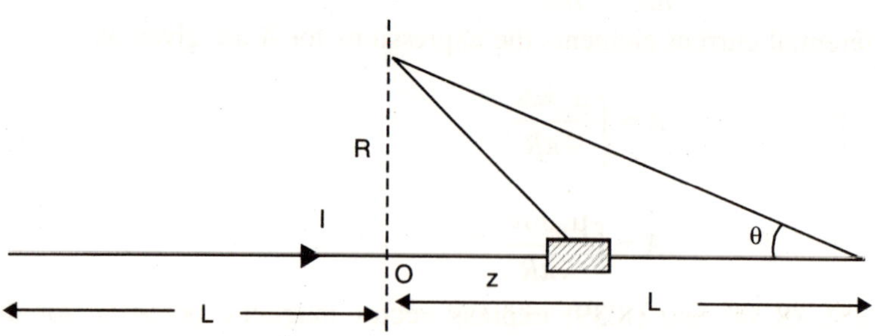

Fig. E 8.9

$$A = \bar{a}_z \frac{\mu_0}{4\pi} I \int_{-L}^{L} \frac{dz}{r} = \bar{a}_z \frac{2\mu_0 I}{4\pi} \int_0^L \frac{dz}{\left(R^2 + z^2\right)^{1/2}}$$

$$A = \frac{\mu_0 I}{2\pi} \bar{a}_z \left[\ln\left(z + \sqrt{R^2 + z^2}\right) \right]$$

$$= \frac{\mu_0 I}{2\pi} \ln \frac{\left(L + \sqrt{R^2 + L^2}\right)}{R} \bar{a}_z \quad \textbf{Ans.}$$

Example 8.10. A radial field $H = \frac{2.4 \times 10^6}{r} \cos\phi \, \bar{a}_r$ A/m exists in free space. Determine the magnetic flux ϕ crossing the surface defined by $-\frac{\pi}{4} < \phi < \frac{\pi}{4}$, $0 \le z \le 1$ m.

Solution.

$$B = \mu_0 H = 4\pi \times 10^{-7} \times \frac{2.4 \times 10^6}{r} \cos\phi \, \bar{a}_r$$

$$\simeq \frac{3}{r} \cos\phi \, \bar{a}_r$$

$$\phi = \int_0^1 \int_{-\pi/4}^{\pi/4} \left[\frac{3}{r} \cos\phi \, \bar{a}_r \right] r \, d\phi \, dz \, \bar{a}_r$$

$$= \int_0^1 \int_{-\pi/4}^{\pi/4} 3\cos\phi \, d\phi \, dz$$

$$= \int_{-\pi/4}^{\pi/4} 3\cos\phi \, d\phi = 3[\sin\phi]_{-\pi/4}^{\pi/4}$$

$$= 3 \cdot \sqrt{2} = 4.24 \, \text{Wb} \quad \textbf{Ans.}$$

Example 8.11. Determine the vector magnetic potential $\bar{A}$ in the region surrounding an infinitely long straight filamentary current I.

Solution. Using equation (8.32)

$$\nabla \times A = B$$

where $B = \frac{\mu_0 I}{2\pi r} \bar{a}_\phi$ for an infinitely long conductor. Since $\bar{B}$ possesses only a ϕ. Component, only the ϕ component of the cylindrical curl is required. Hence

$$\frac{\partial A_r}{\partial z} - \frac{\partial A_z}{\partial r} = \frac{\mu_0 I}{2\pi r} \qquad \text{...(8.11.1)}$$

Since the current filament is placed along z-axis, vector magnetic potential $\bar{A}$ cannot be a function of z.

Therefore equation (8.11.1) reduces to

$$-\frac{\partial A_z}{\partial r} = \frac{\mu_0 I}{2\pi r}$$

or
$$-dA_z = \frac{\mu_0 I}{2\pi} \cdot \frac{dr}{r}$$

or
$$A_z = -\frac{\mu_0 I}{2\pi} \ln r + C$$

Since A_z is vector magnetic potential, it must have a zero reference level. Let this be a distance r_0 *i.e.* as $r \to r_0$ $A_z \to 0$.

$$0 = -\frac{\mu_0 I}{2\pi} \ln r_0 + C$$

or
$$C = \frac{\mu_0 I}{2\pi} \ln r_0$$

Hence
$$A_z = \frac{\mu_0 I}{2\pi}\left(\ln \frac{r_0}{r}\right)\bar{a}_z \quad \textbf{Ans.}$$

Example 8.12. A square loop measuring 1.5 by 1.5 m carries a steady current of 10*A*. The square loop is placed in the *xy* plane as shown in Fig. E 8.12, the origin coincides with a corner of the square. Determine the flux density *B* at a point on the *y*-axis 0.35 m from the origin.

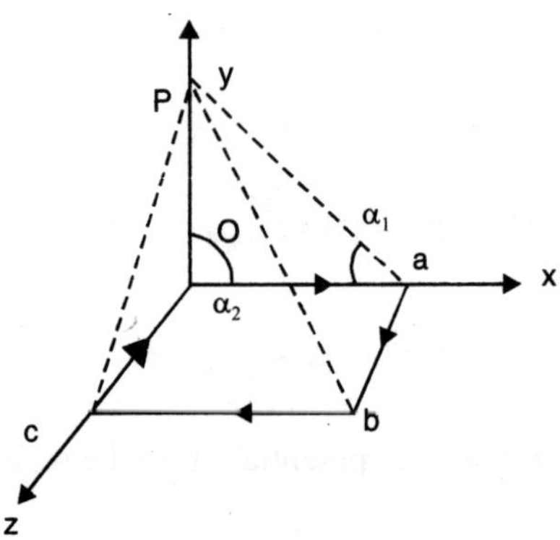

Fig. E 8.12

Solution. From the Fig. E 8.12 $OP = 0.35$ m

$$aP = \sqrt{0.35^2 + 1.5^2} = 1.54 \text{ m} = cP$$

and
$$bP = \sqrt{0.35^2 + \left(1.5\sqrt{2}\right)^2} = \sqrt{0.1225 + 4.45} = 2.15$$

Now using the relation $B = \frac{\mu_0 I}{4\pi R}(\cos\alpha_1 - \cos\alpha_2)$

where α_1 and α_2 are the angle subtended by the line joining the point P and the end of the current element.

$$B_{oa} = B_{co} = \frac{4\pi\times10^{-7}\times10}{4\pi\times0.35}\left(\frac{1.5}{1.54} - \cos\frac{\pi}{2}\right) = 2.783\ \mu\text{T}$$

$$B_{ab} = B_{bc} = \frac{4\pi\times10^{-7}\times10}{4\pi\times1.54}\left(\frac{1.5}{2.15} - 0\right) = 0.45\mu\text{T}$$

In terms of vector components.

$$\bar{B}_{oa} = 2.783\left(a_x\cdot0 + 0\bar{a}_y + 1\bar{a}_z\right)$$

$$B_{co} = 2.783\left(1\cdot\bar{a}_x + 0\bar{a}_y + 0\bar{a}_z\right)$$

$$B_{ab} = 0.45\left(-\frac{0.35}{1.54}\bar{a}_x - \frac{1.5}{1.54}\bar{a}_y + 0\bar{a}_z\right)$$

and

$$B_{bc} = 0.45\left(0\cdot\bar{a}_x - \frac{1.5}{1.54}\bar{a}_y - \frac{0.35}{1.54}\bar{a}_z\right)$$

The resultant flux density is given as

$$3.89\left(0.69\bar{a}_x - 0.226\bar{a}_y + 0.69\bar{a}_z\right)\mu\text{T}\ \textbf{Ans.}$$

Example 8.13. An H due to a current source is given by $H = (y\cos ax)\bar{a}_x + (y + e^x)a_z$. Determine the current density over the yz plane or $x = 0$ plane.

Solution. $\nabla \times H = J$

Now

$$\nabla\times H = \left(\frac{\partial H_z}{\partial y} - \frac{\partial H_y}{\partial z}\right)\bar{a}_x + \left(\frac{\partial H_x}{\partial z} - \frac{\partial H_z}{\partial x}\right)\bar{a}_y + \left(\frac{\partial H_y}{\partial x} - \frac{\partial H_x}{\partial y}\right)\bar{a}_z$$

We are given $H_x = y\cos a\,x$, $H_y = 0$ and $H_z = y + e^x$. Substituting H component values in the curl equations, we have

$$\frac{\partial}{\partial y}\left(y + e^x\right)\bar{a}_x + \frac{\partial}{\partial z}(y\cos ax)\bar{a}_y - \frac{\partial}{\partial x}\left(y + e^x\right)\bar{a}_y - \frac{\partial}{\partial y}(y\cos ax)\bar{a}_z$$

$$= \bar{a}_x + 0 - e^x\bar{a}_y - \cos ax\,\bar{a}_z$$

Now at $x = 0$ $\qquad J = \bar{a}_x - \bar{a}_y - \bar{a}_z$

which is a constant vector of magnitude $\sqrt{3}\ \text{A/m}^2$ **Ans.**

Example 8.14. In a coaxial cable the current density in the inner conductor is given by $J(r) = J_0 e^{-(a-r)}$ where a is the radius of the inner conductor. The outer conductor is an infinitely thin cylindrical shell of radius b. Determine the magnetic field intensity everywhere.

Solution. The current density at the surface of the inner conductor is J_0 when $r = a$.

Using Ampere's law for the inner conductor

$$H = \frac{I}{2\pi r} \text{ or } H \cdot 2\pi r = I = \int_0^r J \cdot 2\pi r\, dr$$

$$= \int_0^r 2\pi J_0 e^{-(a-r)} r\, dr$$

$$= 2\pi J_0 e^{-a} \int_0^r e^r r\, dr$$

$$= 2\pi J_0 e^{-a} \left[re^r - \int e^r dr \right]_0^r$$

$$= 2\pi J_0 e^{-a} \left[re^r - e^r \right]_0^r$$

$$= 2\pi J_0 e^{-a} \left[re^r - e^r + 1 \right] = 2\pi J_0 e^{-a} \left[1 + e^r (r-1) \right]$$

Hence $$H_\phi = \frac{J_0 e^{-a}}{r} \left[1 + e^r (r-1) \right] \textbf{ Ans.}$$

For $a < r < b$

Current $$I = 2\pi J_0 \left[(a-1) + e^{-a} \right]$$

so that $$H_\phi = \frac{2\pi J_0}{2\pi r} (a - 1 + e^{-a})$$

For $r > b$ since $I = 0$ $\quad H_\phi = 0$ **Ans.**

Example 8.15. Determine the scalar magnetic potential from an infinite long current carrying conductor placed along z-axis.

Solution. We know that for such a system $H = \frac{I}{2\pi r} \bar{a}_\phi$

Since $$H = -\nabla V_m = \frac{I}{2\pi r} \bar{a}_\phi$$

$$\frac{1}{r} \frac{\partial V_m}{\partial \phi} = -\frac{I}{2\pi r}$$

$$\frac{\partial V_m}{\partial \phi} = -\frac{I}{2\pi} \text{ or } V_m = -\frac{I}{2\pi} \phi$$ ignoring the constant of integration.

Example 8.16. A cylindrical conductor with current density J_0 has vector magnetic potential as $A = \frac{\mu_0 I_0}{4} (x^2 + y^2) \bar{a}_z$. Determine the magnetic field intensity within the conductor.

Solution.

$$H = \frac{1}{\mu_0}\nabla \times A = -\left\{\frac{\partial}{\partial y}\left(\frac{\mu_0 I_0}{4}\right)(x^2+y^2)\right\}\bar{a}_x - \left\{\frac{\partial}{\partial x}\frac{\mu_0 I_0}{4}(x^2+y^2)\right\}\bar{a}_y$$

$$= \frac{\mu_0 I_0}{4}\left[2y\bar{a}_x - 2x\bar{a}_y\right]$$

$$= -\frac{I_0}{2}\left(y\bar{a}_x - x\bar{a}_y\right)\bar{a}_\phi = \frac{1}{2}J_0 \cdot r\bar{a}_\phi$$

Example 8.17. The vector magnetic potential in the *xy* plane is given by $A = (e^y \cos x)\bar{a}_x + (1+\sin^2 x)\,\bar{a}_z$. Determine the magnetic flux density at the origin.

Solution.

$$B = \frac{\partial}{\partial y}(1+\sin^2 x)\bar{a}_x - \frac{\partial}{\partial x}(1+\sin^2 x)\bar{a}_y - \frac{\partial}{\partial y}(e^y \cos x)$$

$$= -2\sin x \cos x\,\bar{a}_y - e^y \cos x\,\bar{a}_z$$

At $x = y = 0$ $\qquad B(0, 0) = -\bar{a}_z$ **Ans.**

PROBLEMS

P 8.1 State and explain Biot-Savart's law.

P 8.2 Derive an expression for magnetic flux density due to a short length of current carrying conductor at some point *P* when placed along *z*-axis.

P 8.3 Derive an expression for the magnetic field intensity at the centre of a current carrying loop with radius *a* and current *I*.

P 8.4 State and explain Ampere's circuital law.

P 8.5 Derive an expression for the magnetic field intensity at the ends of the solenoid and at the centre of the solenoid. State the assumptions made.

P 8.6 Derive an expression for the magnetic field intensity at the centre of a Toroid.

P 8.7 State and explain Stoke's theorem.

P 8.8 Explain clearly what you mean by scalar and vector magnetic potential. Give their physical significance.

P 8.9 Determine the magnetic field intensity at the centre of a square current loop of side *L*. The centre coincides with the origin of the rectangular coordinate system.

Ans. $H = \frac{2\sqrt{2}}{\pi L} I\bar{a}_z$

P 8.10 The magnetic flux density over the *xy* plane is given by $B = \sin\frac{\pi x}{2}\sin\frac{\pi y}{2}\bar{a}_z$

Determine the flux passing through the square $x > 0$, $y < 2$ m. **Ans.** 1.62 Wb

P 8.11 Determine (*i*) magnetic field intensity in rectangular components at *P*(2, 3, 4) if a current filament on the *z*-axis carries 8 mA in the $\bar{a}_z$ direction (*ii*) Repeat if the filament is located at $x = -1$, $y = 2$ (*iii*) Determine *H* if both filaments are present.

Ans. (*i*) $-294\bar{a}_x + 196\bar{a}_y$ μA/m

(*ii*) $-127\bar{a}_x + 382\bar{a}_y$ μA/m

(*iii*) $-421\bar{a}_x + 578\bar{a}_y$ μA/m

P 8.12 Determine the vector magnetic potential on the axis of a circular current loop.

Ans. Zero

P 8.13 Solve Laplace's equation for the scalar magnetic potential produced by the surface current density $K\bar{a}_z$ flowing on the plane $x = 0$.

Ans. $V_m = -\dfrac{K}{2}y + C$ where *C* is constant.

P 8.14 Determine the magnetic field intensity for a solid cylindrical conductor of radius '*a*' where the current *I* is uniformly distributed over the cross section.

Ans. $r < a \quad H = \dfrac{Ir}{2\pi a^2} a_\phi$

$r > a \quad H = (I/2\pi r)\bar{a}_\phi$

P 8.15 The vector magnetic potential in spherical coordinate is given as $\bar{A} = K \sin\theta \bar{a}_\theta$. Determine the magnetic flux density *B* at the point $\left(K, \dfrac{\pi}{2}, 0\right)$.

Ans. $B = \bar{a}_\phi$

P 8.16 An infinitely long straight wire carries 200 A of current and in its vicinity a circular loop of 50 mm diameter is located with the centre of the loop 0.5 m away from the straight conductor,

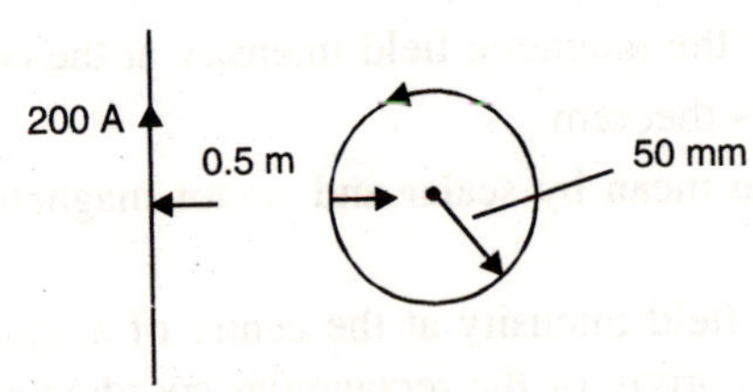

Fig. P 8.16

the wire and the loop are coplanar. Determine the current in the loop when the flux density *B* is zero in the space between the wire and the loop.

Ans. $\dfrac{20}{\pi}$ A

P 8.17 The magnetic field intensity is given in a certain region of space as $H = \left[(x+2y)/z^2\right]\bar{a}_y + \frac{2}{z}\bar{a}_z$ A/m (*i*) Determine $\nabla \times H$ (*ii*) Determine J (*iii*) Use J to find the total current passing through the surface $z = 4$, $1 \le x \le 2$, $3 \le z \le 5$ in the $\bar{a}_z$ direction.

Ans. (*i*) $\left\{2(x+2y)z^3\right\}\bar{a}_x + \frac{1}{z^2}\bar{a}_z$ A/m, (*ii*) Same as part (*i*) (*iii*) $\frac{1}{8}$ A

P 8.18 Assume that $A = 50r^2\bar{a}_z$ Wb/m in a certain region of free space. Determine (*i*) $\bar{H}$ and $\bar{B}$ (*ii*) J (*iii*) use J to find the total current crossing the surface $0 \le r \le 1.0$, $0 \le \phi \le 2\pi$ $z = 0$ (*iv*) Use the value of H_ϕ at $r = 1$ to calculate $\oint H \cdot d\bar{l}$ for $r = 1$, $z = 0$.

Ans. (*i*) $(-100r/\mu_r)\bar{a}_\phi$ A/m, $-100r\,\bar{a}_\phi$ Wb/m^2

(*ii*) $-\frac{200}{\mu_0}\bar{a}_z$ A/m^2 (*iii*) –500 MA (*iv*) –500 MA

P 8.19 Determine H on the axis at a height h of a circular current loop of radius a. Also determine H at the centre of the loop. The radius of the loop is a.

Ans. (*i*) $H = \frac{Ia^2}{2\left(a^2+h^2\right)^{3/2}}\bar{a}_z$, $H_c = \frac{I}{2a}\bar{a}_z$

P 8.20 An infinite current sheet lies in the $z = 0$ plane with $\bar{K} = K\bar{a}_y$. Determine the vector magnetic potential A for the current sheet.

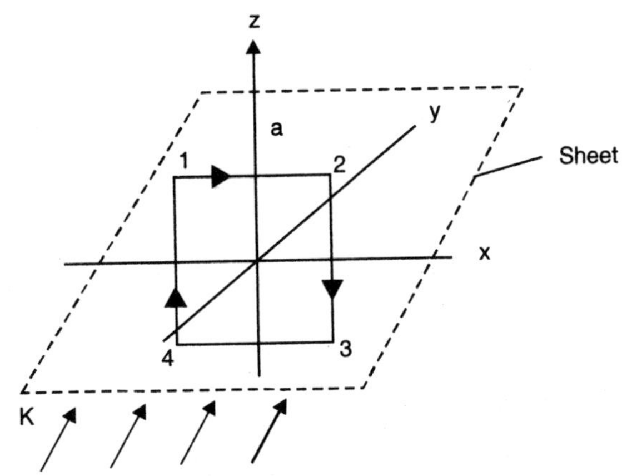

Fig. P 8.20

Ans. $\bar{A} = -\frac{\mu_0}{2}(z - z_0)\bar{K}$ for $z > 0$

$\bar{A} = \frac{\mu_0}{2}(z - z_0)\bar{K}$ for $z < 0$

P 8.21 Two identical circular current loops of radius a = 3 m and I = 20 A are in parallel planes separated on their common axis by 10 m. Determine H at a point midway between the two loops.

Ans. $0.905\bar{a}_n$ A/m

P 8.22 The vector magnetic potential within a cylindrical conductor of radius a is given as

$$A = \frac{\mu_0 I r^2}{4\pi a^2} a_z$$

Determine the magnetic field intensity H.

Ans. $\mu_0 H$ = curl A

P 8.23 Show that in a region with current density J, the vector magnetic potential A satisfies in rectangular coordinates, the equation

$$\nabla^2 A = -\mu_0 J.$$

P 8.24 A 1-cm current element carries 5 A along the x-axis. Determine the magnitude of the magnetic flux density $|B|$ at a radial distance of 5 m in the direction for which the angle θ with respect to the x-axis equals (*i*) 0° (*ii*) 45° (*iii*) 90°.

Ans. (*i*) 0 nT (*ii*) 0.14 nT (*iii*) 0.2 nT

Chapter

9

Magnetic Forces, Materials and Inductance

9.1 INTRODUCTION

We have studied in the previous chapter given a distribution of current sources in space we can find out the magnetic flux on the magnetic field intensity H, the magnetic flux intensity; the scalar magnetic potential V_m and the vector magnetic potential.

Next we consider the forces and torques exerted by the magnetic field on other charges stationary or moving. A magnetic field may be produced by moving charges and may exist force on moving charges. However, magnetic field can not arise due to stationary charges and it cannot produce force on a stationary charge.

Next we consider the properties of various magnetic materials, the analysis of some basic magnetic circuits and the equation of self and mutual inductants of some of the basic components of electric symptoms.

9.2 MAGNETIC FORCE ON CHARGED PARTICLES

When a charged particle is in motion in a magnetic field, it experiences a force at right angles to its velocity and the magnitude of the force is proportional to the charge the velocity and the magnetic flux density (and since the angle between v and B). Mathematically, the force F is given as

$$F = Q\bar{V} \times B \qquad ...(9.1)$$

which shows that the direction of a charged particle in motion can be changed by a magnetic field. The magnitude of the velocity $\bar{V}$ and consequently the kinetic energy will remain unchanged. It is to be noted that this is in contrast to the electric field when the force $F = QE$ does work on the charged particle and, therefore, changes its kinetic energy.

If the magnetic field B is uniform throughout the space and the charged particle has an initial velocity normal to the magnetic field, the path of the charged particle is a circle with certain radius r. The force of the field on the charged particle is of magnitude $F = |Q|VB$ and is directed towards the centre of the circle. The centripetal acceleration is of magnitude

$W^2 r = \dfrac{V^2}{r}$ where W is the angular velocity of the particle.

Thus, by Newton's second law of motion
Force = mass × acceleration

$$QVB = m\frac{V^2}{r} \text{ or } r = \frac{mV}{|Q|B} \quad ...(9.2)$$

It is to be noted that r is directly proportional to the linear momentum of the charged particle.

9.2.1 Electric and Magnetic Fields Combined

In an electric field, electric field intensity at a point is defined as the force experienced by the charge per unit charge *i.e.*

$$E = \frac{F}{Q} \text{ or } F = QE \quad ...(9.3)$$

The force is in the same direction as the electric field intensity (for a positive charge) and is directly proportional to both Q and E. If the charge is in motion, the force at any point in its trajectory is given by equation (9.3).

When both electric and magnetic fields are present in space at the same time, the force on a charged particle is given by the superposition of the total forces *i.e.*

$$F = Q(E + \bar{V} \times B) \quad ...(9.4)$$

Equation (9.4) is known as the Lorentz force equation and its solution is required in determining electron orbits in the magnetrons, protons path B in the cyclotrons or in general, charged particle motion in combined electric and magnetic fields.

Example 9.1. Determine the force on a charged particle of mass 9.1×10^{-27} kg and charge 1.6×10^{-19} C if it enters a magnetic field B of 5mw/m² with an initial velocity of 80 km/sec. Assume V and B are normal to each other. Also determine the radius of the circular path and time required for one revolution.

Solution. Since velocity and B are normal to each other sin θ = 1 and since

$$F = QVB = 1.6 \times 10^{-19} \times 80 \times 1000 \times 5 \times 10^{-3}$$
$$= 640 \times 10^{-19} \text{ Newtons. } \textbf{Ans.}$$

Now Force = m × acceleration

$$640 \times 10^{-19} = 9.1 \times 10^{-27} \times \frac{V^2}{r} = \frac{9.1 \times 10^{-27} \times 80^2 \times 10^6}{r}$$

$$= \frac{582.4 \times 10^{-19}}{r}$$

or
$$r = \frac{582.4 \times 10^{-19}}{640 \times 10^{-19}} = 0.91 \text{ m}$$

Hence time taken for one revolution = $\frac{2\pi r}{v}$

or
$$T = \frac{2\pi \times 0.91}{80 \times 10^3} = 0.071 \text{ msec or } 71 \text{ μsec. } \textbf{Ans.}$$

Example 9.2. The point charge Q = 20 nC has a velocity of 5 × 10^6 m/sec in the direction $\bar{a}_v = 0.6\bar{a}_x + 0.75\bar{a}_y + 0.3\bar{a}_z$. Determine the magnitude of the force existing on the charge by the field (*i*) $B = -3\bar{a}_x + 4\bar{a}_y + 6\bar{a}_z$ mT (*ii*) $E = -3\bar{a}_x + 4\bar{a}_y + 6\bar{a}_z$ kV/m (*iii*) B and E acting simultaneously.

Solution. (*i*) $B = -3\bar{a}_x + 4\bar{a}_y + 6\bar{a}_z$

$$\bar{V} = 5\times10^6\left(0.6\bar{a}_x + 0.75\bar{a}_y + 0.3\bar{a}_z\right)$$

and

$$F = Q\bar{V}\times\bar{B} = 20\times10^{-9}\left(0.6\bar{a}_x + 0.75\bar{a}_y + 0.3\bar{a}_z\right)\times\left(-3\bar{a}_x + 4\bar{a}_y + 6\bar{a}_z\right)$$

$$= 5\times10^6\times20\times10^{-9}\left(2.4\bar{a}_z - 3.6\bar{a}_y + 2.25\bar{a}_z + 4.5\bar{a}_x - 0.9\bar{a}_y - 1.2\bar{a}_x\right)$$

$$= 20\times10^{-9}\left(4.69\bar{a}_z - 4.5\bar{a}_y + 3.2\bar{a}_x\right)\times5\times10^6\times10^{-3}$$

$$\text{Magnitude of } F = 20\times10^{-9}\left(\sqrt{22 + 20.75 + 10.24}\right)$$

$$= 20\times10^{-9}\times7.245\times5\times10^6\times10^{-3}$$

$$= 724.5\ \mu\text{N} \quad \textbf{Ans.}$$

$$F_1 = 100\times10^{-6}\left(4.69\bar{a}_z - 4.5\bar{a}_y + 3.2\bar{a}_x\right)$$

(*ii*) When $E = -3\bar{a}_x + 4\bar{a}_y + 6\bar{a}_z$ kV/m.

and Q = 20 nC, the electrostatic force is

$$F_2 = QE = 20\times10^{-9}\left(-3\bar{a}_x + 4\bar{a}_y + 6\bar{a}_z\right)\times10^3$$

$$= 20\times10^{-9}\left(\sqrt{9+16+36}\right)\times10^3$$

$$= 156\times10^{-9}\times10^3 = 156\ \mu\text{N}$$

(*iii*) Total Force = $F_1 + F_2 = \{(320 - 60)\ \bar{a}_x + (80 - 450)\ \bar{a}_y + (469 + 120)\ \bar{a}_z\}\ 5\times10^{-6}$

$$= (260\bar{a}_x - 370\bar{a}_y + 589\bar{a}_z)10^{-6}$$

$$= (2.6\bar{a}_x - 3.7\bar{a}_y - 5.89\bar{a}_z)10^2\times10^{-6}$$

$$= 742.5\ \mu\text{N} \quad \textbf{Ans.}$$

Example 9.3. The magnetic flux density surrounding the space of the origin is given by B = 4.0 × 10^{-4} $a_z T$ and the electric field intensity E = 5.0 a_z V/m A proton with charge 1.602 × 10^{-9} C and mass 1.673 × 10^{-27} kg enters the field at the origin with an initial velocity V_0 = 2.4 × 10^5 $\bar{a}_x$ m/s. Describe the motion of the electron and give its position after 4 complete revolutions.

Solution. The force due to electric and magnetic field combined is $F = Q\left(E + \bar{v}_0\times\bar{B}\right)$ $= Q_p\left(E\bar{a}_z - V_0 B\bar{a}_y\right)$. Since the electric field is along z-direction, it produces a constant acceleration in the z-direction. Thus the equation of motion in the z-direction is

$z = \dfrac{1}{2}at^2$ where a is the acceleration which is the ratio of force to mass *i.e.* $a = \dfrac{Q_p E}{m}$

Hence, $$z = \frac{1}{2}\left(\frac{Q_p E}{m_p}\right)t^2$$

The other component of force due to magnetic field changes into $-Q_p VB\bar{a}_r$ and produces circular motions perpendicular to the z-axis with period

$$J = \frac{2\pi r}{V}$$

From equation (9.2) $$V = \frac{QBr}{m}$$

Hence, $T = \frac{2\pi r}{QBr}m = \frac{2\pi m}{QB}$.

So these are the two forces one in a straight line due to electric field and another in a circular path due to magnetic field the resultant motion is identical as shown in Fig. E 9.3.

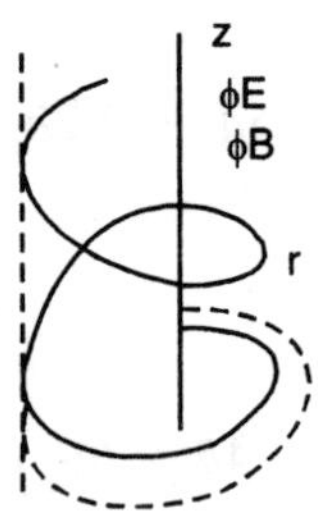

Fig. E 9.3

After four revolutions $x = y = 0$ and

$$Z = \frac{1}{2}\frac{Q_p E}{m_p}(4T)^2 = \frac{1}{2}\frac{1.602\times10^{-19}\times5}{1.673\times10^{-27}}(4T)^2$$

where $$T = \frac{2\pi m_p}{Q_p B} = \frac{2\pi\times1.673\times10^{-27}}{1.602\times10^{-19}\times4\times10^{-4}} = 1.64\times10^{-4}$$

or $$4T = 6.56\times10^{-4} \text{ or } (4T)^2 = 43\times10^{-9}$$

Hence, $$z = 2.4\times10^{+8}\times43\times10^{-8} = 103.2 \text{ m} \quad \textbf{Ans.}$$

9.3 MAGNETIC FORCE ON A DIFFERENTIAL CURRENT ELEMENT

The force on a differential charged particle dQ moving at a uniform velocity and passing through a uniform or steady magnetic field may be written as a differential force as

$$dF = dQ(\bar{v}\times\bar{B}) \quad \ldots(9.5)$$

In actual practice, the differential charge dQ consists of a large number of very small discrete charges occupying a volume which even though is very small yet it is much larger than the average separation between the charges. The differential force expressed in equation (9.5) is thus the sum of the forces on the individual charges. If the charges are electrons in motion is a conductor, the force is transformed to the conductor and it is formed that the sum of the extremely large number of extremely small forces is of practical importance

Now $$dQ = I\,dl.$$

Hence $$dF = I\,dl.\,(V \times B) = I\,(d\bar{L} \times B) \quad ...(9.6)$$

Where $dL = v\,dl$ and is the differential length is the direction of the conventional current I.

If the conductor is straight and the field is constant along it, the differential force may be integrated and the force is given as

$$F = \oint IdL \times B = -I\int B \times d\bar{L} \quad ...(9.7)$$

or $$F = ILB \sin\theta \quad ...(9.8)$$

where θ is the angle between the vectors representing the direction of current flow and the direction of the magnetic flux density. The magnetic force is actually exerted on the electrons that make up the current-I. However, since the electrons are confined to the conductor, the force is effectively transformed to the heavy lattice which in turn is transferred to the conductor as a whole.

However, if the differential charge dQ could be expressed in terms of volume charge density ρ_v, then

$$dQ = \rho_v d_v$$

or $$dF = \rho_0 d_v\,(\bar{v} \times B)$$

or $$dF = (J \times B)\,d_v \quad ...(9.9)$$

If dQ can be expressed as differential surface charge, then

$$dQ = K \cdot d\bar{s} = IdL$$

Hence, $$dF = K \times B\,ds \quad ...(9.10)$$

Since the electrons are moving within the conductor, its almost immobile positive charges, develop coulomb forces between the electron and the positive charges which results into force of attraction between the electrons and positive charges of the crystalline lattice which resist partially the movement of electrons. The charge separation that does result, is discharged by the presence of a slight potential difference across the differential current element in a direction perpendicular to both the magnetic field and velocity of the charges. Thus voltage is known as Hall voltage and this effect is known as Hall effect.

Devices making use of Hall effect can be employed for measurement of magnetic flux density (Flux density meter) and in some cases where current through the devices can be made proportional to the B, it can be used as electronic Wattmeter, Squaring element etc.

Let us consider two differential current elements $I_1 dL_1$ and $I_2 dL_2$ at location 1 and 2 respectively and are at a distance of r_{12}. The magnetic flux density (according to Biot. Savart. Law) due to element 1 at a distance r_{12} in free space is given as

$$dB_2 = \frac{\mu_0 I_1 dL_1 \times \bar{a}\bar{r}_{12}}{4\pi \cdot r_{12}^2}\bar{a}_{r_{12}} \quad ...(9.11)$$

This differential flux density at location 2 will develop a force on current element 2 as

$$d(dF_2) = I_2 dL_2 dB_2 \quad ...(9.12)$$

Replacing dB_2 from equation (9.11) we have

$$d(dF_2) = \frac{\mu_0 I_1 I_2}{4\pi r_{12}^2} dL_2 \times (dL_1 \times \bar{a}_{r_{12}})$$

The total force between two filamentary circuits is thus obtained by integrating two (dL_1 and dL_2)

i.e.
$$F_2 = \frac{\mu_0 I_1 I_2}{4\pi} \oint dL_2 \times \oint \frac{dL_1 \times \bar{a}r_{12}}{r_{12}^2}$$

$$= \frac{\mu_0 I_1 I_2}{4\pi} \oint \left[\oint \frac{r_{12} \times dL_1}{r_{12}^2} \right] \times dL_2 \qquad \ldots(9.13)$$

It is to be noted that the integral within the square gives magnetic flux density B_2 at location 2.

If there are two parallel wires (1– ϕ circuit) separated by a distance d and the current is into the plane of the paper is one of the conductors and coming out of the plane of the paper in the second conductor as shown in Fig. 9.1.

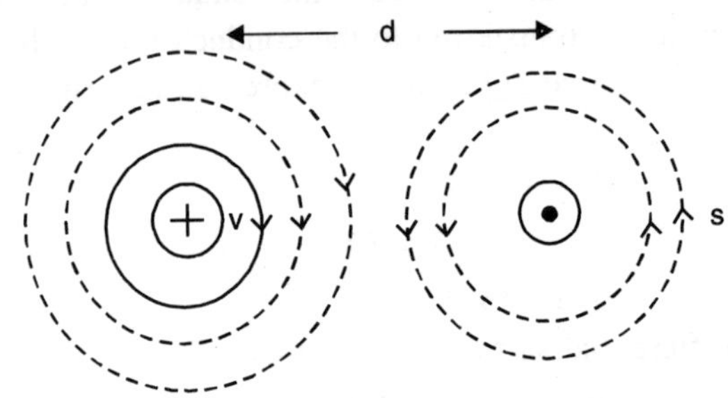

Fig. 9.1

There is addition of flux lines between the space of the two conductors and hence the two conductors will be pulled apart *i.e.* there will be force of repulsion between the two conductors and the magnitude of the force per unit length of the conductor is $\frac{\mu_0 I^2}{2\pi d}$ newton/unit length.

9.4 TORQUE ON A PLANAR COIL

Using equation (9.7) for force on a filamentary closed circuit, we have

$$F = -I \oint B \times dL$$

If we assume B to be uniform equation (9.7) becomes

$$F = -I B \oint dL$$

We have found in electrostatic field that $\oint dL = 0$ *i.e.* the closed time integral is zero irrespective of the path chosen and, therefore, the force on a closed filamentary circuit in a uniform magnetic field is zero. However, it is to be noted that if the field is not uniform the net force need not be zero. Therefore, any actual closed circuit carrying direct current experience a total vector force of zero in a uniform magnetic field. Although the force is zero, the torque is generally not equal to zero.

While defining the torque or moment of a force, it is necessary to consider both an origin at or about which the torque is to be evaluated and the point at which force is applied the moment of a force or torque about a specified point is the cross product of the lever arm about that point and the force. The lever arm $\bar{r}$ is directed from the point about which the torque is to be obtained to the point of application of force. Figure 9.2 shows the force F at point P if the arm $\bar{r}$ between O and P, the torque about O (axis passing through O) is given by

$$T = \bar{r} \times F$$

The direction of the vector torque T is normal to both the force F and the lever arm $\bar{r}$ and is in the direction of progress of r right handed screw as the lever is rotated into the force vector through the smaller angle; we normally choose the origin such that it is usually on the axis of rotation and in the plane containing the applied forces if the several forces are coplanar.

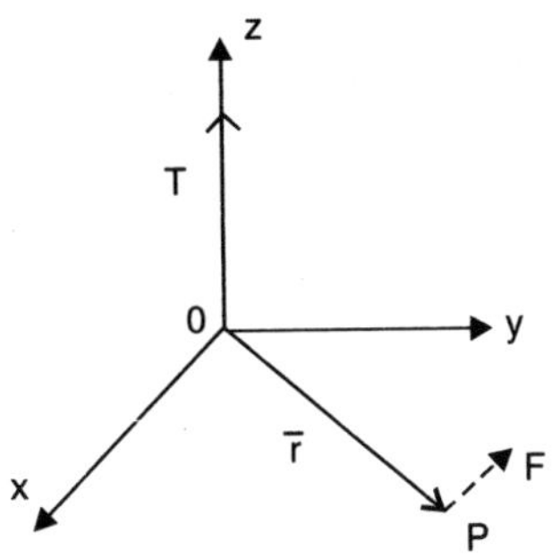

Fig. 9.2 Arm $\bar{r}$ originating from O and terminating at P where F is applied

Next we consider a single turn coil placed in the xy plane or $z = 0$ plane as shown in Fig. 9.3, the width of the coil is dx along $\bar{a}_x$ and length is dy along y direction. The magnetic field B is uniform along $\bar{a}_x$. The currents along $\pm\bar{a}_y$ direction give rise to forces on the coil.

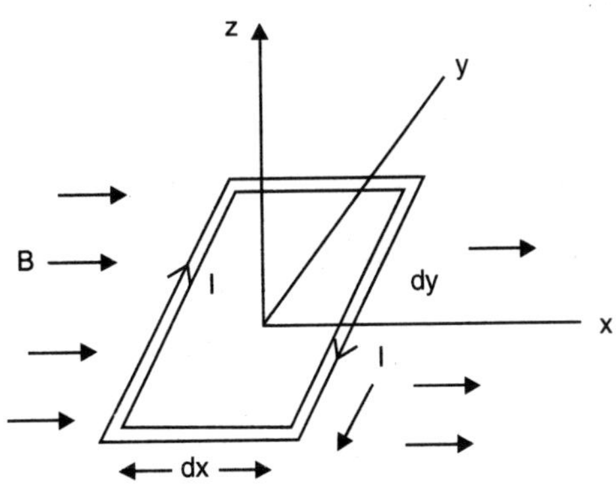

Fig. 9.3 Torque on a planar coil

The force on the left side of the coil

$$F = I\left(dy\,\bar{a}_y + B\bar{a}_x\right) = -BIdy\,\bar{a}_z$$

and force on the right side of the coil

$$F = \left(-I\,dy\,\bar{a}_y\right) \times B\bar{a}_x = B\,I\,dy\,\bar{a}_z$$

The torque due to the left current element requires a lever arm $r = -\frac{dx}{2}\bar{a}_x$ and due to the right element it is $\frac{dx}{2}\bar{a}_x$. The torque due to both the elements along y-axis is

$$T = \left(-\frac{dx}{2}\right)\bar{a}_x \times (-BIdy)\bar{a}_z + \left(\frac{dx}{2}\right)a_x \times (BI\,dy)\bar{a}_z$$

$$= BI\,dx\,dy(-\bar{a}_y) = BIA(-\bar{a}_y) \qquad \text{...(9.14)}$$

where A is the area of the coil. It is to be noted that this expression for the torque holds good for a flat coil of any arbitrary shape and for any axis parallel to the y-axis.

The magnetic moment $\bar{m}$ of a planar current coil or loop is defined as $IA\ \bar{a}_x$ where the unit normal vector $\bar{a}_x$ is determined by the right hand rule. (The right thumb gives the direction of $\bar{a}_x$ when the fingers point in the direction of the current). Therefore, the torque on a planar coil is given by

$$T = \bar{m} \times \bar{B} \qquad \text{...(9.15)}$$

where $\bar{m} = I\,A$ with units A m^2.

If we take loop area as differential vector area dA, the differential magnetic dipole moment $d\bar{m}$ is given as

$$d\bar{m} = I\,d\bar{A}$$

or

$$dT = d\bar{m} \times B$$

or

$$T = m \times B$$

It is to be noted that the torque on the current loop always tends to turn the loop BO as to align the magnetic field produced by the loop with the applied magnetic field which is causing the torque.

Example 9.4. Determine the force on a straight conductor of length 0.5 m carrying a current of 7.5 A in the $-\bar{a}_z - \bar{a}_y$ direction where the magnetic field $B = 3.5 \times 10^{-3}\ (\bar{a}_x - \bar{a}_y)$ Tesla.

Solution. Force $F = I\,(L \times B)$

$$= (7.5 \times 0.5)(-\bar{a}_z) \times 3.5 \times 10^{-3}(\bar{a}_x - \bar{a}_y)$$

$$= 13.125 \times 10^{-3}\left(\frac{-\bar{a}_x - \bar{a}_y}{\sqrt{2}}\right)$$

$$= 18.56 \times 10^{-3}\left(\frac{-\bar{a}_x - \bar{a}_y}{\sqrt{2}}\right) \text{ Newton}$$ **Ans.**

Example 9.5. Determine the vector force experienced by a straight conductor carrying a current of 15 A in a magnetic field $B = -2\bar{a}_x + 3\bar{a}_y + 5\bar{a}_z$ mT in the $\bar{a}_{AB}$ direction when $A(1,1,1)$ and (*i*) $B(2,1,1)$ (*ii*) B (3,5,6).

Solution. (*i*) $\bar{a}_{AB} = \bar{a}_x$

$$F = IL \times B = 15\bar{a}_x \times (-2\bar{a}_x + 3\bar{a}_y + 5\bar{a}_z)$$

$$= -75\bar{a}_y + 45\bar{a}_z$$

(*ii*) $\bar{a}_{AB} = 2\bar{a}_x + 4\bar{a}_y + 5\bar{a}_z$

Hence,

$$F = IL \times B$$

$$= 15(2\bar{a}_x + 4\bar{a}_y + 5\bar{a}_z) \times (-2\bar{a}_x + 3\bar{a}_y + 5\bar{a}_z)$$

$$= 15(6\bar{a}_z - 10\bar{a}_y + 8\bar{a}_z + 20\bar{a}_x + 10\bar{a}_y - 15\bar{a}_x)$$

$$= 15(5\bar{a}_x + 14\bar{a}_z)$$

$$= 75\bar{a}_x + 210\bar{a}_z \text{ mN } \textbf{Ans.}$$

Example 9.6. A conductor is located at $x = 0.5$ m, $y = 0$ and $0 < z < 3$ m which carries a current of 10 A. The magnetic flux density is given as $B = 2.5\bar{a}_x T$. Determine the torque.

Solution. $F = I(L \times B) = 10(3\bar{a}_z \times 2.5\bar{a}_x)$

$$= 75\bar{a}_y \text{N}$$

Now,

$$T = r \times F$$

$$= 0.5\bar{a}_x \times 75\bar{a}_y = 37.5 \text{ Nm } \textbf{Ans.}$$

Example 9.7. Determine the force per unit length on two long straight parallel conductors if each carries a current of 15 A in the same direction and the distance of separation is 0.5 m.

Solution. The force per unit length is given as

$$F = \frac{\mu_0 I^2}{2\pi d} = \frac{4\pi \times 10^{-7} \times 15^2}{2\pi \times 0.5} = 2 \times 10^{-7} \times 225 \times 2$$

$$= 900 \times 10^{-7} \text{ N/m force of attraction.}$$

Example 9.8. A rectangular coil is placed in a magnetic field $B = 0.05\frac{a_x + a_y}{\sqrt{2}}$T as shown in Fig. E 9.8.

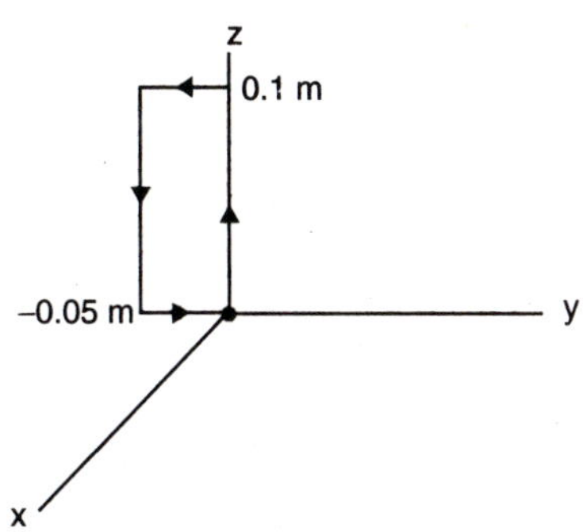

Fig. E 9.8

Determine the torque about the *z*-axis when the coil is in position shown and carries a current of 10 A.

Solution. The magnetic moment $IA\bar{a}_x = 10 \times 0.1 \times 0.05\bar{a}_x$

$$m = 5 \times 10^{-2}\bar{a}_x \text{ Amp-m}^2$$

$$T = m \times B = 5 \times 10^{-2} a_x \times 0.05 \frac{a_x + a_y}{\sqrt{2}} = 17.7 \times 10^{-4}\bar{a}_z$$

It is to be noted that when the coil turns through 45° the direction of *m* and *B* will be same and hence the angle between them will be zero and the torque will be sin 0° = 0.

Example 9.9. Fig. E 9.9 shows a rectangular current loop lying in $z = 0$ plane where the magnetic flux density is given as $B = -0.5\bar{a}_y + 0.9\bar{a}_z$ T. Determine the torque experienced by the loop.

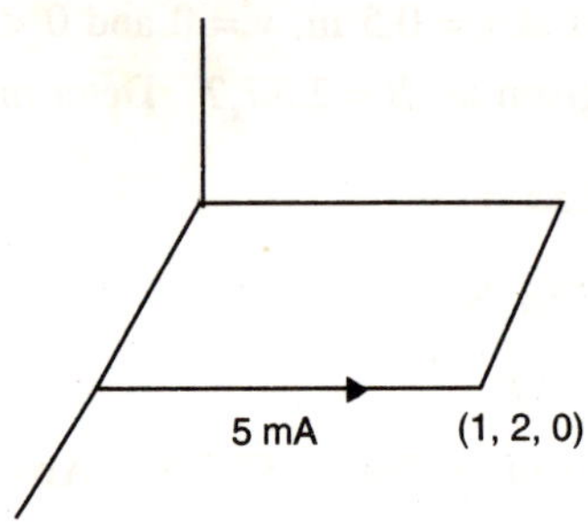

Fig. E 9.9

Solution. Torque $T = m \times B = IA\bar{a}_x \times B$

$$T = 5 \times 10^{-3}\left[(1)(2)a_z\right] \times \left(-0.5\bar{a}_y + 0.9\bar{a}_z\right)$$

Since $a_z \times a_z = 0$ we have $T = 5 \times 10^{-3}\left(2\bar{a}_z\right) \times \left(-0.5\bar{a}_y\right)$

$$= 5\bar{a}_x \text{ mNm} \quad \textbf{Ans.}$$

In order to understand the behaviour of orbiting charged particle, the concept of magnetic moment is essential. Suppose a positive charge *Q* is moving in a circular orbit (or –*Q* (electronic charge moving around a nucleus)) at a velocity *v* or an angular velocity *w* as shown in Fig. 9.4.

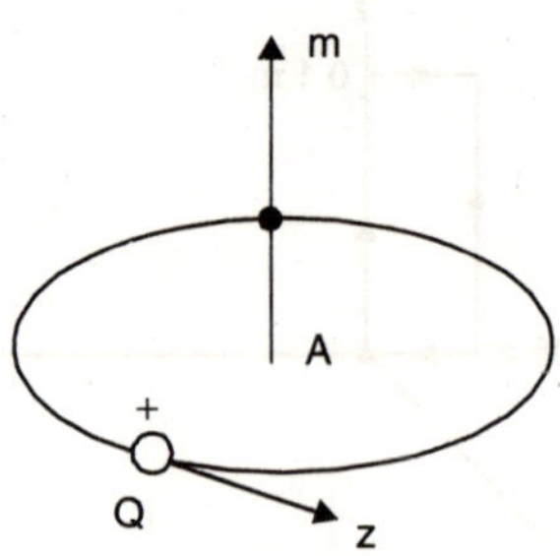

Fig. 9.4

The time taken to compute one circular path is $\frac{2\pi}{\omega}$ and since the charge moved is Q, the equivalent current is $Q\Big/\left(\frac{2\pi}{\omega}\right)$ or $Q\left(\frac{\omega}{2\pi}\right)$ which gives rise to magnetic moment m of IA or $m = \frac{\omega}{2\pi} IA\bar{a}_x$ and if there is an applied field B, the torque developed will be $T = m \times B$ which tends to turn the current loop until m and B are in the same direction and the torque developed will be zero.

It is to be noted that when an orbiting electron moves around a nucleus, it forms a current loop and which will give rise to a magnetic field and if this is an external field present; the field due to the electron motion will add to the externally applied field and hence the net field B will be more than the externally applied field.

Example 9.10. Determine the maximum torque on an orbiting charged particular if the charge is 1.602 $\times 10^{-19}$C, the circular orbit has a radius of 0.6×10^{-10} m, the angular velocity is 4.5×10^{16} rad/sec and $B = 0.5$ mT.

Solution. The orbiting charge has magnetic moment of

$$m = \frac{\omega}{2\pi} QA\bar{a}_x = \frac{4.5\times 10^{16}}{2\pi} \times 1.602 \times 10^{-19} \left(\pi \times 0.36 \times 10^{-20}\right) a_x$$

$$= 1.3 \times 10^{-23} \bar{a}_x \text{ Am}^2$$

The torque developed will be maximum when $\bar{a}_x$ is normal to B and hence

$$T = m \times B = 1.3 \times 10^{-23} \times 0.5 \times 10^{-3}$$
$$= 6.5 \times 10^{-27} \text{ Nm} \quad \textbf{Ans.}$$

9.5 MAGNETIC PROPERTIES OF MATERIALS

We have studied in the preceding paragraph that a magnetic moment is developed due to the orbiting electron.

A second magnetic moment is attributed to electron spin. It is found using the relativistic quantum theory that an electron may have a spin magnetic moment of about $\pm 9 \times 10^{-24}$ Am^2, the plus sign indicates that the external field is enhanced where negative sign shows that the external field is reduced. It is to be noted that in an atom with many electrons present, only the spin of those electrons in shells which are not completely filled will contribute to a magnetic moment for the atom.

A third contribution to the magnetic moment if an atom is due to its nuclear spin. Since the contribution is negligibly small its effect is ignored. However, this effect finds application in nuclear magnetic resonance imaging (MRI) in material sciences.

Thus each atom provides three magnetic moment components and their contribution (relative) determines the magnetic properties of the materials and provides their general magnetic classification. We study here the following types of magnetic materials (*i*) diamagnetic (*ii*) paramagnetic (*iii*) ferromagnetic (*iv*) antiferromagnetic (*v*) ferrimagnetic and (*vi*) superparamagnetic

Diamagnetic materials are those for which the net magnetic moment due to orbital and spin motion combined together is zero. This means that an external field B would not produce any torque as such atom and no realignment of the dipole fields and, therefore, an internal magnetic field is same as the applied external field.

Some of the diamagnetic materials are sodium chloride, copper, gold, silicon, germanium, hydrogen, helium and other inert gases. Metallic bismuth shows a greater diamagnetic effect.

Next we consider materials wherein the magnetic moment due to orbital and spin do not cancel completely. When an external field is applied, there is a small torque on each atomic moment due to orbital motion, as a result, these magnetic moment tend to align with the external field. Thereby enhancing the external filled within the material. However, the diamagnetic effect is still working on the orbiting electron and may contract the increase. However, if there is an increase in B the material is termed as paramagnetic whereas if B decreases it is known as diamagnetic. Some of the paramagnetic materials are, Potassium, tungsten, rare earth elements, oxygen, neodymium oxide etc.

The remaining four classes of materials *e.g.* ferromagnetic, antiferromagnetic, ferrimagnetic and superparamagnetic all have strong magnetic moments. Also, the interaction of adjacent atoms causes an alignment of the magnetic moments of the atoms either in aiding or opposing mode.

In ferromagnetic materials each atom has a large magnetic moment due to orbital motion when spin moments are ignored. Interatomic forces cause these moments to line up in a parallel fashion over regions containing a large number of atoms. These regions are called 'domains'. The size of these domains may vary from one micrometer to several centimeters depending upon the size, shape, material and magnetic history of the sample. For a fresh sample of ferromagnetic material, since the domains are not aligned, the material as a whole has no magnetic moment.

When an external field is applied to a ferromagnetic material, those domains which have moments in the direction of the applied field, increase their size at the expense of their neighbours and the internal magnetic field increases greatly as compared to the external field. When the external field is removed, random domain alignment does not exist and a residual or remanent dipole field remains in the macroscopic structure. This residual field is called hysteresis. Ferromagnetic materials have the property of magnetostriction *i.e.* the charge in dimension of the crystal when it is subjected to a magnetic field. Iron, nickel and cobalt behave as ferromagnetic materials when at room temperature. However, these lose all their ferromagnetic properties above a temperature known as Curie temperature which is 770° C or 1043 K for iron. Some alloys of these metals with each other and with other metals also form ferromagnetic materials *e.g.* alnico, an aluminium - nickel - cobalt alloy with a small amount of copper. It is interesting to note some alloys of non-ferromagnetic metals are ferromagnetic *e.g.* bismuth- manganese and copper-manganese-tin.

In antiferromagnetic materials, the magnetic forces between adjacent atoms cause the magnetic moments to line up in an antiparallel fashion. As a result, the net magnetic moment is zero and these materials are affected only directly by the presence an external magnetic field. Some of the antiferromagnetic materials are manganese oxide, nickel oxides, ferrous sulphide and cobalt chloride. Antiferromagnetism is present at relatively low temperature usually well below room temperature.

The ferrimagnetic materials show an antiparallel alignment of adjacent atomic moments but the moments are not equal. A large response to an external magnetic field, therefore, occurs, although not as large as in case of ferromagnetic materials. The most important group of ferrimagnetic materials are the ferrites which have low electrical conductivity and hence are useful for high frequency transformer as this results in low core losses. Some of the ferrimagnetic materials are iron oxide magnetic (Fe_3O_4) a nickel zinc ferrite and nickel ferrite. Ferrimagnetic also disappears above Curie temperatures.

Superparamagnetic materials consist of an assembly of ferromagnetic particles in a non-ferromagnetic matrix. Even though domains exist within the individual particles the domain walls cannot penetrate the intervening matrix material to the adjacent particle. An important example is the magnetic tape used in audio tape and video tape recorders.

9.6 MAGNETIZATION AND PERMEABILITY

We now consider the effect of current due to bound charges *i.e.* due to orbital electrons, electron spin and nuclear spin. We also take into consideration the externally applied field B_0. The field intensity due to bound charges will have dimensions of H and is called here magnetization M also known as magnetic dipole moments per unit volume and units will be similar to H *i.e.* A/m. Let I_b be the current due to bound charges and as in spheres circuital law $I = \phi\, H.dL$, but also $I_b = \phi.Mdl$.

If B is the flux density due to the total current $I_T = I_b + I$ where I is the current required to the externally applied field. When we take into account I_b we ignore the presence of magnetic material and we assume that now it is any free space. Now, $I_T = \oint \frac{B}{\mu_0} dL$ where B is the net flux density.

Now, $$I = \oint H \cdot dL$$

Hence, $$I_T = I_b + I$$

or $$\oint \frac{B}{\mu_0} dL = \oint M \cdot dL + \oint H \cdot dL$$

or $$\frac{B}{\mu_0} = M + H$$

or $$B = \mu_0 (M + H) \qquad \text{...(9.16)}$$

This relation can be further simplified for linear isotropic media where magnetic susceptibility χ_m can be defined

$$M = \chi_m H \qquad \text{...(9.17)}$$

Thus, $$B = \mu_0 (\chi_m H + H) = \mu_0 H (1 + \chi_m) = \mu_0 \mu_r H$$

Where $\mu_r = 1 + \chi_m$ and is defined as the relative permeability

Finally, $B = \mu H$ where $\mu = \mu_o \mu_r$

As mentioned earlier M is also defined as magnetic dipole moment per unit volume. Mathematically this is expressed as

$$M = \underset{\Delta V \to 0}{\text{Lt}} \frac{1}{\Delta V} \sum_{i=1}^{n\Delta V} m_i \qquad \text{...(9.18)}$$

Example 9.11. Determine the magnetization in a magnetic material which as (*i*) $\mu = 2 \times 10^{-5}$ H/m and H = 150 A/m (*ii*) $\mu_r = 20$ and there are 8.3×10^{28} atoms/m^3 and each atom has a dipole moment of 5×10^{-27} Am^2 (*iii*) $B = 250\ \mu T$ and $\chi_m = 15$.

Solution. (*i*) Since $\mu = \mu_0 \mu_r = 4\pi \times 10^{-7}\ \mu_r = 2 \times 10^{-5}$

Hence, $\mu_r = \frac{22 \times 10^{-7}}{4\pi \times 10^{-7}} \approx 16 = 1 + \chi_m$ or $\chi_m = 15$

Hence, $M = \chi_m H = 15 \times 150 = 2250$ A/m. **Ans.**

(*ii*) Since there are 8.3×10^{28} atoms/m^3 and each atom has dipole moment of 5×10^{-27} Am^2 and by definition magnetization M is the dipole moment per unit volume.

Hence, $$M = 8.3 \times 10^{28} \times 5 \times 10^{-27} = 415 \text{ A/m} \quad \textbf{Ans.}$$

(*iii*) $B = 250\ \mu T\ \chi_m = 15$ Hence $\mu_r = 1 + \chi_m = 16$

$$H = \frac{B}{\mu_0 \mu_r} = \frac{250 \times 10^{-6}}{4\pi \times 10^{-7} \times 16} = 12.4 \text{ A/m.}$$

$$M = \chi_m H = 15 \times 12.4 = 186 \text{ A/m. } \textbf{Ans.}$$

If we utilize several current densities, we have current due to bound charge

$$I_b = \int J_b \cdot ds \qquad ...(9.19)$$

Current due to external applied field $$I = \int J \cdot ds \qquad ...(9.20)$$

Total current $$I_T = \int J_T \cdot ds \qquad ...(9.21)$$

With the help of Stoke's theorem we obtain following relations

$$\nabla \times M = J_b$$

$$\nabla \times H = J \qquad ...(9.22)$$

$$\nabla \times \frac{B}{\mu_0} = J_T$$

9.7 MAGNETIC BOUNDARY CONDITIONS

We have already studied boundary conditions in case of electrostatic fields and therefore, it should not be a big problem to arrive at the proper boundary condition to apply to analogous to electric field $D \leftrightarrow B$, $E \leftrightarrow H$ and M the magnetization at the interface between two different magnetic materials as we have already worked out similar problems for both conducting materials and dielectrics in electric fields.

Consider Fig. 9.5 which shows a boundary between two isotopic homogeneous linear materials with permeabilities μ_1 and μ_2. The boundary condition on the normal components is last case of using Gauss' theorem for a cylindrical gaussian surface. Since the net flux enclosed by the differential surface is zero.

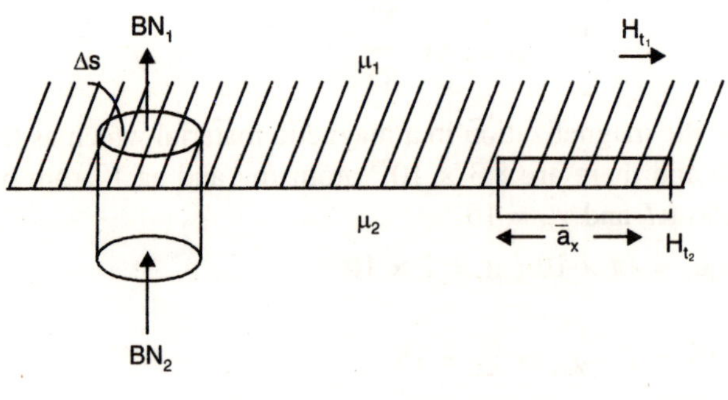

Fig. 9.5 A Gaussian surface and a closed path (For tangential component)

Hence, $$\oint B \cdot ds = 0$$

so that $$B_{N_1} \Delta S - B_{N_2} \Delta S = 0$$

or $$B_{N_1} = B_{N_2} \qquad \text{(In electrical fields } D_{n_1} = D_{n_2}\text{)} \qquad ...(9.23)$$

since $B_{N_1} = \mu_1 H_{N_1}$ and $B_{N_2} = \mu_2 H_{N_2}$

$$H_{N_1} = \frac{\mu_2}{\mu_1} H_{N_2} \qquad ...(9.24)$$

This shows that whereas the normal component of magnetic flux densities are continuous at the boundary, the normal component of magnetic field intensity are discontinuous. Since the relation between H_{N_1} & H_{N_2} is known, we can find the relation between M_{N_1} and M_{N_2} at the boundary

Since $M_{N_1} = x_{m_1} H_{N_1}$ and $M_{N_2} = x_{m_2} H_{N_2}$

$$M_{n_1} = x_{m_1} H_{N_1} = x_{m_1} \frac{\mu_2}{\mu_1} H_{N_2} = \frac{x_{m_1}}{x_{m_2}} H_{N_2} \frac{\mu_2}{\mu_1}$$

$$= \frac{x_{m_1}}{x_2} \frac{\mu_2}{\mu_1} M_{N_2} \qquad ...(9.25)$$

Now to consider the tangential component of H at the boundary we make use of Amperes law about a differential closed path in a plane normal to the boundary surface as shown to the right in Fig. 9.5. Moving along the path clockwise we have

$$H_{t_1} \Delta x - H_{t_2} \Delta x = k \Delta x$$

Where k is the surface current on the boundary

Thus $$H_{t_1} - H_{t_2} = k \qquad ...(9.26)$$

$$\frac{M_{t_1}}{x_{m_1}} - \frac{M_{t_2}}{x_{m_2}} = k$$

or $$M_{t_1} = \frac{x_{m_1}}{x_{m_2}} M_{t_2} + k x_{m_1} \qquad ...(9.27)$$

Normally, $k = 0$ as most of the magnetic materials are bad conductors of electric charges.

For equation (9.26) directions are specified more exactly by using cross product to identify the tangential components

$$(\bar{H}_1 - \bar{H}_2) \times \bar{a}_{N_{12}} = \bar{k}$$

Where $\bar{a}_{N_{12}}$ is the unit vector normal at the boundary directed from region 1 to region 2. An equivalent formulation in terms of the vector tangential components may be more convenient for $\bar{H}$

$$\bar{H}_{t_1} - \bar{H}_{t_2} = \bar{a}_{N_{12}} \times \bar{k} \qquad ...(9.28)$$

For tangential $\bar{B}$ we have

$$\frac{B_{t_1}}{\mu_1} - \frac{B_{t_2}}{\mu_2} = k \qquad ...(9.29)$$

Example 9.12. A ferrite material with relative permittivity $\mu_r = 60$ has magnetic flux density of $B = 0.08\ T$, determine the values of M, H and χ_m.

Solution. Since $\mu_r = 1 + \chi_m$ or $60 = 1 + \chi_m$. Hence $\chi_m = 59$

Since $$B = \mu_0\mu_r H$$

$$0.08 = 4\pi \times 10^{-7} \times 60\ H \text{ or } H = \frac{0.08\times 10^7}{240\bar{d}}$$

or $$H = \frac{8000\times 10^2}{240\pi} = 1061 \text{ A/m}.$$

Hence, $M = \chi_m H = 59 \times 1061 = 62600$ A/m.

9.7.1 Potential Energy and Forces on Magnetic Materials

In electrostatic field we developed an expression for energy in an electric field which is given as

$$W_E = \frac{1}{2}\int_{v.d.} \bar{D}\cdot\bar{E}\,dv \quad \text{...(9.30)}$$

Where D and E are linearity related as $D = tE$

In Magnetic fields we define a magnetostatic field based on assumed magnetic poles. Using the scalar magnetic potential we would develop expression for energy in a magnetic field on the lines similar to the electrostatic field. The quantities in magnetic field analogous to the electrostatic field are used and hence the total energy stored in a steady magnetic field where B is linearly related to H is given as

$$W_H = \frac{1}{2}\int_{vd} B\cdot H\,dv \quad \text{...(9.31)}$$

or $$= \frac{1}{2}\int_{vd} \mu H^2 dv \quad \text{...(9.32)}$$

or $$= \frac{1}{2}\int_{vd} \frac{B^2}{\mu} dv \quad \text{...(9.33)}$$

This energy is being stored throughout the volume with an energy density of 1/2 B.H J/m^3

Next we develop an expression for the force experienced by the magnetic material. Consider a long solenoid with silicon steel core.

The solenoid has a coil of n turns/m length around it which carries a current of I ampere and hence the magnetic field intensity in the core is $nI\,A\cdot t/m$ and using this B can be obtained using the BH curve for the silicon steel. Suppose this is B_{st}. Also suppose that the core is composed of two semi-infinite cylinders which are first touching (A semi-infinite cylinder is one of infinite length having one end located in finite space). We now apply force to separate these cylinders by a differential length dl which results in doing a work $F.dl$. By separating the two cylinders we have created an air gap and this energy is stated in the air gap we have created.

Using equation 9.33, this differential increase in energy is

$$dW_H = F \cdot dl = \frac{1}{2}\frac{B_{st}^2}{\mu_0} A dl \qquad \text{...(9.34)}$$

where A is the area of cross-section of the solenoid

Thus force
$$F = \frac{B_{st}^2}{2\mu_0} A \qquad \text{...(9.35)}$$

9.8 MAGNETIC CIRCUITS

Oersted in 1819 discovered that a current flowing in a straight conductor deflected the needle of a compass and it shows that the paths of magnetic force are concentric circles around the conductor. Similarly if a current is passed through circular (loop) conductor the magnetic fluxes are found to be concentric to the circular conductor and the magnetic flux density at the centre of the circular conductor is given by (loop)

$$B = \frac{\mu I}{2r}$$

where B is the magnetic flux density in Wb/m^2 or in Tesla, I current flowing through the circular loop of the conductor and r the radius of the circular loop, μ the permeability of the material around which the loop is wound.

$$\mu = \mu_0 = 4\pi \times 10^{-7} \text{ H/m}$$

If the coil is wound on a non-magnetic material or it is air cored and $\mu = \mu_0\mu_r$ where μ_r is the relative permeability of the magnetic material. The magnetic flux density is related to magnetic field intensity through μ as given below.

$$B = \mu_0\mu_r H$$

where H is the magnetic field intensity and its units are Ampere turns per meter. μ_r has no units as it is a relative permeability, μ is permeability of the material and its units are Henry/m.

We know that the electrostatic potential V and electric field intensity are related by

$$E = -\nabla V$$

Similarly, the scalar magnetic potential V_m is related to magnetic field intensity H as

$$H = -\nabla V_m$$

In dealing with magnetic circuits, it is convenient to call V_m the magnetomotive force or mmf as it has analogy with electromotive force in electric circuits—just as an electromotive force drives the electric current or conventional current through a conductor, the magnetomotive force similarly creates and drives magnetic flux through a magnetic material or in space. The units of mmf are ampere-turns when current is passed through N no. of coils or turns of a conductor.

The electric potential difference between two points is given as

$$V_{ab} = \int_a^b E \cdot dl$$

Similarly, if we consider a current carrying conductor (current going into the paper), this gives rise to magnetic field or magnetic flux which encircle the conductor clockwise. The direction of flux lines is given by cork-screw rule or Fleming right-hand thumb rule. If the screw is pushed into the cork (driven into the paper) then screw should be moved clockwise and hence the direction of flux line is shown in Fig. 9.6. Similarly according to Fleming right-hand thumb rule means if we hold our thumb in the direction of flow of current; here normal and into the paper the twist in the fingers gives the direction of flux lines on the surface of the paper.

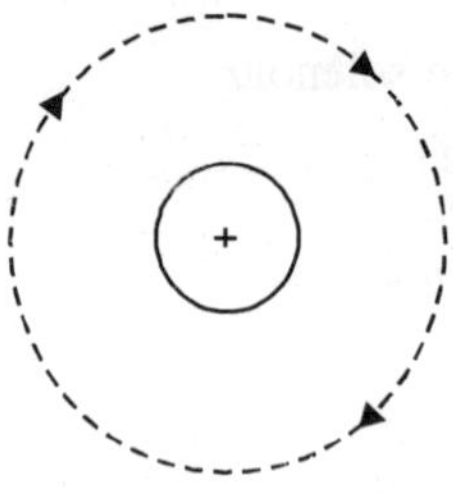

Fig. 9.6 Magnetic field due to a current carrying conductor

Now magnetic field intensity at a distance r from the centre of the conductor is given as

$$H = \frac{I}{2\pi r}$$

and if there are N such conductors, then

$$H = \frac{NI}{2\pi r}$$

The line integral of H over the complete line, here a circle of radius r, will give

$$\oint \frac{NI}{2\pi r} dl = NI = V_{mab}$$

and its units are ampere turns and thus is known as magnetomotive force. The above integrated equation is known as Ampere's circuital law and V_{mab} denotes the ampere turns enclosed by the assumed closed flux line path.

However, in an electric circuit the voltage source is a part of the closed path, in the magnetic circuit the current carrying coil will suround or link the magnetic circuit. In tracing a magnetic circuit we will not be able to identify a pair of terminals at which the magnetomotive force (mmf) is applied.

9.8.1 Analogies between Magnetic and Electric Circuits

As far as analysis of magnetic circuits is concerned, it is analogous to electric circuits. The analogous quantities are listed below.

Electric Circuits	*Magnetic Circuits*
emf (Volt)	mmf (Amp.Turns)
Current (Ampere)	Flux (Weber)
Resistance (Ohm)	Reluctance, R
Conductance G	Permeability (Henry/m)
Electric field intensity, E (V/m)	Magnetic field Intensity, H (AT/m)
Electric flux density, (D Coulomb/m^2)	Magnetic flux density (Wb/m^2)

Simple magnetic circuits can be analyzed based on Ohm's law in electric circuits according to Ohm's law.

$$\frac{V}{I} = R$$

where V is the emf, I the current and R the resistance. Picking up analogous quantities in magnetic circuits from the relations given above, we have

$$\frac{mmf}{flux} = \text{Reluctance}$$

$$\frac{NI}{\phi} = \text{Reluctance}$$

where N is the number of turns in the circular coil.

Now, $\phi = BA$, where A is the area of cross-section of the core on which the coil is wound. Now

$$B = \mu_0 \mu_r H$$

and $H = \dfrac{NI}{l}$, where l is the length of the magnetic path.

Hence reluctance
$$R = \frac{NI}{\mu_0 \mu_r \dfrac{NI}{l} \cdot A}$$

$$= \frac{l}{\mu_0 \mu_r A} \text{ AT/Wb}$$

In electric circuit, the emf drives current to flow which is limited by the resistance of the path of the current, in magnetic circuit, mmf creates and drives flux through a magnetic circuit but the flux value is limited by what is known as reluctance of the magnetic circuit which is defined by the above expression. No specific name is given to the dimension of reluctance except to refer it to as so many units of reluctance.

From the above expression for reluctance it is found that the impediment to the flow of flux which a magnetic circuit presents is directly proportional to the length and inversely proportional to the permeability and cross-sectional area. The results are in consistence with the actual physical conditions.

Series Magnetic Circuit

If there are one or more than one magnetic paths connected in series there could even be an air gap in the path of the same flux (same current in a series electric circuit), the net reluctance is the sum total of the individual magnetic paths.

Parallel Magnetic Circuits

Analogous to electric circuit, the magnetic circuit will be analyzed using Kirchhoff's laws as shown in Fig. 9.7.

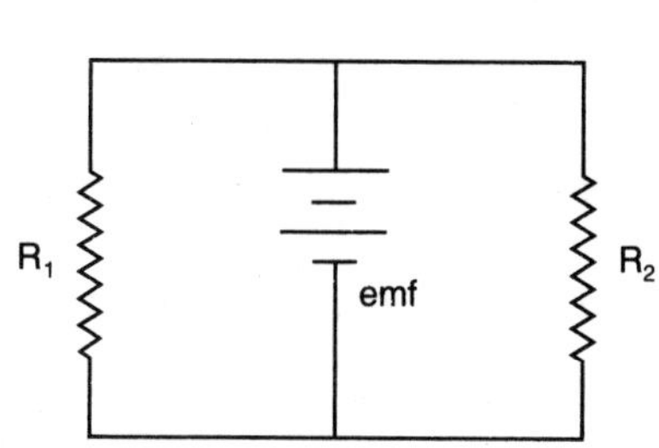

Fig. 9.7 Analogous parallel electric circuits

Let us take a few examples to illustrate the application of what we have studied about the magnetic circuits.

Example 9.13. An iron ring of mean length 50 cm has an air gap of 1mm and a winding of 200 turns. If the relative permeability of air is 300 when a current of 1 A flows through the coil, find the flux density.

Solution. The reluctance of magnetic path R

$$= \frac{50 \times 10^{-2}}{4\pi \times 10^{-7} \times 300\, A} = \frac{50 \times 1000}{12\pi A} = \frac{1326}{A}$$

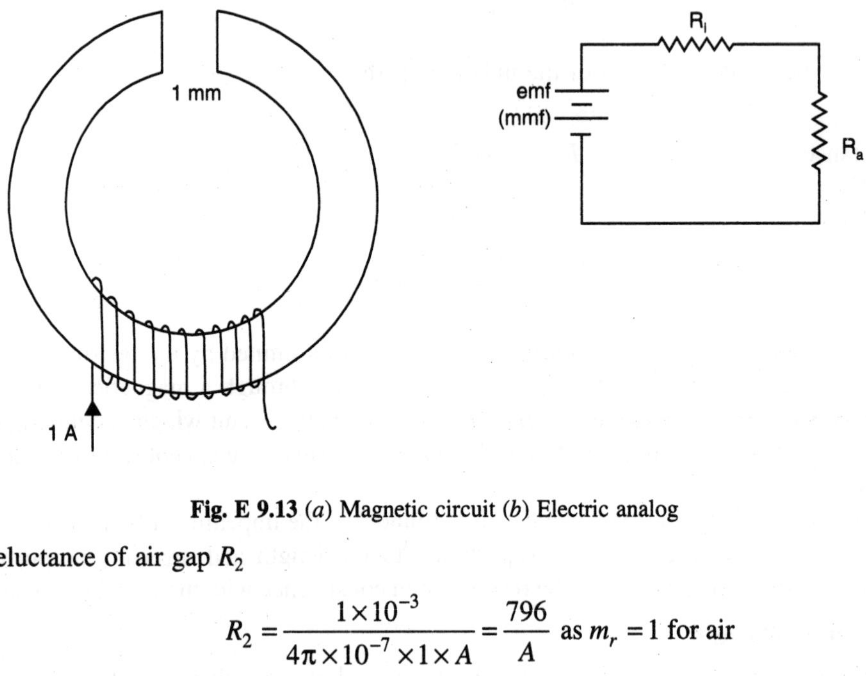

Fig. E 9.13 (*a*) Magnetic circuit (*b*) Electric analog

Now reluctance of air gap R_2

$$R_2 = \frac{1 \times 10^{-3}}{4\pi \times 10^{-7} \times 1 \times A} = \frac{796}{A} \text{ as } m_r = 1 \text{ for air}$$

$$\text{Total reluctance} = \frac{1326}{A} + \frac{796}{A} = \frac{2122}{A}$$

$$\phi = \frac{MMF}{\text{Reluctance}} = \frac{200 \times 1A}{2122} = 94.2 \times 10^{-3} A$$

or, $$B = \frac{\phi}{A} = 94.2 \text{ mWb/m}^2 \textbf{ Ans.}$$

Example 9.14. A cast steel electromagnet has an air gap of length 2 mm and an iron path of length 30 cm. Find the number of ampere turns necessary to produce a flux density of 0.8 Wb/m^2 in the air gap. Assume H = 700 AT/m for the flux density of 0.8 Wb/m^2 for the cast steel.

Solution. Here again there are two reluctances in series, one due to cast steel (say R_1) and the other due to air gap (say R_2).

$$R_1 = \frac{30 \times 10^{-2}}{4\pi \times 10^{-7} \mu_r A}$$

Since $$B = \mu_0 \mu_r H$$

$$0.8 = 4\pi \times 10^{-7} \mu_r \times 700$$

$$\mu_r = 910$$

Hence $$R_1 = \frac{30 \times 10^{-2}}{4\pi \times 10^{-7} \times 910\, A} = \frac{262.4}{A}$$

and $$R_2 = \frac{2 \times 10^{-3}}{4\pi \times 10^{-7} \times 1 \times A} = \frac{1592}{A}$$

Hence total reluctance $$= \frac{1592 + 262.4}{A} = \frac{1854.4}{A}$$

Now $$\phi = \frac{NI}{R}$$

$$0.8 \times A = \frac{NI \cdot A}{1854.4}$$

or $$NI = 1854.4 \times 0.8 \approx 1485 \text{ AT} \quad \textbf{Ans.}$$

Example 9.15. Figure 9.15(*a*) shows a three limb electromagnetic circuit. The central limbs has an air gap of 2 mm. Assume $\mu_r = 800$, determine the current in the 600 turns exciting coil to establish a flux of 100 × 10^{-3} mWb in the air gap.

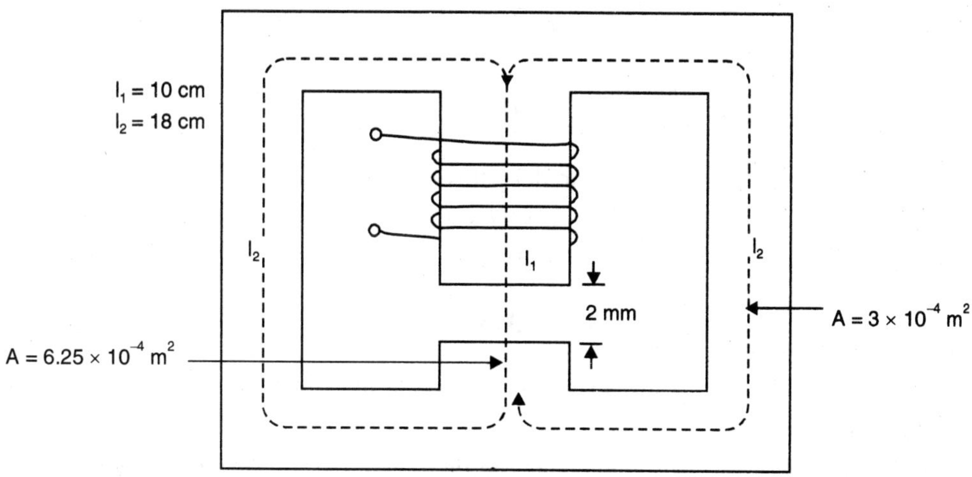

Fig. E 9.15 (*a*)

Solution. Since the flux in the air gap is 100 × 10^{-3} mWb, the flux in the outer limbs will be 50 × 10^{-3} mWb. The analogous electric circuit is as shown.

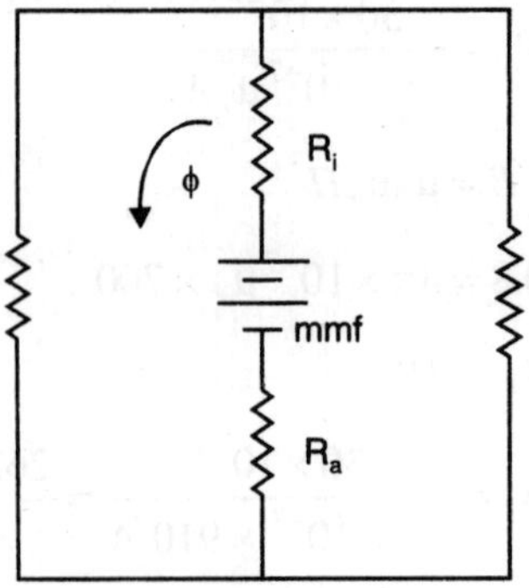

Fig. E 9.15 (*b*)

$$R_a = \frac{2\times10^{-3}}{4\pi\times10^{-7}\times6.25\times10^{-4}}$$

$$= 25.47 \times 10^5$$

$$R_i = \frac{10\times10^{-2}}{4\pi\times10^{-7}\times800\times6.25\times10^{-4}}$$

$$= 159200 = 1.592 \times 10^5$$

Total reluctance = 27.062×10^5

Hence, $\quad \text{mmf} = \phi R = 100 \times 10^{-6} \times 27.062 \times 10^5 = 27.06 \times 10 = 270.6$

Reluctance of one of the parallel paths

$$= \frac{18\times10^{-2}}{4\pi\times10^{-7}\times800\times3\times10^{-4}}$$

$$= \frac{18\times10^{9}}{4\pi\times800\times3} = 5.97\times10^5$$

Hence, mmf required for one of the parallel paths is

$$50 \times 10^{-6} \times 5.97 \times 10^5 = 29.85$$

Hence, total mmf = 270.6 + 29.85 = 300.45 *AT*

$$\text{The current} = \frac{300.45}{600} \simeq 0.5\,\text{A} \quad \textbf{Ans.}$$

We have taken one of the parallel paths as, to find out mmf (emf in electrical analog) we need complete one loop.

Example 9.16. A 680 turns coil is wound on the central limb of the cast steel frame shown. A total flux of 1.6 mWb is required in the gap. Find the current required. Dimensions are in cm. Assume $H = 900$ AT/m for $B = 1$ Wb/m^2.

Solution.

$$B = \mu_0 \mu_r H$$

$$1 = 4\pi \times 10^{-7}\, \mu_r \times 900$$

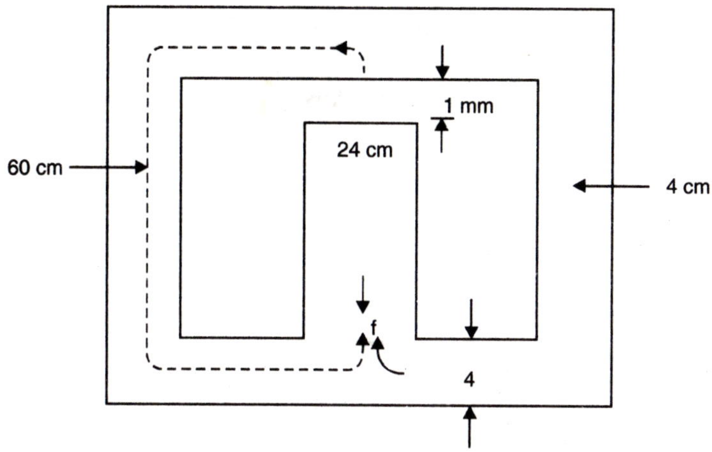

Fig. E 9.16

or $$\mu_r = \frac{10^5}{4\pi \times 9} = 885$$

Reluctance of the air gap $$= \frac{1 \times 10^{-3}}{4\pi \times 10^{-7} \times 1 \times 16 \times 10^{-4}}$$

$$= \frac{10^8}{4\pi \times 16} = 497610 \text{ AT/Wb}$$

Reluctance of the central limb $$= \frac{24 \times 10^{-2}}{4\pi \times 10^{-7} \times 885 \times 16 \times 10^{-4}}$$

$$= \frac{24 \times 10^9}{4\pi \times 885 \times 16} = 13.49 \times 10^4 \text{ AT/Wb}$$

Total reluctance of the central limb

$$= 63.25 \times 10^4 \text{ AT/Wb}$$

mmf required for the central limb

$$= 63.25 \times 10^4 \times 1.6 \times 10^{-3}$$
$$= 1012 \text{ AT}$$

Now reluctance of one of the outer limbs

$$13.49 \times 10^4 \times \frac{60}{24} = 33.72 \times 10^4$$

Hence, AT required $= 33.72 \times 0.8 \times 10^{-3} \times 10^4 = 269.8$

Total AT required $= 1012 + 269.8 = 1281.8$

Since the two paths are identical it is necessary to consider mmf for only one of them.

Hence, total mmf required

$$1012 + 269.8 = 1281.8$$

$$\text{Hence, current required} = \frac{1281.8}{680} = 1.9 \text{ Amp. } \textbf{Ans.}$$

9.8.2 Leakage Flux

In order to understand the concept of leakage flux refer to Fig. 9.8:

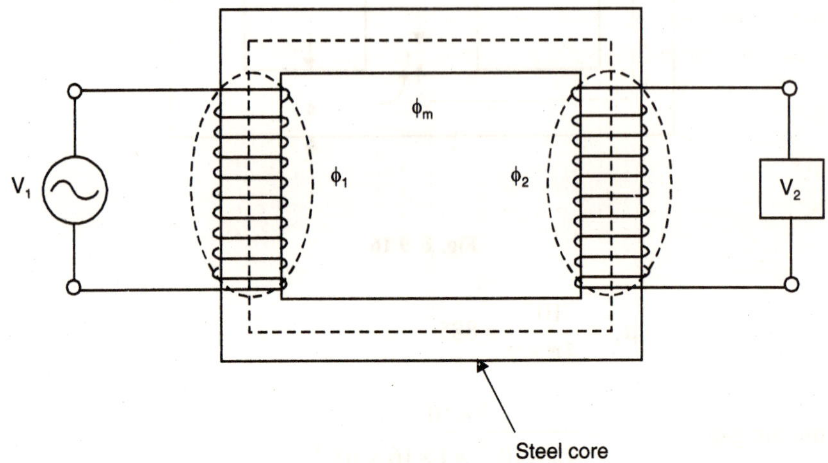

Fig. 9.8 Leakage Flux

As we connect voltage source across primary, current is drawn by the primary winding depending upon the state of secondary. If the secondary winding is open it draws current corresponding to exciting current which gives rise to magnetic flux in steel core and meets the core loss requirements. The steel core is carefully designed to keep the magnetic flux contained within the steel core. Any magnetic flux that is outside the confines of the magnetic steel core is termed as leakage flux. In Fig. 9.8 ϕ_{1l} and ϕ_{2l} are the leakage fluxes of primary and secondary winding respectively, as these fluxes are flowing and completing their path through air rather than steel core, ϕ_m is the flux which is linking both the windings and is known as mutual flux or common flux.

The total flux linking the primary winding is divided into two components: (*i*) the resultant mutual flux confined essentially to the steel core and produced by the combined effect of the primary and secondary currents (if secondary is loaded) and (*ii*) the primary leakage flux which links only the primary. These components have been identified in the basic transformer in Fig. 9.8 where for simplicity the primary and secondary windings are shown on separate (opposite) legs of the core. In an actual transformer with interleaved windings (usually low voltage winding is placed near the core and then sufficient insulation is provided and high voltage is placed above the primary *i.e.* both the limbs have primary as well as secondary winding), the details of the flux map are more complicated but the essential features remain the same. Since the leakage path is mainly through the air, the leakage flux and the voltage induced by it vary linearly with primary current. The effect on the primary circuit is simulated by assigning to the primary a leakage inductance which by definition is equal to leakage flux linkage (leakage flux × No. of turns of primary) per unit primary current or it is equivalent to leakage reactance *i.e.* $2\pi f$ × leakage inductance.

Similarly the effect of secondary leakage flux can be simulated on the secondary winding of the transformer.

9.8.3 Fringing of Flux

To understand fringing of magnetic flux consider Fig. 9.9. In the air gap the magnetic flux line bulge outwards somewhat as shown in Fig. 9.9. This is known as fringing of flux. The effect of the fringing is to increase the cross-sectional area of air gap. The flux density in the air gap reduces as compared to one in the core, therefore a correction due to fringing is required. The usual correction is to add the gap dimension to each dimension of air gap *e.g.* if the core dimensions are 3 cm by 3 cm and air gap g is 0.05 cm, the corrected area would be Ag = (3 + 0.05) (3 + 0.05) = 9.3 cm^2, thereby the gap reluctance decreases by about 3% for the uncorrected gap.

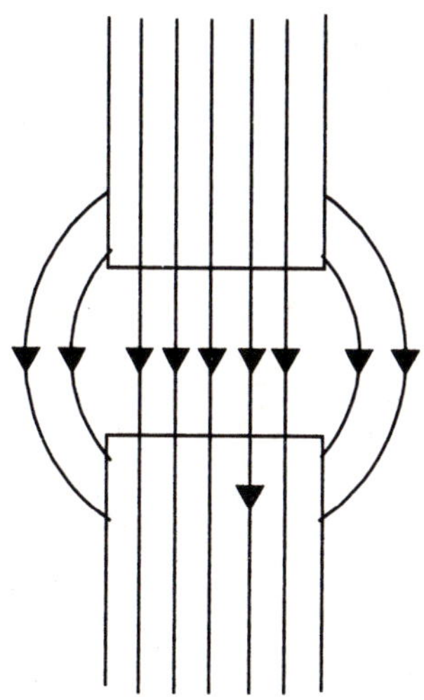

Fig. 9.9 Fringing flux

9.8.4 Dot Convention for Coupled Circuits

The significance of dot convention is explained with the help of Fig. 9.10.

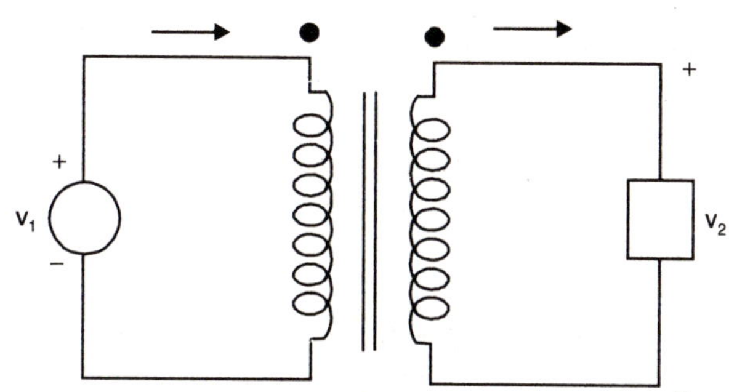

Fig. 9.10 A circuit for dot convention

Suppose a time varying voltage source is connected across the primary of the transformer and at any given instant the voltage source has the polarity shown and current $i(t)$ is in the direction shown by the arrow and is increasing with time entering the dotted terminal, this current induces a voltage in the secondary which is positive at the dotted terminal. Conversely if the current in the primary is increasing and is leaving the dotted terminal, this will induce the voltage polarity in the secondary such that the undotted end will become positive and hence dotted terminals of both the windings will be negative simultaneously. The following experimental procedure can be used to establish dotted ends of the transformer windings. Mark a dot arbitrarily on one end of the primary winding and connect a d.c. source with positive terminal connected to the dotted end and the negative terminal to the undotted end of the winding through a switch. Connect a MC voltmeter across the secondary winding. The end of the secondary winding which goes positive momentarily on closing the switch on the primary side as measured by the voltmeter, is the terminal to be dotted.

If there are more than two windings, similar procedure can be followed for identifying relationship between each pair of windings. For each pair of winding different forms of dots (• ❖ ■ ❁) should be used to avoid any confusion.

Suppose we are required to write loop equations for a mutually coupled circuit shown in Fig. 9.11.

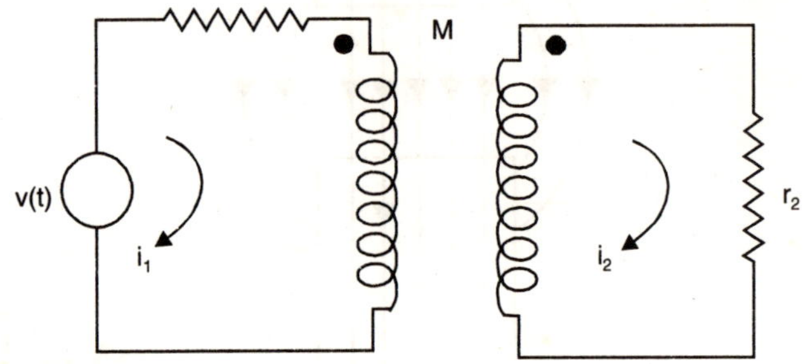

Fig. 9.11 Coupled circuit

For primary loop the drops are $i_1 r_1 + L_1 \dfrac{di_1}{dt}$ + drop in primary loop is due to current in secondary loop. The voltage induced in primary will be determined by the direction of i_2 with respect to its dot in secondary loop. Since current is going away from the dot, the polarity of voltage on the primary side will be positive at the undotted terminal. Therefore, for current i_1 it will be rise in voltage and the equation will be

$$i_1 r_1 + L_1 \frac{di_1}{dt} - M \frac{di_2}{dt} = v(t)$$

and for the secondary loop, similarly

$$i_2 r_2 + L_2 \frac{di_2}{dt} - M \frac{di_1}{dt} = 0$$

The equivalent two loops are as shown in Fig. 9.12(*a*).

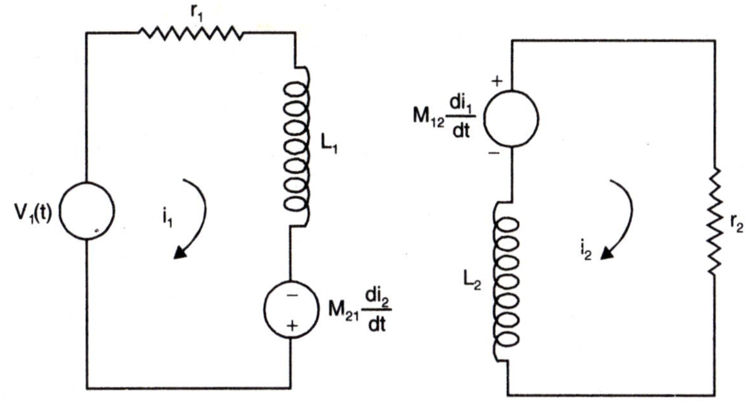

Fig. 9.12 (*a*) Equivalent of Fig. 9.10

9.9 SELF INDUCTANCE

When a coil is connected to an a.c. source, it draws certain current which is changing as a function of time and hence this current produces a flux which changes with time. So it is a case of stationary circuit but the flux is changing as a function of time which induces emf in the coil.

Since the emf is induced in the coil itself in which the circuit/flux is changing, this emf is known as self-induced emf. Since it is mainly due to changing current, the voltage induced is given by

$$e = L\frac{di}{dt} \qquad \text{...(9.35(a))}$$

Here e is the self-induced emf in volts;

$\frac{di}{dt}$ is the time rate of change of current

and L is known as self inductance as it is the quantity of the circuit itself or it is also known as coefficient of self inductance or simply inductance and its unit is Henry. According to Lenz's law the polarity of the emf so induced is such that it opposes the change of current *i.e.* it opposes the external source of voltage as shown in Fig. 9.12(*b*).

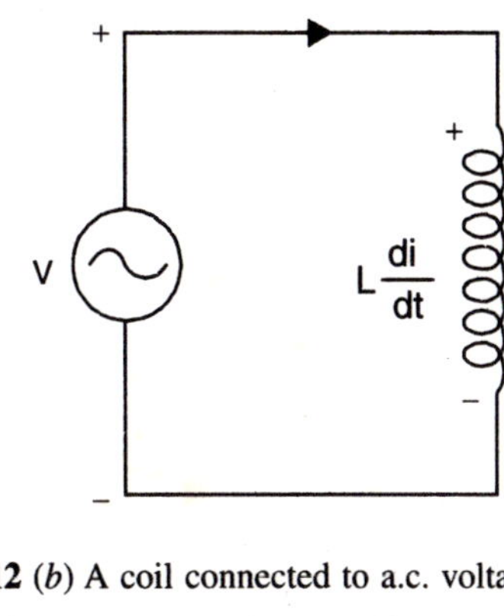

Fig. 9.12 (*b*) A coil connected to a.c. voltage source

We also know, according to Faraday's law of electromagnetic induction that

$$e = N\frac{d\phi}{dt}$$

where N is the no. of turns in the coil and $\frac{d\phi}{dt}$ is the time rate of change of flux.

Equating this equation with equation (9.35(a)), we have

$$L\frac{di}{dt} = N\frac{d\phi}{dt}$$

or

$$L = N\frac{d\phi}{di} \qquad \text{...(9.36)}$$

However, if the B–H curve of the magnetic core over which the coil is wound is a straight line *i.e.* there is no saturation of the core, then ϕ and i are linearily related and equation (9.36) can be rewritten as

$$L = \frac{N\phi}{i} \qquad \text{...(9.37)}$$

This equation defines inductance of a coil and is given as flux linkage per ampere and is a constant quantity. However, we know that B–H curve gets saturated *i.e.* with increase in current after certain value of current, the flux does not increase linearily. Rather the increase in flux is relatively smaller and hence the change in flux linkage per ampere reduces which means the effective inductance of the coil decreases. The variation of inductance as a function of current is shown in Fig. 9.12(c).

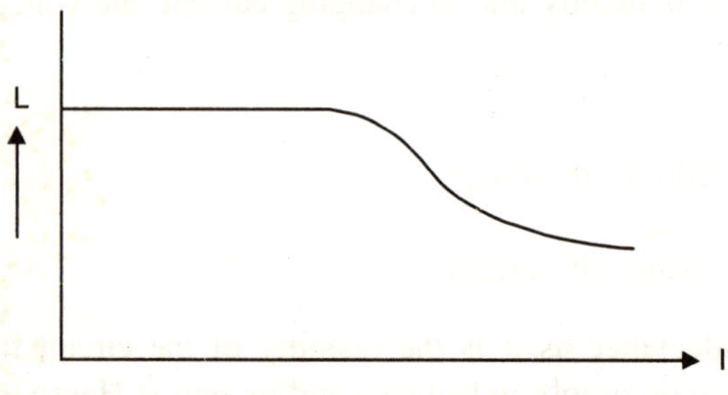

Fig. 9.12 (c) Variation of inductance as I varies.

If we have to find out inductance of a given coil when the saturation of the core takes place we make use of equation (9.36) and we define inductance as the ratio of differential change in flux linkages to differential change in current.

To obtain physical dimensions of the coil we make use of equation (9.37)

Since

$$\phi = \frac{mmf}{\text{Reluctance}} = \frac{Ni}{l/\mu A}$$

Substituting for ϕ in equation (9.37), we have

$$L = \frac{N\{Ni/(l/\mu A)\}}{i} = \frac{N^2\mu A}{l} \qquad \text{...(9.38)}$$

i.e. the inductance of a coil is directly proportional to (*i*) square of number of turns, (*ii*) directly proportional to the permeability of the core material and area of cross-section of the core and inversely proportional to the length of the magnetic circuit. Usually it is the no. of turns which are varied to obtain a desired value of inductance. It is to be noted that for small values of inductances in term of *mH* or μH, usually air-cored coils are used as $\mu_r = 1$ for air and μ_r is quite high for magnetic materials.

9.10 MUTUAL INDUCTANCE

In Fig. 9.13 current I_1 produces a total flux ϕ_{11} and in coil 1. The subscript 11 means one-one and not eleven *i.e.* the flux of coil 1 that links coil 1. The resulting self inductance L_1 of coil 1 is

$$L_1 = \frac{N_1 \phi_{11}}{I_1} H \qquad \text{...(9.39)}$$

Similarly current I_2 produces a total flux ϕ_{22} in coil 2, and therefore, the self-inductance of coil 2 is

$$L_2 = \frac{N_2 \phi_{22}}{I_2} H \qquad \text{...(9.40)}$$

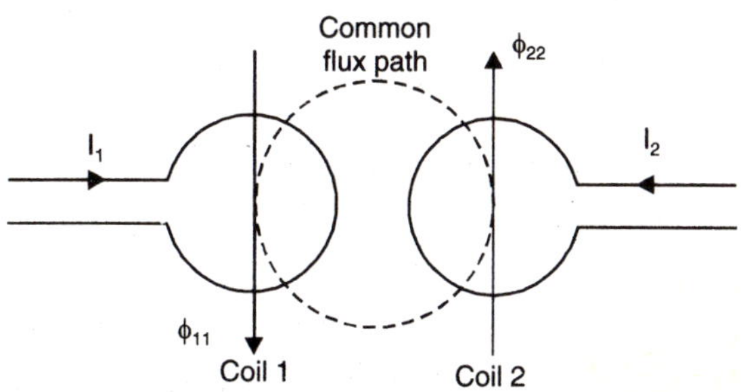

Fig. 9.13 Mutual flux and mutual inductance

The directions of ϕ_{11} and ϕ_{22} have been shown as per the direction of current flows in the respective coils. A part of flux ϕ_{11} links coil 2. Hence, total flux in coil 2 consists of two components, its own flux ϕ_{22} plus the part of ϕ_{11} which is common. The amount of flux of coil 1 which links coil 2 is ϕ_{12}. The subscript "12" means from coil 1 to coil 2. Similarly a part of flux ϕ_{22} of coil 2 which links coil 1 is ϕ_{21}.

In Fig. 9.13, the physical placement of the coils and the direction of currents I_1 and I_2 produce fluxes ϕ_{12} and ϕ_{21} such that these reinforce ϕ_{11} and ϕ_{22} by acting in the same direction as the original ϕ_{11} and ϕ_{22}. Thus these mutual fluxes aid the main fluxes. If, however either I_1 or I_2 is reversed, ϕ_{12} and ϕ_{21} will oppose the fluxes ϕ_{22} and ϕ_{11}. A plus sign is used when the fluxes aid each other, otherwise minus sign is used. However, it should be noted that both ϕ_{12} and ϕ_{21} should be given the same sign either plus or minus. A condition cannot exist when one is plus and the other minus.

The flux ϕ_{21} which links coil 1 having N_1 turns will have flux linkage $N_1 \phi_{21}$ and the flux ϕ_{21} is due to current I_2 in coil 2.

Hence,
$$M_{21} = \frac{N_1 \phi_{21}}{I_2} \qquad \text{...(9.41)}$$

This is the mutual inductance due to flux linkages in the coil 1. Similarly the flux ϕ_{12} which is linking coil 2 of N_2 turns and is due to current I_1 in coil 1.

Hence, $$M_{12} = \frac{N_2 \phi_{12}}{I_1} H \qquad ...(9.42)$$

In the discussion above, we have assumed the mutual fluxes to be additives and hence plus sign is assigned to M_{21} and M_{12}. However, if the mutual fluxes oppose the main fluxes ϕ_{11} and ϕ_{22}, then M is to be given negative sign.

We define the coefficient of magnetic coupling between the two coils as the ratio of common flux to the total flux.

i.e. $$K_1 = \frac{\phi_{12}}{\phi_{11}} \text{ and } K_2 = \frac{\phi_{21}}{\phi_{22}} \qquad ...(9.43)$$

It can be shown that for any two coils, K_1 and K_2 are same and M_{12} and M_{21} are also same. Hence let

$$K_1 = K_2 = K \text{ and } M_{12} = M_{21} = M \qquad ...(9.44)$$

Multiplying equation (9.41) and (9.42) and using equation (9.44), we have

$$M^2 = M_{12} M_{21} = \frac{N_2 \phi_{12}}{I_1} \cdot \frac{N_1 \phi_{21}}{I_2} \qquad ...(9.45)$$

From equation (9.43), $\phi_{12} = K\phi_{11}$ and $\phi_{21} = K\phi_{22}$

Substituting values of ϕ_{12} and ϕ_{21} in equation (9.45), we have

$$M^2 = K^2 \frac{N_1 \phi_{11}}{I_1} \cdot \frac{N_2 \phi_{22}}{I_2} = K^2 L_1 L_2$$

Hence, $$K = \frac{M}{\sqrt{L_1 L_2}} \qquad ...(9.46)$$

or $$M = \pm K \sqrt{L_1 L_2}$$

The coefficient of coupling represents that fraction of the flux of one coil that links a second coil. If a core is of good magnetic material, the coefficient of coupling could be as high as 0.99. However for air-cored coils, this coupling coefficient is as low as 0.05. The mutual inductance induces voltage in these coils. The coil 1 will induce a voltage $M\frac{di_1}{dt}$ in coil 2 whereas coil 2 will induce a voltage $M\frac{di_2}{dt}$ in coil 1 if the mutual fluxes are aiding. If these are opposing the main fluxes then these voltages will be $-M\frac{di_1}{dt}$ and $-M\frac{di_2}{dt}$ respectively.

9.11 HYSTERESIS LOSS

Consider Fig. 9.14 consisting of a coil wound on a magnetic toroid. Assume that the core is initially unmagnetized. If the current i is increased *i.e.* $H = \frac{Ni}{l}$, the magnetic field intensity is increased gradually from zero value. The flux density B starts increasing as shown by *oa* part of the curve in Fig. 9.14 (*c*).

The point a corresponds to a particular value of current say i_1. If i were less than i_1, point a will shift downwards towards o and a smaller final loop will appear.

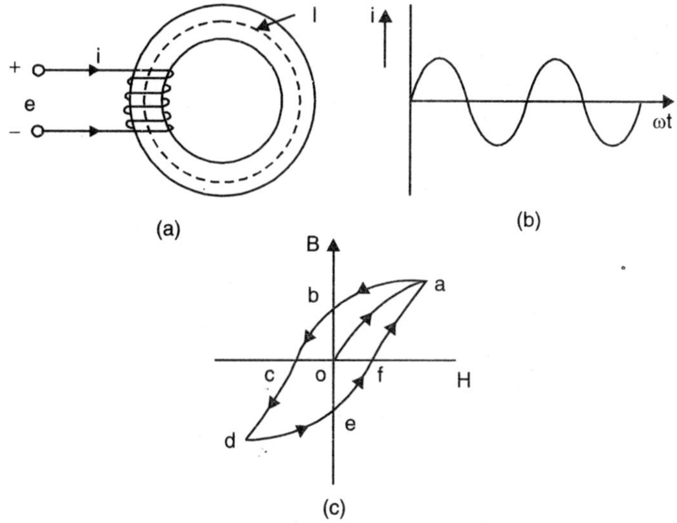

Fig. 9.14 (*a*) Core coil assembly (*b*) Existing current (*c*) Hysteresis loop

Now if we decrease magnetic field intensity *i.e.* decrease current *i.e.* i, *B-H* curve will follow a different path, different from ao. In fact it goes along ab, point b is reached when the field intensity is zero and hence current i is zero. We see that when current through the coil is zero still there is some magnetic flux density in the core of the coil and this is known as residual flux density corresponding to point b in the Fig. 9.14(c). If H is now reversed by reversing the flow of current i, the flux in the core will decrease and the point on the *B-H* curve will move along bc and when H corresponding to H_c is reached, the core is again demagnetized. The magnitude of magnetic field intensity ($-H_c$) is known as coercivity or coercive force of the magnetic core.

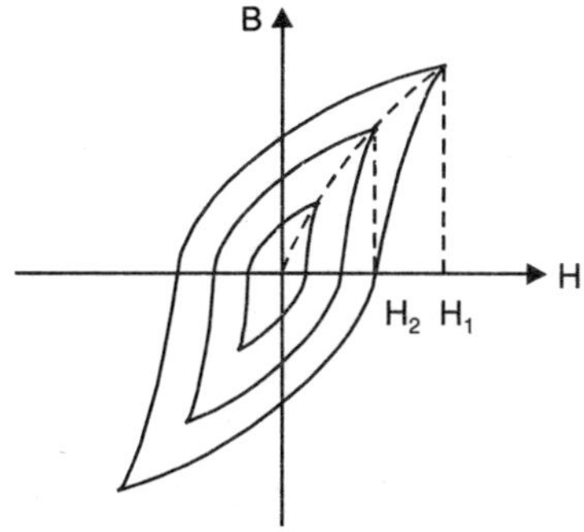

Fig. 9.15 Three hysteresis loops—magnetization characteristics of the magnetic core

If H is further increased in the reverse direction the magnetic flux density will also increase in the reverse direction. *If* H is made corresponding to current $-i_1$ the point on the *B-H* curve moves along the cd path. The point d corresponds to $H_d = -\dfrac{Ni_1}{l}$. If H is now decreased to zero and then increased to the value corresponding to current *i*, the *B-H* curve will follow the path *defa*. Usually the last point *a* is slightly short of the original point but for simplicity we assume that it coincides the original point while moving from *d* through *ef* and towards '*a*'. However, it is found that, if the core is subjected to a few cycles of magnetization, the loop almost closes and this loop is called hysteresis loop. The loop shows that the relationship between *B* and *H* is non-linear and multivalued. It is to be noted that at point *b* the core is magnetized even though the current through the coil is zero. Throughout the complete cycle of magnetization the flux density lags behind the magnetic field intensity. This lagging phenomenon in the magnetic core is known as hysteresis.

Smaller hysteresis loops can be obtained by varying maximum value of current and hence magnetic field intensity. If $i_1 > i_2 > i_3$ where i_1, i_2 and i_3 represent three values of maximum currents and hence H_1, H_2 and H_3. The corresponding hysteresis loops are shown in Fig. 9.15. The locus of the tips of the hysteresis loops shown with dashed line is known as magnetization curve. If the iron core is magnetized from an initial unmagnetized condition, the flux density will follow the magnetization curve.

When current *i* is varied over one cycle, during some part of the cycle, the coil stores energy from the source and during some part, it returns energy back to the source. However, energy flowing into the coil is greater than the energy returned back to the source. Hence, during one cycle of variation of current *i* (Hence *H*), there is a net energy flow from the source to the coil core assembly. The energy difference goes to heat the core. The loss of power in the core due to hysteresis effect is known as hysteresis loss. It can be shown that the area of the hysteresis loop is proportional to the hysteresis loss.

Assuming that the coil has no resistance

$$e = N\frac{d\phi}{dt}$$

and energy transfer during time interval t_1 to t_2 is

$$W = \int_{t_1}^{t_2} ei\,dt \qquad \text{...(9.46(a))}$$

or

$$W = \int N\frac{d\phi}{dt} i\,dt = \int_{\phi_1}^{\phi_2} N_i d\phi \qquad \text{...(9.46(b))}$$

Now, $$\phi = B \cdot A \text{ and } i = \frac{Hl}{N}$$

where l is the length of the magnetic path and A the area of cross-section.
Substituting these values in equation (9.46(*b*)), we have

$$W = \int_{B_1}^{B_2} N \cdot \frac{Hl}{N} \cdot A\,dB = lA \int_{B_1}^{B_2} H \cdot dB$$

Here lA = Volume of the magnetic circuit and the integral terms *HdB* represents the area shown hatched in Fig. 9.16.

The energy received by the core during one circle is

$$W/\text{cycle} = V_{\text{core}} \oint H\,dB$$

$$= V_{core} \times \text{area of } B\text{-}H$$

$$= V_{core} \times W_h$$

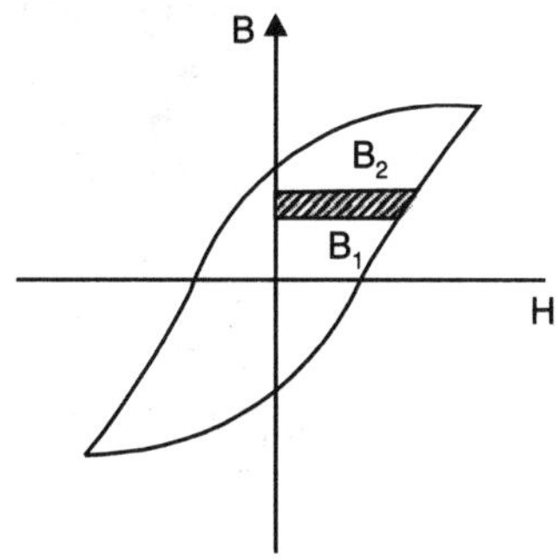

Fig. 9.16 Hysteresis loop

where W_h is the energy density in the core. The power loss in the core due to hysteresis effect is

$$P_h = V_{core} W_h f \qquad \text{...(9.47)}$$

where f is the frequency of current i. Since it is not possible to evaluate the area of the B-H loop, mathematically as the curve cannot be described by some mathematical expression. Steinmetz suggested empirical relation after lot of experimentation to the area of the loop. He suggested that this can be approximated by

Area of B-H loop $= KB^n_{max} f$ ...(9.48)

where B_{max} is the maximum flux density, the value of n varies between 1.5 to 2.5 and k is a constant. Both k and n can be obtained empirically. Using equation (9.47) and (9.48) the hysteresis loss is given by

$$P_n = k_h B^n_{max} f \qquad \text{...(9.49)}$$

Here k_h depends upon the volume and ferromagnetic material of the core. Typical value for grain oriented silicon steel sheet used for electric machines is 1.6. Some typical values for k are: cast steel 0.025, silicon sheet steel 0.001, permalloy 0.0001. k is known as Steinmetz's constant.

9.12 EDDY CURRENT LOSS

When *a.c.* voltage is applied to a coil wound on a magnetic core, the current not only induces voltage within the coil, it induces voltage in the magnetic core also which results in what are known as circulating currents or eddy currents. These eddy currents result in a loss in magnetic materials known as eddy current loss. The circulation of these currents has been shown in Fig. 9.17. Fig. 9.17(*a*) shows flow of eddy current in a solid core whereas (*b*) shows in a laminated core. It has been observed that eddy currents flow cross-wise in the core. The magnitude of eddy current can be reduced substantially by increasing the cross-sectional resistance of the core. The cores are made of the sheets called lamination. The thickness of laminations varies from 0.5 to 5 mm in electrical machines and from 0.01 to 0.5 mm in devices used in electronic circuits which operate at frequencies. Each lamination is insulated from the others by a coating of oxides or varnish or both so that it offers a high resistance to the flow of eddy currents. The eddy current loss can also be reduced by using a high resistivity core material. Addition of a small per cent of silicon say 3% to iron will increase the resistivity significantly. The eddy current

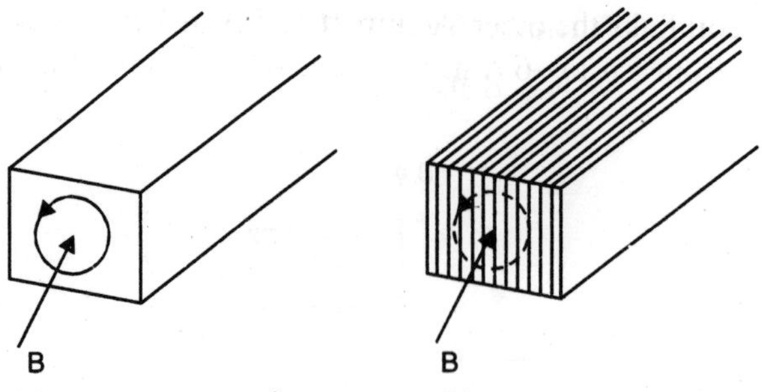

Fig. 9.17 Eddy current loss in core (*a*) Solid core (*b*) Laminated core

loss is given by

$$P_e = K_e f^2 B_m^2 t^2 \times \text{volume}$$

where t is the thickness of laminations and K_e is a constant which depends upon the core material.

The sum of the hysteresis and eddy current loss is known as core loss or iron loss. Since for a machine f and B_m are constant, the core loss is also known as constant loss of the machine.

Eddy currents can be used to the advantage of electric braking purpose *e.g.* in moving iron and moving coil measuring instruments eddy currents are utilized to damp out the final settling of the instrument pointer. Also there are some specially designed machines which utilize eddy current braking.

Example 9.17. In a pair of coupled coils, coil 1 has 600 turns and carries a current of 2.5 A and its fluxes ϕ_{11} and ϕ_{21} are 0.4 and 0.8 m Wb respectively. If coil 2 has 2000 turns, determine L_1, L_2, M and K.

Solution. The total flux due to coil is

$$\phi_1 = \phi_{11} + \phi_{21} = 0.4 + 0.8 = 1.2 \text{ m Wb}$$

Self inductance of coil $1 = \dfrac{N_1\phi_1}{I_1} = \dfrac{600 \times 1.2 \times 10^{-3}}{2.5} = 288$ mH.

The coupling coefficient is $\dfrac{\phi_{21}}{\phi} = \dfrac{0.8}{1.2} = 0.67$

The mutual inductance $\quad M = \dfrac{N_2\phi_{21}}{I_1} = \dfrac{2000 \times 0.8}{2.5} = 640$ mH

and hence

$$K = \frac{M}{\sqrt{L_1 L_2}} = 0.67$$

$$\sqrt{L_1 L_2} = \frac{M}{0.67}$$

or

$$\sqrt{288 L_2} = \frac{640}{0.67}$$

$$L_2 = 3.168 \text{ H } \textbf{Ans.}$$

Example 9.18. Two identical coils of 1000 turns each X and Y lie in parallel planes such that 60% of the flux produced by one coil links the other. A current of 5 A in X produces in it a flux of 0.05 mWb. If the current in X changes from 6 A to –6 A in 10 m sec. What is the emf induced in coil Y? Determine L_x, L_y and M.

Solution. $\phi_x = 0.05$ mWb

$$L_x = \frac{N_x \phi_x}{I_x} = \frac{1000 \times 0.05 \times 10^{-3}}{5} = \frac{0.05}{5} = 10 \text{ mH}$$

Since the two coils are identical $L_x = L_y = 10$ mH **Ans.**

$$\text{Since } K = 0.6 = \frac{M}{\sqrt{L_1 L_2}} = \frac{M}{10} \quad \text{or} \quad M = 6 \text{ mH} \quad \textbf{Ans.}$$

$$\text{Voltage induced in coil } y,\ M\frac{di}{dt} = 6 \times 10^{-3} \frac{12}{10 \times 10^{-13}} = 7.2 \text{ volt } \textbf{Ans.}$$

Example 9.19. Two coupled coils with $L_1 = 0.5$ H and $L_2 = 1.0$ H have a coefficient of coupling 0.8 and the no. of turns in coil 2 is 1500. If the current in coil 1 is $i_1 = 20 \sin 314\,t$ amp, determine the voltage across coil 2 and the maximum flux set up by the coil 1.

Solution. Since $M = k\sqrt{L_1 L_2} = 0.8\sqrt{0.5 \times 1.0} = 0.5657$ H

Now voltage across coil 2 is

$$v_2 = M\frac{di_1}{dt} = 0.5657\frac{d}{dt}(20 \sin 314t)$$

$$= 0.5657 \times 20 \times 314 \cos 314t$$

$$= 3552.5 \cos 314t \textbf{ Ans.}$$

$$v_2 = N_2 \frac{d\phi_{21}}{dt} = 1500 \Rightarrow \frac{d\phi_{21}}{dt} = 3552.5 \cos 314t$$

or $$d\phi_{21} = 2.37 \cos 314t\, dt$$

$$\phi_{21} = \frac{2.37}{314} \sin 314t = 0.0075 \sin 314t$$

or $$\phi_{21} = 7.5 \text{ mWb maximum value } \textbf{Ans.}$$

Example 9.20. Determine the hourly loss of energy in a specimen of iron the hysteresis loop of which is equivalent in area of 250 J/m³, frequency 50 Hz, density of iron 7.5 and weight 10 kg.

Solution. Density is 7.5 which means 7.5 gm/cm³

or $$\frac{7.5 \times 10^6}{10^3 \times \text{m}^3} = \text{kg/m}^3 = 7500 \text{ kg/m}^3$$

$$\text{Hence volume of the specimen} = \frac{10}{7500} = 1.33 \times 10^{-3} \text{ m}^3$$

Therefore, iron loss per cycle = $250 \times 1.33 \times 10^{-3} = 333.33 \times 10^{-3}$ Joule
Therefore, loss per second = $333.33 \times 10^{-3} \times 50 = 16.67$ Joule
Hence, hourly loss = $16.67 \times 3600 = 60{,}000$ Joule **Ans.**

Example 9.21. The maximum flux density in a hysteresis loop is 1.5 Wb/m^2. The abscissa is scaled as 1 cm = 15 A/cm, ordinate is 1 cm = 1 Wb/m^2. The area of the hysteresis loop is 0.6 cm^2.

Determine the hysteresis loss if 1500 cm^3 of the material is subjected to an alternating flux density of 1.5 Wb/m^2 at 50 Hz.

Solution. The area of the loop 0.6 cm^2 gives the hysteresis loss per cycle.

Now 1 cm = 15 AT/cm or AT/m

and 1 cm = 1 Wb/m^2

Therefore, 1 cm^2 area of loop equals $1500 \times 1 = 1500$ J/m^3/cycle

Since the area of the loop given is 0.6 cm^2.

Hence, hysteresis loss = 1500×0.6 J/m^3/cycle = 900

Volume of material = 1500 cm^3 or 1.5×10^{-3} cu.m

Hence, hysteresis loss per cycle = $900 \times 1.5 \times 10^{-3}$

= 1350×10^{-3} Joule

Therefore at 50 Hz the loss per second

$1.35 \times 50 = 67.5$ J/sec or 67.5 watt.

Example 9.22. The hysteresis loss in a sample of iron is found to be 4.9 W/kg at a frequency of 50 Hz and maximum flux density of 1 Wb/m^2. (*a*) Determine the coefficient η in the expression for loss in Joule/m^3/cycle = $\eta B^{1.7}$. The density of iron is 7.5, (*b*) Calculate the loss per kg at 25 Hz and 1.8 Wb/m^2.

Solution. (*a*) The hysteresis loss J/m^3/cycle is

$$\frac{4.9 \times 7500}{50} = 735$$

and this should equal $\eta (B)^{1.7} = \eta(1)^{1.7} = 735$

Hence, $\eta = 735$ **Ans.**

(*b*) To find out loss per kg at 25 Hz and 1.8 Wb/m^2

$$4.9 \times \frac{25}{50} \times (1.8)^{1.7} = 6.65 \text{ watt/kg. } \textbf{Ans.}$$

Example 9.23. Two coils with N_1 and N_2 turns are wound on the same magnetic bar. Show that the mutual inductance between the two coils is $M = \frac{N_1 N_2 K}{R}$ where R is the reluctance of the magnetic circuit.

Solution. $M = K\sqrt{L_1 L_2}$ and $L_1 = N_1^2 \mu_0 \frac{A}{l}$ and $L_2 = N_2^2 \mu_0 \frac{A}{l}$

or $$L_1 = \frac{N_1^2}{R} \text{ and } L_2 = \frac{N_2^2}{R}$$

Hence, $$M = K \cdot \frac{N_1 N_2}{R} \quad \textbf{Ans.}$$

Example 9.24. If a proton is fixed in position and an electron revolves around it in a circular path of radius 0.5×10^{-10} m, determine the magnetic field at the proton.

Solution. The force of attraction between proton and electron is given by

$$F = \frac{Q^2}{4\pi\varepsilon_0 r^2} \text{ which provides the centripetal force for the circular motions.}$$

Thus,
$$m_e W^2 r = \frac{Q^2}{4\pi\varepsilon_0 r^2} \quad \text{or} \quad W^2 = \frac{Q^2}{4\pi\varepsilon_0 m_e r^3}$$

The circular movement of the electron is equivalent to a current loop $I = \frac{W}{2\pi} Q$.

The field due to a current loop with current I and radius r is given as

$$B = \mu_0 H = \mu_0 \frac{I}{2r} = \mu_0 \frac{W}{4\pi r} Q$$

$$B = \frac{\mu_0 Q}{4\pi r} \cdot \frac{Q}{2\sqrt{\pi\varepsilon_0 m_e r^3}}$$

$$= \frac{10^{-7} \times (1.602 \times 10^{-19})^2}{2r^2\sqrt{3 \cdot 14 \times 8.854 \times 10^{-12} \times 9.1 \times 10^{-31} \times 0.5 \times 10^{-10}}}$$

$$= \frac{2.566 \times 10^{-45}}{2 \times 0.25 \times 10^{-20}\sqrt{126.5 \times 10^{-53}}}$$

$$= \frac{5.132 \times 10^{-45}}{10^{-20} \times 10^{-26} \times 3.556} = 14.43 \text{ T}$$

Example 9.25. Consider a plane boundary between two media of permeabilities μ_1 and μ_2 as shown in Fig. E 9.25. Determine the relation between the angles α_1 and α_2. Assume that the media are isotropic with B and H in the same direction.

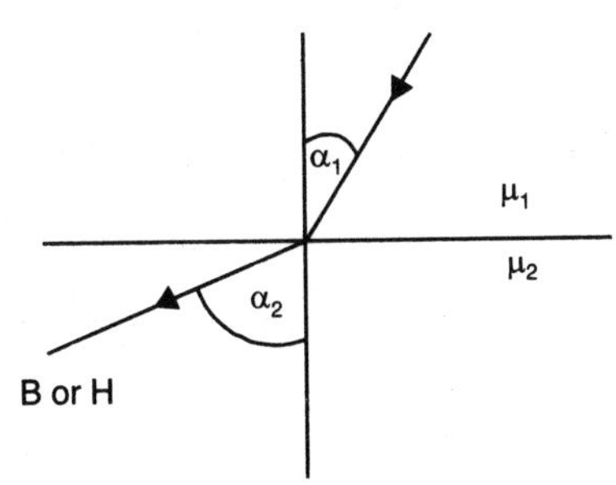

Fig. E 9.25

Solution. From the boundary condition we know that

$$B_{H_1} = B_{H_2} \text{ and } H_{t_1} = H_{t_2}$$

Now $$B_{H_1} = B_1 \cos\alpha_1 \text{ and } B_{H_2} = B_2 \cos\alpha_2$$

or $$H_{t_1} = H_1 \sin\alpha_1 \text{ and } H_{t_2} = H_2 \sin\alpha_2$$

Now $$B_{H_1} = \mu_1 H_1 \cos\alpha_1 \text{ and } B_{H_2} = \mu_2 H_2 \cos\alpha_2$$

or $$\mu_1 H_1 \cos\alpha_1 = \mu_2 H_2 \cos\alpha_2$$

Also $$H_1 \sin\alpha_1 = H_2 \sin\alpha_2$$

$$\frac{H_1 \sin\alpha_1}{\mu_1 H_1 \cos\alpha_1} = \frac{H_2 \sin\alpha_2}{\mu_2 H_2 \cos\alpha_2}$$

$$\frac{\tan\alpha_1}{\mu_1} = \frac{\tan\alpha_2}{\mu_2}$$

or $$\frac{\tan\alpha_1}{\tan\alpha_2} = \frac{\mu_1}{\mu_2} = \frac{\mu_{r_1}}{\mu_{r_2}}$$

This equation gives the relation between angles α_1 and α_2 for B and H lines at the boundary between two magnetic media.

Example 9.26. Refer to Fig. E 9.26 and obtain an expression for the angular deflection in magnetic refraction.

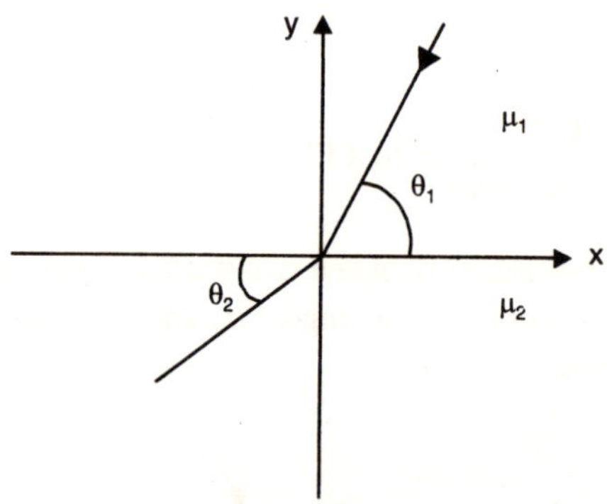

Fig. E 9.26

Solution. At the interface the incident ray in region 1, is refracted through medium by an angle θ_2. Hence the angular deflection is say $\alpha = \theta_1 - \theta_2$

At the boundary $$H_{t_1} = H_{t_2}$$

$$\frac{B_1}{\mu_1}\cos\theta_1 = \frac{B_2}{\mu_2}\cos\theta_2$$

Also $$B_{H_1} = B_{H_2} \text{ or } B_1 \sin\theta_1 = B_2 \sin\theta_2$$

or $$B_1 = B_2 \frac{\sin\theta_2}{\sin\theta_1}$$

$$\frac{B_2}{\mu_1}\frac{\sin\theta_2}{\sin\theta_1}\cos\theta_1 = \frac{B_2}{\mu_2}\cos\theta_2$$

or $$\frac{\cos\theta_1}{\mu_1 \sin\theta_1} = \frac{\cos\theta_2}{\mu_2 \sin\theta_2}$$

or $$\mu_1 \frac{\sin\theta_1}{\cos\theta_1} = \mu_2 \frac{\sin\theta_2}{\cos\theta_2} \text{ or } \mu_1 \tan\theta_1 = \mu_2 \tan\theta_2 = k$$

Now $$\tan\alpha = \tan(\theta_1 - \theta_2) = \frac{\tan\theta_1 - \tan\theta_2}{1 + \tan\theta_1 \tan\theta_2}$$

$$= \frac{\mu_2\mu_1 \tan\theta_1 - \mu_1\mu_2 \tan\theta_2}{\mu_1\mu_2 + (\mu_1 \tan\theta_1)\mu_2(\tan\theta_2)} = \frac{(\mu_2 - \mu_1)k}{\mu_1\mu_2 + k^2}$$

It is to be noted that if $\alpha > 0$ *i.e.* deflection away from the surface normal than $\mu_2 > \mu_1$.

Example 9.27. An iron ring magnet of 0.02 m^2 cross-section area and 300 mm radius has a 1mm gap and a winding of 1200 turns. If the current through the coil is 6A determine the force tending to close the gap. Assume $\mu_r = 1000$ for the iron and nickel fringing.

Solution.

Reluctance of iron path $$\frac{l}{\mu_0\mu_r A} = \frac{2\pi \times 0.3}{4\pi \times 10^{-7} \times 1000 \times 0.02}$$

$$= 7.5 \times 10^{-4}$$

Reluctance of air gap $$= \frac{1 \times 10^{-3}}{4\pi \times 10^{-7} \times 1 \times 0.02} = 3.98 \times 10^4$$

Total reluctance = $(7.5 + 3.98)\ 10^4 = 11.48 \times 10^4$

$$\phi = \frac{MMF}{\text{Reluctance}} = \frac{1200 \times 6}{11.48 \times 10^4} = 62.7 \text{ mWb}$$

$$B = \frac{\phi}{A} = \frac{62.7 \times 10^{-3}}{0.02} = 3.135 \text{ T}$$

Hence $$\text{Force } F = \frac{B^2}{2\mu_0} A = 78.25 \text{ kN} \textbf{ Ans.}$$

Example 9.28. Derive expression for inductance of a Toroid with circular cross-section and a solenoid.

Solution. The expression for inductance is given as

$$L = \frac{\mu_0 N^2 A}{l}$$

(*i*) Toroid: For a toroid if r is the mean radius of the toroid and A is the area of cross section of the toroid then

$$L = \frac{\mu_0 N^2 \cdot A}{2\pi r} H$$

(*ii*) For a solenoid, $L = \dfrac{\mu_0 N^2 A}{l}$ where l is the length of the solenoid and A is the area of the cross section of solenoid.

Example 9.29. Derive an expression for the inductance of a single phase overhead line.

Solution. Inductance of Two-Wire (1-ϕ) Transmission Line

By definition inductance is the flux linkages per ampere (Fig. E 9.29). So the objective is to find out the flux linkages to this system of conductors. Now there are two flux linkages: (*i*) due to internal flux, and (*ii*) due to external flux.

Internal flux linkages: In order to determine the internal flux linkages, we start with the magnetic field intensity H at any distance $r < R$.

$$H = \frac{Ir}{2\pi R^2} \qquad \text{...(9.50)}$$

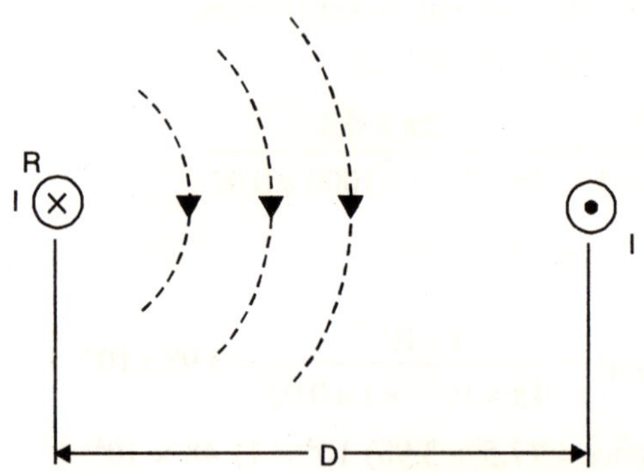

Fig. E 9.29 Magnetic field due to one conductor of a 1-ϕ transmission line

$$\therefore \qquad B = \mu H = \mu_0 H = \frac{\mu_0 I}{2\pi R^2} \cdot r \;(\text{as } \mu_r = 1) \text{ for conductors.}$$

This flux density as we see is varying with r. We can assume this to be constant over an infinitesimal distance dr. The flux lines are in the form of circles concentric to the conductor. Therefore, the flux lines passing through the concentric cylindrical shells of radii r and $r + dr$,

$$d\phi = B \cdot \text{Area normal to flux density } B$$
$$= B \cdot dr \cdot l$$

where l is the length of wire.

In case the inductance per unit is desired, $l = 1$ meter.

$$\therefore \qquad d\phi = B \cdot dr$$

$$= \frac{\mu_0 I}{2\pi R^2} r \cdot dr$$

Now flux linkages = flux × No. of turns

Here since only a part of the conductor ($r < R$) is being enclosed by the flux lines $d\phi$,

$$\therefore \qquad d\lambda = d\phi \cdot \left(\frac{r^2}{R^2}\right)$$

$$= \frac{\mu_0 I}{2\pi R^2} r \cdot dr \cdot \frac{r^2}{R^2}$$

$$\therefore \text{ Total internal flux linkages } \lambda = \int_0^R d\lambda$$

$$= \frac{\mu_0 I}{2\pi R^4} \int_0^R r^3 dr$$

$$= \frac{\mu_0 I}{8\pi} \qquad \ldots(9.51)$$

From this it is clear that the flux linkage due to internal flux is independent of the size of the conductor.

External flux linkages: These flux linkages are due to the flux lines outside the conductor. There will be no flux line that encloses both the conductors. This is because for any distance $r > D$ the total current enclosed is zero (single phase line *i.e.* one conductor is a 'go' conductor and the other 'return'). The magnetic field intensity H due to one conductor at any distance $R \le r < D$,

$$H = \frac{I}{2\pi r}$$

$$B = \mu_0 H = \frac{\mu_0 I}{2\pi r} \left(\mu_r = 1 \text{ as the medium is air}\right)$$

The flux density B can be considered uniform over a distance dr. Therefore, as in the case of internal flux, the flux lines passing through the concentric cylindrical shells with radii r and $(r + dr)$ will be (per unit length).

$$d\phi = B \cdot dr \cdot 1$$

Since this flux encloses only one conductor, therefore, the number of turns enclosed by this flux is one.

$$\therefore \qquad d\lambda = d\phi \cdot 1$$

$$= B \cdot dr \cdot 1 \cdot 1$$

$$= \frac{\mu_0 I}{2\pi r} \cdot dr$$

Therefore, the total external flux linkages due to current flow in one conductor,

$$\lambda = \int_{R}^{D-R} d\lambda$$

The lower limit is because we measure the distances from the centre of the conductor and external flux begins from the surface of the conductor and this extends upto the surface of the other conductor and, therefore, the upper limit $(D - R)$.

$$\lambda = \frac{\mu_0 I}{2\pi}\int_{R}^{D-R}\frac{dr}{r}$$

$$= \frac{\mu_0 I}{2\pi}\ln\frac{D-R}{R}$$

Since R is small as compared to D *i.e.* $R << D$,

$$D - R \simeq D$$

$\therefore$
$$\lambda = \frac{\mu_0 I}{2\pi}\ln\frac{D}{R}$$

$\therefore$ Total flux linkages due to one conductor
= Total internal flux linkages + Total external flux linkages

$$= \frac{\mu_0 I}{8\pi} + \frac{\mu_0 I}{2\pi}\ln\frac{D}{R}$$

$\therefore$ Total flux linkage due to both the conductors $= 2\left[\frac{\mu_0 I}{8\pi} + \frac{\mu_0 I}{2\pi}\ln\frac{D}{R}\right]$

$\therefore$ Inductance L per unit length $= \left[\frac{\mu_0}{4\pi} + \frac{\mu_0}{\pi}\ln\frac{D}{R}\right]$ Henry/meter ...(9.52)

Since $\mu_0 = 4\pi \times 10^{-7}$,

$$L = \left[1 + 4\ln\frac{D}{R}\right]\times 10^{-7}\text{ Henry/metre}$$

$$= 4\times 10^{-7}\times\left[\frac{1}{4} + \ln\frac{D}{R}\right]\text{ Henry/metre} \quad ...(9.53)$$

since $\ln e^{1/4} = 1/4$

$\therefore$
$$L = 4\times 10^{-7}\times\left(\ln e^{1/4} + \ln\frac{D}{R}\right)$$

$$= 4\times 10^{-7}\ln\frac{D}{Re^{-1/4}}$$

$$= 4\times 10^{-7}\ln\frac{D}{R'}\text{ Henry/metre} \quad ...(9.54)$$

The radius R' is that of a fictitious conductor assumed to have no internal flux linkages but with the same inductance as the actual conductor with radius R. The quantity $e^{-1/4} = 0.7788$. The multiplying factor of 0.7788 to adjust the radius in order to account for internal flux linkages applies only to solid round conductors.

9.13 FLUX LINKAGES OF ONE CONDUCTOR IN A GROUP OF CONDUCTORS

Let us now find out the flux linkages of one conductor due to current flowing in the conductor itself and the current flowing in the other conductors. It is assumed here that the sum of the currents in the various conductors is zero. The system of conductors is shown in Fig. 9.18.

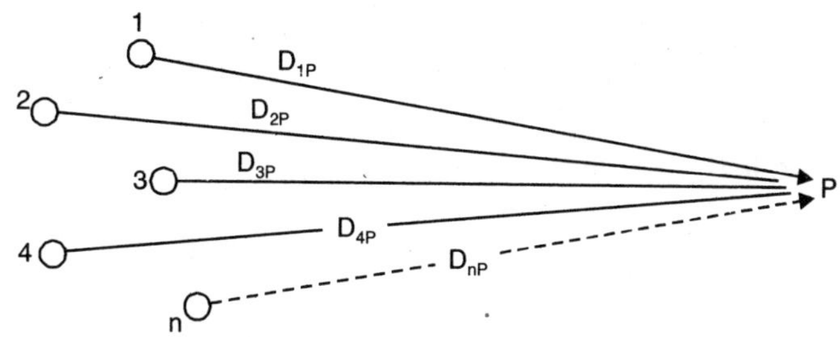

Fig. 9.18 Cross-sectional view of a group of n conductors. Point P is remote from the group of conductors

Theoretically, the flux due to a conductor is extending from the centre of the conductor right upto infinity. We will assume here that P is a point very far from the group of the conductors, the flux linkages will extend upto this point and the distances are as shown in Fig. 9.18. The objective here is to calculate the flux linkages of say conductor 1 due to the current I_1, carried by the conductor itself and flux linkage to conductor 1 due to the current carried by conductors 2,3, ...,n.

The flux linkage of conductor 1 due to the current I_1 including the internal flux linkages

$$\lambda_{1p_1} = \frac{\mu_0 I_1}{8\pi} + \frac{\mu_0 I_1}{2\pi} \ln \frac{D_{1p}}{R_1}$$

$$= 2 \times 10^{-7} \cdot I_1 \ln \frac{D_{1p}}{D_1'} \qquad \text{...(9.55)}$$

The flux linkages λ_{1p_2} to conductor 1 due to current in conductor 2 are

$$\lambda_{1p_2} = 2 \times 10^{-7} \cdot I_2 \ln \frac{D_{2p}}{D_{12}} \qquad \text{...(9.56)}$$

It is to be seen that flux due to conductor 2 that lies between conductor 2 and 1 will not link conductor 1 and therefore the distances involved are D_{2p} and D_{12}.

The flux linkages of conductor 1 due to all the conductors

$$\lambda_{1p} = 2 \times 10^{-7} \left[I_1 \ln \frac{D_{1p}}{R_1'} + I_2 \ln \frac{D_{2p}}{D_{12}} + \ldots + I_n \ln \frac{D_{np}}{D_{\text{ln}}} \right]$$

$$= 2 \times 10^{-7} \left[I_1 \ln \frac{1}{R_1'} + I_2 \ln \frac{1}{D_{12}} + \ldots + I_n \ln \frac{1}{D_{1n}} \right]$$

$$+2\times10^{-7}\left[I_1 \ln D_{1p} + I_2 \ln D_{2p} + ... + I_n \ln D_{np}\right] \quad ...(9.57)$$

Since $\quad I_1 + I_2 + ... + I_n = 0,$

$$\therefore \quad I_n = -(I_1 + I_2 + ... + I_{n-1})$$

Substituting this in the second term of equation (2.21),

$$\lambda_{1p} = 2\times10^{-7}\left[I_1 \ln\frac{1}{R_1'} + I_2 \ln\frac{1}{D_{12}} + ... + I_n \ln\frac{1}{D_{1n}}\right]$$

$$+2\times10^{-7}\left[I_1 \ln D_{1p} + I_2 \ln D_{2p} + ... + I_{n-1} \ln D_{(n-1)p}\right.$$

$$\left. - I_1 \ln D_{np} - I_2 \ln D_{np} ... - I_{n-1} \ln D_{np}\right]$$

$$= 2\times10^{-7}\left[I_1 \ln\frac{1}{R_1'} + I_2 \ln\frac{1}{D_{12}} + ... + I_n \ln\frac{1}{D_{1n}}\right]$$

$$+2\times10^{-7}\left[I_1 \ln\frac{D_{1p}}{D_{np}} + I_2 \ln\frac{D_{2p}}{D_{np}} + ... + I_{n-1} \ln\frac{D_{(n-1)p}}{D_{np}}\right] \quad ...(9.58)$$

PROBLEMS

P 9.1 Starting from first-principles derive Lorenz equation and explain its significance.

P 9.2 Derive expression for magnetic force on a differential current element when placed in a magnetic field due to another differential current element. Hence derive expression for force between two wires per unit length carrying currents of equal magnitude / different magnitude. Explain the nature of force whether attraction or repulsion.

P 9.3 Derive an expression for torque developed on a current carrying planar coil when placed in a magnetic field.

P 9.4 Explain briefly magnetic properties of the following types of materials: (*i*) Diamagnetic (*ii*) paramagnetic (*iii*) Ferromagnetic and (*iv*) Ferrimagnetic.

P 9.5 Explain the terms: Magnetic moment, magnetization permeability, susceptibility of magnetic materials.

P 9.6 The magnetic materials with permeabilities μ_1 and μ_2 form a boundary. Derive relation for B and H at the boundary, angle of incidence and refraction in terms of μ_1 and μ_2.

P 9.7 Derive an expression for inductance of a 1-Q transmission line testing into account both the internal flux and external flux.

P 9.8 Explain the concept and difference between self inductance and mutual inductance of coils.

P 9.9 A conductor 5 m long lies along the *y*-axis with a current of 15 A in the $\bar{a}_y$ direction. Determine the force on the conductor if the field in the region is $B = 0.08\ \bar{a}_x$ Tesla.

Ans. $-6.0\ \bar{a}_z$ Newton

P 9.10 Determine the centripetal force required to hold an electron with mass 9.107×10^{-3} kg in a circular orbit of radius 0.35×10^{-10} m with an angular velocity of 2×10^{16} Rad/sec.

Ans. 1.27×10^{-8} N.

P 9.11 Determine the tangential velocity of a proton in a field $B = 30\ MT$ if the circular path has a diameter of 1 cm.

Ans. 14.4 m/sec

P 9.12 A point charge $Q = 2 \times 10^{-16}$ C and mass 5×10^{-26} kg is moving in the combined fields $E = 100\,\bar{a}_x - 200\,\bar{a}_y + 300\,\bar{a}_z$ V/m and $B = -3\,\bar{a}_x + 2\,\bar{a}_y - \bar{a}_z$ mT. If the charge velocity at $t = 0$ in $v(0) = \left(2\bar{a}_x - 3\bar{a}_y - 4\bar{a}_z\right)10^5$ m/sec. (*i*) Determine the unit vector showing the direction in which the charge is accelerating at $t = 0$ (*ii*) determine the kinetic energy of the charge at $t = 0$.

Ans. (*i*) $0.70\ \bar{a}_x + 0.70\,\bar{a}_y - 0.12\,\bar{a}_z$

(*ii*) 7.25 fJ

P 9.13 An alpha particle and a proton ($Q_\alpha = 2Q_\beta$) enter a magnetic field $B = 1\mu$T with an initial speed $V_0 = 8.5$ m/sec. Assume masses of Alpha and proton particles as 6.68×10^{-27} kg and 1.673×10^{-27} kg respectively. Determine the radii of the circular paths.

Ans. 177 mm, 88.8 mm

P 9.14 Determine the force on a 2 m long conductor located along the *z*-axis carrying a current of 20 A in a magnetic field $B = 0.3\,\bar{a}_x + 0.4\,\bar{a}_y$ T.

Ans. $-16\,\bar{a}_x + 12\,\bar{a}_y$ N.

P 9.15 In Fig. P 9.15 assume that $\mu_2 > \mu_1$, determine the largest angular deflection possible at the interface.

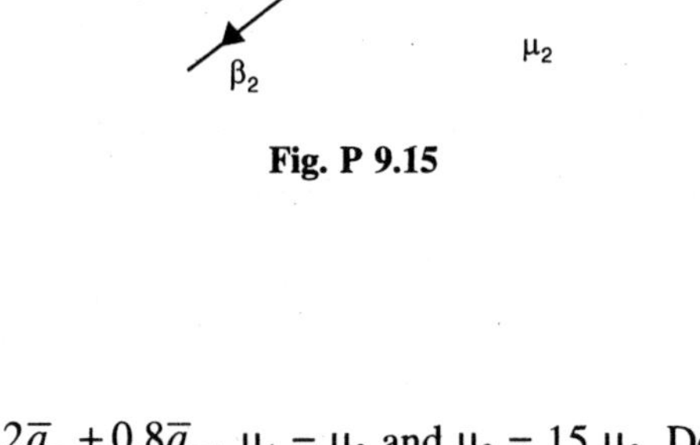

Fig. P 9.15

Ans. $\alpha_{max} = \text{Arc tan}\dfrac{\mu_2 - \mu_1}{2\sqrt{\mu_1\mu_2}}$

P 9.16 In Fig. P 9.15 $B_2 = 1.2\bar{a}_x + 0.8\bar{a}_y$, $\mu_1 = \mu_0$ and $\mu_2 = 15\ \mu_0$. Determine B_1 and H_1.

Ans. $B_1 = 0.08\bar{a}_x + 0.8\bar{a}_y$

$$H_1 = \frac{1.2}{15\mu_0}\bar{a}_x + \frac{0.8}{\mu_0}\bar{a}_y$$

P 9.17 In Fig. P 9.17 show that the angle of emergence θ_4 in region 3 is independent of the permeability of region 2.

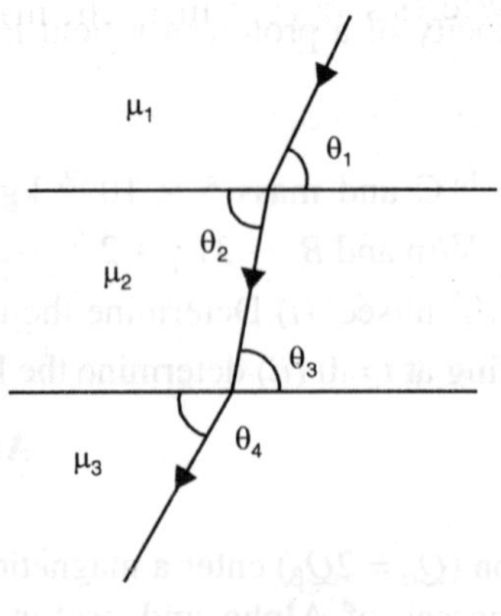

Fig. P 9.17

Ans. $\mu_1 \tan\theta_1 = \mu_3 \tan\theta_4$

P 9.18 A cylinderical conductor of radius '*a*' and permeability carries a *z*-directed uniformly distributed current *I*. Evaluate the magnetization *M* within the conductor.

$$\textbf{Ans. } M = \frac{I}{2\pi a^2}\left(\frac{\mu}{\mu_0} - 1\right) r\bar{a}_\phi$$

P 9.19 Determine the band current density J_b for the *M* evaluated in P 9.18.

P 9.20 A solenoid of length *L* and radius carries *N* turns of tightly wound thin wire. If the current through the coil is *I*, determine the magnetic flux density *B* at the centre of the solenoid. Hence determine *B* if *L* is very large as compared to *R*.

$$\textbf{Ans. } B = \frac{\mu_0 NI}{\sqrt{L^2 + 4R^2}} \text{ T}$$

$$= \frac{\mu_0 NI}{L} \text{ T}$$

P 9.21 A square loop placed in the *xz* plane ($y = 0$ plane) carries a current of *I* loop. This loop is bisected by a very long straight conductor placed parallel to *z*-axis $(\bar{a}_z)$ and carries a current *I*′. Each side of square loop is 2*a* meters. Determine the net force on the loop.

$$\textbf{Ans. } \frac{2\mu_0 II'}{2\pi a}\bar{a}_x$$

P 9.22 Define and explain what you mean by self inductance of coil. How do you compare the inductance of an air-cored coil and iron-cored coil?

P 9.23 Explain what you mean by mutual inductance between coils. Starting from first principles develop an expansion for mutual inductance.

P 9.24 Explain coefficient of magnetic coupling and develop an expression for the same. Discuss its physical significance.

P 9.25 Explain with neat diagram the process of Hysteresis in a magnetic material. How do you obtain magnetization characteristic of a magnetic material from the hysteresis loops?

P 9.26 Develop an expression for hysteresis loss in a magnetic material. Discuss various factors which can be employed to reduce this loss.

P 9.27 Explain with the help of neat diagram eddy current loss in a magnetic material. Discuss various methods to reduce this loss.

P 9.28 Two coils having 30 and 600 turns respectively are wound side by side on a closed iron circuit of section 100 cm^2 and mean length 150 cms. (*a*) Determine the mutual inductance between the coils if the permeability of iron is 2000. (*b*) A current in the first coil increases uniformly from zero to 10 A in 10 m/sec, determine the emf induced in the other coil.

Ans. 0.302 H, 302 V

P 9.29 A circuit has 1000 turns enclosing magnetic circuit 20 cm^2 in section. With 4 A the flux density is 1 Wb/m^2 and with 9 A it is 1.4 Wb/m^2. Determine the mean value of the inductance between these current limits and the induced emf if the current is decreased uniformly from 9 A to 4 A in 50 m sec.

Ans. 0.16 H, 16 V

P 9.30 A magnet has a residual magnetism in the air gap of 1.3 T. The pole face at the air gap is a square section 5 cm × 5 cm. How long (time) must a conductor be moved through the air gap to produce a generated emf of 10 mV. Assume that there is no fringing in the air gap.

Ans. 0.335 sec

P 9.31 An iron magnetic circuit has a uniform cross-sectional area of 5 cm^2 and a length of 25 cm. A coil of 120 turns is wound uniformly over the magnetic circuit. When the current in the coils is 1.5 A, the total flux is 0.3 mWb and when the current is 5 A, the total flux is 0.6 mWb. For each value of current, calculate H and μ_r.

P 9.32 Define the following terms: Magnetic flux density, magnetic field intensity, mmf, reluctance.

P 9.33 Discuss briefly analogies between magnetic and electric circuits.

P 9.34 A two limb magnetic circuit has magnetic path of 30 cm and an air gap of 5 mm. The area of cross-section of magnetic path is 9 cm^2. The relative permeability of core is 5000 and the circuit has 500 turns wound on one limb. The magnetic flux density required is 1 Wb/m^2. Determine (*i*) the current, (*ii*) Flux ϕ and flux linkage.

Ans. (*i*) 0.89 A (*ii*) 9×10^{-4} Wb, 0.45 Wb turns

P 9.35 The total flux in the gap of a permanent magnet is 0.0075 Wb. The diameter of the circular pole faces are 12.7 cm and length of the air gap is 4.57 cm. Determine the values of B and H in the gap. Assume that there is no fringing.

Ans. 0.592 T, 4.7×10^5 A/m

P 9.36 A long wire of radius R carries *d.c.* current I ampere uniformly distributed over the cross-section. Determine magnetic field intensity for all distances $r \leq R$ and $r > R$.

Ans. $\left[H = \dfrac{Ir}{2\pi R^2},\ H = \dfrac{I}{2\pi r}\right]$

P 9.37 A long horizontal wire carrying 20 A is placed crosswise to the earth's magnetic field. At what point does the magnetic field of the wire just cancel terrestrial magnetism? The earth's magnetic field acts downward at an angle of 30° from the vertical with a magnetic flux density of 60 μT.

Ans. 0.067 m

P 9.38 Fig. P 9.38 shows the magnetic circuit. The central limb has a cross-section of 800 mm^2 and each of the outer limb has a cross-section of 500 mm^2.

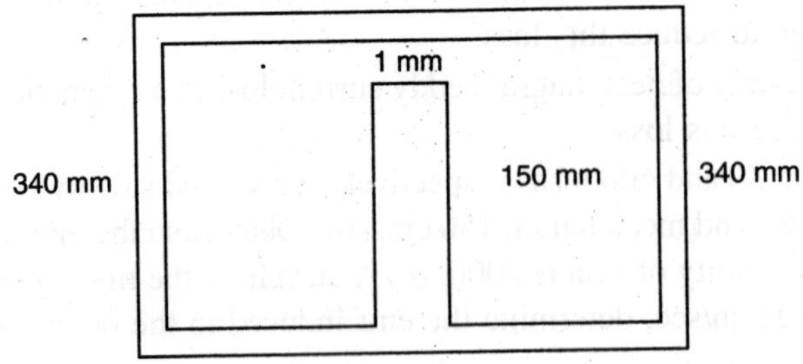

Fig. P 9.38

Determinc the mmf required to produce a flux of 1 m Wb in the central limb. Neglect leakance and assume H for $B = 1.25$ Wb/m^2 as 825 AT/m and for $B = 1$ Wb/m^2 assume $H = 350$ AT/m.

Ans. 1238 AT

P 9.39 A bar magnet of 800 Am^2 magnetic moment has length of 0.1 m and a diameter of 15 mm. Determine

(*i*) the energy density at the centre of the bar and

(*ii*) the total energy in the bar.

Ans. 1.29 GJ/m^3, 22.8 kJ

Chapter

10

Faraday's Law and Maxwell's Equations

10.1 INTRODUCTION

It is to be noted that an electric field is produced by a changing magnetic field and the magnetic field is produced by a changing electric field. In terms of fields it can be said that a time varying magnetic field produces an electromotive force which produces a current in a suitable closed circuit. An electromotive force is nothing but a voltage that arrives in conductors moving in a magnetic field or from changing magnetic field. In simple form Faraday's Law is given as

$$e = -\frac{d\phi}{dt} \text{ volts} \qquad ...(10.1)$$

Equation (10.1) implies a closed path through a net necessarily a closed conducting path. It may contain a capacitor etc. The magnetic flux is that flux which passes through a surface which is enclosed by a perimeter and $\frac{d\phi}{dt}$ is the time rate of change of flux $\frac{d\phi}{dt}$ will have a non-zero value under any of the following conditions.

(*i*) Relative motion between a steady flux and a closed path.
(*ii*) A time changing flux linking a stationary closed path.
(*iii*) A combination of the two.

The minus sign in equation (10.1) shows that the polarity of influenced emf is such that it circulates current which in turn produces flux which opposes the original flux which is due to and it will reduce the magnitude of the emf *e*. This statement that the induced voltage acts to produce an opposing flux is known as Lenz's Law.

If the coil contains N turns, these could be considered as concentrated in one coil, ϕ could be considered as the flux through all turns of the coil and hence

$$e = -N\frac{d\phi}{dt} \qquad ...(10.2)$$

In electrostatic, for the time varying fields, the time integral of electric field leads to an emf or voltage and thus it is expressed as

$$\text{emf} = \oint E \cdot dl \qquad ...(10.3)$$

Replacing ϕ in (1) with the surface integral of B, we have

$$\text{emf} = \oint E \cdot dl = -\frac{d}{dt}\int_s B \cdot ds \qquad ...(10.4)$$

As mentioned earlier, emf can be induced by either moving the path and keeping the magnetic field steady and stationary or by keeping the path fixed and changing the magnetic field (Transform emf)

Let us take a stationary path and time varying magnetic field and therefore in equation (10.4), we take the partial derivative $\frac{\partial}{\partial t}$ inside the integral sign and hence (10.4) becomes

$$\text{emf} = \oint E \cdot dl = -\int_S \frac{\partial B}{\partial t} \cdot dS \qquad ...(10.5)$$

Using Stoke's theorem we first obtain the point form of the equation (10.5) as follows

$$\int_S (\nabla \times E) \cdot dS = -\int \frac{\partial B}{\partial t} \cdot d\bar{S}$$

where surface integral is taken over identical surface and hence we have

$$(\nabla \times E) \cdot d\bar{S} = -\frac{\partial B}{\partial A} d\bar{S}$$

or

$$\nabla \times E = -\frac{\partial B}{\partial t} \qquad ...(10.6)$$

Equation (10.6) is one of Maxwell's four equations in point form or differential form and this is the form which is most generally used. However equation (10.5) is the integral form of the equation (10.6) and represents Faraday's law when applied to fixed points. However, if B is not a function of time equations (10.5) and (10.6) reduce to the electrostatic field equations.

$$\oint E \cdot dl = 0 \qquad \nabla \times E = 0$$

Next we consider a steady constant flux and a moving closed path as shown in Fig. 10.1.

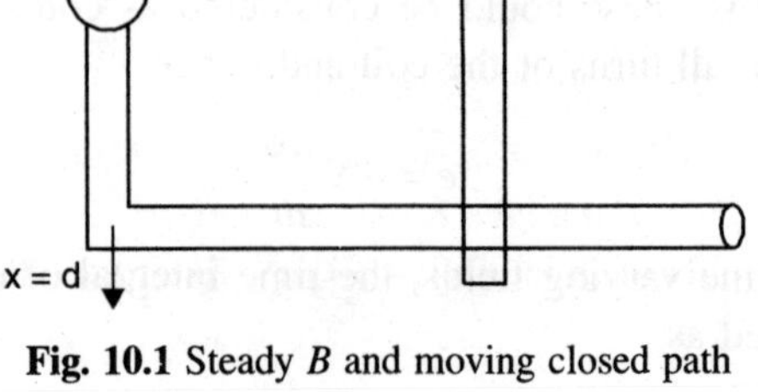

Fig. 10.1 Steady B and moving closed path

The closed circuit consists of the parallel conductor and is fixed and has a high resistance voltmeter and the other conductor is moving with velocity v as shown. Therefore, the conductor intersects flux density normally. Suppose the position of the shorting bar at any time t is at y, and d is the length of the shorting bar, the flux passing through the surface within the closed path, is

$$\phi = Bvd$$

and hence
$$\text{emf} = -\frac{d\phi}{dt} = -Bd\frac{dy}{dt} = -Bvd \qquad ...(10.7)$$

Now let us consider the same example using the concept of motional emf. The force on a charge Q moving at a velocity v in a magnetic field B is

$$F = Q(v \times B)$$

or
$$\frac{F}{Q} = v \times B \qquad ...(10.8)$$

Where force per unit charge is electric field intensity due to motion of charges or E_m. The charges are present in the sliding conducting bar which has both positive and negative charges.

Hence,
$$E_m = \bar{v} \times \bar{B} \qquad ...(10.9)$$

If the moving conductor bar is shifted off the rails E_m would force electrons to one end of the bar (the far end) until the static field due to these charges balances the field induced by the motion of the bar. The resultant tangential electric field intensity would then be zero along the length of the bar.

The motional emf produced by the moving bar (conductor) is,

$$\text{emf} = \oint E_m \cdot d\bar{l} = \oint (v \times B) \cdot d\bar{l} \qquad ...(10.10)$$

The last integral will have non-zero value only if $v \neq 0$. Thus evaluating this integral we have

$$\oint (v \times B) \cdot d\bar{l} = \int_d^0 vBdx = -Bvd \qquad ...(10.11)$$

as before. Since B is not a function of time, this is the total emf. If we have both possibilities *i.e.* the field is changing with time and the path is fixed (transformer induced emf) and field is steady and path is moving (motional emf), we have the total emf as

$$\text{emf} = \oint E \cdot d\bar{l} = \int_S \frac{\partial B}{\partial t} \cdot ds + \oint (v \times B) \cdot d\bar{l} \qquad ...(10.12)$$

This expression is equivalent to the simple statement

$$\text{emf} = -\frac{d\phi}{dt} \qquad ...(10.13)$$

and either can be used to evaluate the induced voltages.

Example 10.1. A 5 turn loop with 0.5 m^2 area situated in air has a uniform magnetic field normal to the plane of the loop. If the flux density changes 8 mT/sec, determine the emf across the terminals of the loop (*ii*) If the emf at the terminals of the loop is 150 mV, determine the rate of change of the magnetic field.

Solution: (*i*) Here B is a function of time

$$\text{emf} = \frac{Nd\phi}{dt} = 5\times 8\times 10^{-3}\times 0.5 = 20 \text{ mV} \quad \textbf{Ans.}$$

$$(ii)\ 150 = -5\frac{dB}{dt}\times 0.5 \text{ or } \frac{dB}{dt} = 6\,\text{mT/sec} \quad \textbf{Ans.}$$

Example 10.2. The circular loop conductor shown in Fig. E 10.2 lies in the $z = 0$ plane has a radius of 0.1 m and a resistance of 6 dm. If the flux density is given to be $B = 0.2 \sin 10^3 t\ \bar{a}_z\ T$ determine the current.

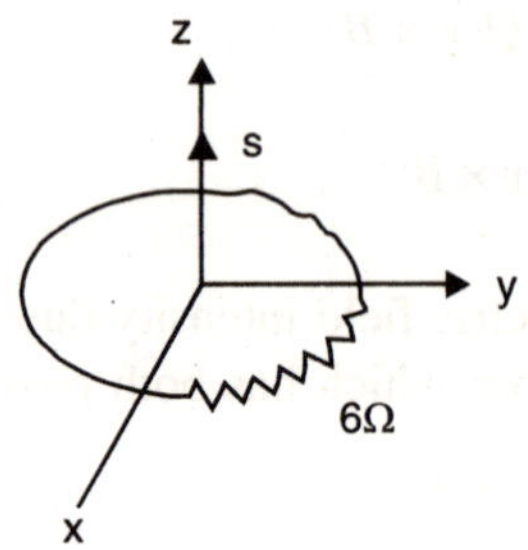

Fig. E 10.2

Solution. $\phi = \text{B.S.} = \pi\times(0.1)^2\times 0.2\sin 10^3 t\bar{a}_z$

$$= \pi\times 1\times 10^{-2}\times 0.2\sin 10^3 t\bar{a}_z$$

$$= 2\pi\times 10^{-3}\sin 10^3 t$$

$$v = -\frac{d\phi}{dt} = -2\pi\cos 10^3 t \text{ volts}$$

$$i = \frac{V}{R} = -\frac{2\pi}{6}\cos 10^3 t = -1.04\cos 10^3 t \quad \textbf{Ans.}$$

Example 10.3. An area of 0.75 m^2 in the $z = 0$ plane is enclosed by a filamentary conductor. If the magnetic field B is given as $B = 0.05\cos 10^3 t\frac{\bar{a}_y + \bar{a}_z}{\sqrt{2}}T$, determine the induced voltage.

Solution. $$v = -\int\frac{\partial B}{\partial t},\ ds\bar{a}_z$$

Now $$\frac{\partial B}{\partial t} = 0.05\times 10^3\sin 10^3 t\frac{\bar{a}_y + \bar{a}_z}{\sqrt{2}}$$

$$= 50\sin 10^3 t\frac{a_y + a_z}{\sqrt{2}}$$

Hence $$V = -\int 50\sin 10^3 t\frac{a_y + a_z}{\sqrt{2}}\cdot ds\bar{a}_z$$

$$= -50 \sin 10^3 t \cdot \frac{0.75}{\sqrt{2}}$$

$$= -26.5 \sin 10^3 t \text{ volt } \textbf{Ans.}$$

Example 10.4. Consider the fixed rectangular loop of area A as shown in Fig. E 10.4. The flux density B is normal to the plane of the loop and is uniform over the area of the loop. However, the magnitude of B varies harmonically with respect to time and is given as

$$B = B_0 \cos \omega t$$

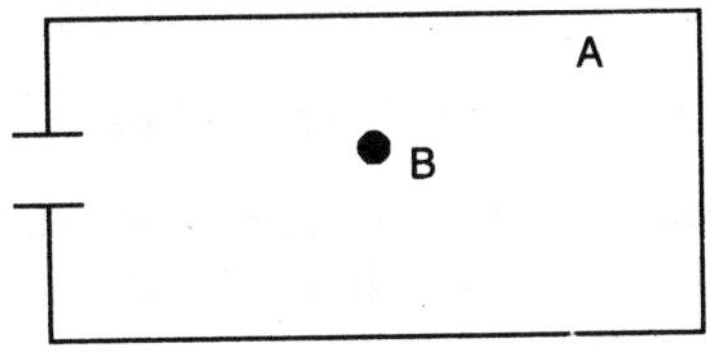

Fig. E 10.4

Determine the total emf induced in the loop.

Solution. This is a case of induced emf due to transfer over action as the coil is fixed in space and B is changing with time. Hence the total emf induced in the coil.

$$v = -\int \frac{\partial B}{\partial t} \cdot dS$$

Now $\frac{\partial B}{\partial t} = B_0 \omega \sin \omega t$ and since $\int dS = A$

Hence $v = AB_0 \omega \sin \omega t$ **Ans.**

Example 10.5. Determine the emf induced when a 50 cm wire aligned with x-axis moves in the plane $z = 3$ m with a velocity $\bar{V} = 2\bar{a}_x + 4\bar{a}_y$ m/s in a magnetic field with flux density

$$B = 3x^2 z\bar{a}_x + 6\bar{a}_y - 3xz^2\bar{a}_z \text{ Wb/m}^2.$$

Solution. Induced emf $= a\int (v \times B) \cdot dL$

$$(2\bar{a}_x + 4\bar{a}_y) \times (3x^2 z\bar{a}_x + 6\bar{a}_y - 3xz^2\bar{a}_z)$$

$$= (12\bar{a}_z + 6xz^2\bar{a}_y - 12x^2 z\bar{a}_z - 12xz^2\bar{a}_x)0.5$$

$$= 6\bar{a}_z + 3xz^2\bar{a}_y - 6x^2 z\bar{a}_z - 6xz^2\bar{a}_x$$

$$= 6\bar{a}_z + 3 \times 0.5 \times 9\bar{a}_y - 6 \times 0.25 \times 3\bar{a}_z - 27\bar{a}_x$$

$$= 6\bar{a}_z + 13.5\bar{a}_y - 4.5\bar{a}_z - 27\bar{a}_x$$

$$= 1.5\bar{a}_z + 13.5\bar{a}_y - 27\bar{a}_x$$

Example 10.6. A circular loop of 10 cm radius is located in the *xy* plane in a *B*-field given by $B = 0.5 \cos 377t \left(3\bar{a}_y + 4\bar{a}_z\right)$J. Determine the voltage induced in the loop.

Solution. We know that voltage induced is given as

$$V = -\int \frac{\partial B}{\partial t} \cdot ds$$

$$= -\int \frac{\partial}{\partial t}(0.5\cos 377t)\left(3\bar{a}_y + 4\bar{a}_z\right)\pi(0.1)^2\,\bar{a}_z$$

$$= 0.5 \times 377 \times 4\pi(0.1)^2 \sin 377t \qquad \left(\text{since } a_y \cdot a_z = 0\right)$$

$$= 23.67 \sin 377t \text{ volts} \quad \textbf{Ans.}$$

Example 10.7. A square loop of side 50 cms is located in free space adjacent to a straight conductor that carries a sinusoidal current of 0.6 A (rms) at 5 kHz. If a small gap is introduced into the loop, determine the induced voltage across the gap.

Solution. Since the magnetic flux density due to a long straight wire at a diameter *r* is given as

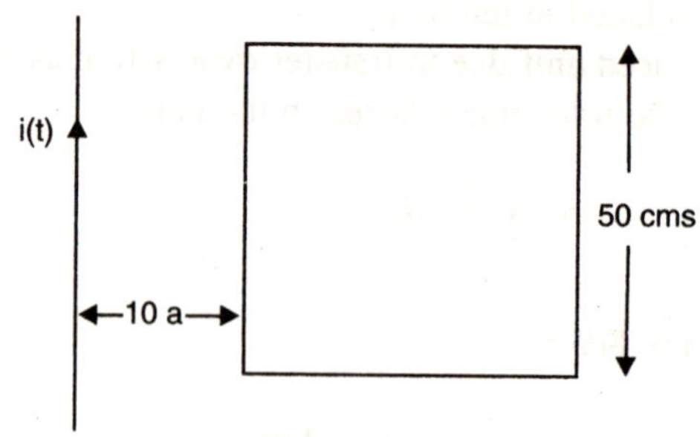

Fig. E 10.7

$$B = \frac{\mu_0 I}{2\pi r}, \text{ hence flux linkages through the square}$$

$$\int B \cdot dr = \frac{\mu_0 I}{2\pi} \cdot \ln\frac{0.6}{0.1}$$

Hence induced emf is $-\dfrac{d\phi}{dt}$

$$v(t) = \frac{\mu_0 \times 0.5 \ln 6}{2\pi} \times i'(t)$$

Let $\quad i(t) = 0.5\sqrt{2}\cos 2\pi ft = 0.5\sqrt{2}\cos 10^4 \pi t$ or $i'(t) = 0.5\sqrt{2} \times 10^4 \times \sin 10^4 \pi t$

$$e = \frac{\mu_0 \times 0.5 \times \ln 6}{2\pi} \times 0.5\sqrt{2} \times 10^4 \pi$$

$$= \frac{4\pi \times 10^{-7} \times 0.5 \times 1.79}{2\pi} \times 0.5\sqrt{2} \times 10^4 \pi$$

$$= 2\pi \times 0.5 \times 1.79 \times 0.5\sqrt{2} \times 10^{-3}$$

$$= 3.97 \text{ mV} \quad \textbf{Ans.}$$

Example 10.8. A square coil 0.60 m on a side rotates about the x-axis at $\omega = 80\ \pi$ rad/sec in a field $B = 0.85\,\bar{a}_z\,T$ as shown. Determine the induced voltage

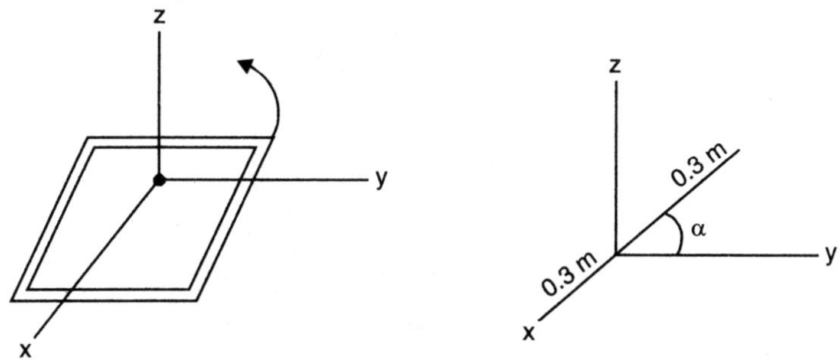

Fig. E 10.8

Solution. Assume that the coil is within $z = 0$ plane at $t = 0$. After time t, it rotates through an angle $\alpha = \omega t\ 80\ \pi t$ rad. The projected area on the $z = 0$ plane is $0.6 \times 0.6 \cos 80\ \pi t$.

Hence $\qquad \phi = BA = 0.85 \times 0.6 \times 0.6 \cos 80\ \pi t.$

and $\qquad v = -\dfrac{d\phi}{dt} = 0.85 \times 0.6 \times 0.6 \times 80\ \pi = 7686 \sin 80\ \pi t$ volts **Ans.**

10.2 DISPLACEMENT CURRENT

Equation (10.6) given below is as a result of Faraday's experimental law and the equation is one of Maxwell's equation in point form or differential form.

$$\nabla \times E = -\frac{\partial B}{\partial t} \qquad \text{...(10.6)}$$

This equation shows that a time varying magnetic field produces an electric field. Current of electric field has the special property of circulation, its time integral about a general closed path is net zero. Next we consider Ampere's circuital law in point form

$$\nabla \times H = J \qquad \text{...(10.14)}$$

This of course shows its inadequately for time varying conditions. We take difference on both the sides and we have

$$\nabla \cdot \nabla \times H = \nabla \cdot J = 0$$

Since difference of a case of a volta is zero $\nabla J = 0$

However, the equation of continuity

$$\nabla \cdot J = -\frac{\partial \rho_v}{\partial t}$$

shows that (10.14) can be true only if $\dfrac{\partial \rho_v}{\partial t} = 0$ which is not true and hence (10.14) must be amended for time varying condition. Suppose we add G to (10.14) we have

$$\nabla \times H = J + G$$

Again taking divergence of the equation, we have

$$\nabla \cdot \nabla \times H = \nabla \cdot J + \nabla \cdot G$$

$$\theta = \nabla \cdot J + \nabla \cdot G$$

Hence
$$\nabla \cdot G = -\nabla \cdot J = \frac{\partial \rho_v}{\partial t}$$

Now
$$\nabla \cdot D = \rho_v$$

Hence
$$\nabla \cdot G = \nabla \cdot \frac{\partial D}{\partial t} \text{ or } G = \frac{\partial D}{\partial t}$$

Therefore, Ampere's Circuital law for time varying condition becomes

$$\nabla \times H = J + \frac{\partial D}{\partial t} \quad ...(10.15)$$

Equation (10.11) is Maxwell's second equation in point form or differential form. The units of $\frac{\partial D}{\partial t}$ are coulombs for m^2 per sec or ampere per second *i.e.* this is also current density since $\frac{\partial D}{\partial t}$ results from a time varying electric flux density or displacement density, Maxwell termed it as a displacement current density and is usually denoted as J_d

$$\nabla \times H = J + J_d \text{ where } J_d = \frac{\partial D}{\partial t}$$

The total displacement current carrying any given surface is expressed by the surface integral

$$I_d = \int_S J_d \cdot d\bar{s} = \int_S \frac{\partial D}{\partial t} \cdot ds$$

$$I + I_d = \int_S (\nabla \times H) \cdot ds = \int_S J \cdot d\bar{s} + \int \frac{\partial D}{\partial t} \cdot ds$$

Displacement current is associated with time varying electric fields and therefore exists in all imperfect conductor or dielectric materials carrying a time varying conductor current. Consider a capacitor connected to an *a.c.* source through conducting leads as shown.

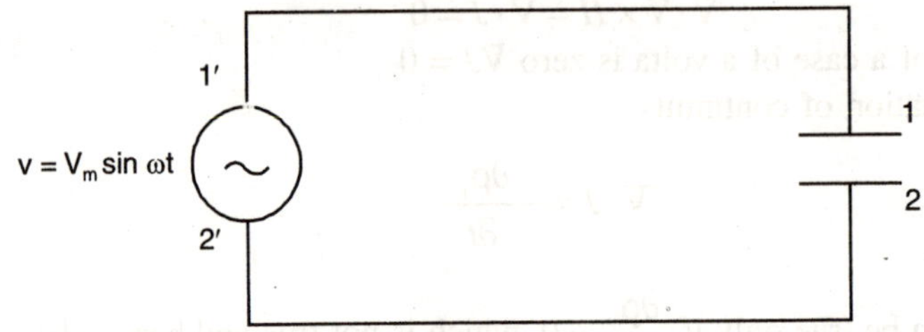

Fig. 10.2

The current that flows in the circuit is the displacement current. The conduction current that passes through the plates of the capacitor is negligibly small. As the current flows from 1′ to 1, the plate 1 of the capacitor gets charged and when potential of plate 1 becomes equal to potential of 1′ of the source, the current ceases to flow. The energy is stored across the plates in the form of electric field. Now since potential of 1 is greater than that of 1′, current flow from 1 to 1′ to 2 the other plate of capacitor. Thus current continues to flow in one half cycle from 1 to 1′ to 2′ to 2 and in another half cycle from 2 to 2′ to 1′ to 1 and this current is called the displacement current. This is the mechanism by which the current (displacement current) $\frac{\partial D}{\partial t}$ is able to flow. It is the turning of the polar molecules to and fro that provides the movement or flow of charge firstly in one direction and then in the reverse direction but the charge never leaves the parent molecule. To indicate this displacement of charges to and fro in the polarized molecules Maxwell gave the name displacement current to the term $\frac{\partial D}{\partial t}$ to distinguish it from the conventional conduction current J in which charges actually drift about in the conductor. The two currents are indistinguishable from each other as both produce I^2R also and will produce magnetic field around them.

Some materials are neither good conductors nor perfect dielectrics, as a result, both conduction current and displacement current exist. Assuming the time dependence $e^{j\omega t}$ for E, the total current density is

$$J_t = J_c + J_d = \sigma E + \frac{\partial}{\partial t}(\varepsilon E) = \sigma E + j\omega\varepsilon E$$

from which $$\frac{J_c}{J_d} = \frac{\sigma}{\omega\varepsilon} \qquad ...(10.16)$$

Therefore, as expected the displacement current becomes increasingly important as the frequency increases.

Example 10.9. A circular cross-section conductor of radius 2 mm carries a current $i_c = 5.5 \sin (4 \times 10^{10}\, t)$. Determine the amplitude of the displacement current density if $\sigma = 35$ Ms/m and $\varepsilon_r = 1$.

Solution. Since $$\frac{J_c}{J_d} = \frac{\sigma}{\omega\varepsilon} = \frac{35\times10^6 \cdot 36\pi}{4\times10^{10}\times10^{-9}} = \frac{35\times36\pi\times10^6}{40}$$

$$J_d = \frac{(5.5\times10^{-6})/\pi(2\times10^{-3})^2}{98.9\times10^6} = 98.9\times10^6$$

$$= \frac{5.5\times10^{-6}}{98.9\times10^6\times4\pi\times10^{-6}} = 4.4\times10^{-3}\ \mu\text{A/m}^2 \textbf{ Ans.}$$

Example 10.10. In a material for which $\sigma = 6$ S/m and $\varepsilon_r = 1$, the electric field intensity $E = 300 \sin 10^{10}\, t$ V/m. Determine the conduction and displacement current densities and the frequency at which they have equal magnitudes.

Solution. $J_c = \sigma E = 6 \times 300 \sin 10^{10}\, t$ A/m^2

Assuming that the field direction does not vary with time

$$J_d = \frac{\partial D}{\partial t} = \frac{\partial}{\partial t}\left(\varepsilon_0 t_r \cdot 300 \sin 10^{10} t\right)$$

$$= \frac{1}{36\pi} \times 10^{-9} \times 1 \times 300 \times 10^{10} \cos 10^{10} t$$

$$= \frac{3000}{36\pi} \cos 10^{10} t = 26.5 \times \cos 10^{10} t$$

For $\quad J_c = J_d,\ \sigma = \omega\varepsilon \text{ or } \omega = \frac{\sigma}{\varepsilon} = \frac{6}{8.854 \times 10^{-12}} = 6.77 \times 10^{11} \text{ rad/sec}$

Now $\quad f = \frac{6.77 \times 10^{11}}{2\pi} = 107.8 \text{ GHz.}$

Example 10.11. A co-axial capacitor with inner radius 6 mm, outer radius 8 mm and length 500 mm has a dielectric for which $\varepsilon_r = 6.7$ and an applied voltage 250 sin 377 t volts. Determine the displacement current i_d and compare with the conductor current. Assume the inner conductor to be at $v = 0$.

Solution. With cylindrical conductor, the potential is constant with ϕ and z, therefore the Laplace's equation reduces to

$$\frac{1}{r}\frac{d}{dr}\left(r\frac{dV}{dr}\right) = 0$$

or $\quad r\frac{dV}{dr} = A$

or $\quad \frac{dV}{dr} = \frac{A}{r}$

or $\quad V = A \ln r + B$

At $r = 0.006$ $v = 0$, $0 = A \ln 0.006 + B$ or $B = -A \ln 0.006$ and at $r = 0.008$ $v = 250 \sin 377\, t$

$$250 \sin 377t = A \ln 0.008 - A \ln 0.006$$

$$= A \ln \frac{8}{6}$$

or $\quad A = \dfrac{250 \sin 377t}{\ln \frac{4}{3}}$

Therefore, $\quad V = \dfrac{250 \sin 377t}{\ln \frac{4}{3}} \ln r - \dfrac{258 \sin 377t}{\ln \frac{4}{3}} \ln 0.006$

$$= \frac{250 \sin 377t}{\ln \frac{4}{3}}\left[\ln \frac{r}{0.006}\right]$$

$$E = -\frac{\partial V}{\partial r} = \frac{876.6 \sin 377t}{r}$$

$$= \frac{876.6 \sin 377t}{r}$$

$$D = \varepsilon_0 \varepsilon_1 E = 8.854 \times 10^{-12} \times 6.7 \times \frac{876.6 \sin 377t}{r}$$

$$= 5.20 \times 10^{-8}/r \ \sin 377t \ \text{C/m}^2.$$

$$J_D = \frac{\partial D}{\partial t} = \frac{5.2 \times 10^{-8}}{r} \cdot 377 \cos 377t$$

$$= \frac{-1.960 \times 10^{-5}}{r}$$

$$i_d = J_D.2\pi rL = 1.96 \times 10^{-5}.\ 2\pi L$$

$$= 1.96 \times 2\pi \times 0.5 \times 10^{-5}$$

$$= 6.15 \times 10^{-5} \cos 377\ t \text{ Amp.}$$

Now capacitance $C = \dfrac{2\pi\varepsilon_0\varepsilon_r}{\ln\dfrac{8}{6}} \times 0.5 = \dfrac{2\pi \times 8.854 \times 10^{-12} \times 6.7 \times 0.5}{0.285}$

$$= 0.653 \text{ nF}$$

$$i_C = C\frac{dV}{dt} = 0.653 \times 10^{-9} \cdot 250 \times 377 \cos 377t$$

$$= 6.15 \times 10^{-5} \cos 377t \text{ Amp.}$$

It can be seen that $i_d = i_C$.

Example 10.12. A lossy dielectric has $\mu_r = 1$, $\varepsilon_r = 10$ and $\sigma = 20$ ns/m. An electric field $E = 200 \sin \omega t\ \bar{a}_z$ V/m exists in the dielectric. Determine the frequency for which the conductor current density and the displacement current density are equal.

Solution. The ratio of $J_C / J_d = \dfrac{\sigma}{\varepsilon\omega}$ and to the two densities to be equal $\sigma = \varepsilon\omega$ or the frequency at which the two densities are equal is $\omega = \dfrac{\sigma}{\varepsilon}$ or $f = \dfrac{\sigma}{2\pi\varepsilon}$.

$$f = \frac{1}{2\pi} \cdot \frac{20 \times 10^{-9} \cdot 36\pi}{10 \times 10^{-9}} = \frac{18 \times 20}{10} = 36 \text{ Hz } \textbf{Ans.}$$

10.3 MAXWELL'S EQUATIONS IN POINT FORM

We have already derived two of Maxwell's equations for time varying fields

$$\nabla \times E = -\frac{\partial B}{\partial t} \quad \text{...(10.17)}$$

and

$$\nabla \times H = J + \frac{\partial D}{\partial t} \quad \text{...(10.18)}$$

The other two equations are unchanged from their non-time varying form

$$\nabla \cdot D = \rho_v \quad \text{...(10.19)}$$

$$\nabla \cdot B = 0 \quad \text{...(10.20)}$$

From equation (10.17) and (10.18) it is clear that when fields are time variable, however H can't exist without an E field nor can E exist without a corresponding H field. However, it is to be noted that a static E field can exist in the absence of a magnetic field H and a conductor with a constant current I has a magnetic field H without an E field. Equation (10.19) states that charge density is a source or sink of electric flux lines.

Equation (10.20) shows that 'magnetic charges' or poles are not known to exist. Magnetic flux is always found in closed loops and never diverges from a point source.

These four equations form the basis of electromagnetic theory. These are partial differential equations and relate the magnetic fields and electric fields to each other and to their sources, charge and current density. The auxiliary equations relating D and E is given as

$$D = \varepsilon E \quad \text{...(10.21)}$$

B and H

$$B = \mu H \quad \text{...(10.22)}$$

condition current density

$$J = \sigma E \quad \text{...(10.23)}$$

The relation between correction current density and volume charge density

$$J = \rho_v V \quad \text{...(10.24)}$$

If we don't have good dielectric or magnetic materials equations (10.21) and (10.22) are modified as

$$D = \varepsilon_0 E + P \quad \text{...(10.25)}$$

and

$$B = \mu_0 (H + M) \quad \text{...(10.26)}$$

For linear materials we may relate P to E as

$$P = \chi_e \varepsilon_0 E \quad \text{...(10.27)}$$

and

$$M = \chi_m H \quad \text{...(10.28)}$$

Finally we include Lorenz Force equation in point form

$$F = \rho_v (E + \overline{v} \times \overline{B}) \quad \text{...(10.29)}$$

Where F is the force per unit volume

10.4 MAXWELL'S EQUATION IN INTEGRAL FORM

First equation is obtained by integrating point form of Maxwell equation (10.17). Integrating (10.17) over a surface and applying Stoke's theorem, we obtain Faraday's law

$$\oint \bar{E} \cdot d\bar{l} = -\int_S \frac{\partial B}{\partial t} d\bar{s} \qquad ...(10.30)$$

Applying same procedure to equation (10.18) we have Amperes' circuital law

$$\oint H \cdot d\bar{l} = I + \int_S \frac{\partial D}{\partial t} \cdot d\bar{s} \qquad ...(10.31)$$

Gauss's law for the electric and magnetic fields are obtained by integrating (10.19) and (10.20) throughout a volume and using the divergence theorem

$$\oint_S D \cdot ds = \int_V \rho_v dv \qquad ...(10.32)$$

and

$$\oint_s B \cdot ds = 0 \qquad ...(10.33)$$

We have seen in previous Chapters that these equations enable us to find boundary conditions on *B*, *D*, *E* and *H* which are required to evaluate constants obtained in solving Maxwell's equations in partial differential form.

Even though the boundary condition have been discussed in earlier chapter a brief view is given here as these are needed for the solution of Maxwell's equation.

Between any two real physical media where $k = 0$ (on bounding surface) equation (10.30) enables us to relate the tangential E – field components *i.e.* $E_{t_1} = E_{t_2}$ and from (10.31) we get $H_{t_1} = H_{t_2}$. The surface integrals produce the boundary conditions on the normal components.

$$D_{n_1} - D_{n_2} = \rho_s \text{ and } B_{n_1} = B_{n_2}$$

Assume a perfect conductor for which σ is infinite but J is finite. From Ohms law, then, in a perfect conductor $E = 0$ and from the point form of Faradays' law $H = 0$ for time varying fields. The point form of Amperes' Circuital law shows that the finite value of $J = 0$ and current must be carried on the conductor surface as a surface current K. Thus if region 2 is a perfect. Conductor various boundary conditions become

$$E_{t_1} = 0, \quad H_{t_1} = k, \quad D_{n_1} = \rho_s, \quad B_{n_1} = 0$$

It is to be noted that surface charge density is considered a physical possibility for either dielectrics, perfect conductors or imperfect conductors but that surface current density is assumed only in case of perfect conductors.

10.5 THE RETARDED POTENTIAL

We know that in electrostatic fields ρ_v is the volume charge density, then scalar electric potential v may be expressed in terms of a static charge distribution.

$$V = \int \frac{\rho_v dv}{4\pi\varepsilon R} \qquad ...(10.34)$$

and the vector magnetic potential is obtained from a steady current distribution and is given as

$$\bar{A} = \int \frac{\mu \bar{J} dv}{4\pi R} \qquad ...(10.35)$$

The differential equations involving V and A are

$$\nabla^2 V = -\frac{\rho_v}{\varepsilon}$$

and $$\nabla^2 A = -\mu J \qquad ...(10.36)$$

Again $$E = -\nabla V \text{ and } \bar{B} = \nabla \times \bar{A}$$

We know that any electromagnetic disturbance will travel at a velocity $v = \frac{1}{\sqrt{\mu\varepsilon}}$ through any homogeneous medium described by μ and t. In case of free space it travels with velocity of light *i.e.* 3×10^8 m/sec. It is, therefore, logical to expect that the potential at any point is due not to the value of the charge density at some distant point at the same instant but to its value at some previous time, because the effect propagates at a finite velocity. Thus equation (10.34) becomes

$$V = \int \frac{[\rho_v]}{4\pi\varepsilon R} dv \qquad ...(10.37)$$

where $[\rho_v]$ indicate that every t appearing in the expression for ρ_v has been replaced by a retarded time

$$t' = t - \frac{R}{C}$$

where R is the point under consideration and c is the velocity of wave.

Thus, if the charge density throughout the space is given by

$$\rho_v = e^{-r} \cos \omega t'$$

then $$[\rho_v] = e^{-r} \cos\left\{\omega\left(t - \frac{R}{C}\right)\right\}$$

where R is the distance between the differential element of charge being considered and the point at which the potential is to be determined.

Similarly retarded vector magnetic potential is given as

$$\bar{A} = \int \frac{\mu[J]}{4\pi R} dV \qquad ...(10.38)$$

The use of retarded time has resulted in the time varying potentials being given the name of retarded potentials.

The current expression $[I] = I_0 \cos \omega\left(t - \frac{R}{C}\right)$ where $[I]$ is called the retarded current. The bracket [] indicates that the current is retarded. The time difference R/C is the interval required for the disturbance to travel the distance R in free space at the velocity of light.

Example 10.13. The electric field intensity E is given as $E = E_m \sin(\omega t - \beta z)\bar{a}_y$ in free space. Determine D, B and H.

Solution. Using Maxwell's equation $\nabla \times E = -\frac{\partial B}{\partial t}$ gives

$$\nabla \times E = \begin{vmatrix} \bar{a}_x & \bar{a}_y & \bar{a}_z \\ \dfrac{\partial}{\partial x} & \dfrac{\partial}{\partial y} & \dfrac{\partial}{\partial z} \\ 0 & E_m \sin(\omega t - \beta z) & 0 \end{vmatrix}$$

$$\frac{\partial B}{\partial t} = -\frac{\partial}{\partial z} E_m \sin(4t - Bz)\bar{a}_x = BE_m \cos(\omega t - \beta z)\bar{a}_x$$

On integral $$B = -\frac{\beta E_m}{\omega} \sin(\omega t - \beta z)\bar{a}_x$$

Since it is a static field constant of integration is neglected.

Now $$\dot{H} = \frac{\beta}{\mu_0} = -\frac{BE_m}{\omega\mu_0} \sin(\omega t - \beta z)\bar{a}_x$$

Also $$D = \varepsilon_0 E = \varepsilon_0 E_m \sin(\omega t - \beta z)\bar{a}_y \text{ **Ans.**}$$

Example 10.14. If the magnetic field intensity H is given as in free space as

$$H = H_m e^{j(\omega t + \beta z)}\bar{a}_z, \text{ determine } E.$$

Solution. Since $\Delta \times H = \dfrac{\partial D}{\partial t}$

$$\begin{vmatrix} \bar{a}_x & \bar{a}_y & \bar{a}_z \\ \dfrac{\partial}{\partial x} & \dfrac{\partial}{\partial y} & \dfrac{\partial}{\partial z} \\ H_m e^{j(\omega t + \beta z)} & 0 & 0 \end{vmatrix} = \frac{\partial D}{\partial t}$$

$$\frac{\partial}{\partial z} H_m e^{j(\omega t + \beta z)}\bar{a}_y = \frac{\partial D}{\partial t}$$

$$j\beta H_m e^{j(\omega t + \beta z)}\bar{a}_y = \frac{\partial D}{\partial t}$$

On integration

$$D = \frac{\beta H_m}{\omega} e^{j(\omega t + \beta z)}\bar{a}_y$$

and $$E = \frac{D}{\varepsilon_0} = \frac{\beta H_m}{\omega \varepsilon_0} e^{j(\omega t + \beta z)}\bar{a}_y \text{ **Ans.**}$$

10.6 PHASOR FORM OF MAXWELL'S EQUATIONS

For harmonic variation, the phasor forms of Maxwell's integral and differential equations can be derived as follows:

For harmonic variation of electric and magnetic fields these are expressed as

$D = De^{j\omega t}$ and $B = Be^{j\omega t}$

$$\frac{\partial D}{\partial k} = j\omega De^{j\omega t}$$

$$\frac{\partial \beta}{\partial t} = j\omega Be^{j\omega t}$$

Integral form of equation

(i) $\int H \cdot dl = \int\left(J + \frac{\partial D}{\partial t}\right) \cdot ds = \int(\sigma E + j\omega\varepsilon E) \cdot ds$

$$= (\sigma + j\omega\varepsilon)\int E \cdot ds$$

(ii) $\int E \cdot dl = -\int \frac{\partial B}{\partial t} \cdot ds = -j\omega\int B \cdot ds = -j\omega\mu\int H \cdot ds$

(iii) $\int D \cdot ds = \int \rho dv = \int \rho dv$

(iv) $\int B \cdot ds = 0 = \int B \cdot ds = 0$

Differential form of equation

$$\nabla \times H = J + \frac{\partial D}{\partial t} = \sigma E + j\omega\varepsilon E = (\sigma + j\omega\varepsilon)E$$

$$\nabla \times E = -\frac{\partial B}{\partial t} = -j\omega\mu H$$

$$\nabla \cdot D = \rho = \rho$$

$$\nabla \cdot B = 0$$

The phasor form of Maxwell's integral and differential equations are given as follows:

$\int H \cdot dl = (\sigma + j\omega\varepsilon)\int E \cdot ds$ $\qquad$ $\int E \cdot dl = -j\omega\mu / H \cdot ds$

$\int D \cdot ds = \int \rho dv$ $\qquad$ $\int B \cdot ds = 0$

Differential form

$\nabla \times H = (\sigma + j\omega\varepsilon)\, E$ $\qquad$ $\nabla \times E = -j\omega\mu H$

$\nabla \cdot D = \rho$ $\qquad$ $\nabla \cdot B = 0$

PROBLEMS

P 10.1 State and explain Faraday's law. Develop Maxwell's equation in both integral form and point form representing Faraday's law.

P 10.2 The rectangular loop shown in Fig. P 10.2 moves towards the origin at a velocity $\bar{v} = -200\bar{a}_y$ m/s in a field $B = 0.8e^{-0.5y}\bar{a}_z$ T. Determine the current at the instant the coil sides are at y = 0.5 m and 0.6 m if $R = 2.5r$.

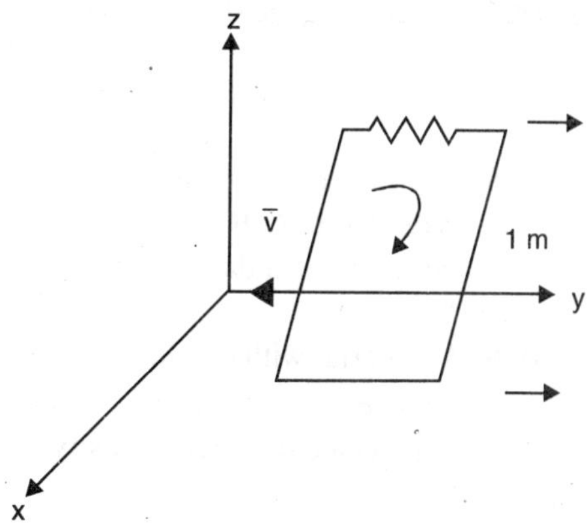

Fig. P 10.2

Ans. 2.43 A

P 10.3 A circular cross-section conductor of radius 1.5 mm carries a current $i_c = 5.5 \sin 4 \times 10^{10} t$ μ A. Determine the amplitude of the displacement of current density if σ = 35 Ms/m and $\varepsilon_r = 1$.

Ans. 7.87×10^{-3} μ A/m²

P 10.4 A parallel plate capacitor of separation 0.6 mm and with a dielectric of $\varepsilon_r = 15.3$ has an applied rms voltage of 25 V at a frequency of 15 GHz. Determine the rms displacement current density. Assuming no fringing effect.

Ans. 5.32×10^5 A/m²

P 10.5 A coil is 75 mm wide and 100 mm long as shown in Fig. P 10.5. Determine the speed of rotation if an rms voltage of 0.25 V is developed in the uniform field $B = 0.45$ $\bar{a}_y$ T.

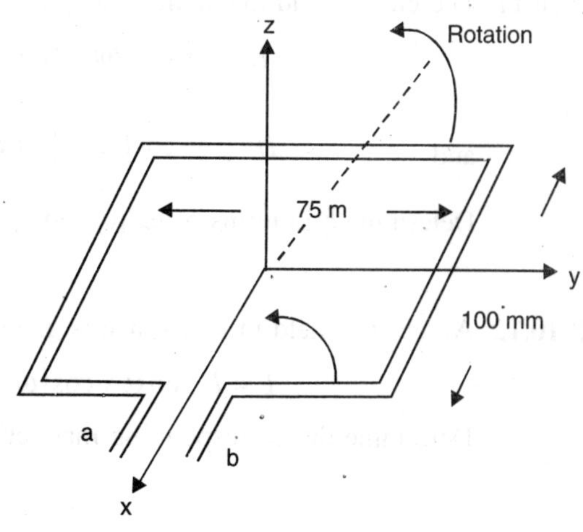

Fig. P 10.5

P 10.6 A uniform and constant field of 10 mWb/m^2 is directed along the z-axis of a rectangular coordinate system. A circular contour in the xy plane centred at the origin has a radius which is decreasing at 200 m/sec. If the initial radius is 150 mm determine the voltage induced in the path.

Ans. 0.6 Volt

P 10.7 A magnetic field $B = 0.5 \sin\frac{\pi x}{2} \cos\frac{\pi y}{2} \sin 377 t \bar{a}_z T$ passes through the stationary loop $0 < x$, $y < 1$ in the xy plane. Determine the voltage induced in the loop.

Ans. $-76.4 \cos 377\, t$

P 10.8 A 20-turn rectangular loop of length 20 cm and width 16 cm rotates at 3000 rpm in a magnetic field $B = 0.5 \sin 377\, t\, T$ directed normal to the axis of rotation. Determine the rms value of the induced voltage.

Ans. 71.1 Volt

P 10.9 A material with conductivity σ and permittivity ε is exposed to a sinusoidally varying electric field $\bar{E}$ with frequency ω. Determine the ratio of conduction current density J_c to the displacement current density J_D.

Ans. $\frac{\sigma}{\omega\varepsilon}$

P 10.10 Write down Maxwell's equations in point form and in integral for time varying magnetic and electric fields.

Ans. $\nabla \times E = -\frac{\partial \dot{B}}{\partial t}, \nabla \times H = J + \frac{\partial D}{\partial t}, \nabla \cdot D = \rho, \nabla \cdot B = 0$

Integral form $\oint_c E \cdot ds = -\frac{d}{dt}\int_s B \cdot ds, \oint_c H \cdot ds = \int_s J \cdot ds + \frac{d}{dt}\int_s D \cdot ds$

$\int_s D \cdot ds = \int_v \rho\, dv, \int_s B \cdot ds = 0$

P 10.11 The electric and magnetic field intensities in free space are given as

$$E = E_0 \cos(\omega t - \beta z)\bar{a}_x$$

and

$$H = \frac{E_0}{\xi} \cos(\omega t - \beta z)\bar{a}_y$$

Determine ξ in terms of ω, μ_0 and ε_0 so that the fields satisfy Maxwell's equations.

Ans. $\xi = \sqrt{\mu_0/\varepsilon_0}$

P 10.12 An electric field in free space is given as

$$\bar{E} = E_0 \sin\alpha\, x \cos(\omega t - \beta z)\bar{a}_y$$

Determine the corresponding magnetic field H.

Ans. $\bar{H} = -\frac{\beta}{\mu_0\omega} E_0 \sin\alpha\, x \cos(\omega t - \beta z)\bar{a}_x$

$-\frac{\alpha}{\mu_0\omega} E_0 \cos\alpha x \sin(\omega t - \beta z)\bar{a}_z$

P 10.13 Four straight conductors form a square with a magnetic field B perpendicular to the square. If all conductors move outward with the same velocity v. While contacting each other at the corners, determine v induced in the square loop at the instant when its area is $2m^2$, $v = 4$ m/sec and $B = \cos(2\pi ft)T$ where $f = 2$kHz. **Ans.** 17.8 kV

P 10.14 If magnetic field intensity H is given as $H = H_0 e^{j(\omega t + \beta z)} \bar{a}_x$ in free space, determine electric field on density D and electric field intensity E.

Ans. $D = H_0 e^{j(\omega t + \beta z)} \bar{a}_y$

$E = D/\varepsilon_0$

P 10.15 In a homogeneous nonconducting medium where $\mu_r = 1$, determine ε_r and W if

$H = 1.0\ e^{j(\omega t - 4/3y)} \bar{a}_x$ and $E = 30\pi e^{j(\omega t - 4/3y)} \bar{a}_x$ V/m.

Ans. $\varepsilon_r = 16$, $\omega = 10^8$ rad/sec.

P 10.16 State and explain what you mean by Retarded Potential. Derive expression for retarded potential for electric and magnetic fields.

P 10.17 Electric flux density in free space is given by $D = D_0 \sin(\omega t + \beta z) \bar{a}_x$. Determine magnetic flux density B.

Ans. $B = \dfrac{-\omega H_0 D_m}{\beta} \sin(\omega t + \beta z) \bar{a}_y$

P 10.18 A magnetic field $H = H_x \cos 2x \cos(\omega t - \beta y) \bar{a}_x$ exists within a dielectric of permittivity ε. Determine (*i*) the displacement current density (*ii*) the electric field intensity and (*iii*) the volume charge density.

Ans. $J_D = -\beta H_x \cos 2x \sin(\omega t - \beta y) \bar{a}_z$

$\bar{E} = \dfrac{\beta}{\omega t} H_x \cos 2x \cos(\omega t - \beta y) \bar{a}_z$

$\rho_v = 0\ C/m^3$

Chapter

11

Transmission Lines

11.1 INTRODUCTION

Transmission lines are the inter-connection between the source (transmitter or generator) and the sink (receiver or loader). These lines may transmit electromagnetic energy or any other form of energy e.g. electric energy from a power plant to a load centre or a subtraction, or light-energy, mechanical work or information e.g. speech, music, data etc. Examples of transmission lines include single phase two wire, 3-wire co-axial strip, microstrip, wave guides and optical fiber lines. The idea of transmission lines comes into effect when the devices to be connected are separated by distances on the order of a wavelength or more. However, in basic circuit analysis, the connection between elements *R, L, C* are of negligible length and hence the voltage or current e.g. is available at the same instant at both the ends of the elements and such elements are known as lumped elements. However, when distances are sufficiently large between the source and the sink (receiver), an inherent time delay appears at different points of the line *i.e.* when the source is switched into the transmission lines, the voltages or currents do not appear instantly at all parts of the network as these signals travel even though with high but finite velocity and hence phase difference appears in these signals. Such large length lines can be analyzed by assuming the parameters *R, L* and *C* of the line as distributed rather than lumped. This means the analysis of such circuits must be made on per unit distance basis. For time harmonic case this condition would lead to a measurable phase difference between each end of the device in question.

11.2 PROPAGATION OF SIGNALS ON TRANSMISSION LINES

For simplicity we assume the transmission line to be loss less *i.e.* since the lines are represented by *R, L, C* and *G* for loss less lines *R* and *G* are zero and therefore we represent a transmission line with series inductance and shunt capacitance as distributed parameters. Suppose the line is represented as a number of π-section as shown in Fig. 11.1.

The section of the line is connected to a battery V_0 through a switch S_1 at one end and at the other end if the line is a resistor *R* connected to the line through switch S_2. Initially suppose the switch S_2 is kept open and S_1 is closed. Since the source is *d.c.* voltage, the inductor L_1, acts as an open circuit *Q* and capacitor C_1, as short circuit. The current in the loop containing L_1, C_1 and the battery V_0 starts increasing. The voltage across the capacitor increases and till such time that the voltage across C_1

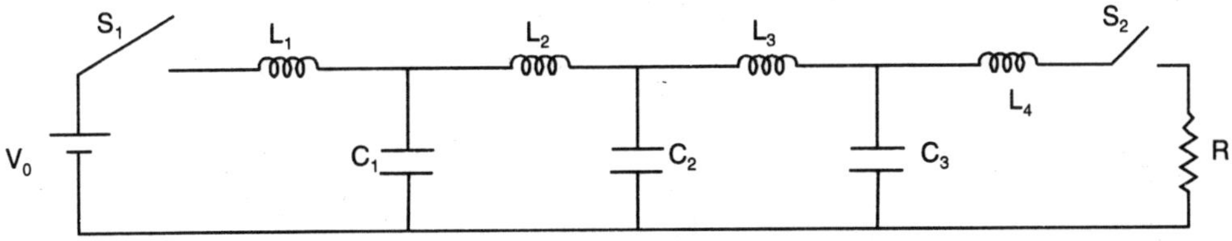

Fig. 11.1 Lumped element model of a transmission line

becomes V_0, no current flows through L_2 and C_2 and hence as the switch S_1 is closed the voltage at the other end of the line is zero at that instant. It takes certain time for the voltage and current to appear at the load end of the line. As soon as C_1 is charged, C_2 starts charging through L_2 and similarly when C_2 is fully charged capacitor C_3 starts charging through L_3 and so on. It is found that smaller the values of L and C in these sections of the line, larger is the velocity of movement of voltage and current waves and it can be seen that the velocity of the wave is since function of inverse of the product of L and C. When the line is fully charged, since the current when S_2 is open, is zero at the resistor end, a current wave is supposed to be reflected back with negative sign at the receiving end, then wiping out the current on the line. Similarly voltage wave is reflected back with positive sign, thus doubling the voltage along the length of the line. These reflection and refraction of voltage and current waves to these reach the end terminals of the line will be treated under transients on intersected lines in the later section.

However, if the resistor R is switched in and the battery is also connected to the left and of the line a shown in Fig. 11.1, capacitor C_3 will first discharge through the resistor, followed by the next nearest capacitor (C_2 and C_1 in that order) and so on. When the network is completely discharged a voltage pulse has been formed across the resistor and so we sometimes call this kind of ladder network a *pulse forming network*. The voltage and current across the line conductors and through the line conductors respectively are associated with the electric and magnetic fields.

We will solve such a network using Maxwell's equation subject to line configuration in a later section. However, we make use of appropriate circuit model of the network and solve for voltage and current in various parts of the network and at any instant of time the distribution of voltage and current will be investigated. We will see, however, that this approach becomes complicated when we include transmission line losses and operate lines of very high frequency and we have to make use of Maxwell's equations for the solution of transmission networks.

11.3 TRANSMISSION LINE EQUATIONS

Next we develop differential equation for determining voltage and currents at any point and at any time on the transmission lines. These equations are also known as wave equations. A transmission line is represented the parameter R, L, C and G where R and G lead to losses with losses and L and C are associated with magnetic and electric fields along the length of the line. G is the shunt conductance of the line spread along the length of the line. Since we are interested in developing differential equation for the transmission line, we consider differential length of the line and write two equations in terms of voltage and currents and the per unit distance parameters R, L, C and G. We assume here the propagation of the wave along $\bar{a}_z$ direction *i.e.* the length of the line is along $\bar{a}_z$ and hence consider a differential length dz of the line at a distance z from the sending end as shown in Fig. 11.2. Since sections of the line look the same from either end we divide the series elements in half to produce a symmetrical network (star equivalent). We could as well make use of the π configuration.

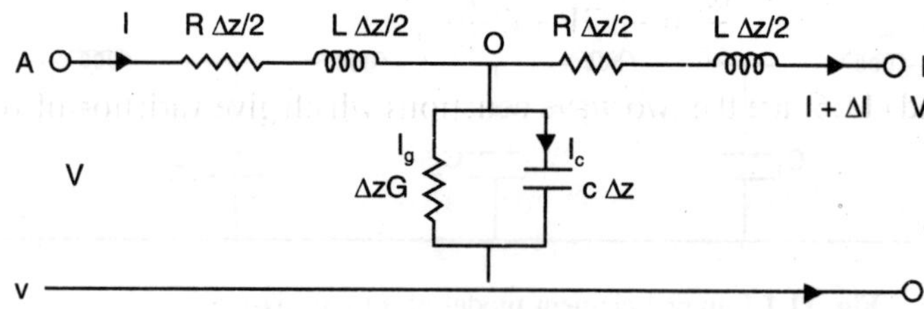

Fig. 11.2 Lumped differential length Δz representation of a transmission line

Writing KVL equation considering the compete length AB for the loop, we have

$$V = \frac{1}{2}RL\Delta Z + \frac{1}{2}L\frac{\partial I}{\partial t}\Delta Z + \frac{1}{2}R(I+\Delta L)\Delta Z + \frac{1}{2}L\left(\frac{\partial I}{\partial t} + \frac{\partial dI}{\partial t}\right)\Delta Z + (V+\Delta V) \quad \ldots(11.1)$$

We now obtain the ratio $\Delta V/\Delta z$ *i.e.* the rate of change of V as the distance z changes. To do this we divide equation (11.1) by Δz and rearranging the terms, we have

$$0 = \frac{1}{2}RI + \frac{1}{2}L\frac{\partial I}{\partial t} + \frac{1}{2}L\left(\frac{\partial I}{\partial t} + \frac{\partial \Delta I}{\partial t}\right) + \frac{1}{2}R(I+\Delta I) + \frac{\Delta V}{\Delta z}$$

or

$$\frac{\Delta V}{\Delta Z} = -\left(RI + L\frac{\partial I}{\partial t} + \frac{1}{2}L\frac{\partial \Delta J}{\partial t} + \frac{1}{2}R\Delta I\right) \quad \ldots(11.2)$$

Next we write $\Delta I = \dfrac{\partial I}{\partial Z}\Delta Z$ and $\Delta V = \dfrac{\partial V}{\partial Z}\Delta Z$ ...(11.2(*a*))

and after substitution in equation (11.2), we have

$$\frac{\partial V}{\partial Z} = -\left(RI + L\frac{\partial I}{\partial t} + \frac{1}{2}L\frac{\partial}{\partial t}\left(\frac{\partial I}{\partial Z}\Delta Z\right) + \frac{1}{2}R\frac{\partial I}{\partial Z}\Delta Z\right)$$

$$\frac{\partial V}{\partial Z} = -\left(1 + \frac{\Delta Z}{2}\frac{\partial}{\partial Z}\right)\left(RI + L\frac{\partial I}{\partial t}\right) \quad \ldots(11.3)$$

Now, in the limiting case when $\Delta Z \to 0$, equation (11.3) reduces to

$$\frac{\partial V}{\partial Z} = -\left(RI + L\frac{\partial I}{\partial t}\right) \quad \ldots(11.4(a))$$

Equation (11.4) is the first of the two wave equations. Now to obtain second wave equation we make use of KCL at node 'O'.

$$I = I_g + I_c + I + \Delta I = G\Delta z\left(V + \frac{\Delta V}{2}\right) + C\Delta Z\frac{\partial}{\partial t}\left(V + \frac{\Delta V}{2}\right) + I + \Delta I \quad \ldots(11.4(b))$$

Now substituting relations 11.2(*a*) in equation (11.4) and simplifying we have

$$\frac{\partial I}{\partial Z} = -\left(1 + \frac{\Delta Z}{2}\frac{\partial}{\partial Z}\right)\left(\partial V + c\frac{\partial V}{\partial t}\right) \quad \ldots(11.5)$$

Again we let $\Delta Z \to 0$ and we have

$$\frac{\partial I}{\partial z} = -\left(GV + C\frac{\partial V}{\partial t}\right) \qquad ...(11.6)$$

Equations (11.4) and (11.6) are the two wave equations which give variation of voltage and current respectively as a function of distance of the transmission line. These equations are also known as telegraphists equations. In order to solve equations (11.4) and (11.6) we differentiate equation (11.4) with respect to z and (11.6) with respect to t and we have

$$\frac{\partial^2 V}{\partial z^2} = -R\frac{\partial I}{\partial z} - L\frac{\partial^2 I}{\partial t\,\partial z} \qquad ...(11.7)$$

and

$$\frac{\partial^2 I}{\partial z\,\partial t} = -G\frac{\partial V}{\partial t} - C\frac{\partial^2 V}{\partial t^2} \qquad ...(11.8)$$

Now substituting for $\frac{\partial I}{\partial z}$ from equation (11.6) into equation (11.7), we have

$$\frac{\partial^2 V}{\partial z^2} = +R\left(GV + C\frac{\partial V}{\partial t}\right) - L\frac{\partial^2 I}{\partial t\,\partial z}$$

Substituting for $\frac{\partial^2 I}{\partial t\,\partial z}$ from (11.8) into above equation, we have

$$\frac{\partial^2 V}{\partial z^2} = R\left(GV + C\frac{\partial V}{\partial t}\right) + LG\frac{\partial V}{\partial t} + LC\frac{\partial^2 V}{\partial t^2}$$

$$= LC\frac{\partial^2 V}{\partial t^2} + (LG + RC)\frac{\partial V}{\partial t} + RGV \qquad ...(11.9)$$

Following similar procedure and differentiating equation (11.4) with respect to t and (11.6) with respect to z we obtain an equation for the current which is identical in form to that of equation (11.9) and we have

$$\frac{\partial^2 I}{\partial z^2} = LC\frac{\partial^2 I}{\partial t^2} + (LG + RC)\frac{\partial I}{\partial t} + RGI \qquad ...(11.10)$$

Equations (11.9) and (11.10) are known as general wave equation for the transmission lines. The solution of these equation under different assumption will be carried out in the following section.

11.4 LOSSLESS TRANSMISSION LINES

Lossless propagation of waves on lines means there is no power dissipation on lines which further means that $R = 0$ and $G = 0$. Under this condition equation (11.9) reduces to

$$\frac{\partial^2 V}{\partial z^2} = LC\frac{\partial^2 V}{\partial t^2} \qquad ...(11.11)$$

At this point, it is prudent to assume a solution of equation (11.11) and we will prove later that the assumed solution is correct. The solution of equation (11.11) is say,

$$V(z,t) = f_1\left(t - \frac{z}{v}\right) + f_2\left(t + \frac{z}{v}\right) = V_f + V_b \qquad \text{...(11.12)}$$

Where v is the velocity of wave on the line which is constant. The arguments of function f_1 and f_2 are $\left(t \mp \frac{z}{v}\right)$. The identities of the functions f_1 and f_2 are not very critical to the solution of equation (11.11) as these can be any functions of $\left(t \mp \frac{z}{v}\right)$. $V_f = f_1\left(t - \frac{z}{v}\right)$ corresponds to forward moving wave along z-axis whereas $V_b = f_2\left(t + \frac{z}{v}\right)$ corresponds to backward movement of wave along z-axis. To understand forward and backward movement of the two components of wave, let us assume that initially the argument of the two waves is zero *i.e.* $t - \frac{z}{v} = 0$ and $\left(t + \frac{z}{v}\right) = 0$ *i.e.* when $t = 0$ and $z = 0$. The values of the two components are $f_1(0)$ and $f_2(0)$. To track these values of the two components we increase t *i.e.* say after time $t + \Delta t$, the value of z will be for $V_{f_1}, \frac{z + \Delta z}{v}$ *i.e.* as t is increased z also increases in order to get another location of $f_1(0)$ and therefore, the wave moves in the positive z direction as time is increased and therefore this would be called as forward component of the wave. However, for f_2, as t is increased to $t + \Delta t$, z must be reduced by Δz to make the argument zero to track $f_2(0)$ after time $(t + \Delta t)$ *i.e.* the movement is in the negative z-axis or direction and, therefore $f_2(0)$ moves in the backward direction and hence it is represented by V_b. It is to be noted that this behaviour occurs irrespective of what be the function f_1 and f_2. Also, it is to be noted that the velocity of propagation v for both the components is the same.

We next show that in functions having the arguments in equation (11.12) are solution to (11.11). We take partial derivations of f_1 with respect to z and t using the chain rule *i.e.*

$$\frac{\partial F}{\partial z} = \frac{\partial F}{\partial x} \cdot \frac{\partial x}{\partial z}$$

Using this rule we have

$$\frac{\partial f_1}{\partial z} = \frac{\partial f_1}{\partial\left(t - \frac{z}{v}\right)} \cdot \frac{\partial (t - z/v)}{\partial z} = -\frac{f_1'}{v} \qquad \text{...(11.13)}$$

as

$$\frac{\partial}{\partial z}\left(t - \frac{z}{v}\right) = -\frac{1}{v}$$

Here f_1' is the derivative of f_1 with respect to its argument. Next, we find partial derivative of f_1 with respect to t.

$$\frac{\partial f_1}{\partial t} = \frac{\partial f_1}{\partial\left(t - \frac{z}{v}\right)} \cdot \frac{\partial (t - z/v)}{\partial t} = f_1' \qquad \text{...(11.14)}$$

as
$$\frac{\partial}{\partial t}\left(t-\frac{z}{v}\right)=1.$$

Next we take second partial derivative of f_1 with respect to z and t and we obtain.

$$\frac{\partial^2 f_1}{\partial z^2}=\frac{f_1''}{v^2} \text{ and } \frac{\partial^2 f_1}{\partial t^2}=f_1'' \qquad \ldots(11.15)$$

Where f_1'' is the second derivative with respect to its argument. Equations (11.15) are substituted in equation (11.11) we have

$$\frac{f_1''}{v^2}=LCf_1'' \text{ or } \frac{1}{v^2}=LC$$

or
$$v=\frac{1}{\sqrt{LC}} \qquad \ldots(11.16)$$

Using the same procedure for f_2 with its argument leads to the same result for velocity v of the wave. The equation 11.16 confirms our assumption that velocity of wave will be inversely proportional to some function of product of L and C. The same result will hold good for the current wave also under lossless condition.

For free space we know that $\frac{1}{\sqrt{LC}}=\frac{1}{\sqrt{\mu_0\varepsilon_0}}=\frac{1}{\sqrt{4\pi\times10^{-7}\times\frac{1}{36\pi}\times10^{-9}}}$

$= 3 \times 10^8$ m/sec.

which is the velocity of light. However, on lossy lines the velocity of wave is less than the velocity of light.

If we use f_2 and make similar operation, we will obtain the same relation as in equation (11.16).

We next find relations between voltage and, currents on the line. We have already found the relation for voltage and current in terms of telegraphist's equation (11.4) and (11.6).

These under lossless condition become when $R = 0$ $G = 0$

$$\frac{\partial V}{\partial z}=-L\frac{\partial I}{\partial t} \qquad \ldots(11.17)$$

and
$$\frac{\partial I}{\partial z}=-C\frac{\partial V}{\partial t} \qquad \ldots(11.18)$$

Now from equation (11.13) and (11.17)

$$\frac{\partial I}{\partial t}=-\frac{1}{L}\frac{\partial V}{\partial z}=-\frac{1}{Lv}\left(f_1'-f_2'\right) \qquad \ldots(11.19)$$

Integrating equation (11.19) with respect to time we obtain the current in terms of its forward and backward propagating components *i.e.*

$$I(z,t)=\frac{1}{LV}\left[f_1\left(t-\frac{z}{v}\right)-f_2\left(t+\frac{z}{v}\right)\right]=I_f+I_b \qquad \ldots(11.20)$$

assuming constant of integration to be zero. From equations (11.17) and (11.18) it is clear that a time varying voltage leads to a time varying current.

The ratio of forward component of voltage to Lv gives the forward component of current. Hence Lv should be an impedance and this impedance is known as characteristic impedance of the line donated by z_0.

$$z_0 = Lv = L \cdot \frac{1}{\sqrt{LC}} = \sqrt{\frac{L}{C}} \qquad \text{...(11.21)}$$

Therefore, $$V_f = I_f z_0 \text{ and } V_b = -I_b z_0 \qquad \text{...(11.22}a\text{)}$$

The negative sign in expression for V_b shows that negative current will be associated with a backward propagating wave that has positive polarity. The direction of forward and backward quantities are shown in Fig. 11.3.

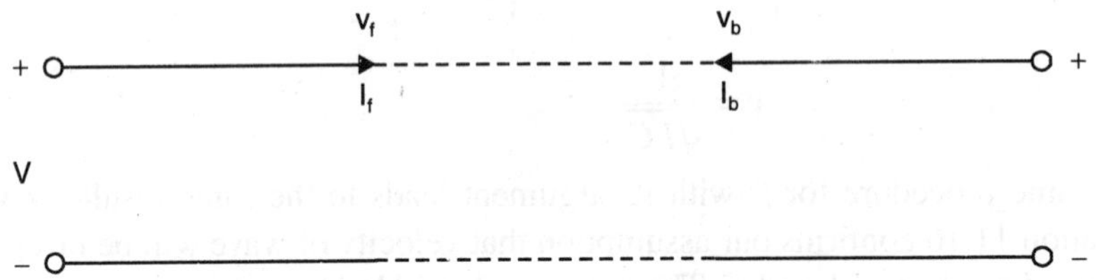

Fig. 11.3 Current direction in waves having positive voltage polarity

This is a general condition applicable to both lossless and lossy lines.

We know that given any time function, it can always be resolved into a sum of series of sinusoidal function having different frequency and therefore, an understanding of behaviour of lines to sinusoidal functions is important. This is the basic theory of frequency domain analysis of signals on transmission lines since the parameters of the line are frequency dependent. The effect of transmission line on every frequency is evaluated and we reassemble the frequency components into the resultant signal in time domain.

In equation (11.14) f_1 and f_2 are general functions. We assign here sinusoidal functions to the voltage function in equation (11.14). We consider here fundamental frequency component of frequency $f = \dfrac{\omega}{2\pi}$ and write $f_1 = f_2 = V_m \cos(\omega t + \phi)$. By convention, the cosine function is chosen here. However sine function can be obtained by taking $\phi = -\pi/2$. We next replace $t = \left(t \mp \dfrac{z}{V_p}\right)$ and we have

$$V(z_1\, t) = V_m \cos\left\{\omega\left(t \mp \frac{z}{V_p}\right) + Q\right\} = V_m \cos\left\{\left(\omega t \mp \frac{\omega z}{v_p}\right) + Q\right\} \qquad \text{...(11.22}b\text{)}$$

Here we have used a new notation for velocity of wave as V_p and is known as phase velocity V_p. This is applicable to fundamental component of sinusoidal wave we let $\phi = 0$ and thus we have the forward and backward components of the wave equation as

$$V_f(z, t) = V_m \text{ Cos } (\omega t - \beta z) \qquad \text{...(11.23}a\text{)}$$

and $$V_b(z, t) = V_m \text{ Cos } (\omega t + \beta z) \qquad \text{...(11.23}b\text{)}$$

where $\beta = \dfrac{\omega}{v_p}$ and V_m is the value of $V(z, t)$ at $z = 0$ and $t = 0$ *i.e.* $V(0,0) = V_m$

$$\beta = \frac{\omega}{v_p} \qquad ...(11.24)$$

Here β is known as the phase constant.

Equations 11.23(a) and 11.23(b) are the solutions of equations (11.11) and represent the real instantaneous forms of the transmission line voltages. The terms ωt and βz have the units of radians. ω is the angular frequency and has units of radians/sec. In a similar way as β is interpreted as a spatial frequency which in this case measures the phase shift per unit distance along the z-direction. Its units are radians/metre.

If we measure these V_f and V_b voltages at $t = 0$, we have

$$V_f(z, 0) = V_b(z, 0) = V_m \text{ Cos } \beta z \qquad ...(11.25)$$

Which is a simple periodic function which repeats every incremental distance λ known as wave length. For equation function the requirement is that $\beta\lambda = 2\pi$ and so

$$\pi = \frac{2\pi}{\beta} = \frac{2\pi}{2\pi f} \cdot V_p = \frac{V_p}{f} \qquad ...(11.26)$$

Crest of the cosine function in equation 11.23 (a) and (b) will occur when the argument of the cosine function is an integral multiple of 2π. However at $t = 0$, equation (11.25) the mth crest occurs when $\beta z = 2m\pi$. However, in general mth crest occurs when

$$\omega t - \beta z = \omega\left(t - \frac{Z}{V_p}\right) = 2m\pi \qquad ...(11.27)$$

To explain these cosine functions also, represent V_f and V_b voltages again for increasing time, the position z must also increase in order to satisfy 11.23(a). Consequently the wave crest travels in the positive z direction at velocity v_p. Similarly for equation 11.23(b) having cosine argument $(\omega t + \beta z)$ describes a wave that travels in the negative z direction, since as time increases z must now decrease to keep the argument constant. The current wave may also be addressed in the same manner. However, complications arise due to the fact that this is a spatial phase difference between voltage and currents.

In order to simplify our analysis it is easier to make use of complex analysis of sinusoidal functions as it greatly simplifies the evaluation and visualization of phase that will be found to accumulate by way of many mechanisms. Also, if it is required to add more than one sinusoidal waves, the task is made simpler if complex analysis (exponential functions) is used.

Sinusoidal functions can be expressed in complex form using Euler's identity

$$e^{\pm jx} = \cos x \pm j \sin x \qquad ...(11.28)$$

From this relation expressions for sinusoidal functions in terms of exponents are derived as follows:

$$\cos x = \text{Re}\left(e^{\pm jx}\right) = \frac{1}{2}\left(e^{jx} + e^{-jx}\right) \qquad ...(11.29)$$

$$\sin x = \text{Im}\left(e^{\pm jx}\right) = \frac{1}{2j}\left(e^{jx} - e^{-jx}\right) \qquad ...(11.30)$$

We next apply (11.29) to equation (11.22) and we have

$$V(z, t) = V_m \cos\{\omega t \pm \beta z + \phi\} = \frac{1}{2}\left(V_m e^{j\phi}\right) e^{j\omega t} e^{+j\beta z} + c.c. \qquad ...(11.31)$$

Here $V_m e^{j\phi}$ is the complex amplitude of the wave. Let this be denoted as V_{mc} *i.e.* $V_m e^{j\phi} = V_{mc}$ *c.c.* represents the complex conjugate of the first term and this conjugate is obtained by negating the exponent j *i.e.* j is to be written as $-j$. V_{mc} is the voltage having both magnitude and phase.

Next we define what is known as complex instantaneous voltage. This is done by multiplying V_{mc} by $e^{j\omega t}$ *i.e.*

$$V_c(z, t) = V_{mc} e^{\pm j\beta z} \cdot e^{j\omega t} \qquad \text{...(11.32)}$$

and the phasor voltage is formed by dropping the factor $e^{j\omega t}$ from V_c and we have

$$V_s(z) = V_{mc} e^{\pm j\beta z} \qquad \text{...(11.33)}$$

and this is the single frequency steady state sinusoidal voltage. Thus V_{mc} is independent of time. This is true otherwise a time varying amplitude would mean the existence of other frequency components in the signal as we are considering only fundamental frequency signals. The significance of the phasor voltage is that we are effectively considering time stand still and observing the stationary wave in space at $t = 0$. The process of evaluating relative phases between various line position (space phase distance) and of combining multiple waves is made much easier in phasor form.

However, this is possible only if all waves under consideration have the same frequency. Using (11.32) and (11.33) the real instantaneous voltage can be constructed using (11.31) as

$$V(z, t) = V_m \cos(\omega t \pm \beta z + \phi) = Re\,[V_s(z)\, e^{j\omega t}] + cc \qquad \text{...(11.34)}$$

In other words we can obtain real sinusoidal voltage wave by multiplying the phasor voltage by $e^{j\omega t}$ and then taking the real part of the resulting expression.

11.5 SOLUTION OF TRANSMISSION LINE EQUATIONS IN PHASOR FORM

The general transmission equation derived earlier as equation (11.9) is reproduced here

$$\frac{\partial^2 V}{\partial z^2} = LC\frac{\partial^2 V}{\partial t^2} + (LG + RC)\frac{\partial V}{\partial t} + RGV \qquad \text{...(11.9)}$$

We try to solve this equation using phasor form. For this we replace $\frac{\partial}{\partial t}$ by $j\omega$ and rewrite equation (11.9) as

$$\frac{\partial^2 V_s}{\partial z^2} = -\omega^2 LCV_s + (LG + RC)\, j\omega V_s + RGV_s$$

$$= V_s\left[-\omega^2 LC + LGj\omega + j\omega RC + RG\right]$$

$$= V_s\left[-\omega^2 LC + j\omega RC + G(R + j\omega L)\right]$$

$$= V_s\left[j\omega C(R + j\omega L) + G(R + j\omega L)\right]$$

$$= (R + j\omega L)(G + j\omega C)V_s$$

Let $\quad z = R + j\omega L$ and $Y = G + j\omega C$

Then, $$\frac{\partial^2 V_s}{\partial z^2} = YZV_s = \gamma^2 V_s \qquad \text{...(11.35)}$$

where $$\gamma^2 = YZ \text{ or } \gamma = \sqrt{YZ}$$

$$\sqrt{(R + j\omega L)(G + j\omega C)} = \alpha + j\beta \qquad \text{...(11.35}a\text{)}$$

where z is the series impedance of the line and Y the shunt admittance of the line per unit length of the line and $\gamma = \alpha + j\beta$ is R propagation constant, α the attenuation constant and the phase constant. The solution of equation (11.35) is

$$V_s(z) = V_f e^{-\gamma z} + V_b e^{\gamma z} \qquad \text{...(11.36)}$$

The equation for current is identical to equation 11.36 for the voltage and is given as

$$I_s(z) = I_f e^{-\gamma z} + I_b e^{\gamma z} \qquad \text{...(11.37)}$$

The relation between the voltage and current waves formed using the telegraphist's equation (11.4) and (11.6). In a manner similar to equation (11.34), equation for current can be written as

$$I(z, t) = I_m \cos(\omega t \pm \beta z + \phi) = \frac{1}{2}\left(I_m e^{je}\right)e^{\pm\beta z} e^{j\omega t} + c.c.$$

$$= \frac{1}{2} I_s(z) e^{j\omega t} + c.c. \qquad \text{...(11.38)}$$

Substituting them for right hand sides of (11.34) and (11.48) into equation (11.4) and (11.6) we obtain

$$\frac{\partial V}{\partial z} = -\left(RI + L\frac{\partial I}{\partial t}\right) \rightleftarrows \frac{dV_s}{dz} = -(R + jwL)I_s = zI_s \qquad \text{...(11.39)}$$

and $$\frac{\partial I}{\partial z} = -\left(GV + C\frac{\partial V}{\partial t}\right) \rightleftarrows \frac{dI_s}{dt} = -(G + j\omega C)V_s = -\gamma V_s \qquad \text{...(11.40)}$$

From equation (11.36)

$$\frac{dV_s}{dz} = -\gamma V_f e^{-\gamma z} + \gamma V_b e^{\gamma x} = -ZI_s = -Z\left(I_f e^{-\gamma z} + I_b e^{\gamma z}\right) \qquad \text{...(11.41)}$$

Equation quotient of $e^{-\gamma z}$ and $e^{\gamma z}$ we have

$$-\gamma V_f = -zI_f \text{ or } \frac{V_f}{I_f} = Z/V = \frac{Z}{\sqrt{\gamma Z}} = \sqrt{\frac{Z}{\gamma}} \qquad \text{...(11.42)}$$

$$Z_0 = \sqrt{\frac{R + j\omega L}{G + j\omega C}} = |z| e^{j\theta} \qquad \text{...(11.43)}$$

Here $\theta =$ is the angle between V_f and I_f

$$\theta = \phi - \xi.$$

Example 11.1. A lossless transmission line is 100 cm long and operates at a frequency of 750 MHz. The line parameters are $L = 0.3$ μH/m and $C = 120$ pF/m. Determine the characteristic impedance, the phase velocity and the phase constant.

Solution. Since the line is lossless $R = G = 0$

The characteristic impedance is $Z_0 = \sqrt{\dfrac{L}{C}} = \sqrt{\dfrac{0.3\times10^{-6}}{120\times10^{-12}}}$

$$= \sqrt{\frac{30\times10^{-8}}{120\times10^{-12}}} = \sqrt{\frac{1}{4}\times10^{4}}$$

$$= \frac{100}{2} = 50\Omega$$

For a lossless line $\gamma = \alpha + j\beta = j\omega\sqrt{LC} \qquad \alpha = 0$

$$\beta = \omega\sqrt{0.3\times120\times10^{-18}}$$
$$= 2\pi \times 750 \times 10^{6} \times 6 \times 10^{-9}$$
$$= 6.28 \times 6 \times 10^{-3} \times 750$$
$$= 28.26 \text{ rad/meter}$$

The phase velocity $= \dfrac{\omega}{\beta} = \dfrac{2\pi\times750\times10^{6}}{28.26} = 1.67\times10^{8}$ m/sec.

From equation (11.46) and the relation $\gamma = \alpha + j\beta$ we have

$$V_s(z) = V_f e^{-\gamma z} + V_b e^{\gamma z} = V_f e^{-\alpha z} e^{-j\beta z} + V_b e^{\alpha z} e^{\beta z} \quad \text{...(11.44)}$$

This is the steady state value of the fundamental frequency sinusoidal wave. In order to obtain real instantaneous value of voltage we multiply equation (11.44) by $e^{j\omega t}$ and thus we have

$$V(z, t) = V_f e^{-\alpha z} \cos(\omega t - \beta z) + V_b e^{\alpha z} \cos(\omega t + \beta z) \quad \text{...(11.45)}$$

Here the forward and backward components of voltage decay exponentially with distance. The backward component also diminishes as z is negative for this component. The phase constant β which is imaginary part of propagation constant γ depends upon both R and G and is still defined by the ratio $\beta = \dfrac{\omega}{v_p}$ and the wavelength $\lambda = \dfrac{2\pi}{\beta}$. Substituting $R = G = 0$ for a lossless line $\alpha = 0$ and $\beta = \omega\sqrt{LC}$ and so $v_p = 1/\sqrt{LC}$.

However, if line is low loss line $R << \omega L$ and $G << \omega C$ conditions apply. Before we make use of this condition we rewrite equation (11.35*a*) as

$$\alpha + j\beta = (R + j\omega L)^{1/2}(G + j\omega C)^{1/2}$$

$$= j\omega\sqrt{LC}\left[1 + \frac{R}{j\omega L}\right]^{1/2}\left[1 + \frac{G}{j\omega L}\right]^{1/2} \quad \text{...(11.46)}$$

Since $\dfrac{R}{\omega L} << 1$ we make use of binomial theorem for expansion of the two terms. We first take all algebraic expression $\sqrt{1+x}$ and write its binomial expansion which is as follows

$$(1+x)^{1/2} = 1 + \frac{x}{2} - \frac{x^2}{8} + \ldots \quad x << 1 \quad \text{...(11.47)}$$

limiting expansion only upto three terms.

$$\gamma = (\alpha + j\beta) = j\omega\sqrt{LC}\left[1 + \frac{R}{j2\omega L} + \frac{R^2}{8\omega^2 L^2}\right]\left[1 + \frac{G}{j2\omega C} + \frac{G^2}{8\omega^2 c^2}\right] \quad \text{...(11.48)}$$

After carrying out the product of the two factors and neglecting the terms like RG^2, R^2G and R^2G^2 as R and G are negligibly small for low loss lines we have

$$\gamma = \alpha + j\beta = j\omega\sqrt{LC}\left[1 + \frac{1}{j2\omega}\left(\frac{R}{L} + \frac{G}{C}\right) + \frac{1}{8\omega^2}\left(\frac{R^2}{L^2} - \frac{2RG}{LC} + \frac{G^2}{C^2}\right)\right] \quad \text{...(11.49)}$$

Separating the real and imaginary components from equation (11.49) we have

$$\alpha = \frac{1}{2}\left(R\sqrt{\frac{C}{L}} + G\sqrt{\frac{L}{C}}\right) \quad \text{...(11.50)}$$

and

$$\beta = \omega\sqrt{LC}\left[1 + \frac{1}{8}\left(\frac{G}{\omega C} - \frac{R}{\omega L}\right)^2\right] \quad \text{...(11.51)}$$

From equation (11.51) it can be seen that the phase velocity $V_p = \frac{\omega}{\beta}$ is frequency dependent. Also the group velocity $V_g = \frac{d\omega}{d\beta}$ depends upon frequency which will lead to signal distortion. However, if $\frac{R}{2} = \frac{G}{C}$ known as Heavisides' condition is satisfied, then the group velocity and phase velocity are independent of frequency and the line is said to be distortionless. Further complications occur when accounting to increase in resistance of the conductor with higher frequencies and the loss increases. This is due to skin effect. As a result, conditions of low loss or distortion free propagation will occur over a limited frequency range.

Next we find characteristic impedance z_0 for a low loss line.

We know

$$z_0 = \sqrt{\frac{R + j\omega L}{G + j\omega C}} = \sqrt{\frac{j\omega L(1 + R/j\omega L)}{j\omega C(1 + G/j\omega C)}}$$

$$= \sqrt{\frac{L}{C}}\left[\frac{\left(1 + \frac{R}{2j\omega L} + \frac{R^2}{8\omega^2 L^2}\right)}{\left(1 + \frac{G}{2j\omega C} + \frac{G^2}{8\omega^2 C^2}\right)}\right]$$

$$= \sqrt{\frac{L}{C}}\left[\frac{1 - j\frac{R}{2\omega L} + R^2/8\omega^2 L^2}{1 - \frac{jG}{2\omega C} + \frac{G^2}{8\omega^2 C^2}}\right]$$

Now multiplying the right hand side with $\dfrac{1+\dfrac{G^2}{8\omega^2C^2}+\dfrac{jG}{2\omega C}}{1+\dfrac{G^2}{8\omega^2C^2}+j\dfrac{G}{2\omega C}}$

and assuming that x is small as compared to units and hence $\dfrac{1}{1+x}\approx 1-x$

We have

$$z_0=\sqrt{\frac{L}{C}}\left[1+\frac{1}{2\omega^2}\left\{\frac{1}{4}\left(\frac{R}{L}+\frac{G}{C}\right)^2-\frac{G^2}{C^2}\right\}+\frac{1}{2\omega}\left(\frac{G}{C}-\frac{R}{L}\right)\right] \quad \ldots(11.52)$$

If however, we consider, distortionless line when $\dfrac{R}{L}=\dfrac{G}{C}$ the above expression for z_0 reduces

$Z_0=\sqrt{\dfrac{L}{C}}$ as is true when both R and G are zero.

Group velocity: Consider a uniform plane wave travelling in the positive z direction as shown in Fig. 12.1. Let this total electric field be given by

$$E_x = E_0 \cos(\omega t-\beta z) \quad \ldots(1)$$

Suppose that the wave has not one but two frequencies of equal magnitude given as

$$\omega_0+\Delta\omega \text{ and } \omega_0-\Delta\omega$$

The β values corresponding to these two frequencies are

$$\beta_0+\Delta\beta \text{ corresponding to } \omega_0+\Delta\omega$$

and

$$\beta_0-\Delta\beta \text{ corresponding to } \omega_0-\Delta\omega$$

For frequency $\omega_0+\Delta\omega$

$$E_x' = E_0\cos\{(\omega_0+\Delta\omega)t-(\beta_0+\Delta\beta)z\} \quad \ldots(2)$$

and for frequency $\Delta\omega_0-\Delta\omega$

$$E_x'' = E_0\cos\{(\omega_0-\Delta\omega)t-(\beta_0-\Delta\beta)z\} \quad \ldots(3)$$

The total field is given by after rearranging

$$E_x = E_x'+E_x'' = 2E_0\cos(\omega_0 t-\beta_0 z)\cos(\Delta\omega t-\Delta\beta z) \quad \ldots(4)$$

The two cosine factors indicate the presence of beats *i.e.* a slow variation superimposed on a more rapid one.

For a constant phase point.

$$\omega_0 t-\beta_0 z = \text{constant}$$

and

$$\omega_0-\beta_0\frac{dz}{dt}=0$$

or

$$\frac{dz}{dt}=\frac{\omega_0}{\beta_0}=v=f_0\lambda_0 \text{ the phase velocity} \quad \ldots(5)$$

If the argument of the second cosine term is set to a constant we have

$$\Delta\omega t - \Delta\beta z = \text{constant}$$

$$\Delta\omega - \Delta\beta\frac{dz}{dt} = 0$$

or
$$\frac{dz}{dt} = \frac{\Delta\omega}{\Delta\beta} = \Delta\lambda\Delta f = u \text{ group velocity} \quad \ldots(6)$$

Where u is the phase velocity of the wave envelope and is termed as group velocity. In the above analysis we consider $(\omega_0 + \Delta\omega)$ and $(\omega_0 - \Delta\omega)$ as the two side band frequencies caused by the modulation of a carrier frequency ω_0 by a frequency $\Delta\omega$, the carrier frequency being suppressed.

In nondispersive media the group velocity and phase velocity are identical. Free space is an example of nondispersive lossless medium and in free space $u = v = c$. However, in dispersive medium, phase and group velocities are different. A dispersive medium is one where the phase velocity is a function of the frequency and hence of the free space wavelength. Dispersive media are of two types.

(*i*) Normally dispersive: Here change in phase velocity with wavelength is positive *i.e.* $\frac{dv}{d\lambda} > 0$ and $u < v$.

(*ii*) Anomalously dispersive: Here $\frac{dv}{d\lambda} < 0$ and $u > v$.

For a particular frequency (bandwidth negligibly small)

$$u = v + \beta\frac{dv}{d\beta} \quad \text{group velocity} \quad \ldots(7)$$

$$u = v - \lambda\frac{dv}{d\lambda} \quad \text{group velocity} \quad \ldots(8)$$

Equations (7) and (8) are useful for finding the group velocity for a given phase velocity.

Example 11.2. A low loss transmission line has $G = 0$ but R is finite such that $R << \omega L$. From equation (11.52) determine the characteristic impedance of the line.

Solution. From equation 11.52 we have

$$Z_0 = \sqrt{\frac{L}{C}}\left[1 + \frac{1}{2\omega^2}\left\{\frac{1}{4}\left(\frac{R}{L} + \frac{G}{C}\right)^2 - \frac{G^2}{C^2}\right\} + \frac{j}{2\omega}\left(\frac{G}{C} - \frac{R}{L}\right)\right]$$

Substituting $G = 0$ we have

$$Z_0 = \sqrt{\frac{L}{C}}\left[1 + \frac{1}{2\omega^2}\frac{R^2}{4L^2} - \frac{jR}{2\omega L}\right]$$

Since $R << \omega L$, the expression simplifies to

$$Z_0 = \sqrt{\frac{L}{C}}\left(1 - j\frac{R}{2\omega L}\right)$$

Since $R << \omega L$, the magnitude of $Z_0 = \sqrt{\frac{L}{C}}$ and angle $\theta = \tan^{-1}\left(-\frac{R}{2\omega L}\right)$

Example 11.3. A transmission line operating at 500 Mrad/sec has line parameters as follows

$R = 0.3\ \Omega/\text{m}$ $L = 0.375\ \mu\text{H/m}$, $G = 15\ \mu\text{s/m}$ and $C = 150$ pF/m.

Determine (*i*) α (*ii*) β (*iii*) λ (*iv*) Z_0.

Solution. From equation (11.50) we have

$$(i)\quad \alpha = \frac{1}{2}\left(R\sqrt{\frac{C}{L}} + G\sqrt{\frac{L}{C}}\right) = \frac{1}{2}\left(0.3\sqrt{\frac{150\times10^{-12}}{0.375\times10^{-6}}} + 15\times10^{-6}\sqrt{\frac{0.375\times10^{-6}}{150\times10^{-12}}}\right)$$

$$= \frac{1}{2}\left(0.3\times20\times10^{-3} + 15\times10^{-6}\times\frac{1}{20}\times10^{3}\right)$$

$$= \frac{1}{2}(6+0.75)\times10^{-3} = 3.375 \text{ mNP/m} \quad \textbf{Ans.}$$

$$(ii)\quad \beta = \omega\sqrt{LC}\left[1+\frac{1}{8}\left(\frac{G}{\omega C}-\frac{R}{\omega L}\right)^2\right]$$

$$= 500\times10^{6}\times\sqrt{0.375\times10^{-6}\times150\times10^{-12}}\left[1+\frac{1}{8}\left(\frac{15\times10^{-6}}{500\times150\times10^{-6}}\right)\right]$$

$$= 500\times10^{6}\times7.5\times10^{-9}\left(1+\left(\frac{200}{8}-\frac{0.3}{500\times0.375\times10^{-6}}\right)^2\right)$$

$$= 500\times10^{6}\times7.5\times10^{-9}\left[1+\frac{1}{8}\left(200-\frac{300}{500\times0.375}\right)^2\right]$$

$$= 500\times10^{6}\times7.5\times10^{-9} = 3.75 \text{ rad/m}.$$

$$(iii)\quad \lambda = \frac{2\pi}{\beta} = \frac{6.28}{3.75} = 1.67 \text{ m}$$

$$(iv)\quad Z_0 = \sqrt{\frac{L}{C}}\left(1-j\frac{R}{2\omega L}\right)$$

$$= \sqrt{\frac{0.375\times10^{-6}}{150\times10^{-12}}}\left(1-j\frac{0.3}{500\times0.375}\right)$$

$$= \sqrt{\frac{37.5}{150}\times10^{4}}\ (1-j0.0016)$$

$$= (50-j0.08)$$

When switch S is closed, a voltage wave of magnitude V and current wave of magnitude I start travelling towards the shorted end. Consider again the last element dx where the electrostatic energy stored by the element is $\frac{1}{2}CdxV^2$ and electromagnetic energy $\frac{1}{2}LdxI^2$. Since the voltage at the shorted end is zero, the electrostatic energy vanishes and is transformed into electromagnetic energy. As a result, let the change in the current be i; then

$$\frac{1}{2}CdxV^2 = \frac{1}{2}LdxI^2$$

or

$$V = iZ$$

or

$$i = \frac{V}{Z} = I$$

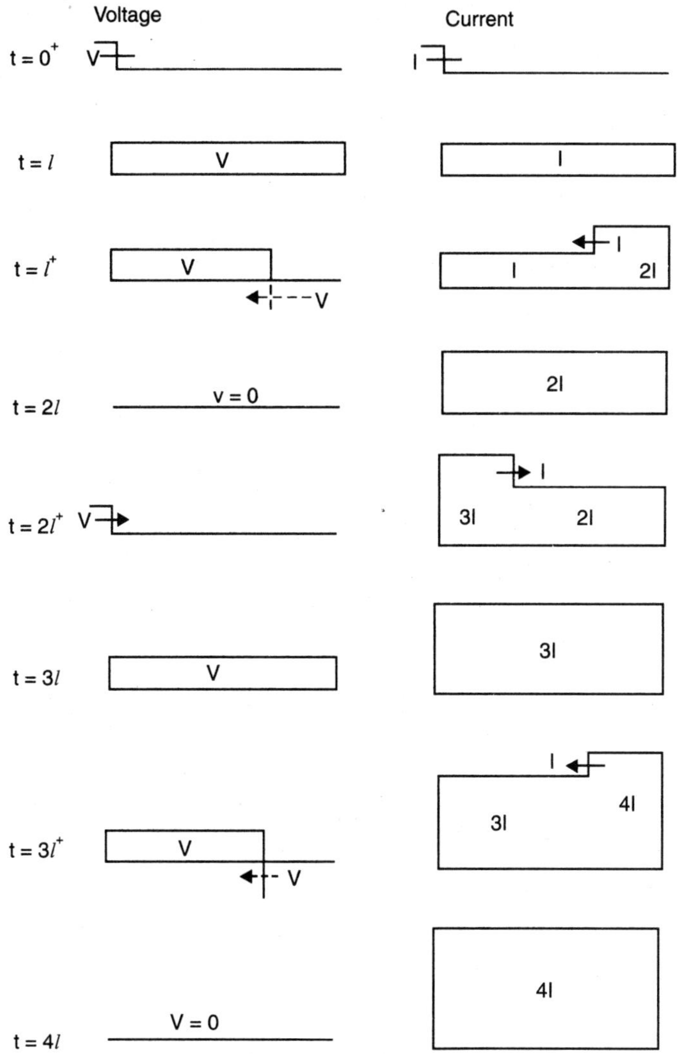

Fig. 11.7 Variation of voltage and current in a short ended line

This means the increase in current is I amperes. As a result the total current at shorted end, when the current wave reaches the end is $(I + I) = 2I$ amperes. This could be considered due to a reflected current wave of magnitude I amperes. Therefore for a short-circuited end the current wave is reflected back with positive sign and coefficient of reflection is unity. Since the voltage at the shorted end is zero, a voltage wave of $-V$ could be considered to have been reflected back into the line, *i.e.* the current wave in case of short-circuited end is reflected back with positive sign and with coefficient of reflection as unity, whereas the voltage wave is reflected back with negative sign and unity coefficient of reflection. The variation of voltage and current over the line is explained in Fig. 11.7.

It is seen from above that the voltage wave periodically reduces to zero after it has travelled through a distance of twice the length of the line whereas after each reflection at either end the current is built up by an amount $V/Z_n = I$. Theoretically, the reflection will be infinite and therefore the current will reach infinite value. But practically in an actual system the current will be limited by the resistance of the line and the final value of the current will be $I' = V/R$, where R is the resistance of transmission line.

Line Terminated Through a Resistance

Let Z be the surge impedance of the line terminated through a resistance R (Fig. 11.8). It has been seen in the previous sections that whatever be the value of the terminating impedance whether it is open or short circuited, one of the two voltage or current waves is reflected back with negative sign. Also, since the reflected wave travels along the overhead line or over the line along which the incident wave travelled, therefore, the following relation holds good for reflected voltage and current waves.

$$I' = -\frac{V'}{Z}$$

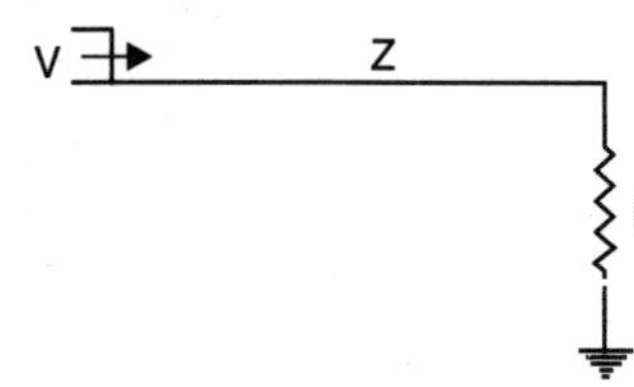

Fig. 11.8 Line terminated through a resistance

where V' and I' are the reflected voltage and current waves. Also,

Refracted or transmitted wave = Incident wave + Reflected wave

Let V'' and I'' be the refracted voltage and current waves into the resistor R when the incident waves V and I reach the resistance R. The following relations hold good:

$$I = \frac{V}{Z}$$

$$I' = -\frac{V'}{Z}$$

$$I'' = \frac{V''}{R}$$

Since $I'' = I + I'$ and $V'' = V + V'$, using these relations, we have

$$\frac{V''}{R} = \frac{V}{Z} - \frac{V'}{Z} \quad \text{...(11.58)}$$

$$= \frac{V}{Z} - \frac{V'' - V}{Z}$$

$$= \frac{2V}{Z} - \frac{V''}{Z}$$

or
$$V'' = \frac{2VR}{Z + R} \qquad \text{...(11.59)}$$

and current
$$I'' = \frac{2V}{R + Z} = \frac{V}{Z} \cdot \frac{2Z}{R + Z}$$

$$= I \cdot \frac{2Z}{R + Z} \qquad \text{...(11.60)}$$

Similarly substituting for V'' in terms of $(V + V')$, equation (11.58) becomes

$$\frac{V + V'}{R} = \frac{V}{Z} - \frac{V'}{Z}$$

or
$$V' = V \cdot \frac{R - Z}{R + Z} \qquad \text{...(11.61)}$$

and
$$I' = -\frac{V'}{Z} = -\frac{V}{Z} \cdot \frac{(R - Z)}{R + Z} \qquad \text{...(11.62)}$$

From the relations above, the coefficient of refraction for current waves

$$= \frac{2Z}{R + Z}$$

and for voltage waves

$$= \frac{2R}{R + Z}$$

Similarly, the coefficient of reflection for current waves

$$= -\frac{R - Z}{R + Z}$$

and for voltage waves

$$= +\frac{R - Z}{R + Z}$$

Now the two extreme cases can be derived out of this general expression. For open circuit,

$$R \to \infty$$

and coefficient of refraction for current waves

$$\frac{2Z}{\infty + Z} = 0$$

and coefficient of refraction for voltage waves

$$= \frac{2R}{R + Z} = \frac{2}{1 + Z/R} = \frac{2}{1 + Z/\infty} = 2$$

Similarly, coefficient of refraction for current waves

$$= -\frac{R-Z}{R+Z} = -\frac{1-Z/R}{1+Z/R} = -1$$

and coefficient of reflection for voltage waves

$$= \frac{R-Z}{R+Z} = 1$$

Similarly, to find out the coefficients of reflection and refraction for current and voltage waves for the short circuit case, the value of $R = 0$ is to be substituted in the corresponding relations as derived in this section.

It is, therefore, seen here that whenever a travelling wave looks into a change in impedance, it suffers reflection and refraction. It is shown below that in case $Z = R$ *i.e.* the line is terminated through a resistance whose value equals the surge impedance of the line (*i.e.* no change in the impedance) there will be no reflection and the wave will enter fully into the resistance, *i.e.* the coefficient of refraction will be unity whereas the coefficient of reflection will be zero.
when $R = Z$, substituting this, the coefficient of reflection for current wave

$$= -\frac{R-Z}{R+Z} = \frac{Z-Z}{Z+Z} = 0$$

and for voltage wave

$$= \frac{R-Z}{R+Z} = 1$$

The coefficient of refraction for current wave

$$= \frac{2Z}{R+Z} = \frac{2Z}{2Z} = 1$$

and for voltage wave

$$= \frac{2R}{R+Z} = 1$$

It is seen that when a transmission line is terminated through a resistance equal to its surge impedance the wave does not suffer reflection and, therefore, such lines could be said to be of infinite length. Such lines are also called as matched lines and the load corresponding to this is known as surge impedance loading or natural impedance loading.

Line Connected to a Cable

A wave travels over the line and enters the cable (Fig. 11.9). Since the wave looks into a different impedance, it suffers reflection and refraction at the junction and the refracted voltage wave is given by

$$V'' = V \cdot \frac{2Z_2}{Z_1 + Z_2}$$

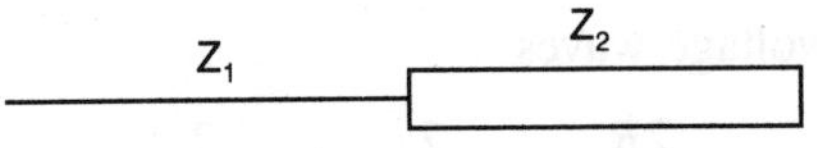

Fig. 11.9 Line connected to a cable

The other waves can be obtained by using the relations (11.60–11.62). The impedance of the overhead line and cable are approximately 400 ohms and 40 ohms respectively. With these values it can be seen that the voltage entering the cable will be

$$V'' = V \cdot \frac{2 \times 40}{40 + 400} = \frac{2}{11} V$$

i.e. it is about 20% of the incident voltage *V*. It is for this reason that an overhead line is terminated near a station by connecting the station equipment to the overhead line through a short length of underground cable. Besides the reduction in the magnitude of the voltage wave, the steepness is also reduced because of the capacitance of the cable. This is explained in the next section. The reduction in steepness is very important because this is one of the factors for reducing the voltage distribution along the windings of the equipment. While connecting the overhead line to a station equipment through a cable it is important to note that the length of the cable should not be very short (should not be shorter than the expected length of the wave) otherwise successive reflections at the junction may result in piling up of voltage and the voltage at the junction may reach the incident voltage.

Reflection and Refraction at a T-junction

A voltage wave V is travelling over the line with surge impedance Z_1 as shown in Fig. 11.10. When it reaches the junction, it looks a change in impedance and, therefore, suffers reflection and refraction. Let V_2'', I_2'' and V_3'', I_3'' be the voltages and currents in the lines having surge impedances Z_2 and Z_3 respectively. Since Z_2 and Z_3 form a parallel path as far as the surge wave is concerned, $V_2'' = V_3'' = V''$. Therefore, the following relations hold good:

$$V + V' = V''$$

$$I = \frac{V}{Z_1}, \; I' = -\frac{V'}{Z_1}$$

$$I_2'' = \frac{V''}{Z_2} \text{ and } I_3'' = \frac{V''}{Z_3}$$

and

$$I + I' = I_2'' + I_3'' \qquad \ldots(11.63)$$

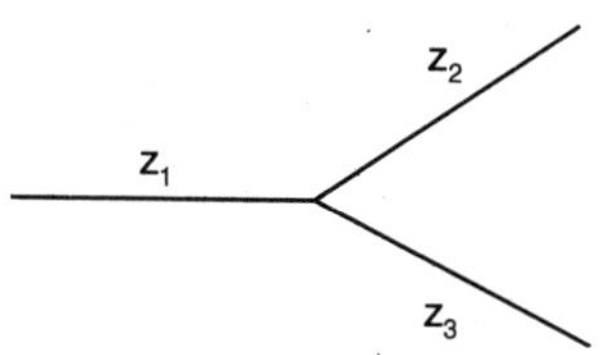

Fig. 11.10 A bifurcated line

Substituting in equation (11.63) the values of currents

$$\frac{V}{Z_1} - \frac{V'}{Z_1} = \frac{V''}{Z_2} + \frac{V''}{Z_3}$$

Substituting for $V' = V'' - V$,

$$\frac{V}{Z_1} - \frac{V'' - V}{Z_1} = \frac{V''}{Z_2} + \frac{V''}{Z_3}$$

or

$$\frac{2V}{Z_1} = V''\left[\frac{1}{Z_1} + \frac{1}{Z_2} + \frac{1}{Z_3}\right]$$

$$V'' = \frac{2V/Z_1}{\frac{1}{Z_1} + \frac{1}{Z_2} + \frac{1}{Z_3}} \quad ...(11.64)$$

Similarly other quantities can be derived.

Example 11.5. A 3-phase transmission line has conductors 1.5 cms in diameter spaced 1 metre apart in equilateral formation. The resistance and leakance are negligible. Calculate (*i*) the natural impedance of the line, (*ii*) the line current if a voltage wave of 11 kV travels along the line, (*iii*) the rate of energy absorption, the rate of reflection and the state and the form of reflection if the line is terminated through a star connected load of 1000 ohm per phase, (*iv*) the value of the terminating resistance for no reflection and (*v*) the amount of reflected and transmitted power if the line is connected to a cable extension with inductance and capacitance per phase per cm of 0.5×10^{-8} H and 1×10^{-6} μF respectively.

Solution. (*i*) The inductance per unit length

$$= 2 \times 10^{-7} \ln \frac{d}{r} \text{ H/meter}$$

$$= 2 \times 10^{-7} \ln \frac{100}{0.75}$$

$$= 2 \times 10^{-7} \ln 133.3$$

$$= 2 \times 10^{-7} \times 4.89$$

$$= 9.78 \times 10^{-7} \text{ Hm}^{-1}$$

The capacitance per phase per unit length

$$= \frac{2\pi\varepsilon}{\ln d/r} \text{ Fm}^{-1}$$

$$= \frac{2\pi \times 10^{-9}}{36\pi \ln d/r}$$

$$= \frac{1}{18} \times \frac{10^{-9}}{4.89} = 1.136 \times 10^{-11}$$

∴ The natural impedance $= \sqrt{\frac{L}{C}}$ ohms

$$= \sqrt{\frac{9.78 \times 10^{-7}}{1.136 \times 10^{-11}}} = 294\,\Omega \textbf{ Ans.}$$

(*ii*) The line current $= \dfrac{11000}{\sqrt{3} \times 294} = 21.6$ amps. **Ans.**

(*iii*) Since the terminating resistance is of higher value as compared to the value of the surge impedance of the line, the reflection is with a positive sign.

The voltage across the terminating resistance

$$E'' = \frac{2Z_2 E}{Z_1 + Z_2}$$

where Z_1 = line surge impedance, Z_2 = terminating impedance and E = incident voltage.

$$E'' = 2 \times \frac{11000}{\sqrt{3}} \frac{1000}{1294} = 9.8 \text{ kV}$$

$\therefore$ The rate of power consumption $= \dfrac{2E''^2}{R}$ MW

$$= \frac{3 \times 9.8 \times 9.8}{1000} \times 1000 \text{ kW}$$

$$= 288 \text{ kW } \textbf{Ans.}$$

The reflected voltage $E' = \dfrac{Z_2 - Z_1}{Z_2 + Z_1} E = \dfrac{1000 - 294}{1294} \times \dfrac{11}{\sqrt{3}}$ kV

$$= \frac{706}{1294} \times \frac{11}{\sqrt{3}} = 3.465 \text{ kV}$$

$\therefore$ The rate of reflected energy $= \dfrac{3 \times 3.465^2}{294} \times 1000$ kW

$$= 121.8 \text{ kW } \textbf{Ans.}$$

(*iv*) In order that the incident wave when reaches the terminating resistance, does not suffer reflection, the terminating resistance should be equal to the surge impedance of the line, *i.e.* 294 ohms.

(*v*) The surge impedance of the cable $= \sqrt{\dfrac{L}{C}} = \sqrt{\dfrac{0.5 \times 10^{-8}}{10^{-12}}}$

$$= 70.7 \text{ ohm}$$

The refracted voltage $= \dfrac{2 \times 70.7}{294 + 70.7} \times \dfrac{11}{\sqrt{3}}$

$$= \frac{2 \times 70.7 \times 11}{\sqrt{3} \times 364.7} = 2.46 \text{ kV}$$

The reflected voltage $= \dfrac{70.7 - 294}{364.7} \times \dfrac{11}{\sqrt{3}}$

$$= \frac{-223.3 \times 11}{\sqrt{3} \times 364.7} = -3.9 \text{ kV}$$

∴ The refracted and reflected powers are respectively

$$\frac{3\times 2.46^2}{70.7}\times 1000 = 256 \text{ kW and } \frac{3\times 3.9^2}{294}\times 1000 = 155 \text{ kW } \textbf{Ans.}$$

Example 11.6. A surge of 15 kV magnitude travels along a cable towards its junction with an overhead line. The inductance and capacitance of the cable and overhead line are respectively 0.3 mH, 0.4 μF and 1.5 mH, 0.012 μF per km. Find the voltage rise at the junction due to the surge.

Solution. In this problem the surge travels from the cable towards the overhead line and hence there will be positive voltage reflection at the junction.

$$\text{The natural impedance of the cable} = \sqrt{\frac{0.3\times 10^{-3}}{0.4\times 10^{-6}}}$$

$$= \sqrt{\frac{3\times 10^{-4}}{0.4\times 10^{-6}}} = 27.38.$$

$$\text{The natural impedance of the line} = \sqrt{\frac{1.5\times 10^{-3}}{0.012\times 10^{-6}}}$$

$$= \sqrt{\frac{1.5\times 10^{-3}}{0.12\times 10^{-7}}} = 353 \text{ ohms.}$$

The voltage rise at the junction is the voltage transmitted into the overhead line as the voltage is zero before the surge reaches the junction.

$$E'' = \frac{2\times 353\times 15}{353+27} = \frac{2\times 353\times 15}{380} = 27.87 \text{ kV } \textbf{Ans.}$$

Example 11.7. A surge of 100 kV travelling in a line of natural impedance 600 ohms arrives at a junction with two lines of impedances 800 ohms and 200 ohms respectively. Find the surge voltages and currents transmitted into each branch line.

Solution. The problem deals with a reflection at a T-joint. The various natural impedances are: $Z_1 = 600$ ohms, $Z_2 = 800$ ohms, $Z_3 = 200$ ohms. The surge magnitude is 100 kV.

The surge as it reaches the joint suffers reflection and here the two lines are in parallel; therefore, the transmitted voltage will have the same magnitude and is given by

$$E'' = \frac{2E/Z_1}{\frac{1}{Z_1}+\frac{1}{Z_2}+\frac{1}{Z_3}} = \frac{2\times 100/600}{\frac{1}{600}+\frac{1}{800}+\frac{1}{200}}$$

$$= \frac{0.333}{(1.67+1.25+5.0)\times 10^{-3}} = \frac{0.333\times 10^3}{7.92}$$

$$= \frac{33.3}{7.92}\times 10 = 42.04 \text{ kV } \textbf{Ans.}$$

The transmitted current in line $Z_2 = \frac{42.04 \times 1000}{800}$ amps = 52.55 amps **Ans.**

The transmitted current in line $Z_3 = \frac{42.04 \times 1000}{200}$ amps = 210.2 amps **Ans.**

Line Terminated Through a Capacitance

We consider here that a d.c. surge of infinite length travels over the line of surge impedance *Z* and is incident on the capacitor as shown in Fig. 12.17. We are interested in finding out the voltage across the capacitor.

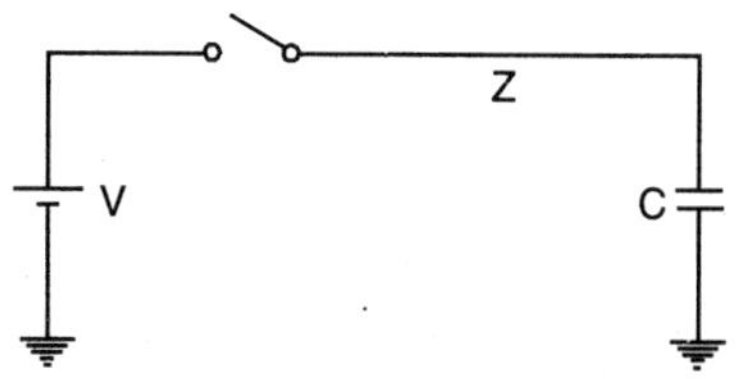

Fig. 11.12 Line terminated through a capacitance

i.e. the refracted voltage. The refracted voltage, using equation (11.59)

$$V''(s) = \frac{2 \cdot 1/Cs}{Z + 1/Cs} \cdot \frac{V}{s} = \frac{2V}{s} \cdot \frac{1}{ZCs + 1}$$

$$= \frac{2V}{s} \cdot \frac{1/ZC}{s + 1/ZC} = 2V\left[\frac{1}{s} - \frac{1}{s + 1/ZC}\right]$$

$$v''(t) = 2V\left[1 - e^{-t/ZC}\right] \qquad \ldots(11.65)$$

The variation of voltage is shown in Fig. 11.13(*b*).

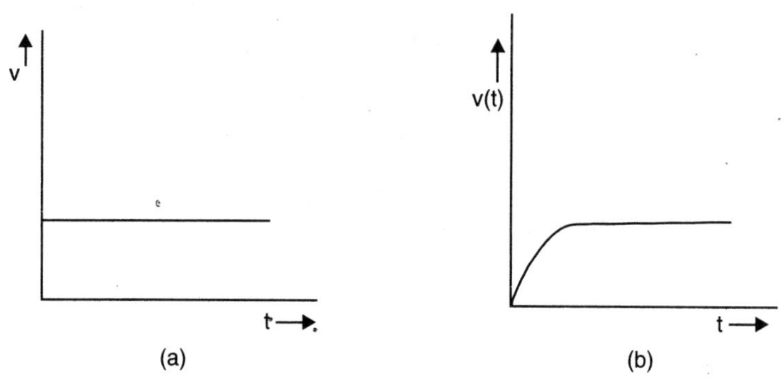

Fig. 11.13 (*a*) Incident voltage and (*b*) Voltage across capacitor

It is to be noted that since terminating impedance is not a transmission line, therefore, $V''(s)$ is not a traveling wave but it is the voltage across the capacitor *C*.

Capacitor Connection at a T

The voltage across capacitor is given by the equation

$$V''(s) = \frac{2V/Z_1 s}{\frac{1}{Z_1} + \frac{1}{Z_2} + Cs} = \frac{2VZ_2}{s} \cdot \frac{(1/Z_1 Z_2 C)}{\frac{(Z_1 + Z_2)}{Z_1 Z_2 C} + s}$$

$$= \frac{2V}{sZ_1 C} \cdot \frac{1}{s + \frac{Z_1 + Z_2}{Z_1 Z_2 C}}$$

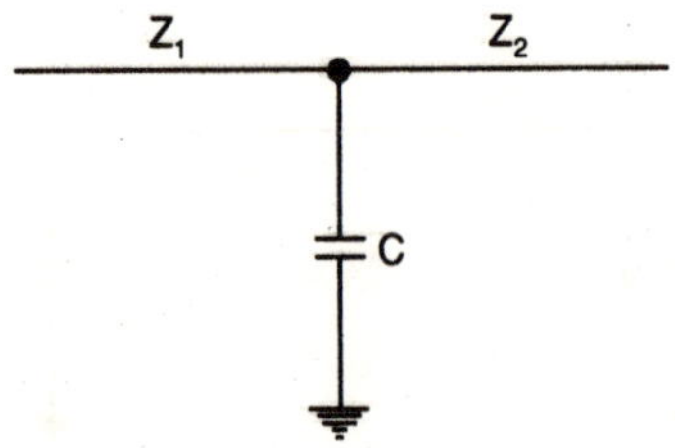

Fig. 11.14 Capacitor connected at *T*

Let $\frac{Z_1 + Z_2}{Z_1 Z_2 C} = \alpha$; then

$$V''(s) = \frac{2V}{s} \cdot \frac{1/Z_1 C}{s + \alpha}$$

or
$$V''(s) = \frac{2V}{s} \cdot \frac{Z_2}{Z_1 + Z_2} \cdot \frac{(Z_1 + Z_2)/Z_1 Z_2 C}{(s + \alpha)}$$

$$= \frac{2V}{s} \cdot \frac{Z_2}{Z_1 + Z_2} \cdot \frac{\alpha}{s + \alpha} = \frac{2VZ_2}{Z_1 + Z_2}\left[\frac{1}{s} - \frac{1}{s + \alpha}\right]$$

or
$$v''(t) = \frac{2V \cdot Z_2}{Z_1 + Z_2}\left[1 - \exp\left(-\frac{Z_1 + Z_2}{Z_1 Z_2 C} t\right)\right] \qquad ...(11.66)$$

The variation of the wave is shown in Fig. 11.15.

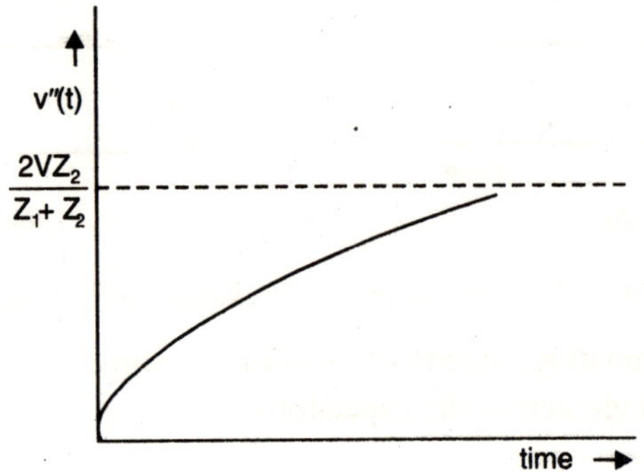

Fig. 11.15 Variation of voltage across the capacitor

We have assumed in the derivation of the expression for voltage across the capacitor in the previous section that the travelling surge is of infinite length. Let us now derive the expression when the surge is of finite duration say τ (Fig. 11.16). Also, let the magnitude of this wave be V units. The wave could be decomposed into two waves.

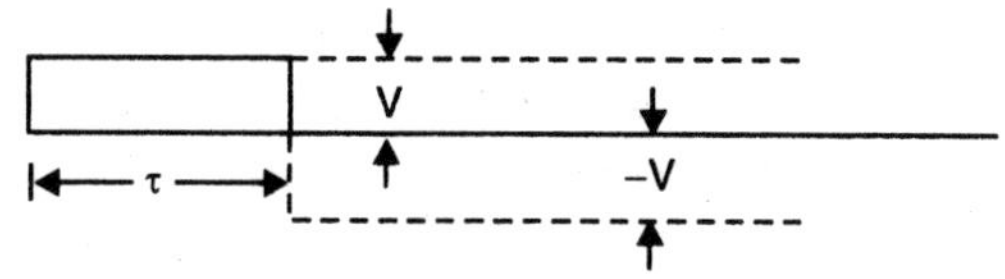

Fig. 11.16 Surge of finite length τ

$$f(t) = Vu(t) - Vu(t-\tau)$$

Here $$Vu(t-\tau) = V \text{ for } t \ge \tau$$
$$= 0 \text{ for } t < \tau$$

With this, voltage across the capacitor is given by

$$V''(s) = \mathcal{L}\{f(t)\} \cdot \frac{2/Z_1}{\frac{1}{Z_1} + \frac{1}{Z_2} + Cs} = \frac{2V/Z_1 s}{\frac{1}{Z_1} + \frac{1}{Z_2} + Cs} - \frac{(2V/Z_1 s) \cdot e^{-\tau s}}{\frac{1}{Z_1} + \frac{1}{Z_2} + Cs}$$

$$v''(t) = 2V \cdot \frac{Z_2}{Z_1 + Z_2}\left[1 - \exp\left(-\frac{Z_1 + Z_2}{Z_1 Z_2 C} t\right)\right] - \frac{2VZ_2}{Z_1 + Z_2}\left[1 - \exp\left\{-\frac{Z_1 + Z_2}{Z_1 Z_2 C}(t - \tau)\right\}\right]$$

The variation of voltage is shown in Fig. 11.17.

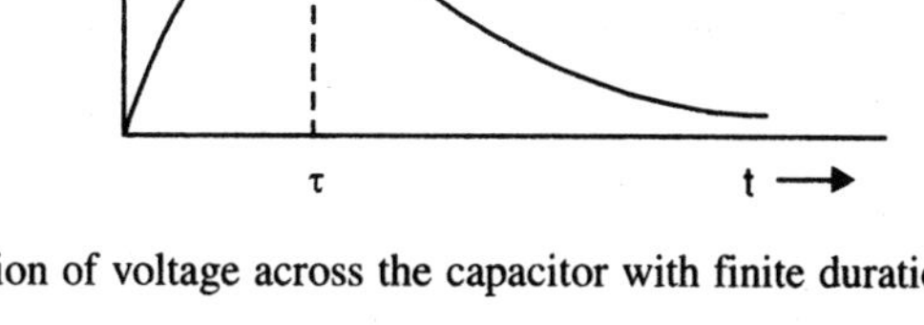

Fig. 11.17 Variation of voltage across the capacitor with finite duration incident surge

Thus for time $0 < t < \tau$ only the first term in the expression is active and for $t \ge \tau$ both the terms are active. The rise in voltage is maximum at $t = \tau$ when the value will be

$$v''(t) = \frac{2VZ_2}{Z_1 + Z_2}\left[1 - \exp\left(-\frac{Z_1 + Z_2}{Z_1 Z_2 C}\tau\right)\right] - \frac{2VZ_2}{Z_1 + Z_2}\left[1 - e^0\right]$$

$$= \frac{2VZ_2}{Z_1 + Z_2}\left[1 - \exp\left(-\frac{Z_1 + Z_2}{Z_1 Z_2 C}\tau\right)\right] \quad ...(11.67)$$

It is, therefore, clear that the attenuation in the magnitude of voltage for a short wave is much more rapid than for long wave.

We have seen that the effect of a shunt capacitor is to reduce the steepness and magnitude of the wave reaching an equipment. Since an inductor is dual to a capacitor, an inductor in series of the lines should give the same effect.

Example 11.8. A 500 kV 2 μsec rectangular surge on a line having a surge impedance of 350 ohms approaches a station at which the concentrated earth capacitance is 3000 pF. Determine the maximum value of the transmitted wave.

Solution. The diagram corresponding to the problem is as follows:

The maximum value of voltage will be

$$E'' = 2E\left[1 - \exp\left(-\frac{\tau}{ZC}\right)\right]$$

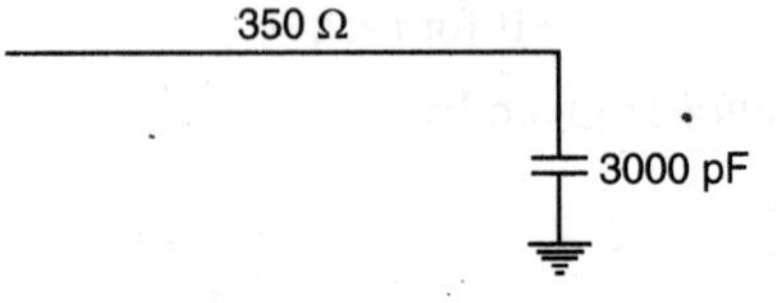

Fig. E 11.8

$$= 2\times 500\left[1 - \exp\left(-\frac{2\times 10^{-6}\times 10^{12}}{350\times 3000}\right)\right]$$

$$= 2\times 500\left[1 - \exp\left(-\frac{2\times 10^{3}}{350\times 3}\right)\right]$$

$$= 2\times 500\left[1 - e^{-1.9}\right]$$

$$= 2\times 500\left[1 - 0.15\right]$$

= 850 kV **Ans.**

Example 11.9. An inductance of 800 μH connects two sections of a transmission line each having a surge impedance of 350 ohms. A 500 kV 2 μs rectangular surge travels along the line towards the inductance. Determine the maximum value of the transmitted wave.

Solution. The maximum value of the transmitted surge is given by

$$E'' = E\left[1 - \exp\left(-\frac{2Z}{L}\tau\right)\right].$$

$$= 500\left[1 - \exp\left(-\frac{2\times 350}{800}\times 2\right)\right]$$

$$= 500\left[1 - e^{-0.875\times 2}\right]$$

$$= 500\left[1 - e^{-1.750}\right]$$

$$= 500[1 - 0.173]$$

$$= 413.5 \text{ kV} \quad \textbf{Ans.}$$

11.8 VOLTAGE STANDING WAVE RATIO (VSWR)

The abbreviation VSWR stands for voltage standing wave ratio: When a wave suffers reflection at the other end of the line, the incident wave and reflected wave are travelling on the same line. As a result, there will be superposition of the two voltages. The superposition of the two waves may form a combination of a standing wave and a travelling wave. Depending upon the terminal condition *i.e.* whether the terminal point is open or shorted or it has impedance greater than the impedance of the line or less than the impedance of the line over which the wave is incident. We define voltage standing wave ratio as

$$VSWR = \frac{V_{max}}{V_{min}} = \frac{I_{max}}{I_{min}} \quad \text{...(11.68)}$$

Where V_{max} is the maximum amplitude of standing voltage wave and is given as $V_{oi} + V_{or} = V_{max}$ where V_{oi} is the incident voltage and V_{or} is the reflected voltage

Hence
$$VSWR = \frac{V_{max}}{V_{min}} = \frac{V_{oi} + V_{or}}{V_{oi} - V_{or}}$$

$$= \frac{1 + \dfrac{|V_{or}|}{|V_{oi}|}}{1 - \dfrac{|V_{or}|}{|V_{oi}|}} = \frac{1 + \rho_v}{1 - \rho_v} \quad \text{...(11.69)}$$

Where ρ_v is the coefficient of reflection for voltage.

Therefore,
$$VSWR = \frac{1 + |\rho_v|}{1 - |\rho_v|} \quad \text{...(11.70)}$$

From (11.70), the coefficient of reflection for voltage is given as

$$\rho_v = \frac{VSWR - 1}{VSWR + 1}$$

From this it is clear that VSWR is an indication of the quality of termination *i.e.* how the line is terminated. If a line is perfectly matched, no reflection takes place and hence $\rho_v = 0$ and VSWR = 1. However, if the wave is fully reflected *i.e.* $\rho_v = 1$, VSWR $= \dfrac{1+1}{1-1} = \infty$ *i.e.* VSWR is infinity. If the impedance of the line and the terminating impedance are complex, then reflection coefficient for voltage is given as

$\rho_v = \dfrac{Z_L - Z_0}{Z_L + Z_0}$ where Z_0 is the impedance of the line. It is to be noted that while evaluating VSWR, only magnitude of the reflection coefficient is to be used and in that case VSWR is same whether reflection coefficient for certain termination is positive or negative but has the same magnitude.

Figure 11.18 shows a terminated line with incident and reflected waves.

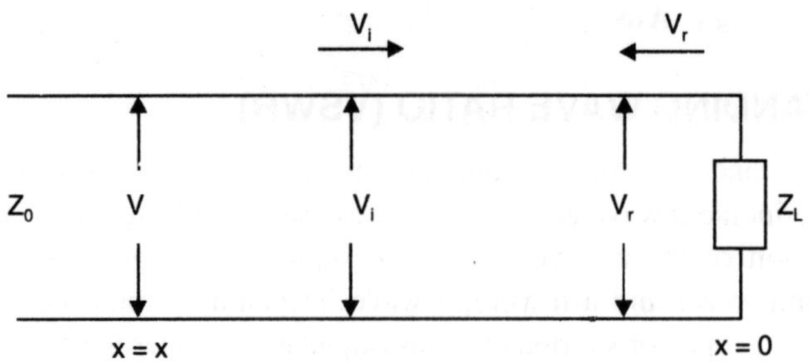

Fig. 11.18 Terminated

Here V_i is the incident voltage, V_r the reflected voltage and V as resultant voltage *i.e.* $V = V_i + V_r$ at a distance x from the load impedance Z_L and Z_0 is the characteristic impedance of the line.

$$V_i = |V_i|e^{\gamma x},\ V_r = |V_r|e^{-vx + j\xi}$$

where ξ is phase shift at the load.

Therefore, at the load point $x = 0$, we have $V_i = |V_i|$ and $V_r = |V_r|e^{j\xi}$

Therefore, at the load point the ratio of reflected and the incident voltages is

$$\frac{V_r}{V_i} = \frac{|V_r|}{|V_i|}\underline{|\xi} = \rho_v$$

Therefore, $$V = V_i\left(e^{\gamma x} + e^{-\gamma x}\cdot\rho_v\right) \qquad ...(11.71)$$

Similarly, the resultant current I at a point on the line is the sum of I_i and I_r where

$$I_i = |I_i|e^{\gamma x - i\delta} \qquad I_r = |I_r|e^{-vx + j(\xi - \delta)}$$

where δ is the phase difference between current and voltage. At this load point, the ratio of reflected and incident current is

$$\frac{I_r}{I_i} = \frac{|I_r|}{|I_i|}\underline{|\xi} = \rho_i$$

Thus, $$I = I_i + I_r = I_i e^{-j\delta}\left(e^{\gamma x} + \rho_i e^{-\gamma x}\right) \qquad ...(11.72)$$

We know that $\rho_v = -\rho_i$

Let Z_x be the impedance of the line at any point x when looking towards the load. Z_x is the ratio of V to I at that point at a distance x from the load. Hence using equation (11.71) and (11.72) we have

$$Z_x = \frac{V}{I} = \frac{|V_i|}{|I_i|}\underline{|\delta}\ \frac{e^{\gamma x} + \rho_v e^{-\gamma x}}{e^{\gamma x} - \rho_v e^{-\gamma x}}$$

Substituting for $\rho_v = \dfrac{Z_L - Z_0}{Z_L + Z_v}$ and $\rho_v = -\rho_i$ we have after simplification

$$Z_x = Z_0 \frac{Z_L + Z_0 \tanh \gamma x}{Z_0 + Z_L \tanh \gamma x} \qquad \text{...(11.73)}$$

This is the general expression for the line impedance Z_x as a function of the distance x from the load.

Next we consider the following conditions at the terminating point (*i*) Open Circuited (*ii*) Short Circuited.

(*i*) When load end is open circuited $Z_L = \infty$, hence equation (11.73) reduces to

$$Z_{sc} = Z_x = \frac{Z_0}{\tanh \gamma x} = z_0 \coth \gamma x \qquad \text{...(11.74)}$$

(*ii*) If line is short circuited $Z_L = 0$ hence equation (11.73) reduces to

$$Z_{sc} = Z_x = Z_0 \tanh \gamma x \qquad \text{...(11.75)}$$

It is to be noted that γ in general is complex and thus

$$\tanh \gamma x = \frac{\sinh \alpha x \cos \beta x + j \cosh \alpha x \sin \beta x}{\cosh \alpha x \cos \beta x + j \sinh \alpha x \sin \beta x}$$

$$= \frac{\tanh \alpha x + j \tan \beta x}{1 + j \tanh \alpha x \tan \beta x}$$

Next we find the product of Z_{oc} and Z_{sc} from equation (11.74) and (11.75) we have

$$Z_{oc} Z_{sc} = Z_0 \cosh \gamma x \cdot Z_0 \tanh \gamma x$$

$$= Z_0^2$$

or

$$Z_0 = \sqrt{Z_{sc} Z_{oc}} \qquad \text{...(11.76)}$$

For the lossless line $\alpha = 0$ and $\tan h\gamma x = j \tan\beta x$ and hence equation (11.73) which is an expression for Z_x for a general line, reduces to

$$Z_x = Z_0 \frac{Z_L + jZ_0 \tan \beta x}{Z_0 + jZ_L \tan \beta x} \qquad \text{...(11.77)}$$

Equation (11.77) gives impedance of the lossless line. It is to be noted that when the line is O.C. or S.C., its equivalent Z_x is a pure reactance. The impedance relations developed above apply to all uniform two conductor lines e.g. co-axial lines, strip lines, micro strip lines and two wire lines. They give the input impedance Z_x of a uniform transmission line if length x and characteristic impedance Z_o terminated in load Z_L. Z_x is also known as wave impedance.

From equation (11.56) we have the expression for incident power as

$$P_i = \frac{1}{2} \frac{|V_i|^2}{|Z_0|} e^{-2\alpha x} \cos \theta$$

However, the reflected power can be obtained by substituting V_r for V_i and we thus have

$$P_r = \frac{1}{2}\frac{|\rho_v V_i|^2}{|Z_0|} e^{-2\alpha x} \cos\theta$$

Therefore, the fraction of the incident power that is transmitted into the load or dissipated in the load is given as

$$\frac{P_t}{P_i} = 1 - |\rho_v|^2 \qquad \ldots(11.78)$$

Example 11.10. Determine the VSWR when the reflection coefficient for voltage is 0.8.

Solution.
$$VSWR = \frac{1+\rho_v}{1-\rho_v} = \frac{1.8}{0.2} = 9$$

Example 11.11. A line with $Z_0 = 100\ \Omega$ is terminated in a load impedance of $Z_L = (100 - j150)\Omega$. If the incident power is 200 mW, determine the power transmitted into the load.

Solution. The given quantity is $P_i = 200$ mW and the required one is P_t the transmitted power. From equation (11.78) we have

$$\frac{P_t}{P_i} = 1 - \rho_v^2 \quad \text{or} \quad P_t = \left(1 - \rho_v^2\right) P_i$$

So, to find out P_t we need evaluate reflection coefficient

$$\rho_v = \frac{Z_L - Z_0}{Z_L + Z_0} = \frac{100 - j150 - 100}{100 + j150 + 100} = \frac{-j150}{200 + j150} = \frac{150\angle -90°}{250\angle 36.86} = 0.6\angle -126.86$$

$$|\rho_v| = 0.6$$

Hence, $\quad P_t = (1 - 0.6)^2 \times 200 = 128$ mW **Ans.**

Example 11.12. Determine the line impedance Z_0 for a $x/4$ length line to match a load Z_L to a desired value Z_x.

Solution. In general $Z_x = Z_0 \dfrac{Z_L + jZ_0 \tan\beta x}{Z_0 + jZ_L \tan\beta x}$

Now substituting $x = \lambda/4$ we have $\beta x = \dfrac{\beta\lambda}{4}$

and $\beta = \dfrac{2\pi}{\lambda}$ Hence $\beta x = \dfrac{\lambda}{4} \cdot \dfrac{2\pi}{\lambda} = \dfrac{\pi}{2}$ and $\tan\dfrac{\pi}{2} = \infty$

Hence
$$Z_{\lambda/4} = Z_0 \cdot \frac{+jZ_0}{jZ_L} = \frac{Z_0^2}{Z_L}$$

or
$$Z_0 = \sqrt{Z_{\lambda/4} Z_L} \quad \textbf{Ans.}$$

If $Z_L = 100\ \Omega$ and $Z_x = 50\ \Omega$, determine Z_0.

From above $Z_0 = \sqrt{Z_x Z_L} = \sqrt{100 \times 50} = 70.7\,\Omega$ **Ans.**

Example 11.13. A plane wave is lossless medium 1 impringes normally on lossy medium 2. Determine the ratio of transmitted to incident power in terms of the voltage standing wave ratio. Also determine the ratio of reflected to incident power.

Solution. (*i*) Since we know that $\dfrac{P_t}{P_i} = 1 - |\rho_v|^2$

and
$$S = \frac{1+|\rho_v|}{1-|\rho_v|} \quad \text{or} \quad \rho_v = \frac{S-1}{S+1}$$

Hence
$$\frac{P_t}{P_i} = 1 - \left(\frac{S-1}{S+1}\right)^2 = \frac{4S}{(S+1)^2} \quad \textbf{Ans.}$$

(*ii*) Now $\dfrac{P_r}{P_i} = 1 - \dfrac{P_t}{P_i} = 1 - 1 + \rho_v^2 = \rho_v^2 = \left(\dfrac{S-1}{S+1}\right)^2$ **Ans.**

Example 11.14. The electric field $\bar{E}$ of a plane wave in a lossless dielectric is given by $E = 25 \cos(10^9 t - 25z)\ \bar{a}_y$ V/m. Determine the (*i*) amplitude (*ii*) the frequency (*iii*) the direction in which the wave is travelling (*iv*) wavelength and (*v*) phase velocity of the wave.

Solution. The general expression for $E = E_m \cos(\omega t - \beta z)\ \bar{a}_y$

Hence

(*i*) The amplitude is E_m V/m

(*ii*) The frequency $f = \dfrac{10^9}{2\pi} = 159.15$ MHz.

(*iii*) The direction of motion of the wave is along positive z-axis as the argument is $(\omega t - \beta z)$.

(*iv*) Wavelength $\lambda = \dfrac{2\pi}{\beta} = \dfrac{2\pi}{25} = 0.2512$ m

(*v*) Phase velocity $V_p = \lambda f = 0.2512 \times 159.15 \times 10^6 = 3.99 \times 10^7$ m/sec $= 4 \times 10^7$ ms^{-1}.

Example 11.15. A transmission line with characteristic resistance R_e is connected at one end to a d.c. source of voltage V_S and internal resistance R_c and the other end it has a load resistance R_L. The d.c. source is connected to the line at $t = 0$. Determine (*i*) V_b (reflected voltage at the first-reflection). (*ii*) The steady state voltage across the load.

Solution. Let reflection coefficient at the source junction be ρ_S and that at the load end be ρ_L.

The first incident voltage at the load point will be

$$V_{fL} = \frac{R_c V_s}{R_c + R_s}$$

(*i*) Hence first reflected voltage will be $V_{fL}\,\rho_L$

(*ii*) and under steady state condition the voltage at any point is given by the superposition of the first forward (incident) first reflected, second forward, second reflected wave and so on. This requires

$$V_L = V_{f1} + V_{b1} + V_{f2} + V_{b2} + \ldots$$

$$= V_{f1}\left[1 + \rho_L + \rho_s\rho_L + \rho_s\rho_L^2 + \rho_s^2\rho_L^2 + \rho_{s2}^2\rho_L^3 + \ldots\right]$$

$$= V_{f1}\left[\left(1 + \rho_s\rho_L + \rho_s^2\rho_L^2 + \ldots\right) + \rho_L\left(1 + \rho_s\rho_L + \rho_s^2\rho_L^2 + \ldots\right)\right]$$

Since ρ_S and ρ_L are smaller than unity

$$V_L = V_{f1}\left[\frac{1}{1-\rho_s\rho_L} + \frac{\rho_L}{1-\rho_s\rho_L}\right] = V_{f_1}\left(\frac{1+\rho_L}{1-\rho_s\rho_L}\right) \textbf{ Ans.}$$

Example 11.16. If in example 11.14 $R_C = 100\Omega$, $R_L = 150\ \Omega$, $R_S = 50\ \Omega$ and $V_S = 200$ volts. Determine V_L under steady state condition. Also determine the steady state load current.

Solution. The incident voltage travelling from transmission line towards the source is

$$V_{f1} = \frac{200 \times 100}{100 + 50} = \frac{200 \times 100}{150} = 133.3 \text{ volt}$$

$$\rho_s = \frac{100 - 50}{150} = 0.33$$

$$\rho_L = \frac{150 - 100}{250} = 0.2$$

$$V_L = 133.3\left[\frac{1+\rho_L}{1-\rho_s\rho_L}\right] = 133.3 \times \frac{1.2}{1 - 0.33 \times 0.2}$$

$$= \frac{133.3 \times 1.2}{0.934}$$

$$= 171.26 \text{ volts } \textbf{Ans.}$$

The steady state load current is

$$I_L = \frac{V_L}{R_L} = \frac{171.26}{150} = 1.14 \text{ A } \textbf{Ans.}$$

Example 11.17. A common type of transmission line for microwave application has an open circuit impedance of $150\underline{|75°}\,\Omega$ and a short circuit impedance of $50\underline{|-30°}\,\Omega$. Determine the characteristic impedance of the line.

Solution. Since $Z_C = \sqrt{Z_{oc}Z_{sc}}$

$$= \sqrt{150 \times 50\underline{|-5°}} = 86.6\underline{|-2.5°} \textbf{ Ans.}$$

Example 11.18. Determine the impedance on a 100 Ω transmission line at a distance of $\frac{\lambda}{8}$ from a 400 Ω load.

Solution. Using equation 11.77, we have

$$Z_x = Z_0 \frac{Z_L + jZ_0 \tan \beta x}{Z_0 + jZ_L \tan \beta x}$$

Since $$\frac{\lambda\beta}{8} = \frac{2\pi}{8} = \frac{\pi}{4} = \beta x$$

Hence $$Z_x = 100 \frac{400 + j100 \tan \pi/4}{100 + j400 \tan \pi/4} = \frac{400 + j100}{1 + j4}$$

$$= \frac{400 + j100}{1 + j4} = \frac{4.123\underline{|14^\circ} \times 100}{4.123\,\underline{|75.9}}$$

$$= 100\,\underline{|-61.96^\circ}. \textbf{ Ans.}$$

Example 11.19. A lossless 50 ohm transmission line is terminated in (35 + j 65) ohm. Determine (*i*) Voltage reflection coefficient ρ_V (*ii*) VSWR (*iii*) Impedance Z_x at 0.35λ from load.

Solution. (*i*) $$\rho_v = \frac{Z_L - Z_0}{Z_L + Z_0} = \frac{35 + j65 - 50}{35 + j65 + 50} = \frac{-15 + j65}{85 + j65}$$

$$= \frac{-3 + j13}{17 + j13} \cdot \frac{13.34\underline{|103}}{21.4\underline{|37}} = 0.623\underline{|66^\circ} \quad \textbf{Ans.}$$

(*ii*) $$VSWR = \frac{1 + |\rho_v|}{1 - \rho_v} = \frac{1.623}{0.377} = 4.3 \quad \textbf{Ans.}$$

(*iii*) $$Z_x = 50 \left[\frac{(35 + j65)\cos 126 - j50 \sin 126}{50 \cos 126 - (35 + j65)\sin 126} \right]$$

since $$\beta x = \frac{2\pi}{\lambda} \cdot 0.35\lambda = 0.7\pi \text{ or } \beta x = 0.7 \times 180 = 126^\circ$$

$$= 50 \times \frac{(35 + j65)(-0.58778) - j50 \times 0.809}{50(-0.58778) + j(35 + j65)(0.809)}$$

$$= 50 \times \frac{-205.7 - j38.20 - j40.45}{-29.39 + j28.315 - 52.58} = 50 \times \frac{-20.57 - j2.75}{-81.97 + j28.315}$$

$$= \frac{(-20.57 + j2.25)\ (-81.97 - j28.315)}{(-81.97 + j28.315)\ (-81.97 - j28.315)}$$

$$= 50 \times \frac{(-20.57 + j2.25)(-81.97 - j28.315)}{7520}$$

$$= 50 \times \frac{1686 + 63.7 + j582.44 - j184.4}{7520} = 11.63 + j2.66\,\Omega \ \textbf{Ans.}$$

PROBLEMS

P 11.1 Explain with a neat diagram what you mean by Pulse forming networks.

P 11.2 Develop transmission line equations for determining voltage and current at any time and at any point of the transmission line.

P 11.3 What are the telegraphists equation? Starting with telegraphists equation develop general wave equations for transmission line.

P 11.4 Using wave equations for a loss line, explain the concept of forward component of wave and backward component of wave on the line and show that the velocity of both the components is given by $v = 1/\sqrt{LC}$.

P 11.5 Explain the terms phase velocity and phase constant and give relation between the two quantities.

P 11.6 Write forward and backward component of voltage wave solution in terms of cosine function and explain their significance.

P 11.7 Starting with the general transmission line equation obtain its solution in phasor form.

P 11.8 Differentiate between phase velocity and group velocity of wave. Explain the concept of distortionless lines.

P 11.9 Derive an expression for the characteristic impedance of a low-loss line.

P 11.10 Starting from first principle derive an expression for power transmitted over a lossy transmission line. Also derive an expression for power loss when a wave has travelled through a distance z over the line.

P 11.11 Derive expression for coefficient of reflection and transmission whenever a wave looks into a change in impedance. Derive an expression for both voltage and currents and derive relation between the coefficients.

P 11.12 Explain what do you mean by Voltage Standing Wave Ratio. Discuss clearly its significance. Derive a relation between VSWR and reflection coefficient and hence express reflection coefficient in terms of VSWR.

P 11.13 Assuming load to be located at $x = 0$, derive an expression for Z_x the impedance of the line at any point x when looking towards the load in terms of the load and characteristic impedance of the line and parameters of the line.

P 11.14 Derive expression for Z_x when (*i*) line is open circuited (*ii*) Line is short circuited (*iii*) When the line is lossless. Show that

$$Z_0 = \sqrt{Z_{oc} Z_{sc}}\,.$$

P 11.15 Show that the ratio of transmitted power to the incident power at the load is given as $\frac{P_t}{P_i} = 1 - |\rho_v|^2$.

P 11.16 An overhead transmission line with characteristic impedance 400 ohms is 300 kms long. One end of the line is short circuited and at the other end a source of 11 kV is suddenly switched in. Calculate the current at the source end 0.005 seconds after the voltage is applied.

Ans. 79.38 A

P 11.17 An overhead line with surge impedance 400 ohm bifurcates into two lines of surge impedance 400 ohms and 40 ohms respectively. If a surge of 20 kV is incident on the overhead line, determine the magnitudes of voltage and current which enter the bifurcated line.

Ans. 3.33 kV, 8.33 A, 83.33 A

P 11.18 A co-axial line has an open circuit impedance $Z_{OC} = 105\angle 17°$ and a short circuit impedance $Z_{SC} = 24\angle -3°$ ohms. Determine the characteristic impedance of the line.

Ans. $75\angle -5°$ ohms

P 11.19 A signal of 100 volt is applied to a 50 ohm co-axial transmission line terminated in a 200 ohm load. Determine (*i*) the coefficient of reflection for voltage (*ii*) the magnitude of the reflected voltage and (*iii*) the magnitude of the reflected current.

Ans. (*i*) 0.6 (*ii*) 60 V (*iii*) 1.2 A

P 11.20 Two transmission lines are to be joined end to end. Line 1 is 30 m long and is rated at 0.1dB/m. Line 2 is 45 m long and is rated at 0.15 dB/m. The joint is not perfect and has a loss of 3 dB. Determine the percentage of input power that reaches the output of the combination.

Ans. 5.3%

P 11.21 A 10 m section of lossless transmission line having $R_C = 50\ \Omega$ and $V = 200$ m/μs^{-1} is connected to a voltage source $V_S = 100\angle 0°$, 26 MHz and internal impedance $Z_S = 50 + j\,0\ \Omega$. If the line is terminated in a load of impedance $100 + j\,50\ \Omega$, determine the input impedance of the line.

Ans. $Z_{in}(0) = 19.21 + j\,3.52\ \Omega$

P 11.22 A transmission line with characteristic impedance 30 Ω has a load of 90 Ω connected to it. The source voltage is 100 V *d.c.* with internal resistance 60 Ω. The source is connected to the line through a switch. Determine the steady state voltage across the load when the switch is closed at $t = 0$.

Ans. $V_L = 60$ V

P 11.23 A transmission line with characteristic impedance Z_0 is terminated with a complex load $Z_L = \sqrt{2}z_0 e^{j\pi/4}$. At what distances from the load (in wavelength) is the impedance a pure resistance?

Ans. $l = 0.089\ \lambda$ plus $n\lambda/4$

P 11.24 Replacing the unknown load of a 100 ohm transmission line with an open circuit yields an input impedance to the line of $-j\,80$ ohms. With Z_L attached, the input impedance is $30 + j\,40$ ohm. Determine the load impedance Z_L.

Ans. $Z_L = 322 - j\,486$ ohm.

P 11.25 A 0.8 km long line operating at 5 kHz has characteristic impedance of $94\angle -23.2°$ ohm and total attenuation of 0 06 nP and the phase shift between input and output is 8°. Determine *R*, *L*, *G* and *C* per km for the line, the phase velocity on the line and the power lost on the line when the sending end power is 3 W and the load is matched.

Ans. $R = 13\ \Omega$ km^{-1}, $L = 393.75$ μHm^{-1} $G = 0$, $i = 0.0644$ μF km^{-1}, $V_p = 7 \times 10^4$ km s^{-1} $\rho_L = 0.34$ watts

P 11.26 A 600 ohm transmission line is 150 m long operates at 400 kHz with $\alpha = 2.4 \times 10^{-3}$ nPm^{-1} and $\beta = 0.0212$ rad m^{-1} and supplies a load impedance $Z_L = 424.3\ \angle 45°$ ohm. Determine the length of the line in wavelengths, ρ_R, ρ_S and Z_S. For a received voltage $V_R = 50\angle 0° \, V$ determine V_S.

Ans. length 0.51 λ, $\rho_R = 0.45 \angle 116.6°$, $\rho_S = 0.22 \angle 113.6°$

$Z_S = 502.7 \angle 22.8°$ $V_S = 75 \angle 167.3°$ V.

P 11.27 For a parallel-wire line with $R = 2.59 \times 10^{-3}$ ohm m^{-1} $L = 2$ μH m^{-1}, $C = 5.56$ pF m^{-1} and $G = 0$, determine the characteristic impedance, propagation constant, velocity of propagation and wavelength for operation at 5 kHz.

Ans. $Z_0 = 600\angle -12$ ohm, $\alpha = 2.19 \times 10^{-6}$ nP m^{-1}

$\beta = 1.048 \times 10^{-4}$ rad m^{-1}, $V_P = 2.998 \times 10^8$ m s^{-1}, $\lambda = 59.96$ km.

P 11.28 A 50 ohm lossless line is terminated by a load impedance $Z_L = 50 - j\,75\ \Omega$. If the incident power is 100 mW, determine the power dissipated by the load. **Ans.** 64 mW.

P 11.29 Two characteristics of a certain lossless transmission line are $Z_0 = 50\ \Omega$ and $\gamma = 0 + j\,0.2\,\pi$m^{-1} at $f = 60$ MHz.

(*a*) Determine L and C for the line (*b*) A load $Z_L = 60 + j\,80$ ohm is located at $z = 0$. Determine the shortest distance from the load to a point at which the impedance is resistant.

Ans. (*a*) 83.3 nH m^{-1}, 33.3 pF m^{-1} (*b*) 65 cms.

P 11.30 A lossy cable has $R = 2.25\ \Omega$m^{-1}, $L = 1$ μHm^{-1} $C = 100$ pFm^{-1} and $G = 0$ at 500 MHz. Determine (*i*) the attenuation constant of the line at 500 MHz. Also determine (*ii*) the attenuation per 30.5 m of length in decibels. (*iii*) Determine the input impedance to a 10 m section of the cable at 500 MHz, assuming that the load impedance is an (*a*) Open circuit, (*b*) a short circuit.

Ans. (*i*) $\alpha = 1.125 \times 10^{-2}$ nP m^{-1} (*ii*) 3 dB (*iii*) (*a*) $Z_{in} = 892.56 \angle 0°$ (*b*) $11.24 \angle 0°$ ohms.

Chapter

12

The Uniform Plane Wave

12.1 INTRODUCTION

The application of Maxwell's equations to the solution of electromagnetic wave propagation has been studied. In an electromagnetic wave a changing electric field produces a changing magnetic field which in turn generates an electric field and so on, with a resulting propagation of energy. The uniform plane wave represents the simplest case yet it is of great practical importance for waves normally encountered.

We have studied uniform plane waves propagating a transmission in the Chapter 11. We have seen how voltages and currents propagate as waves in transmission lines and we know the existence of electric and magnetic fields. So here, transmission line is the structure where the fields are confined and it allows them to travel along its length as waves. However, it is to be noted that even when transmission line does not exist, fields still exist and will propagate in free space and so they may assume any magnitude and direction.

12.2 THE WAVE EQUATION

While considering electromagnetic wave equations in free space following assumptions are made

(*i*) The free charge density in the field is zero *i.e.* $\rho_v = 0$.

(*ii*) The medium is assumed to be both magnetically and electrically homogeneous and isotropic. Subject to these conditions, Maxwell's equation may be given in terms of E and H as

$$\nabla \times H = \varepsilon_0 \frac{\partial E}{\partial t} \quad \text{...(12.1)}$$

$$\nabla \times E = -\mu_0 \frac{\partial H}{\partial t} \quad \text{...(12.2)}$$

$$\nabla \cdot E = 0 \quad \text{...(12,3)}$$

$$\nabla \cdot H = 0 \quad \text{...(12.4)}$$

From equation (12.1) it is clear that if electric field E is changing with time at some point (space) then magnetic field H has curl at that point, therefore, H varies spatially in a direction normal to its orientation (as it is cross product). Also, if E is changing with time then H will also change with time though not necessarily in the same way as E. Similarly from equation (12.2) it is found that a time varying H generates E which having curl, varies spatially in the direction normal to its orientation (as it is cross product $\nabla \times E$).

From the above it is clear that if a set of initial values of E and H at all points in the electromagnetic field are given, their values at all future times may be found. Also, if values of E and H are specified as a function of time over any surface e.g. the surface of the radio antenna, it is possible to find their values at all other points in the field. Summarizing the above, the solution of the Maxwell's equation subject to appropriate initial and boundary condition will provide the following answers:

(*i*) Given a set of initial values of E and H at all points in an electromagnetic field, to find their values at all future times.

(*ii*) Given E and H as functions of time over any surface, to find their values at all other points in the field.

We assume the existence of a uniform plane wave wherein both fields E and H lie in the transverse plane *i.e.* the plane whose normal is the direction of propagation as shown in Fig. 12.1.

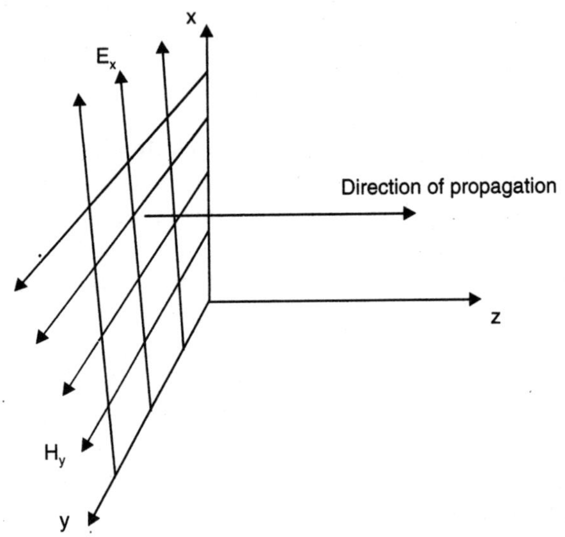

Fig. 12.1 Transverse Electromagnetic (TEM) waves propagating in the z direction with E in x-direction and H in y direction

Also, both the fields are constant in magnitude in the transverse plane. For this reason such a wave is called Transverse Electromagnetic (TEM) wave. The desired spatial variation of both fields in the direction normal to their orientation will occur only in the direction of travel or normal to the transverse plane (as it is $E \times H$).

Now writing $E \times H$ in full form we have in rectangular form

$$\nabla \times H = \left(\frac{\partial H_z}{\partial y} - \frac{\partial H_y}{\partial z}\right)\bar{a}_x + \left(\frac{\partial H_x}{\partial z} - \frac{\partial H_z}{\partial x}\right)\bar{a}_y + \left(\frac{\partial H_y}{\partial x} - \frac{\partial H_x}{\partial y}\right)\bar{a}_z$$

Since all the partial derivative of H are zero except $\frac{\partial H_y}{\partial z}\bar{a}_x$, we thus have

$$\nabla \times H = -\frac{\partial H_y}{\partial z}\bar{a}_x$$

Now $\nabla \times H = \varepsilon_0 \dfrac{\partial E}{\partial t}$ and since E is along x-direction only, therefore

$$\nabla \times H = -\frac{\partial H_y}{\partial z}\bar{a}_x = \varepsilon_0 \frac{\partial E_x}{\partial t}\bar{a}_x$$

or
$$\frac{\partial H_y}{\partial z} = -\varepsilon_0 \frac{\partial E_x}{\partial t} \qquad \text{...(12.5)}$$

And using equation (12.2), we can write

$$\nabla \times E = \frac{\partial E_x}{\partial z} a_y = -\mu_0 \frac{\partial H_y}{\partial t}\bar{a}_y \qquad \text{...(12.6)}$$

Equations (12.5) and (12.6) are rewritten as

$$\frac{\partial H_y}{\partial z} = -\varepsilon_0 \frac{\partial E_x}{\partial t} \qquad \text{...(12.7)}$$

and
$$\frac{\partial E_x}{\partial z} = -\mu_0 \frac{\partial H_y}{\partial t} \qquad \text{...(12.8)}$$

These equations can be compared with the telegraphists' equations of Chapter 11.

Now differentiating (12.8) with respect to z and (12.7) and (12.8) with respect to t, we have

$$\frac{\partial^2 E_x}{\partial z^2} = -\mu_0 \frac{\partial^2 H_y}{\partial z \partial t} \qquad \text{...(12.9)}$$

and
$$\frac{\partial^2 H_y}{\partial z \partial t} = -\varepsilon_0 \frac{\partial^2 E_x}{\partial t^2} \qquad \text{...(12.10)}$$

Substituting equation (12.10) into equation (12.9), we have

$$\frac{\partial^2 E_x}{\partial z^2} = \mu_0 \varepsilon_0 \frac{\partial^2 E_x}{\partial t^2} \qquad \text{...(12.11)}$$

Equation (12.11) is analogous to equation (11.11). We identify this equation as the wave equation for the x-polarized TEM electric field in free space and from this equation we identify the propagation velocity.

$$V = \frac{1}{\sqrt{\mu_0 \varepsilon_0}} = \frac{1}{\sqrt{4\pi \times 10^{-7} \times \dfrac{1}{36\pi} \times 10^{-9}}} = 3 \times 10^8 \text{ ms}^{-1}$$

which is the velocity of light.

Similarly, if we differentiate equation (12.8) with t and equation (12.7) with z and simplifying further as we did with equation (12.9) and (12.10) we have the wave equation for the magnetic field as

$$\frac{\partial^2 H_y}{\partial z^2} = \mu_0 H_0 \frac{\partial^2 H_y}{\partial t^2} \qquad \text{...(12.12)}$$

The solution of equations (12.11) and (12.12) is sought on the same line as we did for equation (11.11). The solution is of general form and will have forward and backward propagation in general form. For equation (12.11) the general solution is

$$E_x(z, t) = f_1\left(t - \frac{z}{v}\right) + f_2\left(t + \frac{z}{v}\right) \quad \text{...(12.13)}$$

where f_1 and f_2 can be any function whose argument is for $\left(t \pm \frac{z}{v}\right)$. We assume cosine function of a particular frequency and write the solution to (12.13) in the form of forward and backward propagating cosines. Since the waves are sinusoidal we denote their velocity as the phase velocity V_p. The solution is given as

$$E_x(z, t) = E_{xf}(z, t) + E_{xb}(z, t)$$

$$= |E_{mf}|\cos\left\{\omega\left(t - \frac{z}{v_p}\right) + \phi_1\right\} + |E_{mb}|\cos\left\{\omega\left(t + \frac{z}{v_p}\right) + \phi_2\right\} \quad \text{...(12.14)}$$

$$= |E_{mf}|\cos(\omega t - k_0 z + \phi_1) + |E_{mb}|\cos\{(\omega t + k_0 z) + \phi_2\} \quad \text{...(12.15)}$$

The first part of the solution is for the forward wave whereas the second part represents the backward component of the wave. Since these waves are travelling in free space wherein the phase velocity is $V_p = c$ the velocity of light. Also K_0 is the wave number and is given as $K_0 = \frac{\omega}{c}$ rad s^{-1}. Here K_0 is represented as a spatial frequency which in the present situation measures the phase shift per unit distance along the z-direction and is measured in rad m^{-1}. K_0 is also the phase constant for lossless propagation of uniform plane waves in free space. The wavelength in free space is the distance over which the spatial phase shifts by 2π radians, assuming fixed time or

$$K_0 z = K_0 \lambda = 2\pi \text{ or } \lambda = \frac{2\pi}{k_0} \text{ in free space} \quad \text{...(12.16)}$$

It is to be noted that the solution in equation (12.15) represents the real instantaneous forms of electric field.

We now express the real instantaneous fields of equation (12.15) in terms of their phasor form picking out only the forward moving component from the solution in equation (12.15) we write phasor form as

$$E_x(z, t) = \frac{1}{2}\underbrace{|E_{sf}|e^{j\phi_1}e^{-jk_0 z}}_{E_{x0}} \cdot e^{j\omega t} + c.c. = \frac{1}{2}E_{xs}\,e^{j\omega t} + c.c.$$

$$= \text{Re}\left[E_{xs}\right]e^{j\omega t} \quad \text{...(12.17)}$$

Here E_{x0} is given as $E_{x0} = E_{sf} + e^{j\phi_1}$ and is the complex amplitude. We know that taking the partial derivative of any field quantity with respect to time is equivalent to multiplying the corresponding phasor by $j\omega$ e.g., we can express the equation 12.7 given as

$$\frac{\partial H_y}{\partial z} = -\varepsilon_0 \frac{\partial E_x}{\partial t} \qquad ...(12.18)$$

to

$$\frac{\partial H_y}{\partial z} = -j\varepsilon_0 \omega E_{xs}(z) \qquad ...(12.19)$$

The intermediate step is as follows

$$E_x(z, t) = \frac{1}{2} E_{xs}(z) e^{j\phi} + c.c. \text{ and } H_y(z, t) = \frac{1}{2} H_{ys}(z) e^{j\omega t} + c.c. \qquad ...(12.19a)$$

After substituting the fields in 12.19*a* into 12.18, the latter equation simplifies to equation (12.19). In this process we note that (*i*) the complex conjugate term in (12.19*a*) produce their own equation redundant with (12.19) (*ii*) The $e^{j\omega t}$ factors common to both sides cancel out (*iii*) the partial derivatives are consistent to total derivatives with respect to z as the phasor H depends only on z. Converting the Maxwell's equations (12.1) to (12.4) in phasor form we have

$$\nabla \times H_s = j\omega\, \varepsilon_0\, E_s \qquad ...(12.20)$$

$$\nabla \times E_s = -j\omega\, \mu_0\, H_s \qquad ...(12.21)$$

$$\nabla \cdot E_s = 0 \qquad ...(12.22)$$

$$\nabla \cdot H_s = 0 \qquad ...(12.23)$$

It is to be noted that equations (12.22) and (12.23) are no longer independent relations as these can be determined by taking the divergence of (11.20) and (11.21).

Equations (11.20) to (11.23) can be used to obtain the sinusoidal steady state vector form of wave equations in free space. Taking curl of both sides of equation (11.21) we have

$$\nabla \times \nabla \times E_s = \nabla\, (\nabla . E_s) - \nabla^2 E_s = -j\omega\mu_0 \nabla \times H \qquad ...(12.24)$$

Since $\nabla . E_s = 0$ from equation (12.22), the above equation becomes

$$\nabla^2 E_{xs} = j\omega\mu_0\, \nabla \times H = -\, K_0{}^2 E_{xs} \qquad ...(12.25)$$

after substituting $\nabla \times H = j\omega\varepsilon_0 E_s$. Here $k_0 = \frac{\omega}{c} = \omega\sqrt{\mu_0 \varepsilon_0}$.

Equation (12.25) is known as the vector Helmholtz equation in free space

Writing equation (12.25) in rectangular coordinates we have

$$\frac{\partial^2 E_{xs}}{\partial x^2} + \frac{\partial^2 E_{xs}}{\partial y^2} + \frac{\partial^2 E_{xs}}{\partial z^2} = -k_0^2 E_{xs} \qquad ...(12.26)$$

Since E_{xs} is not a function if x and y, equation (12.26) reduces to

$$\frac{d^2 \cdot E_{xs}}{dz^2} = -k_0^2 E_{xs} \qquad ...(12.27)$$

The solution of equation (12.27) is known to be

$$E_{xs}(z) = E_{x_0 f} e^{-jk_0 z} + E_{x_0 b} e^{jk_0 z} \qquad ...(12.28)$$

To find out H field we refer to equation (12.21) we have

$$\nabla \times E_s = -j\omega\mu_0 H_s$$

Since E_s is only along x-axis, hence only this component will be considered as the partial derivative, if E_s along y and z axis are zero. Similarly since H_s is along y-axis only hence the above equation reduces to

$$\frac{dE_{xs}}{dz} = -j\omega\mu_0 H_{sy} \qquad \text{...(12.29)}$$

Now differentiating (12.28) with respect to z and substituting it in equation (12.29) we have

$$H_{ys} = \frac{1}{j\omega\mu_0}\left[(-jK_0)E_{x_0f}e^{-jk_0z} + (jK_0)E_{x_0b}e^{jk_0z}\right] \qquad \text{...(12.30)}$$

Since $$K_0 = \frac{\omega}{c} = \omega\sqrt{\mu_0\varepsilon_0}$$

$$H_{ys} = E_{x_0f}\sqrt{\frac{\varepsilon_0}{\mu_0}}e^{-jk_0z} - E_{x_0b}\sqrt{\frac{\varepsilon_0}{\mu_0}}e^{jk_0z} \qquad \text{...(12.31)}$$

$$= H_{y_0f}e^{-jk_0z} - H_{y_0b}e^{jk_0z} \qquad \text{...(12.32)}$$

Constituting the above expression (12.31) in real instantaneous form *i.e.* in cosine form we have

$$H_y(z,t) = E_{xf}\sqrt{\frac{\varepsilon_0}{\mu_0}}\cos(\omega t - k_0z) - E_{xb}\sqrt{\frac{\varepsilon_0}{\mu_0}}\cos(\omega t + k_0z) \qquad \text{...(12.33)}$$

Where E_{xf} and E_{xb} are assumed real.

From equations (12.31) and (12.32) we find that the electric and magnetic field amplitudes of the forward propagating wave in free space are related through.

$$E_{x_0f}\sqrt{\frac{\varepsilon_0}{\mu_0}} = H_{y_0f}$$

or $$E_{x_0f} = H_{y_0f}\sqrt{\frac{\mu_0}{\varepsilon_0}} = \eta_0 H_{y_0f} \qquad \text{...(12.34a)}$$

Similarly the backward propagating wave amplitudes are also related

$$E_{x_0b} = H_{y_0b}\sqrt{\frac{\mu_0}{\varepsilon_0}} = -\eta_0 H_{y_0b} \qquad \text{...(12.34b)}$$

Here η_0 is the intrinsic impedance of free space and is given as

$$\eta_0 = \sqrt{\frac{\mu_0}{\varepsilon_0}} = \sqrt{\frac{4\pi\times10^{-7}}{10^{-9}}\cdot 36\pi} = 12\pi\times 10$$

$$= 376.8 \approx 377\,\Omega$$

η_0 is impedance as $\eta_0 = \dfrac{E}{H} = \dfrac{V/m}{I/m} = \dfrac{V}{I} = Z$.

12.3 WAVE PROPAGATION IN DIELECTRICS

We next study the propagation of uniform plane wave through dielectric material with permittivity ε and permeability μ_0. We analyze this problem assuming again as in the previous article that the medium (here dielectric material) is assumed to be homogeneous and isotropic. Homogeneous means μ and ε are constant and isotropic means μ and ε are independent of the orientation of the fields. The Helmholtz equation is

$$\nabla^2 E_s = -k^2 E_s \qquad \ldots(12.35)$$

Where wave number k is a function of μ and ε and is given by

$$k = \omega\sqrt{\mu\varepsilon} = k_0\sqrt{\mu_r \varepsilon_r} \qquad \ldots(12.36)$$

As in equation (12.27) for free space, for medium like a dielectric this equation can be written as

$$\frac{d^2 E_{xs}}{dz^2} = -k^2 E_{xs} \qquad \ldots(12.37)$$

Here the wave number k can be a complex for a dielectric and, therefore, it is referred to as the complex propagation constant. A general solution of equation (12.37) usually results in a complex k, and therefore, it is usually written in terms of its real and imaginary parts as suggested below

$$jk = \alpha + j\beta \qquad \ldots(12.37a)$$

We know that a solution to second order differential equation (12.37) is given as

$$E_{xs} = E_{xf} e^{-jkz} = E_{xf} e^{-\alpha z} . e^{-j\beta z} \qquad \ldots(12.38)$$

To get this solution in cosine form, multiply equation (11.38) by $e^{j\omega t}$ and we consider only the real part of the field and we have

$$E_x = E_{xf} e^{-\alpha z} \cos(\omega t - \beta z) \qquad \ldots(12.39)$$

This expression represents a uniform plane wave that propagates in the forward z-direction with phase constant β, but for positive α, the amplitude of the forward wave decays with its propagation. Thus, the general effect of the complex valued k is to have a travelling wave which changes its amplitude as it propagates in the forward path. If α is positive, it is called the attenuation constant as the amplitude of the wave decays with distance and if α is negative, the amplitude of the wave builds up with the increase in distance and then α is known as gain co-efficient. The latter effect will usually occur in laser amplifiers. Here ofcourse we will consider situations where α will be positive.

The manner in which physical processes in a material can affect the wave electric field can be explained using a complex permittivity of the form.

$$\varepsilon = \varepsilon' - j\varepsilon'' = \varepsilon_0\left(\varepsilon_r' - j\varepsilon_r''\right) \qquad \ldots(12.40)$$

Bound electrons or ion oscillation and dipole relaxation, both of these give rise to complex permittivity and result in wave losses. Also, the permittivity is affected by the conduction of free electrons or holes.

In magnetic field also losses may occur and thus complex permeability may be used instead of simple permeability. The complex permeability is given as

$$\mu = \mu' - j\mu'' = \mu_0\left(\mu_r' - j\mu_r''\right) \qquad \ldots(12.41)$$

Examples of such materials are ferrimagnetic materials or ferrites. Here of course we will avoid using materials with complex permeabilities. Substituting equation (12.40) into equation (12.36) we have

$$k = \omega\sqrt{\mu\varepsilon} = \omega\sqrt{\mu(\varepsilon' - \varepsilon'')} = \omega\sqrt{\mu\varepsilon'}\left(\sqrt{1 - j\frac{\varepsilon''}{\varepsilon'}}\right) \qquad \ldots(12.42)$$

When $\varepsilon'' << \varepsilon'$, the second radical in equation (12.42) becomes real and equals unity. With non zero ε'', k is complex and so losses occur which are evaluated using the attenuation co-efficient α as in 12.37(*a*).

The phase constant β and hence wavelength and phase velocity will also be affected by ε''. α and β are obtained by taking the real and imaginary parts of jk from equation (12.42).

$$\alpha = \text{Re}(jk) = \omega\sqrt{\frac{\mu\varepsilon'}{2}}\left\{\sqrt{1+\left(\frac{\varepsilon''}{\varepsilon'}\right)^2} - 1\right\}^{1/2} \quad \text{...(12.43)}$$

and

$$\beta = \text{Im}(jk) = \omega\sqrt{\frac{\mu\varepsilon'}{2}}\left\{\sqrt{1+\left(\frac{\varepsilon''}{\varepsilon'}\right)^2} + 1\right\}^{1/2} \quad \text{...(12.44)}$$

From equation (12.43) we find that α is non-zero if ε'' is nonzero. The ratio $\frac{\varepsilon''}{\varepsilon'}$ is known as loss tangent $\tan\delta$ where δ is the angle between the displacement current and the total current.

From equation (12.39) phase velocity is given by

$$v_p = \frac{\omega}{\beta} \quad \text{...(12.45)}$$

The wavelength as we know is the distance required to effect a phase change of 2π radians *i.e.*

$$\beta\lambda = 2\pi$$

which gives the fundamental definition of wavelength

$$\lambda = \frac{2\pi}{\beta} \quad \text{...(12.46)}$$

Since we have uniform plane wave the magnetic field is obtained as

$$H_{ys} = \frac{E_{xf}}{\eta} e^{-\alpha z} e^{-j\beta z}$$

where η the intrinsic impedance is a complex quantity and is given as

$$\eta = \sqrt{\frac{\mu}{\varepsilon' - j\varepsilon''}} = \sqrt{\frac{\mu}{1-\varepsilon''/\varepsilon'}}\cdot\sqrt{\frac{\mu}{\varepsilon'}} \quad \text{...(12.47)}$$

Therefore, in a dielectric media the electric and magnetic fields are not in phase. In free space, however, the two fields are in phase.

If the dielectric medium is loss less or it is a perfect dielectric then $\varepsilon'' = 0$ and hence $\varepsilon = \varepsilon'$. From equation (12.43) $\alpha = 0$ and from equation (12.44).

$$\beta = \omega\sqrt{\mu\varepsilon'} \text{ (For lossless medium)} \quad \text{...(12.48)}$$

with $\alpha = 0$, the real electric field becomes

$$E_x = E_{xf}\cos(\omega t - \beta z) \quad \text{...(12.49)}$$

Equation (12.49) represents a forward propagating wave at a phase velocity V_p, where

$$V_p = \frac{\omega}{\beta} = \frac{1}{\sqrt{\mu\varepsilon'}} = \frac{c}{\sqrt{\mu_r\varepsilon'_r}} \quad \text{...(12.50)}$$

The wavelength for the wave is

$$\lambda = \frac{2\pi}{\beta} = \frac{2\pi}{\omega\sqrt{\mu\varepsilon'}} = \frac{1}{f\sqrt{\mu\varepsilon'}} = \frac{c}{f\sqrt{\mu_r\varepsilon_r'}} = \frac{\lambda_0}{\sqrt{\mu_r\varepsilon_r'}} \text{(lossless medium)}$$

where λ_0 is the wave'ength in free space. Since μ_r, $\varepsilon'_r > 1$ the phase velocity and the wavelength in media are smaller as compared to free space.

Associated with electric field E_x, H_y is the magnetic field intensity and is given as

$$H_y = \frac{E_{xf}}{\eta}\cos(\omega t - \beta z) \qquad ...(12.51)$$

where η, the intrinsic impedance is given as $\eta = \sqrt{\frac{\mu}{\varepsilon}}$

We next consider the conducting materials where $J = \sigma E$. Where σ is the conductivity of the material. We next consider Maxwell's curl equation (12.20) and using equation (12.40) we have

$$\nabla \times H_s = j\omega(\varepsilon' - j\varepsilon'')E_s = \omega\varepsilon'' E_s + j\omega\varepsilon' E_s \qquad ...(12.52)$$

$$= J_s + j\omega\varepsilon E_s$$

$$= \sigma E_s + j\omega\varepsilon' E_s$$

$$= (\sigma + j\omega\varepsilon')E_s$$

$$\nabla \times H_s = J_{\sigma s} + J_{ds} \qquad ...(12.53)$$

Here we have interpreted ε as ε'. $J_{\sigma s}$ represents the conduction current density and J_{ds} the displacement current density. Comparing equations (12.52) and (12.53) we find that in a conductive medium.

$$\varepsilon'' = \frac{\sigma}{\omega} \qquad ...(12.54)$$

From equation (12.53) we have the ratio of conduction current density to displacement current density as

$$\frac{J_{\sigma s}}{J_{ds}} = \frac{\varepsilon''}{j\varepsilon'} = \frac{\sigma}{j\omega\varepsilon'} \qquad ...(12.55)$$

This shows that these two vectors point in the same direction in space but these are 90° out of phase in time reference frame. Conduction current and displacement current are the current in the resistor with a current in parallel capacitor as displacement current and the angle between the total current and displacement is denoted by δ and $\tan\delta$ is known as the loss tangent. This is shown in Fig. 12.2.

From the phasor diagram $\tan\delta$ = loss angle $= \frac{\sigma E_s}{\omega\varepsilon' E_s}$

$$= \frac{\sigma}{\omega\varepsilon'} \qquad ...(12.56)$$

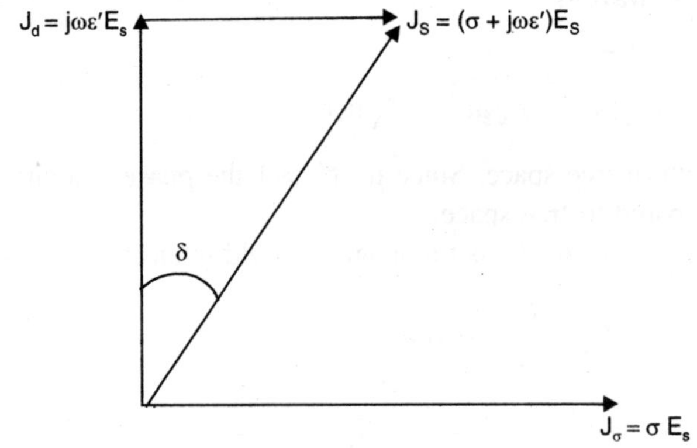

Fig. 12.2 Conduction and displacement current densities and the loss angle tan δ

If the loss tangent is small, useful approximation for the attenuation and phase constant can be obtained. We have already found that for small loss in the dielectric $\varepsilon''/\varepsilon' << 1$ and if this condition is met thus it is a good dielectric.

Now from equation (12.54) $\varepsilon'' = \frac{\sigma}{\omega}$ then (12.42) becomes

$$jk = j\omega\sqrt{\mu\varepsilon'}\sqrt{1 - j\frac{\sigma}{\omega\varepsilon'}} \qquad ...(12.57)$$

We expand second radical term $\sqrt{1 - j\frac{\sigma}{\omega\varepsilon'}}$ using binomial expansion

$$(1+x)^n = 1 + nx + \frac{n(n-1)}{2!}n^2 + ... \text{ where } x << y.$$

Here $x = -\frac{j\sigma}{\omega\varepsilon'}$, and $n = {}^1\!/_2$ and thus we have limiting the expansion to first-three terms.

$$jk = j\omega\sqrt{\mu\varepsilon'}\left[1 - j\frac{\sigma}{2\omega\varepsilon'} + \frac{1}{8}\left(\frac{\sigma}{\omega\varepsilon'}\right)^2 + ...\right] = \alpha + j\beta \qquad ...(12.58)$$

Now, for a good dielectric

$$\alpha = \text{Re}(jk) = j\omega\sqrt{\mu\varepsilon'}\left(-j\frac{\sigma}{2\omega\varepsilon'}\right) = \frac{\sigma}{2}\sqrt{\frac{\mu}{\varepsilon'}} \qquad ...(12.59)$$

and

$$\beta = I_m(jk) = \omega\sqrt{\mu\varepsilon'}\left[1 + \frac{1}{8}\left(\frac{\sigma}{\omega\varepsilon'}\right)^2\right] \qquad ...(12.60)$$

In many situation $\frac{\sigma}{\omega\varepsilon'} << 1$ is negligibly small and hence equation (12.60) reduces to

$$\beta = \omega\sqrt{\mu\varepsilon'} \qquad ...(12.61)$$

Similarly using binomial expansion to equation (12.47) we have the expression for the intrinsic impedance η as

$$\eta = \sqrt{\frac{\mu}{\varepsilon'}}\left[1 - \frac{3}{8}\left(\frac{\sigma}{w\varepsilon'}\right)^2 + j\frac{\sigma}{2w\varepsilon'}\right]$$

$$= \sqrt{\mu/\varepsilon'}\left[1 + j\frac{\sigma}{2\omega\varepsilon'}\right] \qquad \text{...(12.62)}$$

These approximations hold good when $\frac{\sigma}{\omega\varepsilon'} < 0.1$.

12.4 POYNTING THEOREM

We now develop an expression for the power associated with the electromagnetic waves and theorem associated with power is known as POYNTING theorem. To develop this, we start with Maxwell's curl equation.

$$\nabla \times H = J + \frac{\partial D}{\partial t} \qquad \text{...(12.63)}$$

Taking scalar product on both sides of equation (12.63) with *E*, we have

$$E \cdot \nabla \times H = E \cdot J + E \cdot \frac{\partial D}{\partial t} \qquad \text{...(12.64)}$$

The vector identity for $E.(\nabla \times H) = -E.\ \nabla \times H + H.\ \nabla \times E$ ...(12.65)
Hence (12.64) is rewritten as

$$H \cdot \nabla \times H - \nabla \cdot (E \times H) = J.E + E \cdot \frac{\partial D}{\partial t} \qquad \text{...(12.66)}$$

Since
$$\nabla \times E = -\frac{\partial B}{\partial t}$$

$$-H \cdot \frac{\partial B}{\partial t} - \nabla \cdot (E \times H) = J.E. + E \cdot \frac{\partial D}{\partial t}$$

or
$$-\nabla \cdot (E \times H) = J.E + E \cdot \frac{\partial D}{\partial t} + H \cdot \frac{\partial B}{\partial t}$$

$$= J.E + \varepsilon E \cdot \frac{\partial E}{\partial t} + \mu H \cdot \frac{\partial H}{\partial t} \qquad \text{...(12.67)}$$

The two time derivatives in (12.67) can be rearranged as

$$\varepsilon E \cdot \frac{\partial E}{\partial t} = \frac{\partial}{\partial t}\left(\frac{1}{2}D.E.\right) \qquad \text{...(12.68}a\text{)}$$

$$\mu H \cdot \frac{\partial H}{\partial t} = \frac{\partial}{\partial t}\left(\frac{1}{2}B.H.\right) \qquad \text{...(12.68}b\text{)}$$

Equation (12.67) is rewritten as

$$-\nabla \cdot (E \times H) = J.E. + \frac{\partial}{\partial t}\left(\frac{1}{2}D.E.\right) + \frac{\partial}{\partial t}\left(\frac{1}{2}B.H.\right) \quad ...(12.69)$$

Now we integrate equation (12.69) throughout a volume

$$-\int_v (E \times H) ds = \oint_v J \cdot E\, dv + \frac{d}{dt}\int \frac{1}{2} D \cdot E\, dv + \frac{d}{dt}\int \frac{1}{2} B \cdot H\, dv \quad ...(12.70)$$

Equation (12.70) is known as Poyntings' Theorem. On the right hand side the first integral represents total power dissipated in the volume. The second term represents the total energy stored in the electric field in the volume and the third term represents the total energy stored in the magnetic field in the volume. Since the second and third integrals have time derivatives, therefore, these terms in effect represent the instantaneous power flowing to increase the stored energy in the total volume.

The sum total of powers on the right hand side represent the total power flowing into the volume and so the total power flowing out of the volume is

$$\oint_s (E \times H) \cdot ds \text{ watts.} \quad ...(12.71)$$

Where the integral is to be carried over the surface surrounding the volume. The cross product $E \times H$ is known as the Poynting vector S *i.e.*,

$$S = E \times H \text{ watts/m}^2 \quad ...(12.72)$$

which can be interpreted as the instantaneous power density in watts/m^2. The direction of the vector S indicates the direction of flow of power at a point. In our case since E_x and H_y are along x-axis and y-axis respectively and these lie in the same plane and are orthogonal to each other, the direction flow of power will be along z-axis *i.e.*

$$E_x \bar{a}_x \times H_y \bar{a}_y = s_z \bar{a}_z \quad ...(12.73)$$

In a perfect dielectric the E and H field amplitudes are given by

$$E_x = E_{xf} \cos(\omega t - \beta z)$$

$$H_y = \frac{E_{xf}}{\eta} \cos(\omega t - \beta z)$$

where η is the intrinsic impedance and is real (resistive) in this case. Therefore, the power density is given by

$$S_z = \frac{E_{xf}^2}{\eta} \cos^2(\omega t - \beta z) \quad ...(12.74)$$

In case of a lossy dielectric E_x and H_y are not in time phase hence we have

$$E_x = E_{xf} e^{-\alpha z} \cos(\omega t - \beta z)$$

and if $\eta = \eta \angle\theta$, then $H_y = \dfrac{E_{xf}}{|\eta|} e^{-\alpha z} \cos(\omega t - \beta z - \theta)$

Thus,
$$S_z = E_x H_y = \frac{E_{xf}^2}{|\eta|} e^{-2dz} \cos(\omega t - \beta z)\cos(\omega t - \beta z - \theta) \quad ...(12.75)$$

Finding the time average power density we have

$$S_z = \frac{1}{T}\int_0^T \frac{E_{xf}^2}{|\eta|} e^{-2\alpha z} \cos(\omega t - \beta z)\cos(\omega t - \beta z - \theta)\,dt$$

$$= \frac{1}{T}\int_0^T \frac{E_x^2}{|\eta|} e^{-2\alpha z} \frac{\cos(2\omega t - 2\beta z - \theta)\cos\theta}{2}\,dt$$

The average of the second harmonic component over complete cycle is zero and hence

$$S_z = \frac{1}{2}\frac{E_{xf}^2}{|\eta|} e^{-2\alpha z} \cos\theta \quad ...(12.76)$$

If we take E and H in phasor forms, the pointing vector S is given as

$$S = \frac{1}{2}\mathrm{Re}\left[E_x \times H_s\right] \mathrm{W/m^2} \quad ...(12.77)$$

Here
$$E_s = E_{xf} e^{-j\beta} \bar{a}_x \text{ and } H_s = \frac{E_{xf}}{|\eta|} e^{j\theta} e^{j\beta} \bar{a}_y$$

where E_{xf} is assumed to be real.

12.5 PROPAGATION IN GOOD CONDUCTORS - SKIN EFFECT

In Chapter 11 we studied rigorously the behaviour of transmission line when uniform plane waves are travelling assuming the lines to be lossless or low loss line. We obtained approximate values of attenuation coefficient, phase constant and intrinsic impedance. Good conductor transmission lines satisfy the criterion $\sigma/(\omega\varepsilon') >> 1$ and thus these are high loss lines. We are interested in finding out the losses that occur on wave transmission into a good conductor. The plane waves associated with electromagnetic fields exist in an external dielectric that adjoins the conductor surface and the waves travel along the surface of the line conductor. A portion of the overall field will exist within the conductor which will produce power loss due to conduction currents. As the wave travels along the surface of the conductor the loss increases with length of travel along the surface.

Therefore, the energy associated with the wave decreases continuously as the wave travels through the conductor. In the previous article we have found that the ratio of conduction current density to the displacement current density in a conducting material is given by $\sigma/(\omega\varepsilon') >> 1$ and we try to find out good approximation to find α, β and η for a good conductor.

From equation (12.57) the general expression for the propagation constant is given as

$$jk = j\omega\sqrt{\mu\varepsilon'}\sqrt{1 - j\sigma/\omega\varepsilon'}$$

Since $\sigma/\omega\varepsilon' >> 1$ we simplify the above expressions as

$$jk = j\omega\sqrt{\mu\varepsilon'}\sqrt{-j\frac{\sigma}{\omega\varepsilon'}}$$

$$= j\sqrt{-j\omega^2\mu\varepsilon'\cdot\frac{\sigma}{\omega\varepsilon'}} = j\sqrt{-j\omega\sigma\mu}$$

Now since $$-j = 1\angle -90°, \sqrt{-j} = 1\angle -45° = \frac{1}{\sqrt{2}}(1 - j1)$$

Therefore, $$jk = j(1-j)\sqrt{\frac{\omega\mu\sigma}{2}}$$

$$= (1+j)\sqrt{\pi f\mu\sigma} = \alpha + j\beta \qquad ...(12.78)$$

Hence $$\alpha = \beta = \sqrt{\pi f\mu\sigma} \qquad ...(12.79)$$

From equation (12.79) it is clear that irrespective of the values of μ and σ of the conductor frequency *f* of the fields, α and β are equal. Assuming the electric field applied is along *x*-axis as in Fig. 12.1 and E_x component is propagating along *z*-axis, then

$$E_x = E_{xf}e^{-\alpha z}\cos(\omega t - \beta z)$$

$$= E_{xf}e^{-z\sqrt{\pi f\mu\sigma}}\cos\left(\omega t - z\sqrt{\alpha f\mu\sigma}\right) \qquad ...(12.80)$$

Let $z > 0$ be the region within the conductor and $z = 0$ is the surface of the conductor where $z < 0$ is the region for a perfect dielectric. At the boundary where $z = 0$ equation (12.80) becomes

$$E_x = E_{xf}\cos\omega t \qquad ...(12.81)$$

We consider this field as the source field that produces the fields within the conductor. Since displacement current is negligible

$$J = \sigma E$$

which shows that the conduction current density at any point within the conductor is directly proportional to electric field intensity at that point.

$$J_x = \sigma E_x = \sigma E_{xf}e^{-z\sqrt{\pi f\mu\sigma}}\cos\left(\omega t - z\sqrt{\pi f\mu\sigma}\right) \qquad ...(12.82)$$

From equation (12.80) it is found that there is exponential decay or decrease in the electric field intensity and from (12.82) it is found that there is exponential decrease in the conduction current density with the penetration into the conductor. The exponential factor is unity at $z = 0$ and decreases to $e^{-1} = 0.368$ when $z = \dfrac{1}{\sqrt{\pi f\mu\sigma}}$.

This distance is denoted by δ or *d* and is known as depth of penetration or the skin depth. Skin here means the surface of the conductor and depth into the conductor from its skin is known as skin depth

$$d = \delta = \frac{1}{\sqrt{\pi f\mu\sigma}} = \frac{1}{\alpha} = \frac{1}{\beta} \qquad ...(12.83)$$

Depth of penetration is an important parameter in describing conductor behaviour in electromagnetic fields. To get some idea of the magnitude of the skin depth we consider copper as the conductor material which has $\sigma = 5.8 \times 10^7$ S/m, $\mu = \mu_0 = 4\pi \times 10^{-7}$ and for 50 Hz source.

Hence $$\delta = \frac{1}{\sqrt{\pi f \times 4\pi \times 10^{-7} \times 5.8 \times 10^{-7}}} = \frac{1}{2\pi\sqrt{f5.8}} = \frac{0.066}{\sqrt{f}} = 9.33 \text{ mm.}$$

It is to be noted that power density carries an exponential term $e^{-2\alpha z}$ and thus we see that the power density is multiplied by a factor of $(0.368)^z = 0.135$ for every 9.33 mm of distance into copper. However, at microwave frequency of 10,000 MHz $\delta = 6.61 \times 10^{-4}$ mm which means all fields in a good conductor such as copper are essentially zero at distances greater than a few skin depths from the surface. Electromagnetic energy when operating at thousand of MHz is not transmitted in the interior of a conductor. However, it travels into the region surrounding the conductor whereas the conductor only guides the wave. It is for this reason that it is desirable to have hollow conductors than solid conductors especially when operating at higher frequency so the current or electric fields are limited only upto a few skin depths and rest of the cross-section of the conductor is not being utilized. In fact at microwave frequencies, the skin depth being very small, only surface coating of the guiding conductor is important. A piece of glass with an evaporated silver surface 3μm thick is an excellent conductor at these frequencies.

Next we develop expressions for phase velocity and wavelength within a good conductor. From equation (12.83) we have

$$\alpha = \beta = \frac{1}{\delta} = \sqrt{\pi f \mu \sigma}$$

Since $$\beta = \frac{2\pi}{\lambda} \quad \text{or} \quad \lambda = \frac{2\pi}{\beta} = 2\pi\delta \qquad \text{...(12.84)}$$

Also we know that $$V_p = \frac{\omega}{\beta} = \omega\delta \qquad \text{...(12.85)}$$

We next find out H_y the magnetic field associated with electric field E_x and then we can find out expression for intrinsic impedance. To achieve this we begin with expression (12.47) for intrinsic impedance for a good conductor and replacing $\varepsilon'' = \sigma/\omega$ we have

$$\eta = \sqrt{\frac{j\omega\mu}{\sigma + j\omega\varepsilon'}}$$

Since $\sigma >> \omega\varepsilon'$ $$\eta = \sqrt{\frac{j\omega\mu}{\sigma}} = \frac{\sqrt{2}\angle 45^\circ}{\sigma\delta} = \frac{(1+j)}{\sigma\delta} \qquad \text{...(12.86)}$$

Writing equation (12.81) in terms of skin depth we have

$$E_x = E_{xf} e^{-z/\delta} \cos\left(\omega t - \frac{z}{\delta}\right) \qquad \text{...(12.87)}$$

Hence $$H_y = \frac{E_x}{\eta} = \frac{\sigma\delta E_{xf}}{\sqrt{2}} e^{-z/\delta} \cos\left(\omega t - \frac{z}{\delta} - \frac{\pi}{4}\right) \qquad \text{...(12.88)}$$

Comparing (12.87) and (12.88) we find that electric field maximum amplitude leads the magnetic field maximum amplitude by one eighth $\left(\frac{\pi}{4}\right)$ of a cycle at every point. Using equations (12.87) and

(12.88) we may obtain the time average poynting vector by applying the relation given in equation (12.77)

$$\left(S_z\right) = \frac{1}{2}\frac{\sigma\delta E_{xf}^2}{\sqrt{2}} e^{-2z/\delta} \cos\frac{\pi}{4} = \frac{1}{4}\sigma\delta E_{xf}^2 e^{-2z/\delta} \qquad \ldots(12.89)$$

When $z = \delta$ *i.e.* single skin depth, the power density is only $e^{-2} = 0.135$ of its value at the surface.

Consider a cuboid conductor with length L and width b as shown in Fig. 12.3. The objective is to find average power loss in a width $0 < y < b$ and length $0 < x < L$ when the conduction current density is given as $J_x = J_{xf}e^{-z/\delta}.\ e^{-jz/\delta}$

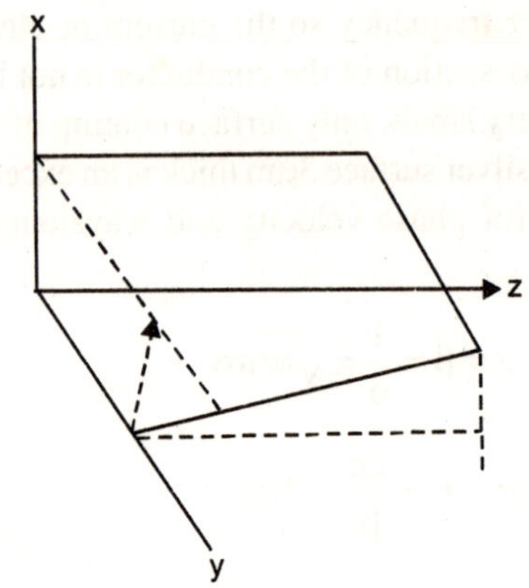

Fig. 12.3

The power crossing the conductor surface within this area

$$P_L = \int_A \left(S_z\right) da = \int_0^b\int_0^L \frac{1}{4}\sigma\delta E_{xf}^2\, e^{-2z/\delta} dx\, dy$$

$$= \frac{1}{4}\sigma\delta\, E_{xf}^2 \cdot b \cdot L = \frac{1}{4}\sigma\delta b L E_{xf}^2$$

However, $J_x = \sigma E_x$

Hence,

$$P_L = \frac{1}{4\sigma}\delta b L J_{x0}^2 \qquad \ldots(12.90)$$

From equation (12.90) it is clear that the average power loss in a conductor with skin effect present may be calculated by assuming that the total current is uniformly distributed over skin depth δ. The resistance of a slab of width b and length L and infinite thickness is same as slab of width b, length L and thickness δ as the current is assumed to be uniformly distributed over thickness δ. The concept of skin effect can be applied to conductor of circular cross section provided that radius a of the conductor is much greater than the skin depth. The resistance of such a conductor can be obtained at higher frequency where skin effect is present by considering a slab of width equal to the circumference $2\pi a$ and thickness δ. Hence

$$R = \frac{L}{\sigma S} = \frac{L}{\sigma \cdot 2\pi a\delta} \qquad \ldots(12.91)$$

We compare the resistance of copper conductor of 1 mm radius and 1 km length for *d.c.* current and for megafrequency currents where skin effect is significant whereas for *d.c.* current skin effect is zero.

For *d.c.* current

$$R_{dc} = \frac{10^3}{\pi(10^{-2})^2\, 5.8\times10^7} = \frac{10^3}{\pi\times5.8\times10} = 5.48\,\Omega$$

For 1 MHz where skin depth is 0.066 mm, and then $a >> \delta$ and using equation (12.91)

$$R = \frac{10^3}{5.8\times10^7\times2\pi\times10^{-3}\times0.066\times10^{-3}} = 41.5\,\Omega$$

12.6 WAVE POLARIZATION

In the previous sections we have assumed E and H field to lie along $\bar{a}_x$ and $\bar{a}_y$ direction and the propagation is along z-axis. This may be possible in case of uniform plane waves in free space. However, the directions of E and H within the plane perpendicular to $\bar{a}_z$ may change as function of time and position, depending upon how the wave was generated or what type of medium it is propagating through. Therefore, a complete description of the wave would require a statement of the orientation of its field vectors besides its wavelength, phase velocity and power. Wave polarization of an electric field is defined as the electric field vector orientation as a function of time, at a fixed point in space. A better way of specifying the polarization will be to specify the field orientation at all points as some waves demonstrate spatial variations in their polarization. It is to be noted that if only field E is specified, orientation of field H can be obtained using Maxwell's equations.

When electric field E is specified in a fixed straight orientation for all times and positions, the wave is said to be linearly polarized. We have assumed E to lie along x-axis, however, it may lie in any orientation in the xy plane, yet it is said to be linearily polarized. It lies in xy plane, its two components along $\bar{a}_x$ and $\bar{a}_y$ are say respectively E_{x0} and E_{y0} and E can be represented in time and space as

$$E_s = \left(E_{y0}\bar{a}_x + E_{y0}\bar{a}_y\right)e^{-\alpha z}e^{-j\beta z} \quad \text{...(12.92)}$$

The expression for magnetic field H_s can be formed by determining its $\bar{a}_x$ and $\bar{a}_y$ components directly from those of E_s. H_s for the E_s wave of equation (12.92) is given as

$$H_s = \left[H_{x0}\bar{a}_x + H_{y0}\bar{a}_y\right]e^{-\alpha z}\,e^{j\beta z}$$

$$= \left[-\frac{E_{x0}}{\eta}\bar{a}_x + \frac{E_{x0}}{\eta}\bar{a}_y\right]e^{-az}e^{-j\beta z} \quad \text{...(12.93)}$$

The two fields E and H have been sketched in Fig. 12.4.

The negative sign in the expression for H_s is clear from the figure. E_{y0} and H_{x0} are normal to each other and since H_{x0} is along $-\bar{a}_x$, negative sign has been assigned. The direction of power flow given by $E \times H$ is in the positive z-direction. Using equation (12.92) and (12.93), the power density in the wave is obtained using equation (12.77).

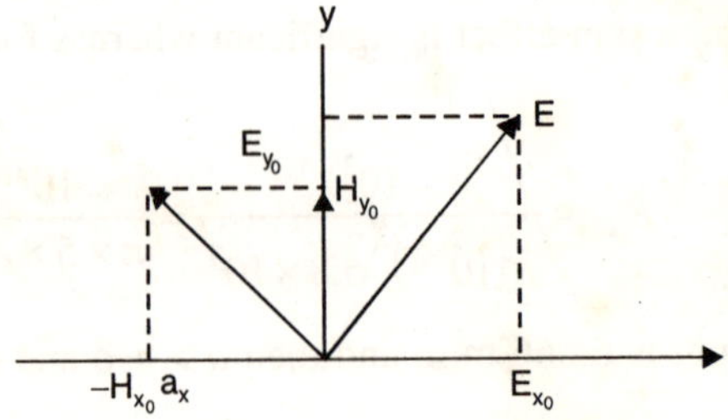

Fig. 12.4 *E* and *H* field anywhere in the *xy* plane

$$(S_z) = \frac{1}{2}\text{Re}\left[E \times H^*\right] = \frac{1}{2}\text{Re}\left[E_{xf}H_{yf}^*\left(\bar{a}_x \times \bar{a}_y\right) + E_{yf}H_{xf}^*\left(\bar{a}_y \times \bar{a}_x\right)\right]e^{-2\alpha z}$$

$$= \frac{1}{2}\text{Re}\left[\frac{E_{xf}E_{xf}^*}{\eta^*} + \frac{E_{yf}E_{yf}^*}{\eta^*}\right]e^{-2\alpha z}\bar{a}_z$$

$$= \frac{1}{2}\text{Re}\left(\frac{1}{\eta^*}\right)\left\{\left(E_{xf}\right)^2 + \left(E_{yf}\right)^2\right\}e^{-2\alpha z}\ \text{W/m}^2 \qquad \ldots(12.94)$$

For power flow we need to have product of mutually perpendicular components of *E* and *H*. A linearly polarized wave can be considered as two distinct plane waves having *x* and *y* polarization whose electric fields are combining in phase to produce the net *E* field. The same procedure holds good for the magnetic field *H* also. It is to be noted that any polarization state can be described (linear, elliptic or circular) in terms of mutually perpendicular components of the electric field and their relative phasing.

Next we consider a phase difference of ϕ between E_{xf} and E_{yf} such that $\phi < \pi/2$. For simplicity we consider propagation in a lossless medium. The total field in phasor form is

$$E_s = \left(E_{xf}\bar{a}_x + E_{yf}e^{j\phi}\bar{a}_y\right)e^{-j\beta z} \qquad \ldots(12.95)$$

To understand easily we convert this wave to real instantaneous form by multiplying by $e^{j\omega t}$ and taking only the real part, we have

$$E(z, t) = E_{xf}\cos(\omega t - \beta z)\bar{a}_x + E_{xf}\cos(\omega t - \beta z + \phi)\bar{a}_y \qquad \ldots(12.96)$$

Whose E_{xf} and E_{yf} are assumed to be real. Suppose we let $t = 0$ in which case (12.96) becomes

$$E(z, 0) = E_{xf}\cos\beta z\bar{a}_x + E_{yf}\cos(\beta z - \phi)\bar{a}_y \qquad \ldots(12.97)$$

We now plot the components of $E(z, 0)$ as a function of *z* as shown in Fig. 12.5.

Since time is fixed at zero, the wave is frozen in position. An observer moving along *z*-axis can measure the components of *E* and thus the orientation of the total electric field at each point. Consider point '*a*' the crest of $E_{xf}(0)$ and point *b* the crest of $E_{yf}(0)$. If ϕ were zero, point *b* would have coincided with point (*a*). Since $\phi \neq 0$ the crest of E_{xf} is now displaced to point *b* farther down *z*-axis. The two points are separated by a distance (ϕ/β) where E_{yf} thus lags behind E_{xf} when we consider the spatial dimension.

Now suppose the observer stops at some location on the *z*-axis and time is allowed to move forward. Both the field now move in the positive *z* direction as in (12.96) indicates. This time the point *b* reaches the observer first and then it reaches point *a*. Thus E_y leads E_x when we consider the time

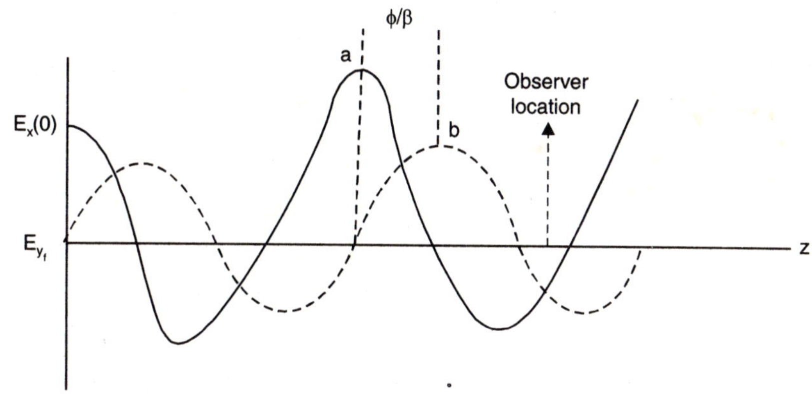

Fig. 12.5 Plot of equation (12.97). As time increases from zero both waves travel to the right as per equation (12.96). Thus to an observer at a fixed location, the *y*-component leads in time.

dimension. In either situation (fixed t and z varying or vice versa) the observer finds that the net field rotates about the z-axis while its magnitude changes. Considering a starting point in z and t at which the field has a given orientation and magnitude, the wave will return to its original orientation and magnitude at a distance of one wavelength for fixed t or at a time $t = \frac{2\pi}{\omega}$ later at a fixed z.

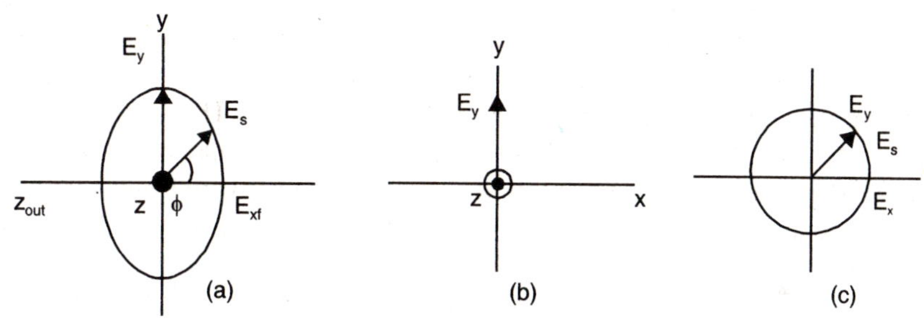

Fig. 12.6 Wave polarization (*a*) Elliptical (*b*) Linear (*c*) Circular

For illustration consider Fig. 12.6. If we take the length of the field vector as a measure of its magnitude, we find that at a fixed position the tip of the field vector traces out the shape of the ellipse over time ($2\pi/\omega$). The wave is said to be elliptically polarised. Elliptical polarization is of course the general polarization state of a wave since it can assume any magnitude and phase difference between E_x and E_y. Linear polarization is a special case of elliptical polarization where the phase difference is zero. However, if $E_y = E_x = E_0$ and when $\phi = \pm\frac{\pi}{2}$ the wave is said to be circularly polarized as shown in Fig. 12.6(*c*). Mathematically if we substitute these conditions for circular polarization in equation (12.96) we have.

$$E(z, t) = E_{xf}\left[\cos(\cos\omega t - \beta z)\bar{a}_x + \cos\left(\omega t - \beta z \pm \frac{\pi}{2}\right)\bar{a}_y\right]$$

$$= E_{xf}\left[\cos(\omega t - \beta z)\bar{a}_x \mp \sin(\omega t - \beta z)\bar{a}_y\right] \qquad \ldots(12.98)$$

If we fix a position along z such as $z = 0$ and allow time to vary, (12.98) with $\phi = \frac{\pi}{2}$ becomes

$$E(0, t) = E_{fx}\left[\cos(\omega t)\bar{a}_x - \sin(\omega t)a_y\right] \quad \text{...(12.99)}$$

However if we let $\phi = -\pi/2$ in (12.98) we have

$$E(0, t) = E_{xf}\left[\cos(\omega t)\bar{a}_x + \sin(\omega t)\bar{a}_y\right] \quad \text{...(12.100)}$$

The field vector of equation (12.100) rotates in the counter clockwise direction in the xy plane while maintaining constant amplitude E_{fx} and so the tip of the vector traces out a circle. However, if we let $\phi = \frac{\pi}{2}$ equation (12.99) whose field vector rotates in the clockwise direction. With forward z propagation (12.99) describes a left circularly polarized wave and (12.100) describes a right circularly polarized wave. The same convention applies to elliptical polarization, where the description left elliptical polarization and right elliptical polarization are used.

The most important advantage of the circularly polarized wave is that the reception of this type of wave does not depend on the antenna orientation in the plane normal to the propagation direction. Dipole antennas e.g. are required to be oriented along the electric field direction of the signal they receive. However, if circularly polarized signals are transmitted, the receiver orientation requirements are relaxed considerably.

Circularly polarized light can be generated using an anisotropic medium—a material whose permittivity is a function of electric field direction. A crystal orientation can be found such that along one direction e.g. x-axis, the permittivity is the lowest while along the orthogonal direction say y-axis the permittivity is highest. The strategy is to input a linearly polarized wave with its field vector at 45° to the x and y axes of the crystal. It will thus have equal amplitudes x and y component in the crystal and these will now propagate in the z-direction at different speeds. A phase difference accumulates between the two components and if the length of the crystal is long this may add upto $\pi/2$. The wave at the output thus becomes circularly polarized. Such a crystal, cut to the right length and used in this manner is called a *quarter wave plate* since it introduces a relative phase shift of $\pi/2$ between E_x and E_y which is equivalent to $\lambda/4$.

Circularly polarized waves can be expressed in phasor form as follows. Equation (12.98) can be expressed as

$$E(z, t) = \text{Re}\left[E_{xf}e^{j\omega t}e^{-j\beta z}\left\{\bar{a}_x + e^{\pm j\pi/2}\bar{a}_y\right\}\right]$$

Since $e^{\pm j\pi/2} = \pm j$, the above equation becomes

$$E_s = E_{xf}\left(\bar{a}_x \pm ja_y\right)e^{-j\beta z} \quad \text{...(12.101)}$$

Here plus sign is for the left circular polarization and the minus sign for right circular polarization. If the wave propagates in the negative z direction, the wave equation becomes

$$E_s = E_{xb}\left(\bar{a}_x \pm ja_y\right)e^{j\beta z} \quad \text{...(12.102)}$$

Here plus sign is for right circular polarization and the minus sign for the left polarization.

Example 12.1. Determine the phase velocity, attenuation constant, phase constant and intrinsic impedance for a 1 MHz plane wave in a large block of copper when $\sigma = 5.8 \times 10^7$ Sm^{-1}, $\varepsilon_r = 1$, $\mu_r = 1$.

Solution. The propagation constant is $\gamma = \sqrt{j\omega\mu(\sigma + j\omega\varepsilon)}$

$$\gamma = \sqrt{j(2\pi\times10^6)(4\pi\times10^{-7})\left(5.8\times10^7 + j2\pi\times10^6\times\frac{1}{36\pi}\times10^{-9}\right)}$$

$$= \sqrt{j8\pi^2\times10^{-1}\left(5.8\times10^7 + j\frac{10^{-3}}{18}\right)} = 2\pi\sqrt{(j0.2)(5.8\times10^7)}$$

$$= 2\pi\times10^3\times3.4\angle45^\circ$$

$$= 2.135\times10^4\angle45^\circ$$

$$\alpha + j\beta = \left(1.51\times10^4 + j1.51\times10^4\right) \text{ per meter. } \textbf{Ans.}$$

Hence $\alpha = 1.51\times10^4$ nP m^{-1} and $\beta = 1.51\times10^4$ rad m^{-1} **Ans.**

Hence wavelength in copper $\lambda = \dfrac{2\pi}{\beta} = \dfrac{2\pi}{1.51\times10^4} = 415.8\ \mu\text{m}$ **Ans.**

and phase velocity $v = \dfrac{\omega}{\beta} = \dfrac{2\pi\times10^6}{1.51\times10^4} = 415.8\,\text{m}\,\text{s}^{-1}$ **Ans.**

The intrinsic impedance $\eta = \sqrt{\dfrac{j\omega\mu}{\sigma + j\omega t}}$

$$= \sqrt{\frac{j(2\pi\times10^6)(4\pi\times10^{-7})}{5.8\times10^7 + j(2\pi\times10^6)\cdot\frac{1}{36\pi}\times10^{-9}}}$$

$$= 2\pi\sqrt{\frac{j2\times10^{-1}}{5.8\times10^7}} = \frac{2\pi}{10^4}\times0.587$$

$$= 3.68\times10^{-4}\angle45^\circ\,\Omega \ \textbf{Ans.}$$

Example 12.2. Determine the wavelength of an electromagnetic wave travelling in free space at 30 GHz.

Solution. Since the phase velocity in free space

$$u = \frac{1}{\sqrt{\mu_0\varepsilon_0}} = \sqrt{\frac{1}{4\pi\times10^{-7}\times\frac{1}{36\pi}\times10^{-9}}} = 3\times10^8\,\text{m}\,\text{s}^{-1}$$

Hence $$\lambda = \frac{u}{f} = \frac{3\times10^8}{30\times10^9} = 0.01\ \text{m} \ \textbf{Ans.}$$

Example 12.3. A 1 MHz plane wave travelling in a normally dispersive lossless medium has a phase velocity at this frequency of 250 mm s^{-1}. The phase velocity as a function of wavelength is given by $v = k\sqrt{\lambda}$ where k is a constant. Determine the group velocity.

Solution. The wavelength of the plane wave $= \dfrac{c}{f} = \dfrac{3\times 10^8}{10^6} = 300$ m

Since here $\dfrac{dv}{d\lambda} > 0$. Hence the group velocity is given by equation

$$u = v - \lambda\frac{dv}{d\lambda} = v - \frac{k}{2}\sqrt{\lambda}$$

Since $\qquad v = k\sqrt{\lambda}\,,\quad \dfrac{dv}{d\lambda} = \dfrac{k}{2}\sqrt{\lambda}$

Since $\qquad k = \dfrac{v}{\sqrt{\lambda}}$

Hence $\qquad u = v - \dfrac{v}{\sqrt{\lambda}}\cdot\dfrac{1}{2}\sqrt{\lambda} = u = v - \dfrac{v}{2} = \dfrac{v}{2} = 125 \text{ mm s}^{-1}$ **Ans.**

Example 12.4. A plane 1 GHz travelling wave in air with peak electric field intensity of 2Vm^{-1} is incident normally on a large copper sheet. Determine the average power absorbed by the sheet per sq. meter of area.

Solution. We know that the intrinsic impedance of the media is given as

$$\eta = \sqrt{\frac{j\omega\mu}{\sigma + j\omega t}} \quad \text{Here } \mu = \mu_0 \text{ as } \mu_r = 1 \text{ for copper}$$

Since copper has $\sigma >> \omega\varepsilon$ hence $\eta = \sqrt{\dfrac{j\omega\mu}{\sigma}} = \sqrt{\dfrac{\omega\mu}{\sigma}}\angle 45°$

Now σ for copper is 5.8×10^7 Sm^{-1} and $\mu_r = 1$

$$\eta = \sqrt{\frac{2\pi\times 10^9 \times 4\pi\times 10^{-7}}{5.8\times 10^7}}\cdot\cos 45° = 8.2 \text{ m}\Omega$$

Since the incident electric field is 2Vm^{-1} which results into incident current of $\dfrac{2V}{\eta}$. But since copper plate is a short circuit, the transmitted current will be $2\cdot\dfrac{2V}{\eta}$ and this is the value of H.

i.e. $H = \dfrac{2\times 2}{\eta}$ A m^{-1}. Here η is intrinsic impedance and its value for free space is 377 Ω and hence

$$\text{Power loss} = \frac{1}{2}|H_z|^2 \operatorname{Re} I_0$$

$$= \frac{1}{2}\left(\frac{4}{377}\right)^2 \times 8.2 \times 10^{-3}$$

$$= 0.5 \times 1.1 \times 10^{-4} \times 8.2 \times 10^{-3} = 451 \text{nw m}^{-2} \textbf{ Ans.}$$

Example 12.5. Determine the power that is flowing along a co-axial cable with inner conductor radius of *'a'* and outer radius *'b'* when a *dc* current I flows outward along the inner conductor and returns along the sheath. Assume the resistance of the conductor to be negligible.

Solution. The electric field intensity is radial to the surface of the conductor and is given as

$$E = \frac{V}{r \log \frac{b}{a}}$$

and the magnetic field intensity is circumferential and is given by

$$H_r = \frac{I}{2\pi r}$$

Since E and H are normal to each other, hence

$$S = S_z = \frac{VI}{2\pi r^2 \log b/a}$$

and direction of flow of power is normal to the plane containing E and H and is, therefore along the direction of flow of current in the conductor.

The power flow is confined to the region between the conductors and the sheath and is for any plane perpendicular to the axis of the cable.

$$P = \oint s \cdot \bar{a}_n ds = \int_0^b s_z \cdot 2\pi r\, dr = \int_0^b \frac{VI}{2\pi r^2 / n\, b/a} \cdot 2\pi r\, dr$$

$$= VI$$

Example 12.6. Express the electric field

$$E_y(z, t) = 150 \cos(10^6 t - 0.5z + 30°)\,\text{Vm}^{-1} \text{ as phasor.}$$

Solution. We write the expression in exponential notation

$$E_y(z, t) = \operatorname{Re}\left[150 e^{j10^6 t} \cdot e^{-j0.52} \cdot e^{j30}\right]$$

The quantity within the bracket will have real (cos component) and imaginary (sin component) components. Since our original expression has only cos component we have taken *Re* part of it.

Again dropping *Re* and $e^{j10^6 t}$ where $10^6 = w$, we have the expression in phasor λ form

$$E_{ys}(z) = 150 e^{-j0.5z} e^{j30°} = 150 e^{-j0.5z + j30°}$$

Example 12.7. The electric field of a uniform plane wave is given by $E_0 = 150\bar{a}_x + 30\angle 30°\bar{a}_y$. Determine the phasor and real instantaneous fields if the wave is propagating at 1 MHz in the positive z direction in free space.

Solution The general phasor expression for the given electric field is

$$E_s(z) = \left[150\bar{a}_x + 30e^{j30}\bar{a}_y\right]e^{-x_0 z}$$

where $$k_0 = \frac{w}{c} = \frac{2\pi \times 10^6}{3 \times 10^8} = 0.021 \text{ rad m}^{-1}$$

The real instantaneous form is then obtained using equation (12.17)

$$E(z, t) = \text{Re}\left[150e^{-j0.021}\, e^{j2\pi N_0^6 t}\,\bar{a}_x + 30e^{j30°} \cdot e^{-j0.0212}\, e^{j2\pi\times10^6 t}\bar{a}_y\right]$$

$$= \text{Re}\left[150e^{(j2\pi\times10^6 t - j0.021z)}\,\bar{a}_x + 30e^{j(2\pi\times10^6 t - 0.021z + 30°)}\,\bar{a}_y\right]$$

$$= 150\cos(2\pi\times10^6 t - 0.021z)\bar{a}_x + 30\cos(2\pi\times10^6 t - 0.021z + 30°)\bar{a}_y$$

Example 12.8. A uniform plane wave of an electric field with amplitude 300 V m^{-1} is propagating along z-axis. If $\bar{E} = E\bar{a}_x$ and ω = 1 M rad s^{-1}, determine (*i*) the frequency (*ii*) the wavelength (*iii*) the period (*iv*) the amplitude of H.

Solution. (*i*) Given $2\pi f = 1 \times 10^6 \text{ rad s}^{-1}$

Hence $$f = \frac{10^6}{2\pi} = 159 \text{ kHz} \quad \textbf{Ans.}$$

(*ii*) Since $f\lambda = C$, $$\lambda = \frac{3\times10^8}{159\times10^3} = \frac{300\times10^6}{159\times10^3} = 1.88 \text{ km} \quad \textbf{Ans.}$$

(*iii*) $$T = \frac{1}{f} = \frac{1}{159\times10^3} = \frac{1}{0.159\times10^6} = 6.28\ \mu\text{sec} \quad \textbf{Ans.}$$

(*iv*) Since $E_y = \sqrt{\frac{\mu_0}{\varepsilon_0}}H_x$. Hence $H_x = E_y\sqrt{\frac{\varepsilon_0}{\mu_0}}$

$$= 300\sqrt{\frac{(1/36\pi)10^{-9}}{4\pi\times10^{-7}}} = \frac{300}{120\pi} = 0.796 \text{A m}^{-1} \quad \textbf{Ans.}$$

Example 12.9. Express the phase velocity v, the intrinsic impedance η, the wave length λ and the wave number or the phasor number or phase constant $\beta = \frac{2\pi}{\lambda}$ for a wave travelling in a lossless material medium characterized by μ_r, ε_r in terms of the free space values.

Solution. $u = \frac{\mu_0}{\sqrt{\mu_r t_r}}$, $\eta = \eta_0 \sqrt{\frac{\mu_r}{\varepsilon_r}}$, $\lambda = \frac{\lambda_0}{\sqrt{\mu_r \varepsilon_r}}$, $\beta = \beta_0 \sqrt{\mu_0 \varepsilon_r}$

Example 12.10. A uniform plane wave is propagating in a material that is characterized by $\varepsilon_r = 2.53$, $\mu_r = 1$ and $\sigma = 0$. If the electric field is given by

$$E = 100 \cos\left(10\pi \times 10^9 t - \beta z\right) \bar{a}_y .$$

Determine (*i*) The phase velocity (*ii*) the wavelength (*iii*) the magnitude of the magnetic field (*iv*) the time domain expression for the magnetic field.

Solution. (*i*) From example 12.9 $v = \frac{\mu_0}{\sqrt{\mu_r \varepsilon_r}} = \frac{3 \times 10^8}{\sqrt{2.53 \times 1}} = 1.886 \times 10^8 \text{ m s}^{-1}$ **Ans.**

(*ii*) $2\pi f = 10\pi \times 10^9$ or $f = 5 \times 10^9$ and $\lambda_0 = \frac{3 \times 10^8}{5 \times 10^9} = 0.6 \times 10^{-1}$

$$\lambda = \frac{\lambda_0}{\sqrt{2.53}} = \frac{0.6 \times 10^{-1}}{1.59} = 0.0377 \quad \textbf{Ans.}$$

(*iii*) $H = E\sqrt{\frac{\varepsilon}{\mu}} = 100 \times \frac{1}{120\pi} \sqrt{2.53} = 0.422 \text{ A m}^{-1}$

$$\text{Now } \beta = \frac{\omega}{v_p} = \frac{\pi \times 10^{10}}{1.886 \times 10^8} = 166.48 \quad \textbf{Ans.}$$

(*iv*) Hence, expression for $H = 0.42 \cos(10\pi \times 10^9 t - 166.48z)\bar{a}_y$ **Ans.**

Example 12.11. The electric field of a plane wave in a dielectric is given by

$$E = 3.77 \times 10^{-2} \cos\left(2\pi \times 10^8 t - \frac{4\pi}{3} z\right)$$

(*i*) Show that $\varepsilon_r = 4$ (*ii*) Write the expression for the vector magnetic field.

Solution. (*i*) Since $\frac{\omega}{\beta} = \frac{3 \times 10^8}{\sqrt{\mu_r}} = \frac{2\pi \times 10^8}{4\pi} \times 3 = \frac{3 \times 10^8}{\sqrt{\mu_r}}$

or $$\frac{1}{2} = \frac{1}{\sqrt{\mu_r}} \quad \text{or} \quad \mu_r = 4 \quad \textbf{Ans.}$$

(*ii*) Magnitude of $H = E \cdot \sqrt{\frac{\varepsilon}{\mu}} = E\sqrt{\frac{\varepsilon_0 \varepsilon_r}{\mu_0 \mu_r}} = \frac{E\sqrt{\varepsilon_r}}{120\pi} = \frac{2E}{120\pi} = \frac{3.77 \times 10^{-2} \times 2}{120\pi} = 2 \times 10^{-4}$

$$H = -2 \times 10^{-4} \cos\left\{2\pi \times 10^8 t - \frac{4\pi}{3} z\right\} \text{ A m}^{-1} \quad \textbf{Ans.}$$

Example 12.12. An H field travels in the $-\bar{a}_z$ direction in free space with phase shift constant β of 20 rad m^{-1} and an amplitude of $\left(\frac{1}{4\pi}\right)$A m^{-1}. If the field has direction $-\bar{a}_y$ when $t = 0$ and $z = 0$. Write expression for E and H. Determine frequency and wavelength.

Solution. For free space η the intrinsic impedance = 120 π

and E and H are related through $\frac{E}{H} = 120\pi = \eta$

Now, it is known that $H(z, t) = -\frac{1}{4\pi}\cos(\omega t - \beta z)\bar{a}_y$

For propagation in $-z$ direction $\frac{E}{H} = -120\pi$

or amplitude of $E = -120\pi\left(-\frac{1}{4\pi}\right) = 30$

Hence $E_x = 30 \cos(\omega t + 20z)$ V m^{-1}

Thus, $E(z, t) = 30 \cos(\omega t + 20z)\ \bar{a}_x$ V m^{-1} **Ans.**

Since β is given to be 20 rad m^{-1} and $\lambda = \frac{2\pi}{\beta} = \frac{2\pi}{20} = \frac{1}{10}$m **Ans.**

$$f = \frac{c}{\lambda} = \frac{3 \times 10^8}{\pi} \times 10 = \frac{30}{\pi} \times 10^8 \text{ Hz} \quad \textbf{Ans.}$$

Example 12.13. Determine the skin depth δ at a frequency of 2.5 MHz in aluminium where σ = 38.2 MS/m and μ_r = 1. Also determine γ and the wave velocity.

Solution. We know skin depth $\delta = \frac{1}{\sqrt{\pi f \mu_\sigma}} = \frac{1}{\sqrt{\pi \times 2.5 \times 10^6 \times 4\pi \times 10^{-7} \times 38.2 \times 10^6}}$

$$= \frac{1}{2\pi\sqrt{2.5 \times 38.2 \times 10^5}}$$

$$= 0.0515 \times 10^{-3} = 51.5\ \mu\text{m}.$$

From equation (12.84) $\alpha = \beta = \frac{1}{\delta} = \frac{10^6}{51.5} = \frac{100}{51.5} \times 10^4 = 1.94 \times 10^4$

Hence $\gamma = \alpha + j\beta = 1.94 \times 10^4 + j1.94 \times 10^4$ **Ans.**

and wave velocity $= \frac{\omega}{\beta} = \frac{2\pi \times 2.5 \times 10^6}{1.94 \times 10^4} = 810$ m s^{-1} **Ans.**

Example 12.14. A voltage source V is connected to a pure resistance R by a length of a co-axial cable. Show that use of the poynting vector P in the dielectric results in the same instantaneous power in the resistor as in circuit analysis.

Solution. Let V be the voltage of the source and I the current in the resistor through the co-axial cable. Let a be the radius of the inner conductor b the radius of the sheet and r is any distance between a and b. The electric field intensity radial to the inner conductor is given as

$$E = \frac{V}{r \ln b/a} \bar{a}_r \text{ and } H = \frac{I}{2\pi r} a_\phi$$

Since H field is circumferential, E and H are normal to each other. Hence instantaneous power is

$$P = E \times H = \frac{VI}{2\pi r^2 \ln b/a} a_z \sin \bar{a}_r \times a_\phi = a_z$$

Since these are in cylindrical co-ordinates the differential area will be $dr\ r\ d\phi$. Hence total power is given by $\int_{\partial}^{2\pi} \int_a^b \frac{VI}{2\pi r^2 /n b/a} r dr\, d\phi$

$$P(t) = \int_a^b \frac{VI}{r \ln b/a} dr = \frac{VI \ln b/a}{\ln b/a} = VI$$

which is the power consumed by the resistor with voltage V across its terminals and I the current through the resistor.

Example 12.15. In free space $E(z, t) = 150 \sin(\omega t - \beta z)\ \bar{a}_x$ V m^{-1}. Determine the total power passing through a rectangular area of sides 30 mm and 15 mm in the $z = 0$ plane.

Solution. $E(0, t) = 150$ and since it is free space

$$H = \frac{E}{\eta_p} = \frac{150}{120\pi}$$

$$P = \frac{1}{2}(E \times H^*) \text{ area} = \frac{1}{2}\frac{150 \times 150}{120\pi} \times 30 \times 15 \times 10^{-6} = \frac{1}{2}\frac{150 \times 150}{4\pi} \times 15 \times 10^{-6}$$

$$= \frac{1}{2}\frac{75 \times 75 \times 15}{\pi} \times 10^{-6} = 13.4 \text{ mW} \textbf{ Ans.}$$

Example 12.16. A 10 GHz uniform plane wave is propagating in polyethylene which has $\varepsilon' = 2.26$ and $\varepsilon'' = 2 \times 10^{-4}$. If the amplitude of the electric field intensity is 500 V m^{-1} and the material is assumed to be lossless, determine (*i*) phase constant (*ii*) the wavelength in the polyethylene (*iii*) the velocity of propagation (*iv*) The intrinsic impedance.

Solution. (*i*) Since $\beta = \omega\sqrt{\mu\varepsilon'} = \omega\sqrt{\mu_0\varepsilon_0}\sqrt{\varepsilon_r'} = \frac{\omega\sqrt{\varepsilon_r'}}{c}$

$$\text{Hence } \beta = \frac{2\pi \times 10^{10}\sqrt{2.26}}{3 \times 10^8} = \frac{2\pi \times 1.5 \times 10^{10}}{3 \times 10^8} = 314 \text{rad m}^{-1} \textbf{ Ans.}$$

(*ii*) Wavelength $\lambda = \frac{2\pi}{\beta} = \frac{2\pi}{314} = 2$ cms. **Ans.**

(*iii*) Velocity of propagation $f \cdot \lambda = 10 \times 10^9 \times 2 \times 10^{-2} = 2 \times 10^8 \, \text{m s}^{-1}$ **Ans.**

(*iv*) The intrinsic impedance $= \sqrt{\frac{\mu}{\varepsilon_0'}} = \sqrt{\frac{\mu_r \mu_0}{\varepsilon_0 \varepsilon_r'}} = \sqrt{\frac{\mu_0}{\varepsilon_0}} \cdot \frac{1}{\sqrt{\varepsilon_r'}} = \frac{\eta_0}{\sqrt{\varepsilon_r'}}$

$$= \frac{377}{\sqrt{2.26}} = \frac{377}{1.5} = 251\Omega$$

The amplitude of magnetic field intensity $\frac{E}{\eta} = \frac{500}{251} = 1.99 \text{A m}^{-1}$

Hence $|H| = 1.99 \text{ A m}^{-1}$ **Ans.**

Example 12.17. An electric field of 100 V m^{-1} plane wave of frequency 300 MHz travels in the positive z direction in an infinite lossless medium having $\varepsilon_r = 9$, $\mu_r = 1$ and $\sigma = 0$. Write complete time domain expressions for E and H fields. Also calculate average power density.

Solution. Since $\eta = \eta_0 \sqrt{\frac{\mu_r}{\varepsilon_r}} = 120\pi \sqrt{\frac{1}{9}} = 40 \ \pi\Omega$

$$\beta_0 = \frac{\omega}{c} = \frac{2\pi \times 300 \times 10^6}{3 \times 10^8} = 2\pi \, \text{rad m}^{-1}$$

and $$\beta = \beta_0 \sqrt{\mu_r \varepsilon_r} = 2\pi \sqrt{1 \times 9} = 6\pi \text{ rad m}^{-1}$$

Hence $$E = 100 \cos(\omega t - \beta z) \bar{a}_x$$

$$= 100 \cos(2\pi \times 300 \times 10^6 t - 6\pi z) \bar{a}_x \text{ V m}^{-1}$$

and $$H = \frac{E}{\eta} = \frac{100}{40\pi} \cos(2\pi \times 300 \times 10^6 t - 6\pi z) \bar{a}_y \text{ A m}^{-1}$$

The average power density is given as $\frac{1}{2} \frac{E^2}{|\eta|}$

$$S = \frac{1}{2} \frac{100^2}{40\pi} = 39.8 \text{W m}^{-2} \ \textbf{Ans.}$$

Example 12.18. Assume a conductivity of the material in example 12.17 as $\sigma = 10^5$ S m^{-1}, determine (*i*) intrinsic impedance (*ii*) propagation constant and average power density.

Solution. (*i*) If the material is conductive

$$\eta = \sqrt{\frac{j\omega\mu}{\sigma + j\omega\varepsilon}} = \sqrt{\frac{j2\pi \times 300 \times 10^6 \times 4\pi \times 10^{-7}}{10 + j2\pi \times 300 \times 10^6 \times \frac{1}{36\pi} \times 10^{-9} \times 9}}$$

$$= \sqrt{\frac{j2\pi \times 30 \times 4\pi}{10 + j2\pi \times 3 \times \frac{9}{36\pi} \times 10^{-1}}} = \frac{2\pi\sqrt{j60}}{\sqrt{10 + j\frac{3}{2} \times 10^{-1}}}$$

$$= \frac{2\pi \times 7.746\angle 45°}{\sqrt{10 + j0.15}} = \frac{2\pi \times 7.746\angle 45°}{3.16\angle 0.43°} = 15.39\angle 44.57°$$

$$= 15.39\angle 44.57° \ \textbf{Ans.}$$

(*ii*) Now propagation constant $\gamma = \sqrt{j\omega\mu(\sigma + j\omega\varepsilon)}$

$$2\pi\sqrt{j60(10 + j0.15)} = 2\pi \times 7.746\angle 45° \times 3.16\angle 0.43°$$

$$= 153.72\angle 45.43°$$

$$= 107.87 + j109.51 \text{rad m}^{-1} \ \textbf{Ans.}$$

The average power density

$$S^* = \frac{1}{2}\frac{E^2}{|\eta|} = \frac{100^2}{2 \times (15.4)} e^{-2\times 107.87z} \cos 44.57\,\bar{a}_e$$

$$= 231.3e^{-215z}\bar{a}_x \text{ watt m}^{-2} \ \textbf{Ans.}$$

PROBLEMS

P 12.1 Starting with Maxwell's equation derive wave equation in free space. State the assumptions made. What information do you get from the solution of these equations?

P 12.2 What are forward and backward waves? Explain.

P 12.3 Explain the terms: Wave number, wavelength, phase constant, attenuation constant, propagation constant, phase velocity, group velocity, intrinsic impedance.

P 12.4 Starting with suitable set of Maxwell's equations obtain the sinusoidal steady state vector form of wave equations in free space.

P 12.5 Explain complex permittivities and complex permeability of material. Derive expressions of attenuation constant and phase constant for such materials. Also derive expression for intrinsic impedance of such media.

P 12.6 What are conduction current density and displacement current density? For conducting materials show that

$$\nabla \times H_s = J_{\sigma s} + J_{ds}$$

P 12.7 Derive expression for attenuation constant and phase constant for a good dielectric material. Hence derive expression for intrinsic impedance.

P 12.8 Show that the intrinsic impedance of free space is 120π ohms.

P 12.9 Starting with Maxwell's equation, develop expression for Poynting theorem and explain the significance of each term in the expression.

P 12.10 Explain what you mean by Poynting vector and explain its significance. Derive expression for power density for a perfect dielectric. Also derive expression for average power density for a lossy dielectric.

P 12.11 Explain the phenomenon of electromagnetic wave propagation in good conductor. Derive expression for attenuation constant and phase constant for a good conducting medium.

P 12.12 Explain the phenomenon of skin effect when electromagnetic wave travel over a good conducting material. Also develop an expression for depth of penetration in such a medium.

P 12.13 Explain how concept of skin depth can be utilized for selecting size of the conductor for electromagnetic wave propagation.

P 12.14 For a good conductor show that the electric field maximum amplitude leads to the magnetic field maximum amplitude by one eighth of a cycle at every point of propagation.

P 12.15 Explain what you mean by wave polarization of electromagnetic waves. Discuss advantages and disadvantages of wave polarization.

P 12.16 Explain clearly what you mean by left circularly polarized wave and right circularly polarized wave.

P 12.17 Show that a linearly polarized wave can be obtained by the superposition of two circularly polarized waves rotating in opposite direction but at the same angular rate.

P 12.18 Determine the polarizations of the following plane waves

(*i*) $E = 1\cos(\omega t + \beta z)\bar{a}_x + 1\sin(\omega t + \beta z)\bar{a}_y$

(*ii*) $E = 1\cos(\omega t + \beta z)\bar{a}_x - 1\sin(\omega t + \beta z)\bar{a}_y$

(*iii*) $E = 1\cos(\omega t + \beta z)\bar{a}_x - 2\sin(\omega t + \beta z - \pi/4)\bar{a}_y$

Ans. (*i*) Left circular (*ii*) right circular (*iii*) right elliptical

P 12.19 Show that a plane wave with

$$E = E_{m_1}\cos(wt - \beta z)\bar{a}_x + E_{m_2}\cos(wt - \beta z + \theta)\bar{a}_y$$

is plane polarized if and only if $\theta = 0$.

P 12.20 Explain what is a "quarter wave plate" when referred to wave polarization.

P 12.21 An elliptically polarized wave traveling in the positive z-direction in air has x and y components.

$E_x = 5\sin(\omega t - \beta z)$ V/m

$E_y = 10\sin(\omega t - \beta z + 75°)$ V/m

Determine the average power per unit area covered by the wave.

Ans. 165 mW m^{-2}

P 12.22 An elliptically polarized wave in a medium with $\sigma = 0$ $\mu_r = 2$, $\varepsilon_r = 5$ has H field components normal to the direction of propagation and normal to each other of amplitude 3 and 4 Am^{-1}. Determine the average power conveyed through an area of 5m^2 normal to the direction of propagation.

Ans. 14.9 kW

P 12.23 A plane 1 GHz wave in air with peak electric field intensity of 2 V/m is incident normally on a large copper sheet. Determine the average power absorbed by the sheet per square meter of area. **Ans.** 460 nWm^{-2}

P 12.24 Determine the group velocity for a 100 MHz wave in a normally dispersive lossless medium for which the phase velocity $v = 2 \times 10^7\ \lambda^{2/3}$ m s^{-1}.

Ans. 13.9 Mm s^{-1}

P 12.25 A parallel plate capacitor is filled with a dielectric with *p.f.* 0.004 and $\varepsilon_r = 14$. The plates are 350 mm square and the distance of separation between them is 12 mm. If 300 V rms at 3 MHz is applied to the capacitor, determine the power dissipated as heat.

Ans. 8.6 Watts

P 12.26 A plane wave is given by $E = E_0 \cos(\omega t - \beta z)$.

(*a*) Show that the resultant superposition of two waves of this type, of equal magnitude and of angular frequencies $\omega_0 + \Delta\omega$ and $(\omega_0 - \Delta\omega)$ and two corresponding wave numbers $\beta_0 + \Delta\beta$ and $(\beta_0 - \Delta\beta)$ can be expressed as $E = 2\, E_0 \cos(\omega_0 t - \beta_0 z) \cos(\Delta\omega - \Delta\beta z)$.

(*b*) Discuss the statement that the group velocity of a wave packet (a group of waves with a small range of ω and β, which move together) in a dielectric medium is given by $v_g = d\omega/d\beta$.

P 12.27 The electric field intensity of a plane wave

$$E = (a_x + a_y)10^{-3} \cos(2\pi \times 10^7 t - \beta z) \text{ V m}^{-1}$$

propagates in ferrite material that has $\mu_r = 1000$ $\varepsilon_r = 3$ and $\sigma = 10^{-5}$ S m^{-1}. Determine β and λ of the wave (*ii*) Determine its average intensity (*iii*) Write expression for H of the wave.

Ans. (*i*) $\lambda = 0.548$ m, $\beta = 11.47$/m, (*ii*) 1.45×10^{-10} W m^{-2}

(*iii*) $H = (a_y - \bar{a}_x)1.45 \times 10^{-7} \cos(2\pi \times 10^7 t - 11.47z)$ A m^{-1}

P 12.28 A plane wave of 1 MHz is propagating in fresh water. Assume $\varepsilon'' = 0$, $\mu_r = 1$ and at 1 MHz $\varepsilon_r' = 81$. Determine phase constant β, wavelength λ, phase velocity, the intrinsic impedance and expressions for field intensity if its maximum amplitude is 1 V m^{-1}.

Ans. $\beta = 0.19$ rad m^{-1}, $\lambda = 33$ m, $v_p = 3.3 \times 10^7$ m s^{-1}

$\eta = 42\ \Omega$ $E = 1 \cos(2\pi \times 10^6 t - 0.19z)$ V m^{-1}

P 12.29 A plane wave of 1 MHz is propagating in sea water for which $\sigma = 4$ S m^{-1} and $\varepsilon'_r = 81$. Determine loss tangent, skin depth, the wavelength and the phase velocity.

Ans. Tan $\theta = 8.9 \times 10^2$, $\delta = 25$ cms, $\lambda = 1.6$ m, $v_p = 1.6 \times 10^6$ m s^{-1}

P 12.30 A uniform plane wave propagating along positive z-direction has the propagation constant $jk = (0.2 + j1.5)$ per meter and intrinsic impedance $\eta = 450 + j60\ \Omega$. If $\omega = 300$ M rad sec^{-1} Determine μ, ε', ε'' for the medium.

Ans. $\mu = 2.28 \times 10^{-6}$ H m^{-1}, $\varepsilon' = 1.07 \times 10^{-11}$ F m^{-1}

and $\varepsilon'' = 2.9 \times 10^{-12}$ F m^{-1}

P 12.31 A good conductor is planar in form and it carries a uniform plane wave that has a wavelength of 100.3 mm and a velocity of 3×10^5 m s^{-1}. Assume the conductor to be nonmagnetic, determine the frequency and the conductivity. **Ans.** 1 GHz, 1.1×10^5 S m^{-1}

P 12.32 In free space $E(z, t) = 10^3 \sin(\omega t - \beta z)\bar{a}_y$ V m^{-1}. Determine $H(z, t)$.

Ans. $H(z, t) = -\dfrac{10^3}{377}\sin(\omega t - \beta z)\bar{a}_x \text{ A m}^{-1}$

P 12.33 Determine the intrinsic impedance η, the propagation constant γ and the wave velocity u for a conducting medium where $\sigma = 58$ MS m^{-1} $\mu_r = 1$ at a frequency 100 MHz.

Ans. $\gamma = 2.14 \times 10^5 \angle 45°$, $\eta = 3.69 \times 10^{-3} \angle 45°$, $u = 4.15 \times 10^3$ m s^{-1}

Multiple Choice Questions

1. A lossy capacitor with a dielectric of permittivity ε and conductivity σ operates at a frequency ω, the loss tangent for the capacitor is given by

(a) $\frac{\omega\sigma}{\varepsilon}$ (b) $\omega\varepsilon/\sigma$

(c) $\sigma/\omega\varepsilon$ (d) $\sigma\omega\varepsilon$

2. The ratio of velocity of propagation of electromagnetic waves in an overhead line and in a cable with relative permittivity 4 is

(a) 0.25 (b) 0.5

(c) 2.0 (d) 4.0

3. If an electromagnetic wave travelling in a dissipative medium in the x direction has peak electric field E_1, E_2, E_3 at points x, $x + 1$, $x + 2$ respectively, then

(a) $E_2 = \sqrt{E_1 E_3}$ (b) $E_1 = \sqrt{E_2 E_3}$

(c) $E_3 = \sqrt{E_1 E_2}$ (d) $\frac{E_1 + E_3}{2} = E_2$

4. The relation between the wave impedance Z_ω and the intrinsic impedance η for a TEM wave is given by

(a) $Z_\omega = 0.5\eta$ (b) $Z_\omega = \frac{1}{\eta}$

(c) $Z_\omega = 2\eta$ (d) $Z_\omega = \eta$

5. The voltage reflection coefficient for a wave travelling over a transmission line terminated in an impedance equal to the surge impedance of the line is

(a) zero (b) 1

(c) −1 (d) ∞

6. Match List I and List II and select the correct answer using the code given below in the lists.

List I	**List II**
A. Ampere's law	1. $F = Q[\bar{E} + (\bar{V} \times \bar{B})]$
B. Biot Savarts' law	2. $B = \frac{\mu_0 I}{4\pi} \int \frac{dl \times R}{R^2}$
C. Lorentz Force law	3. $\nabla \times B = \mu_0 J$
D. Magnetic flux law	4. $\nabla \cdot B = 0$

Codes

	A	B	C	D
(*a*)	3	2	1	4
(*b*)	2	3	4	1
(*c*)	3	2	4	1
(*d*)	2	3	1	4

7. Match list I with list II

List I	**List II**
A. $\oint_s B \cdot ds$	1. $\int \rho dv$
B. $\oint_l E \cdot dl$	2. zero
C. $\oint_s D \cdot ds$	3. $(\sigma + j\omega\varepsilon)\int E \cdot ds$
D. $\oint_l H \cdot dl$	4. $-j\omega\mu \int H \cdot ds$

Codes

	A	B	C	D
(*a*)	2	4	3	1
(*b*)	2	4	1	3
(*c*)	4	2	1	3
(*d*)	4	2	3	1

8. Match list I with list II and select the correct answer using the codes given below the list

List I	**List II**
A. $\nabla \times \bar{H} = \bar{J} + \frac{\partial \bar{D}}{\partial t}$	1. Faraday's law
B. $\nabla \times \bar{E} = \frac{\partial \bar{D}}{\partial t}$	2. Gauss' law
C. $\nabla \cdot D = \rho$	3. Ampere's law

Codes

	A	B	C
(a)	3	1	2
(b)	2	1	3
(c)	3	2	1
(d)	1	2	3

9. A circular current carrying loop is shown in the given figure. The vector potential at point P on the axis of the loop will be

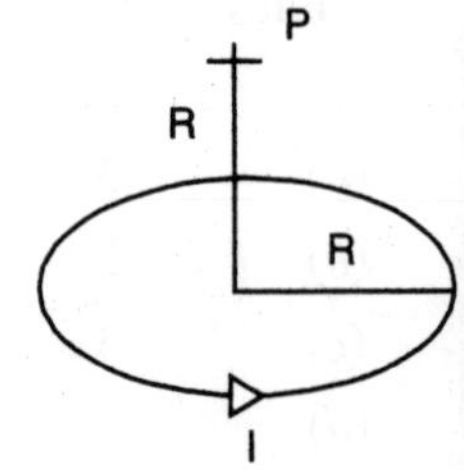

Fig. Q. 9

(a) zero

(b) $\mu_0 \frac{(2\pi R)I}{R}\bar{a}_z$

(c) $\mu_0 \frac{(2\pi R)I}{\sqrt{2}R}\bar{a}_z$

(d) $\mu_0 \frac{(\pi R^2)I}{R}\bar{a}_z$

10. The electric field of a uniform plane wave is given by $\bar{E} = 15\sin(\omega t - \pi z)\bar{a}_x + 15\cos(\omega t - \pi z)$ $\bar{a}_y$ V/m. The nature of polarization of the wave is.

(a) Linear

(b) Elliptical

(c) Right-circular

(d) Left-circular

11. The electric field of a uniform plane wave is given by

$$E = 10\cos(3\pi \times 10^8 t - \pi z)\bar{a}_x$$

List I	List II
A. Phase velocity	1. 2
B. Wavelength	2. 3.14
C. Frequency	3. 377
D. Phase constant	4. 1.5×10^8
	5. 3×10^8

	A	B	C	D
(a)	5	4	3	2
(b)	3	4	2	1
(c)	4	3	1	2
(d)	5	1	4	2

12. The electric field intensity of a uniform plane wave in air is

$$E = 20e^{-jsz}\bar{a}_x - 25e^{jsz}\bar{a}_y \text{ V/m}$$

The components $\bar{H}_x$ and $\bar{H}_y$ of $\bar{H}$ are

(a) $H_x = \frac{20}{120\pi}\bar{a}_x$ $\qquad H_y = -\frac{25}{120\pi}\bar{a}_y$

(*b*) $H_x = -\frac{25}{120\pi}\bar{a}_x$ $\quad H_y = -\frac{20}{120\pi}\bar{a}_y$

(*c*) $H_x = \frac{25}{120\pi}\bar{a}_x$ $\quad H_y = -\frac{20}{120\pi}\bar{a}_y$

(*d*) $H_x = -\frac{20}{120\pi}\bar{a}_x$ $\quad H_y = -\frac{25}{120\pi}$

13. Conservation of charge implies that

(*a*) $\oint_S \bar{J}\cdot d\bar{s} = -\frac{d}{dt}\oint_v \rho\, dv$ $\quad$ (*b*) $\oint \bar{J}\cdot d\bar{s} = 0$

(*c*) $\oint \bar{J}\cdot d\bar{s} = \oint_v \rho dv$ $\quad$ (*d*) $\oint \bar{J}\cdot d\bar{s} = \rho$

14. Two vectors $\vec{A}$ and $\vec{B}$ are such that $\bar{A}+\bar{B} = n\bar{A}$, where *n* is a positive scalar. The angle between $\bar{A}$ and $\bar{B}$ is

(*a*) $\frac{\pi}{2}$ $\quad$ (*b*) $\frac{3\pi}{4}$

(*c*) π $\quad$ (*d*) 2π

15. List II gives mathematical expressions for the variables given in list I with list II and select the correct answer using the code given below the lists.

List I	**List II**
A. Intrinsic impedance	1. $1/\sqrt{\mu\varepsilon}$
B. velocity of wave propagation	2. $\sqrt{\frac{\mu}{\varepsilon}}$
C. Skin depth	3. $\frac{1}{\sqrt{\pi f \mu\sigma}}$
D. Attenuation constant	4. $\frac{1}{f\sqrt{\mu\omega}}$
	5. $\sqrt{\omega}\,\mu\sigma$

Codes

	A	B	C	D
(*a*)	1	2	3	4
(*b*)	2	1	4	5
(*c*)	2	1	3	5
(*d*)	1	2	5	3

16. A plane wave propagates in z direction with field components E_y and H_x with a time and z-dependence of the form $e^{(j\omega t - j\beta z)}$, E_y/H_x is given by

(a) $\frac{\omega\mu}{\beta}$ (b) $\frac{\omega\varepsilon}{\beta}$

(c) $-\frac{\beta}{\omega\mu}$ (d) $-\frac{\beta}{\omega\varepsilon}$

17. Match list I and list II and select the correct answer using the codes given below the lists.

List I	List II (Intrinsic Impedance)
A. Lossless dielectric	1. $\sqrt{\frac{j\omega\mu}{\sigma + j\omega\varepsilon}}$
B. Good conductor	2. $\sqrt{\frac{\mu}{\varepsilon}}\left(1 + j\frac{\sigma}{2\omega\varepsilon}\right)$
C. Poor conductor	3. $\sqrt{\frac{\mu}{\varepsilon}}$
D. Lossy	4. $(1+j)\sqrt{\frac{\omega\mu}{2\sigma}}$

Codes

	A	B	C	D
(a)	4	3	2	1
(b)	4	3	1	2
(c)	3	4	2	1
(d)	3	4	1	2

18. In a dielectric medium of relative permittivity 4, the electric field intensity is 20 sin $(10^8 t - \beta z)$ $\bar{a}_y$ V/m and z being in meters. The phase shift constant β is

(a) $\frac{1}{3}$ rad m^{-1} (b) $\frac{2}{3}$ rad m^{-1}

(c) $\frac{2\pi}{3}$ rad m^{-1} (d) $\frac{1}{6\pi}$ rad m^{-1}

19. The unit of poynting vector is

(a) power (b) power density

(c) energy (d) energy density

20. The equation $\bar{E} = E\sin(\omega t - \beta z)\bar{a}_x + E\sin(\omega t - \beta z)\bar{a}_y$ represents

(a) Left-circularly (b) Right-circularly

(c) Linearly (d) Elliptically polarized wave

21. The intrinsic impedance of free space is
   (*a*) 377 Ω   (*b*) $\sqrt{\mu_0 \varepsilon_0}$
   (*c*) $j\sqrt{\dfrac{\varepsilon_0}{\mu_0}}$ (*d*) $\sqrt{\dfrac{\varepsilon_0}{\mu_0}}$

22. If the electric field components of a wave is $E = \cos(6 \times 10^8 \pi t + 50z)\ \bar{a}_x$ V/m then the wave
   (*a*) propagates in $-x$ direction   (*b*) amplitude is $\sqrt{2}$ V/m.
   (*c*) is not travelling in free space   (*d*) propagation in $+z$ direction.

23. A plane electromagnetic wave is travelling in a highly dissipative medium in the direction *ABC* as shown in the figure. The electric field E_A, E_B and E_C at points *A*, *B* and *C* respectively are related as

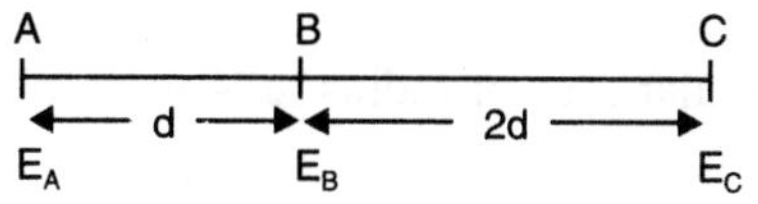

Fig. Q. 23

   (*a*) $E_A^2 E_C = E_B^2$ (*b*) $E_A E_C^2 = E_B^2$
   (*c*) $E_A E_C = E_B^2$ (*d*) $E_A E_B = E_C^2$

24. In the source free wave equation

$$\nabla^2 E - \mu_0 \varepsilon_0 \mu \varepsilon \frac{\partial^2 E}{\partial t^2} - \mu_0 \mu_\sigma \frac{\partial E}{\partial t} = 0$$

   The term responsible for the attenuation of the wave is
   (*a*) $\mu_0 \mu \sigma \dfrac{\partial E}{\partial t}$ (*b*) $\mu_0 \varepsilon_0 \mu \varepsilon \dfrac{\partial^2 E}{\partial t^2}$
   (*c*) $\nabla^2 E$ (*d*) (*a*) and (*b*)

25. Three media are characterized by
   1. $\varepsilon_r = 8$, $\mu_r = 2$, $\sigma = 0$
   2. $\varepsilon_r = 1$, $\mu_r = 9$, $\sigma = 0$
   3. $\varepsilon_r = 4$, $\mu_r = 4$, $\sigma = 0$

   The values of the intrinsic impedances of the media 1, 2, 3 respectively are
   (*a*) 188, 377, 1131 Ω   (*b*) 377, 1131, 188 Ω
   (*c*) 188, 1131, 377 Ω   (*d*) 1131, 188, 377 Ω

26. Ohm's law in point form in field theory can be expressed as
   (*a*) $V = RI$ (*b*) $J = E/\sigma$
   (*c*) $J = \sigma E$ (*d*) $R = \dfrac{\rho L}{A}$

27. If a field vector $\bar{B}$ is solenoidal, which of these is true?

(*a*) $\oint_L B \cdot d\bar{l} = 0$ (*b*) $\oint_S B \cdot d\bar{s} = 0$

(*c*) $\nabla \times B = 0$ (*d*) $\nabla . B = 0$

28. For linear isotropic materials both $\bar{E}$ and $\bar{H}$ have the time dependence $e^{j\omega t}$ and regions of interest are free of charge. The value of $\nabla \times H$ is given by

(*a*) $\sigma\bar{E}$ (*b*) $j\omega\varepsilon\bar{E}$

(*c*) $\sigma\bar{E} + j\omega\varepsilon E$ (*d*) $\sigma E - j\omega\varepsilon E$

29. In free space $E(z, t) = 50\cos(\omega t - \beta z)\bar{a}_x$ V m^{-1} and $H(z, t) = \frac{5}{12\pi}\cos(\omega t - \beta z)\bar{a}_y$ A m^{-1}. The average power crossing a circular area of radius $\sqrt{24}$ m in plane z = constant is

(*a*) 200 W (*b*) 250 W
(*c*) 300 W (*d*) 350 W

30. If $\bar{A}$ and $\bar{J}$ are the vector potential and current density vectors associated with a coil, then $\int A \cdot J \, dV$ has the units of

(*a*) Flux linkage (*b*) Power
(*c*) Energy (*d*) Inductance

31. The various symbols have their usual meaning, which of the following expression represents characteristic impedance of the transmission line

(*a*) $\frac{\gamma}{R + j\omega L}$ (*b*) $\frac{R + j\omega L}{\gamma}$

(*c*) $\frac{G + j\omega C}{\gamma}$ (*d*) $\sqrt{\frac{G + j\omega C}{R + j\omega L}}$

32. A plane electromagnetic wave $(\bar{E}_i, \bar{H}_i)$ travelling in a perfect dielectric medium of surge impedance z strikes normally an infinite perfect dielectric medium of surge impedance $2z$. If the refracted *e.m.* wave is $(\bar{E}_r, H_r)$, the ratio of $\frac{E_i}{E_r}$ and $\frac{H_i}{H_r}$ are respectively

(*a*) 1/3 and 3/2 (*b*) 3/2 and 1/3

(*c*) 3/4 and 3/2 (*d*) 3/4 and 2/3

33. For a perfect conductor, the field strength at a distance equal to the skin depth is x% of the field strength at its surface. The value of x% is

(*a*) zero (*b*) 50%
(*c*) 36% (*d*) 26%

34. The characteristic impedance of a transmission line increases with increase in

(a) resistance and capacitance per unit length.
(b) resistance and inductance per unit length.
(c) capacitance and conductance per unit length.
(d) capacitance and inductance per unit length.

35. Which of the equations are not Maxwell's equation?

1. $\nabla \cdot J = -\frac{\partial \rho v}{\partial t}$ 2. $\nabla \cdot D = \rho v$

3. $\nabla \cdot E = -\frac{\partial B}{\partial t}$ 4. $\oint H \cdot dl = \oint_S \left\{ \sigma E + \varepsilon \frac{\partial E}{\partial t} \right\} \cdot ds$

Select the correct answer using the codes given below

Codes

(a) 2 and 4 (b) 1 alone
(c) 1 and 3 (d) 1 and 4

36. In free space $H(z, t) = 0.1\cos(4 \times 10^7 t - \beta z)\bar{a}_x$ A/m. The expression for $E(z, t)$ is

(a) $E(z, t) = 37.7\cos(4 \times 10^7 t - \beta z)\bar{a}_y$ (b) $E(z, t) = 2.65 \times 10\cos(4 \times 10^7 t - \beta z)\bar{a}_z$
(c) $E(z, t) = 37.7\cos(4 \times 10^7 t - \beta z)\bar{a}_x$ (d) $E(z, t) = -37.7\cos(4 \times 10^7 t - \beta z)\bar{a}_y$

37. A charge Q is uniformly distributed throughout the sphere of radius a. Assuming the potential at infinity to zero the potential at $r = b < a$ is

(a) $-\int_{\infty}^{b} \frac{Qr}{4\pi\varepsilon_0 a^3} dr$ (b) $-\int_{\infty}^{b} \frac{Q}{4\pi\varepsilon_0 r^2} dr$

(c) $-\int_{\infty}^{a} \frac{Q}{4\pi\varepsilon_0 r^2} dr - \int_{\infty}^{b} \frac{Qr}{4\pi\varepsilon_0 a^3} dr$ (d) $-\int_{\infty}^{a} \frac{Q}{4\pi\varepsilon_0 r^2} dr$

38. Which one of the following pairs is not correctly matched?

(a) $\int_S D \cdot ds = \int_v \nabla \cdot D\, dv$ Gauss' Theorem

(b) $\int \bar{D} \cdot d\bar{s} = \int_v \rho dv$ Gauss' law

(c) $\varepsilon_r v = -\frac{d\phi_m}{dt}$ Coulombs' law

(d) $\oint_l E \cdot dl = \int_S (\nabla \times E) ds$ Strokes Theorem

39. Interface of two regions of two magnetic materials is current free. The region 1 for which $\mu_{r_1} = 2$ is defined by $z < 0$ and region 2 $z > 0$ has $\mu_{r_2} = 1$. If $B_{a_1} = 1.2\bar{a}_x + 0.8\bar{a}_y + 0.4\bar{a}_z$ then $\bar{H}_1$ is

(*a*) $\frac{1}{\mu_0}\left[0.6\bar{a}_x + 0.4\bar{a}_y + 0.4\bar{a}_z\right]$A/m (*b*) $\frac{1}{\mu_0}\left[1.2\bar{a}_x + 0.8\bar{a}_y + 0.8\bar{a}_z\right]$A/m

(*c*) $\frac{1}{\mu_0}\left[1.2\bar{a}_x + 0.4\bar{a}_y + 0.4\bar{a}_z\right]$A/m (*d*) $\frac{1}{\mu_0}\left[0.6\bar{a}_x + 0.4\bar{a}_y + 0.8\bar{a}_z\right]$A/m

40. A plane slab of dielectric having dielectric constant-5, placed normal to a uniform field with a flux density of 2 C/m^2 is uniformly polarized. The polarization of the slab is
(*a*) 0.4 C/m^2 (*b*) 1.6 C/m^2
(*c*) 2.0 C/m^2 (*d*) 6.4 C/m^2

41. Given that $H = 0.5\ e^{-0.1\,x}\sin(10^6 t - 2x)\,\bar{a}_x$ A/m, which one of the following statements is not correct?
(*a*) Wave is linearly polarized along $\bar{a}_z$
(*b*) The velocity of the wave is 5×10^5 m/s
(*c*) The complex propagation constant is $(0.1 - j^2)$
(*d*) The wave is traveling along $\bar{a}_x$.

42. A lossless transmission line of characteristic impedance z_0 and $l < \lambda/4$ is terminated at the load end by a short circuit. Its input impedance z_{in} is
(*a*) $z_{in} = -jz_0 \tan \beta l$ (*b*) $z_{in} = jz_0 \cot \beta l$
(*c*) $z_{in} = j\, z_0 \tan \beta l$ (*d*) $z_{in} = -jz_0 \cot \beta l$

43. What will be the reflected wave for an elliptically polarized wave incident on the interface of a dielectric at the Brewester angle?
(*a*) Elliptically polarized (*b*) Linearly polarized
(*c*) Right circularly polarized (*d*) Left circularly polarized

44. In a loss less medium the intrinsic impedance $\eta = 60\ \pi$ and $\mu_r = 1$. The value of the dielectric constant ε_r is
(*a*) 2 (*b*) 1
(*c*) 4 (*d*) 8

45. A flat slab of dielectric $\varepsilon_r = 5$ is placed normal to a uniform field with a flux density $D = 1$ C/m^2. The slab is uniformly polarized. The polarization of the slab in C/m^2 is
(*a*) 0.8 (*b*) 1.2
(*c*) 4 (*d*) 6

46. The wavelength of a wave propagating in a wave guide is
(*a*) smaller than the free space wavelength.
(*b*) greater than the free space wavelength.
(*c*) directly proportional to the group velocity.
(*d*) inversely proportional to the group velocity.

47. For a lossless dielectric medium, the phase constant for a traveling wave, β is proportional to
(*a*) ε_r (*b*) $\sqrt{\varepsilon_r}$
(*c*) $\frac{1}{\varepsilon_r}$ (*d*) $\frac{1}{\sqrt{\varepsilon_r}}$

48. Match list I with list II and select the correct answer using the codes given below in the lists.

List I	List II
A. Continuity equation	1. $\nabla \times H = J + \frac{\partial D}{\partial t}$
B. Amperes' law	2. $J = \frac{\partial D}{\partial t}$
C. Displacement current	3. $\nabla \times E = -\frac{\partial B}{\partial t}$
D. Faraday's law	4. $\nabla \cdot J = -\frac{\partial \rho_v}{\partial t}$

Codes

	A	B	C	D
(*a*)	4	3	2	1
(*b*)	4	1	2	3
(*c*)	2	3	4	1
(*d*)	2	1	4	3

49. The vector $\bar{A} = 5x^2 \sin \pi x \bar{a}_x$, the divergence of vector A at $x = \frac{1}{2}$ is

(*a*) 2 (*b*) 3
(*c*) 5 (*d*) 6

50. The divergence of a field directed radially and varying inversely with the radius within a cylinder is

(*a*) π (*b*) zero
(*c*) π/2 (*d*) ∞

51. A scalar function *V* is independent of *x* and *z* and the gradient of *V* is $20y\bar{a}_y$. If $V = 10$ at $y = 0$, the general expression for *V* is given as

(*a*) $10y^2$ (*b*) $5y^2 + 10$
(*c*) $10y^2 + 10$ (*d*) $10y^2 - 10$

52. Two positive charges *Q* and 4*Q* are separated by a distance *d* in air. A third charge is so placed that the entire system is in equilibrium. The magnitude and location of the charge is

(*a*) $-\frac{3Q}{8}, \frac{9}{64}d$ (*b*) $-\frac{4Q}{9}, \frac{2}{3}d$
(*c*) $-\frac{2Q}{5}\sqrt{\frac{2}{5}}d$ (*d*) None of these

53. Two infinite uniform sheets of charge each with densities ρ_s are located at $x = \pm 1$. Electric field intensity between the plates is given as

(*a*) $\frac{\rho_s}{\varepsilon_0}\bar{a}_x$ (*b*) $-\frac{\rho_s}{\varepsilon_0}\bar{a}_y$
(*c*) zero (*d*) $\frac{2\rho_s}{\varepsilon_0}\bar{a}_x$

54. Within a sphere of radius a the charge density is $\rho_v = 1 - \left(\frac{r}{a}\right)^3$, the electric field intensity E at $r = 4a$ is

(a) $\frac{a^3}{96\varepsilon_0}$ (b) $\frac{a^2}{48\varepsilon_0}$

(c) $\frac{a^2}{96\varepsilon_0}$ (d) $\frac{a}{96\varepsilon_0}$

55. An electric field $E = xy^2\bar{a}_x - xz^2\bar{a}_y + xyz\bar{a}_z$. The x-component of its curl at (2, 1, 3) is given as

(a) 6 (b) 12

(c) 18 (d) 24

56. A hollow sphere of radius R has a surface charge density ρ_s C/m^2. The electric field intensity at a distance $r > R$ from the centre of the sphere is

(a) $\frac{4\rho_s}{\varepsilon_0}\left(\frac{R}{r}\right)^2 \bar{a}_r$ (b) $\frac{\rho_s}{\varepsilon_0}\left(\frac{R}{r}\right)^2 \bar{a}_r$

(c) $\frac{4}{3}\pi\left(\frac{R}{r}\right)^2 \frac{\rho_s}{\varepsilon_0}$ (d) None of these

57. The volume charge density corresponding to $D = 10\sin^2\phi a_r + ra_\phi + \frac{10z}{r}\cos^2\phi\, a_z$ depends upon:

(a) ϕ only (b) z only

(c) ϕ and z (d) neither ϕ nor z

58. A circular loop 6 m in diameter carries a uniform charge density of 1μ C/m. The potential at a point on its axis 4 m away from the centre of the loop is

(a) 33 kV (b) 34 kV

(c) 45 kV (d) 48 kV

59. For the electric fields E shown at the interface of two dielectrics ε_1 and ε_2, the angles θ_1 and θ_2 are related as

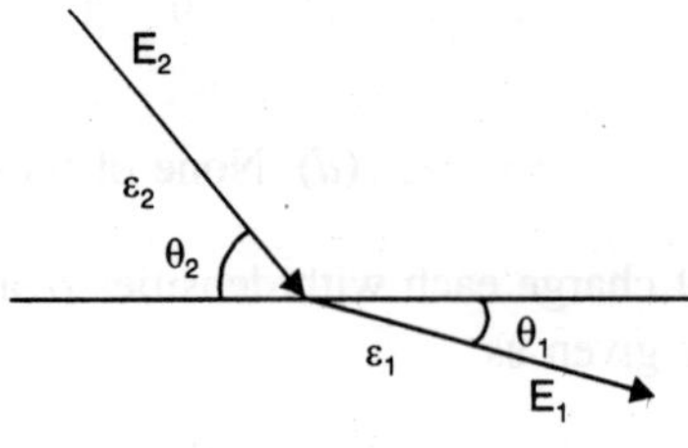

Fig. Q. 59

(a) $\varepsilon_1 \cos\theta_1 = \varepsilon_2 \cos\theta_2$ (b) $\varepsilon_2 \cos\theta_1 = \varepsilon_1 \cos\theta_2$

(c) $\varepsilon_1 \tan\theta_1 = \varepsilon_2 \tan\theta_2$ (d) $\varepsilon_1 \tan\theta_2 = \varepsilon_2 \tan\theta_1$

60. A circular disc of radius 4 m with a charge density $\rho_s = 12 \sin\phi\ \mu C/m^2$ is enclosed by surface S. The net flux that crosses the surface S is
    (*a*) 2π coulomb (*b*) 4π coulomb
    (*c*) 1 coulomb (*d*) zero coulomb

61. The electric flux density $D = 15\, x\, \bar{a}_x$ C/m^2. The flux crossing a 1 m^2 area that is normal to the x-axis at $x = 3$ m is
    (*a*) 30C (*b*) 45C
    (*c*) 5C (*d*) None of these

62. In spherical coordinates electric flux density $D = \frac{5r^2}{4}\bar{a}_r$ when $r \le 2$. The charge density in the region is
    (*a*) zero (*b*) 5
    (*c*) $5r$ (*d*) $\frac{5}{4}r$

63. A conductor of uniform cross section and 150 m long has a voltage drop of 1.3 V and a current density of 5×10^5 C/m^2. The conductivity of the conductor material is (approx.)
    (*a*) 8.5×10^7 S m^{-1} (*b*) 5.8×10^7 S m^{-1}
    (*c*) 7.6×10^7 S m^{-1} (*d*) 10^8 S m^{-1}

64. If the current density in a conductor varies with r as $J = \frac{10^3}{r}$. The total current in a circular conductor of radius 2 mm is
    (*a*) 4 A (*b*) 4π A
    (*c*) 8 A (*d*) 8π A

65. At the interface of two dielectrics with no free charges at the boundary
    (*a*) The tangential component of E is continuous across the interface.
    (*b*) The normal component of flux density is continuous.
    (*c*) The tangential component of flux density is continuous.
    (*d*) (*a*) and (*b*)

66. For a solid conductor (cylindrical) of radius a, if the current I is uniformly distributed over the cross-section, the magnetic field intensity H
    (*a*) inside the conductor is inversely proportional to r.
    (*b*) inside the conductor is directly proportional to r.
    (*c*) outside the conductor is inversely proportional to r^2.
    (*d*) none of the above is correct.

67. Tick out the correct statement
    (*a*) The divergence of a curl is a zero vector.
    (*b*) The curl of a gradient is never zero.
    (*c*) The curl of a gradient is a zero vector.
    (*d*) None of the above is correct.

68. The maximum torque on a 100 turn rectangular coil 0.2 m by 0.3 m carrying a current of 2 A in a field $B = 6$ Tesla is
    (*a*) 66 Nm (*b*) 68 Nm
    (*c*) 72 Nm (*d*) 74 Nm

69. Three insulating materials with same maximum working stress and permittivity 2.5, 3.0, 4.0 are used in a single core cable. The location of the materials with respect to the core of the cable will be
    (*a*) 2.5, 3.0, 4.0 (*b*) 3.0, 2.5, 4.0
    (*c*) 4.0, 3.0, 2.5 (*d*) 4.0, 2.5, 3.0

70. If δ is the loss angle of the cable, its p.f. is
    (*a*) sin δ (*b*) cos δ
    (*c*) p.f. is independent of δ (*d*) none of these

71. In a lossy dielectric material, the ratio of conduction current density to displacement current density is given as
    (*a*) $\frac{\sigma\varepsilon}{\omega}$ (*b*) $\frac{\omega}{\sigma\varepsilon}$
    (*c*) $\frac{\sigma}{\omega\varepsilon}$ (*d*) $\frac{\sigma\omega}{\varepsilon}$

72. An interface (current free) separating material 1 with σ_1 and μ_1 and material 2 with σ_2 and μ_2 parameters. Tick out the correct one
    (*a*) $\mu_1 B_{n_1} = \mu_2 B_{n_2}$ (*b*) $\frac{\tan\theta_1}{\tan\theta_2} = \frac{\mu_{r_1}}{\mu_{r_2}}$
    (*c*) $\mu_1 H_{t_1} = \mu_2 H_{t_2}$ (*d*) $H_{t_1} = H_{t_2}$

73. Inside a right circular cylinder $\mu_{r_1} = 100$. The exterior is free space. If $\overline{B}_1 = 2.5\overline{a}_\phi$ Tesla inside the cylinder, B_2 just outside the cylinder is
    (*a*) $2.5\overline{a}_\phi$ (*b*) $2.5\times10^{-2}\, a_r$
    (*c*) $2.5\overline{a}_r$ (*d*) $250\overline{a}_z$

74. For a good conductor, attenuation constant α, phase constant β and depth of penetration are related as
    (*a*) $\alpha = \beta = 1/\delta$ (*b*) $\alpha = \delta$
    (*c*) $\beta > \alpha$ (*d*) $\alpha\delta = \beta$

75. E_i and H_i waves in free space are travelling and are normally incident on the interface with a perfect dielectric for which $\varepsilon_r = 4$. The ratio of reflected to incident E_i/E_r, H_i/H_r.
    (*a*) + 0.333, – 0.333 (*b*) – 0.333, 0.333
    (*c*) 0.333, 0.333 (*d*) – 0.333, – 0.333

76. In the above question the ratio of the transmitted to incident waves are
    (*a*) 0.666, –1.333 (*b*) 0.666, 1.333
    (*c*) – 0.666, –1.333 (*d*) – 0.666, 1.333

77. An electric field in free space is given as $E(z, t) = 10^3 \sin(\omega t - \beta z)\bar{a}_y$ V/m. The magnetic field $H(z, t)$ is given as

(a) $H(z, t) = \frac{10^3}{120\pi}\sin(\omega t - \beta z)\bar{a}_y$

(b) $H(z, t) = \frac{10^3}{120\pi}\cos(\omega t - \beta z)\bar{a}_x$

(c) $H(z, t) = -\frac{10^3}{120\pi}\sin(\omega t - \beta z)\bar{a}_x$

(d) $H(z, t) = -\frac{10^3}{12\pi}\sin(\omega t - \beta z)\bar{a}_y$

78. In free space $E(z, t) = 50 \cos(\omega t - \beta z)\bar{a}_x$ V/m. The average power crossing a circular area of radius 2 m in the plane z = constant is

(a) 65.1 W
(b) 41.67 W
(c) 130 W
(d) 83 W

79. The inductance of free space is

(a) 40 nH/m
(b) 50 nH/m
(c) 60.6 nH/m
(d) 31.4 nH/m

80. The magnitude of polarization P for a dielectric material in which $E = \pi$ MV/m and $k_e = 6$ is

(a) 0.167 mC/m^2
(b) 16.7 mC/m^2
(c) 0.5 mC/m^2
(d) none of these.

81. The electric and magnetic fields in free space in spherical coordinates system are given by

$$E = \frac{10}{r}\sin\theta\cos\left(\omega t - \frac{4\pi}{3}\right)\bar{a}_\theta \text{ V/m}, \; H = \frac{10}{120\pi r}\sin\theta\cos\left(\omega t - \frac{4\pi}{3}\right)a_\phi \text{ W/m}^2$$

the instantaneous power flow is given as

(a) $\frac{100}{120\pi r^2}\sin^2\theta\cos^2\left(\omega t - \frac{4\pi}{3}\right)\bar{a}_\phi$

(b) $\frac{50}{120\pi r^2}\sin^2\theta\cos^2\left(\omega t - \frac{4\pi}{3}\right)\bar{a}_r$

(c) $\frac{100}{120\pi r^2}\sin^2\theta\cos^2\left(4t - \frac{4\pi}{3}\right)\bar{a}_r$

(d) $\frac{50}{120\pi r^2}\sin^2\theta\cos^2\left(\omega t - \frac{4\pi}{3}\right)\bar{a}_r$

82. The angle of the intrinsic impedance for a good conductor is

(a) $\frac{\pi}{4}$
(b) $\frac{\pi}{8}$
(c) $\frac{3\pi}{8}$
(d) none of these

83. The electric field E in a lossless, linear, homogeneous, isotropic charge free medium characterized by μ, ε is given by

$$E_1 = E_{10}\cos(\omega t - \beta z)\bar{a}_x \quad \left(\beta = \omega\sqrt{\mu\varepsilon}\right)$$

The magnetic field H_1 is given as

(a) $H_1 = \frac{E_{10}}{\eta}\sin(\omega t - \beta z)\bar{a}_x$ (b) $H_1 = \frac{E_{10}}{\eta}\cos(\omega t - \beta z)\bar{a}_y$

(c) $H_1 = \frac{E_{10}}{\eta}\sin\left(\omega t - \beta z - \frac{\pi}{4}\right)\bar{a}_z$ (d) $H_1 = \frac{E_{10}}{\eta}\sin\left(\omega t - \beta z - \frac{3\pi}{4}\right)\bar{a}_y$ where $\eta = \sqrt{\frac{\mu}{\varepsilon}}$

84. A 5 GHz plane wave is propagating in a material with $\varepsilon_r = 4$, $\mu_r = 1$ and $\sigma = 0$. If the electric field $E = 10\cos(10\pi \times 10^9 t - \beta z)\,\bar{a}_x$ V/m, phase velocity of the wave is given as
(a) 1.89×10^8 ms^{-1} (b) 1.5×10^8 ms^{-1}
(c) 1.5×10^7 ms^{-1} (d) 3×10^7 ms^{-1}

85. The wavelength of the wave in the previous question is
(a) 0.3 cm (b) 0.45 cm
(c) 3 cm (d) 4.5 cm

86. The electric field E plane wave is given by

$$E = E_{m_1}\cos(\omega t - \beta z)\bar{a}_x + E_{m_2}\cos(\omega t - \beta z + \theta)\bar{a}_y$$

This is said to be linearly polarized if θ equals
(a) $\theta = 90°$ (b) $\theta = -90°$
(c) $\theta = 0°$ (d) $\theta = \frac{\pi}{4}$

87. If in the above question; $E_{m_1} = E_{m_2} = E_m$, the wave E represents a right circular polarization if the angle θ is
(a) $\theta = 90°$ (b) $\theta = -90°$
(c) $\theta = 0°$ (d) $\theta = \frac{\pi}{4}$

88. If in expression for E in above question $\theta = 90°$ and $E_{m_1} = E_{m_2} = E_m$, the wave represents
(a) right circular polarization (b) left circular polarization
(c) elliptically polarized wave (d) linear polarized wave

89. If in expression for E in question 86 $E_{m_1} \neq E_{m_2}$, the wave is
(a) right circular polarization (b) left circular polarization
(c) elliptically polarized wave (d) linear polarized wave

90. The superposition of two circularly polarized waves rotating in opposite direction but at the same angular rate give rise to
(a) circularly polarized wave (b) elliptically polarized wave
(c) linearly polarized wave (d) none of these

91. An electric field plane wave is given as

$$E = 1\cos(\omega t + \beta z)\bar{a}_x + 1\sin(\omega t + \beta z)\bar{a}_y$$

Polarization of the given wave is

(*a*) linearly (*b*) right circular
(*c*) left circular (*d*) elliptical

92. An electric field plane wave is given as

$$E = 1\cos(\omega t + \beta z)\bar{a}_x - 1\sin(\omega t + \beta z)\bar{a}_y$$

Polarization of the given wave is

(*a*) linearly (*b*) right circular
(*c*) left circular (*d*) elliptical

93. An electric field plane wave is given as

$$E = 1\cos(\omega t + \beta z)\bar{a}_x - 2\sin(\omega t + \beta z - 45°)\bar{a}_y$$

Polarization of the wave is

(*a*) linearly (*b*) right circular
(*c*) left circular (*d*) elliptical

94. There is a planar interface at $z = 0$ between free space and glass with $\sigma = 0$, $\varepsilon_r = 4$, $\mu_r = 1$. A plane wave with electric field $E = 1\cos(\omega t - \beta z)\bar{a}_x$ V/m is incident from free space (medium 1) normal to the glass, the transmission and reflection coefficient are

(*a*) $\frac{2}{3}, \frac{1}{3}$ (*b*) $-\frac{2}{3}, \frac{1}{3}$

(*c*) $\frac{2}{3}, -\frac{1}{3}$ (*d*) $-\frac{2}{3}, -\frac{1}{3}$

95. In the above question 94 the time average power transmitted through a 1 m^2 patch of the interface is

(*a*) 2.36 mW (*b*) 1.18 mW
(*c*) 3.54 mW (*d*) 0.59 mW

96. The SWR for medium 1 of Question 94 is

(*a*) 2 (*b*) 1/2
(*c*) 4 (*d*) 1/4

97. The electric and magnetic fields in a medium are given by

$$E = -2jE_m \sin\beta_1 z\bar{a}_x \quad H_1 = \frac{2E_m}{\eta_1}\cos\beta_1 z\bar{a}_y \quad \eta_1 > 0$$

The two fields in time domain are

(*a*) $2E_m \sin\beta_1 z\cos\omega t\, a_x, \; \frac{2E_m}{\eta_1}\cos\beta_1 z\cos\omega t\,\bar{a}_y$

(*b*) $2E_m \sin\beta_1 z\sin\omega t\,\bar{a}_x, \; \frac{2E_n}{\eta_1}\cos\beta_1 z\cos\omega t\,\bar{a}_y$

(c) $2E_m \sin\beta_1 z \sin\omega t\, \bar{A}_x, \ \dfrac{2E_m}{\eta_1}\cos\beta_1 z \sin\omega t\, \bar{a}_y$

(d) $2E_m \sin\beta_1 z \sin\omega t\, \bar{A}_x, \ \dfrac{2E_m}{\eta_1}\sin\beta_1 z \sin\omega t\, \bar{a}_y$

98. The time domain expression for the electric field

$$E = -\left(0.3\bar{a}_y + 0.2\bar{a}_x\right)e^{j(-0.5\pi y + 0.86\pi x)}\ \text{V/m}$$

is given as

(a) $-0.3\cos(\omega t - 0.5\pi y + 0.86\pi x)\bar{a}_y - 0.2\cos(\omega t - 0.5\pi y + 0.86\pi x)\bar{a}_x$

(b) $-0.6\cos(\omega t - 0.5\pi y + 0.86\pi x)\bar{a}_y - 0.4\cos(\omega t - 0.5xy + 0.86\pi x)\bar{a}_x$

(c) $-0.6\sin(\omega t - 0.5\pi y + 0.866\pi x)\bar{a}_y - 0.4\sin(\omega t - 0.5xy + 0.86\pi x)\bar{a}_x$

(d) $-0.3\sin(\omega t - 0.5\pi y + 0.866\pi x)\bar{a}_y - 0.2\sin(\omega t - 0.5\pi x + 0.86\pi x)\bar{a}_x$

99. A 10 MHz plane wave $E = E_m\cos(\omega t + \beta x)\bar{a}_z$ travels in a non-magnetic medium ($x > 0$) with permittivity $\varepsilon_r = 9$. The wavelength in the dielectric is

(a) 1 m (b) 0.5 m

(c) 5 m (d) 10 m

100. The magnetic field for a linearly polarized wave in a lossless magnetic material with free space permittivity is given as

$$H = 5e^{j(\omega t - \beta y)}\bar{a}_x\ \text{A/m}$$

For the material, intrinsic impedance is 1000 Ω, the electric field in time domain is given as

(a) $6750\cos(\omega t - \beta y)\,\bar{a}_z\ \text{Vm}^{-1}$ (b) $5000\cos(\omega t - \beta y)\,\bar{a}_x\ \text{Vm}^{-1}$

(c) $5000\cos(\omega t - \beta y)\,\bar{a}_z\ \text{Vm}^{-1}$ (d) $5000\cos(\omega t - \beta y)\,\bar{a}_y\ \text{Vm}^{-1}$

101. The time average poynting vector for the wave in question 100 is

(a) $15\bar{a}_y\ \text{kWm}^{-2}$ (b) $15.9\bar{a}_x\text{kWm}^{-2}$

(c) $12.5\bar{a}_y\text{kWm}^{-2}$ (d) $12.5\bar{a}_x\text{kWm}^{-2}$

102. If a lossless cable has $L = 0.25$ μH/m and $C = 100$ pF/m, its velocity of propagation and characteristic impedance are

(a) $3 \times 10^8\ \text{ms}^{-1}$, 50 ohm (b) $2 \times 10^8\ \text{ms}^{-1}$, 50Ω

(c) $5 \times 10^8\ \text{ms}^{-1}$, 40Ω (d) $2.5 \times 10^8\ \text{ms}^{-1}$, 50Ω

103. A transmission line with a characteristic resistance of 50 ohm is connected to a 100 ohm resistance load. A d.c. source of 30 V and zero internal resistance is connected to the line. The voltage reflection coefficient at the source end and load end are

(a) 0, –1/3 (b) –1, – 1/3

(c) –1, 1/3 (d) 1, 1/3

104. A vector field F is given as $\overline{F} = 25\overline{a}_r + 12a_\phi - 20\overline{a}_z$ at the point (8, 120°, 5). The vector field component of F perpendicular to the cylinder $r = 8$ is:

(*a*) $25\overline{a}_r$ (*b*) $12\overline{a}_r$

(*c*) $\frac{2\pi}{3}\overline{a}_r$ (*d*) $5\overline{a}_r$

105. A loss-less transmission line of length $\lambda/8$ is short circuited at one end. The impedance measured at the other end is jz at a frequency of 13.5 MHz. If the frequency is raised to 27 MHz, the impedance measured will be

(*a*) $j\,2z$ (*b*) $-jz$

(*c*) zero (*d*) infinity.

106. The line integral of the vector field $F(x, y, z) = xy\overline{a}_x + 2\overline{a}_y$ around the closed path is

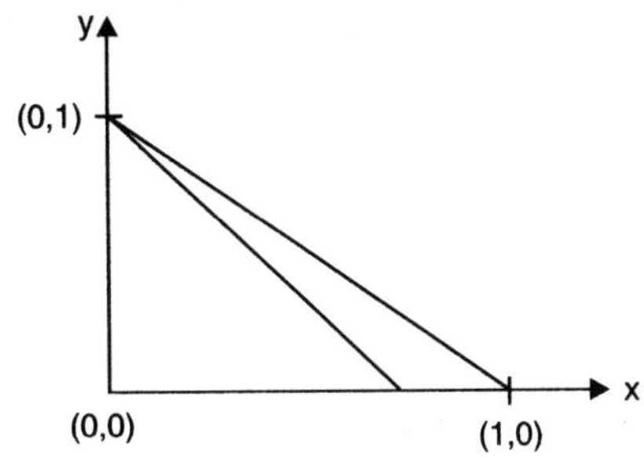

Fig. Q. 106

(*a*) 1/3 (*b*) 1/6

(*c*) – 1/6 (*d*) zero

107. The electric flux density within a sphere of radius a is given by $D = \frac{10}{r^2}\overline{a}_r - r^2\phi\sin\theta\overline{a}_\phi$. The volume charge density in the sphere is given as

(*a*) r (*b*) $-r$

(*c*) r^2 (*d*) 10

108. Two perfect dielectrics media with relative permittivity ε_{r_1} and ε_{r_2}, and $\rho_s = 0$, the normal and tangential component of D at the surface are

(*a*) $D_{n_2} = D_{n_1}, D_{t_2} = D_{t_1}$ (*b*) $D_{n_2} = D_{n_1}, \frac{D_{t_2}}{\varepsilon_2} = \frac{D_{t_1}}{\varepsilon_1}$

(*c*) $\frac{D_{n_2}}{\varepsilon_2} = \frac{D_{n_1}}{\varepsilon_1}, D_{t_2} = D_{t_1}$ (*d*) none of the above is correct

109. A conductor of radius a carries a current of I uniformly distributed over the conductor cross section. The magnetic field intensity within and outside the conductor is given as

(a) $\frac{Ir^2}{2\pi a^2}, \frac{I}{2\pi r}$ (b) $\frac{Ir}{2\pi a^2}, \frac{I}{2\pi r}$

(c) $\frac{Ir}{2\pi a^2}, \frac{Ir^2}{2\pi a^2}$ (d) $\frac{Ir^2}{2\pi a^2}, \frac{Ir}{2\pi a^2}$

Answers to MCQ

1. (c)	2. (c)	3. (a)	4. (d)	5. (a)
6. (a)	7. (b)	8. (a)	9. (c)	10. (d)
11. (d)	12. (a)	13. (a)	14. (d)	15. (c)
16. (a)	17. (d)	18. (b)	19. (b)	20. (b)
21. (a)	22. (b)	23. (c)	24. (a)	25. (c)
26. (c)	27. (a)	28. (c)	29. (b)	30. (c)
31. (b)	32. (c)	33. (c)	34. (b)	35. (c)
36. (d)	37. (c)	38. (c)	39. (a)	40. (b)
41. (a)	42. (c)	43. (a)	44. (c)	45. (a)
46. (a)	47. (b)	48. (b)	49. (c)	50. (b)
51. (c)	52. (b)	53. (c)	54. (d)	55. (c)
56. (b)	57. (d)	58. (b)	59. (c)	60. (d)
61. (b)	62. (c)	63. (b)	64. (b)	65. (d)
66. (b)	67. (c)	68. (c)	69. (c)	70. (a)
71. (c)	72. (d)	73. (a)	74. (a)	75. (b)
76. (b)	77. (c)	78. (b)	79. (b)	80. (a)
81. (c)	82. (a)	83. (b)	84. (b)	85. (c)
86. (c)	87. (b)	88. (b)	89. (c)	90. (c)
91. (c)	92. (b)	93. (d)	94. (c)	95. (b)
96. (a)	97. (b)	98. (a)	99. (d)	100. (c)
101. (c)	102. (b)	103. (c)	104. (a)	105. (d)
106. (c)	107. (b)	108. (b)	109. (b)	

REFERENCES

1. KRAUS J.D. and Fleisch D.A. *Electromagnetics with Application,* V Edition McGraw Hill, New York 1999.
2. Hayt W.H. and Buck J.A. *Engineering Electromagnetics*, VII Edition Tata McGraw Hill, New Delhi 2006.
3. R. Plonsey and Collin R.E. *Principles and Applications of Electromagnetic Fields*, McGraw Hill, New York 1961.
4. H.H. Skilling, *Fundamentals of Electric Waves*, 2nd Edition, John Wiley & Sons, 1948.
5. Rao N.N., *Basic Electromagnetics with Applications*, Prentice Hall of India New Delhi 1974.
6. Joseph A. Edminister *Electromagnetics Schaum's series*, 2nd Edition McGraw Hill, 1993.

7. Syed A. Nasar, *Solved Problems in Electromagnetics*, Schaum's solved problem series, McGraw Hill, New York, 1992.
8. Seely S, *Introduction to Electromagnetic Fields*, McGraw Hill, 1958.
9. Jordon, Edward C, *Electromagnetic Waves and Radiating Systems*, Prentice Hall Inc. 1950.
10. Dekker A.J., *Electrical Engineering Materials*, Englewood Cliffs N.J. Prentice Hall, 1959.
11. Maxwell J.C., *A Treatise on Electricity and magnetism*, 3rd Edition Oxford University Press, New York, Also Dover Publication, New York 1954.
12. Seely S. and Poularikas, A.D., *Electromagnetics: Classical and Modern Theory and Applications*, Marcel Dekker, New York 1979.
13. Jordan E.C. and Balmain, K.G., *Electromagnetic Waves and Radiating Systems*, 2nd Edition Prentice Hall, 1968.
14. Paul C.R., Whites K.W. and Nasar S.Y., *Introduction to Electromagnetic Fields*, 3rd Edition McGraw Hill, New York 1998.
15. Holiday D.R., Resnick and D. Walker J., *Fundamentals of Physics*, 5th Edition John Wiley & Sons, New York 1997.
16. Nussbaum, A., *Electromagnetic Theory for Engineers and Scientists*, Prentice Hall, 1965.
17. Seshadari, S.R., *Fundamentals of Transmission Lines and Electromagnetic Fields*, Addison Wesley, 1971.
18. Chang, D.K., *Field and Wave Electromagnetics*, 2nd Edition Addison Wesley, 1989.

IMPORTANT INFORMATIONS

Intrinsic Impedance for various materials

Partially conducting medium $\eta = \sqrt{\dfrac{j\omega\mu}{\sigma + j\omega\varepsilon}}$

Conducting medium $\eta = \sqrt{\dfrac{\omega\mu}{\sigma}}\angle 45°$

Perfect dielectric $\eta = \sqrt{\dfrac{\mu}{\varepsilon}}$

Free space $\eta_0 = \sqrt{\dfrac{\mu_0}{\varepsilon_0}} = 120\pi\Omega$

Maxwell's Equations

Law	Integral form	Differential form
Ampere's	$\oint H \cdot dl = \int_S \left(J + \dfrac{\partial D}{\partial t}\right) \cdot ds = I$	$\nabla \times H = J + \dfrac{\partial D}{\partial t}$
Faraday	$\oint E \cdot dl = -\int \dfrac{\partial B}{\partial t} \cdot ds = V$	$\nabla \times E = -\dfrac{\partial B}{\partial t}$

Gauss' law for electric fields	$\oint D \cdot ds = \int_V \rho \cdot dv = Q$	$\nabla \cdot D = \rho$
Gauss' law for magnetic fields	$\oint B \cdot ds = 0$	$\nabla \cdot B = 0$

Maxwell's Equation for Harmonic Variation or Phasor Form

$$\oint H \cdot dl = (\sigma + j\omega E)\int E \cdot ds \qquad \nabla \times H = (\sigma + j\omega\varepsilon)E$$

$$\oint E \cdot dl = -j\omega\mu \int H \cdot ds \qquad \nabla \times E = -j\omega\mu H$$

$$\oint D \cdot ds = \int_V \rho \cdot dv \qquad \nabla . D = \rho$$

$$\oint_s B \cdot ds = 0 \qquad \nabla . B = 0$$

If u_0, η_0, λ_0 and β_0 are the quantities in free space for a plane wave then for a loss less medium with ε_r, μ_r, these quantities are

$$u = \frac{u_0}{\sqrt{\mu_r \varepsilon_r}}, \eta = \eta_0 \sqrt{\frac{\mu_r}{\varepsilon_r}}, \lambda = \frac{\lambda_0}{\sqrt{\mu_r \varepsilon_r}}, \beta = \beta_0 \sqrt{\mu_r \varepsilon_r}$$

Continuity equation $\nabla \cdot J = -\dfrac{\partial \rho_v}{\partial t}$

For transmission lines

$$v_p = \frac{1}{\sqrt{LC}}, \; \gamma = \sqrt{(R + j\omega L)(G + j\omega E)} = \left(\frac{R}{2}\sqrt{\frac{C}{L}} + \frac{G}{2}\sqrt{\frac{L}{C}}\right) + j\omega\sqrt{LC}$$

$$= \alpha + j\beta$$

$$\lambda = \frac{2\pi}{\beta} = \frac{v_p}{f}$$

$$VSWR = \frac{|V_{max}|}{|V_{min}|} = \frac{1 + |\rho_v|}{1 - |\rho_v|}$$

Good conductor: At high frequencies electromagnetic energy is not transmitted in the interior of a conductor, it travels in the region surrounding the conductor, while the conductor merely guides the waves.

For a good conductor

$$\alpha = \beta = \frac{1}{\delta} = \sqrt{\pi f \mu \sigma}$$

$$\beta = \frac{2\pi}{\lambda}, \lambda = 2\pi\delta, v_p = \frac{\omega}{\beta}, v_p = \omega\delta$$

Standing waves: When waves travelling in a perfect dielectric ($\sigma_1 = 0$, $\alpha_1 = 0$) are normally incident on the interface with a perfect conductor ($\sigma_2 = \infty$, $\eta_2 = 0$), the reflected wave in combination with the incident wave produces a standing wave. In this wave oscillations at all points of a half wavelength interval are in phase. The combination of incident and reflected waves may be written.

$$E(z, t) = \left[E_{0i}e^{j(\omega t - \beta z)} + E_{0r}e^{j(\omega t + \beta z)}\right]\bar{a}_x = e^{j\omega t}\left(E_{0i}e^{-j\beta z} + E_{0r}e^{j\beta z}\right)\bar{a}_x$$

Since $\qquad \eta_2 = 0, \dfrac{E_{or}}{E_{oi}} = -1$ and

$$E(z, t) = e^{j\omega t}\left(E_{oi}e^{-j\beta z} - E_{oi}e^{j\beta z}\right)\bar{a}_x$$

$$= e^{j\omega t}E_{oi}\left(-j2\sin\beta z\right)\bar{a}_x$$

$$= -2jE_{oi}\sin\beta z\, e^{j\omega t}\bar{a}_x$$

or taking the real part

$$E(z, t) = 2E_{oi}\sin\beta z \sin\omega t\bar{a}_x$$

Poynting Vector: $P = E \times H$ is the poynting vector, the instantaneous rate of energy flow per unit area at a point.

In the cross product that defines the poynting vector, the fields are supposed to be in real form. If instead E and H are expressed in complex form and have the common time dependence e^{jwt}, then the time average of $\underline{P}$ is given as

$$p_{av} = \frac{1}{2}\mathrm{Re}\left[E \times H^*\right]$$

Divergence

Rectangular $\qquad \nabla \cdot D = \dfrac{\partial D_x}{\partial x} + \dfrac{\partial D_y}{\partial y} + \dfrac{\partial D_z}{\partial z}$

Cylindrical $\qquad \nabla \cdot D = \dfrac{1}{r}\dfrac{\partial (D_r)}{\partial r} + \dfrac{1}{r}\dfrac{\partial D_\phi}{\partial \phi} + \dfrac{\partial D_z}{\partial z}$

Spherical $\qquad \nabla \cdot D = \dfrac{1}{r^2}\dfrac{\partial}{\partial r}\left(r^2 D_r\right) + \dfrac{1}{r\sin\theta}\dfrac{\partial}{\partial \theta}\left(D_\theta \sin\theta\right) + \dfrac{1}{r\sin\theta}\dfrac{\partial D_\phi}{\partial \phi}$

Gradient

Rectangular $\qquad \nabla V = \dfrac{\partial V}{\partial x}\bar{a}_x + \dfrac{\partial V}{\partial y}\bar{a}_y + \dfrac{\partial V}{\partial z}\bar{a}_z$

Cylindrical $\qquad \nabla V = \dfrac{\partial V}{\partial r}\bar{a}_r + \dfrac{1}{r}\dfrac{\partial V}{\partial \phi}\bar{a}_\phi + \dfrac{\partial V}{\partial z}\bar{a}_z$

Spherical $\qquad \nabla V = \dfrac{\partial V}{\partial r}\bar{a}_r + \dfrac{1}{r}\dfrac{\partial V}{\partial \theta}\bar{a}_\theta + \dfrac{1}{r\sin\theta}\dfrac{\partial V}{\partial \phi}\bar{a}_\phi$

Curl

Rectangular
$$\nabla \times H = \left(\frac{\partial H_z}{\partial y} - \frac{\partial H_y}{\partial z}\right)\bar{a}_x + \left(\frac{\partial H_x}{\partial z} - \frac{\partial H_z}{\partial x}\right)\bar{a}_y + \left(\frac{\partial H_y}{\partial x} - \frac{\partial H_x}{\partial y}\right)\bar{a}_y$$

Cylindrical
$$\nabla \times H = \left(\frac{1}{r}\frac{\partial H_z}{\partial \phi} - \frac{\partial H_\phi}{\partial z}\right)\bar{a}_r + \left(\frac{\partial H_r}{\partial z} - \frac{\partial H_z}{\partial r}\right)\bar{a}_\phi + \frac{1}{r}\left[\frac{\partial r H_\phi}{\partial r} - \frac{\partial H_r}{\partial \phi}\right]\bar{a}_z$$

Spherical
$$\nabla \times H = \frac{1}{r\sin\theta}\left[\frac{\partial H_\varphi \sin\theta}{\partial \theta} - \frac{\partial H_\theta}{\partial \phi}\right]a_r + \frac{1}{r}\left[\frac{1}{\sin\theta}\frac{\partial H_r}{\partial \phi} - \frac{\partial (rH_\phi)}{\partial r}\right]a_\theta$$
$$+\frac{1}{r}\left[\frac{\partial (rH_\theta)}{\partial \phi} - \frac{\partial H_r}{\partial \theta}\right]\bar{a}_\phi$$

Laplacian

Rectangular
$$\nabla^2 V = \frac{\partial^2 V}{\partial x^2} + \frac{\partial^2 V}{\partial y^2} + \frac{\partial^2 V}{\partial z^2}$$

Cylindrical
$$\nabla^2 V = \frac{1}{r}\frac{\partial}{\partial r}\left(r\frac{\partial V}{\partial r}\right) + \frac{1}{r^2}\frac{\partial^2 V}{\partial \phi^2} + \frac{\partial^2 V}{\partial z^2}$$

Spherical
$$\nabla^2 V = \frac{1}{r^2}\cdot\frac{\partial}{\partial r}\left(r^2\frac{\partial V}{\partial r}\right) + \frac{1}{r^2\sin\theta}\frac{\partial}{\partial \theta}\left(\sin\theta\frac{\partial V}{\partial \theta}\right) + \frac{1}{r^2\sin^2\theta}\cdot\frac{\partial^2 V}{\partial \phi^2}$$

	Differential length	**Differential area**	**Differential volume**
Rectangular	$(dx\bar{a}_x + dy\bar{a}_y + dz\bar{a}_z)$	$dxdy$, $dydz$, $dzdx$	$dxdydz$
Cylindrical	dr, $rd\phi$, dz	$dr\, rd\phi$, $drdz$, $rd\phi\, dz$	$rdrd\phi dz$
Spherical	dr, $rd\theta$, $r\sin\theta d\phi$	$rdr\, d\theta$, $r\sin\theta\, drd\phi$, $r^2\sin\theta d\theta d\phi$	$r^2\sin\theta drd\theta d\phi$

$D = \varepsilon_0 E + P$, $P = \chi_e \varepsilon_0 E$

$D = \varepsilon_0 E + \chi_e \varepsilon_0 E = \varepsilon_0 E(1 + \chi_e)$ and $\varepsilon_r = 1 + \chi_e$

$D = \varepsilon_0 \varepsilon_r E$, $\varepsilon = \varepsilon_0 \varepsilon_r$

Index

A

B

C

D

N

O

P

Q

R

S